AF480048

Recent Advances in Design, Development and Operation of Micro Air Vehicles

Recent Advances in Design, Development & Operation of Micro Air Vehicles

Editors

Dr. G.K. Viswanadh

Professor of Civil Engineering,
JNTUH College of Engineering,
JNTUH, Kukatpally, Hyderabad.

and

Dr. M.V.S.S. Giridhar

Assistant Professor in Water Resources,

Institute of Science and Technology,
JNTUH, Kukatpally, Hyderabad.

JAWAHARLAL NEHRU TECHNOLOGICAL UNIVERSITY HYDERABAD
Kukatpally, Hyderabad – 500 085, (A.P.) INDIA.

Published by :

 **BS Publications**
A unit of **BSP Books Pvt., Ltd.**

4-4-309, Giriraj Lane, Sultan Bazar,
Hyderabad - 500 095 - A.P.
Phone : 040 - 23445605, 23445688
e-mail : info@bspbooks.net
www.bspublications.net

ISBN : 978-93-90211-10-4

Contents

TECHNICAL SESSION – II

GUIDANCE, NAVIGATION AND CONTROL

TECHNICAL SESSION – III
MAV APPLICATIONS

TECHNICAL SESSION – IV

MODELLING AND STIMULATIONS AND FUSELAGE AND PROPULSION SYSTEMS

(x)

Session I

Sensors and Avionics

A ROBUST APPROACH TO OBTAIN PANORAMIC VIEW OF UAV CAPTURED IMAGES

Athilingam, .R[1]   **A. Nuzrath Hameedha**[2]   **K.Senthilkumar**[3]

[1]Research scholar, [2]PG student, [3]Associate Professor

Avionics Division, Department of Aerospace Engineering, Madras Institute of Technology,
Anna University, Chennai -600044
Contact id: aathikannan@gmail.com

ABSTRACT

Panoramic view is an important requirement in the field of aerial surveillance. Multiple images taken by Unmanned Aerial Vehicles (UAV) with mutual object of interest is to be stitched together to obtain the panoramic view of the surveyed area. Panoramic image stitching used to create virtual environment for many applications is a key technology in 3D realization, Geo referenced mapping and target localization and lots of stitching algorithms are developed in recent years.

This paper proposes a panorama image stitching system which combines an image matching algorithm based on corner detection and backward image blending algorithm.   The application is focused towards generation of panoramic view of the images captured by the UAV – DAKSHA*. The algorithm is designed to be extremely efficient and fast in its execution and is intended for use in stitching images captured by Unmanned Aerial Vehicles. We present full details of how the extraction of the heuristic is done from the inputs and how it drastically stitches images to provide panoramic view of the UAV -DAKSHA image data.

The algorithm works by extracting the corners of the two images to be stitched. To get the scene image of wide view field, firstly block matching based on corner detection is applied to estimate the motion vector field, and then the parameters of transformation model can be calculated with the backward algorithm to implement image sequence stitching. Because this method is not sensitive to ordering, orientation, scaling and illumination, its stitching precision is much higher than many other methods.

Keywords : Image Stitching, Panorama view , UAV , corner detection, block matching, Aerial Surveillance

INTRODUCTION

Image stitching is a technology that carries on image matching and blending to image sequences which are overlapped with each other, and finally builds a seamless and high quality panorama photo with high resolution and wide view field. Image stitching is an important research field of image processing, and has widely applications in the fields of photogrammetric, computer vision, remote sensing image processing, medical image analysis, computer graphic and so on[1,2] The image stitching process mainly includes image matching and image blending, and the key problem of image stitching is how to realize the image matching, by which find the matching parameters among the images. At present, image matching can be categorized two methods: region-based method and feature-based method. Region-based method [3, 4] is to use the image grey-scale information to determine the matching parameters between images, which make full use of the image information, but often be with large amount of calculation. Moreover, in the presence of image noise, image distortion and large scaling and rotation between images, the obtained stitching results are often not the correct. Feature-based method [5, 6] makes use of the distinct image features such as image contours and corners to realize the image matching with less calculation, further this method have higher robustness in case of the scaling and rotation between images, the correct feature extraction of images is very important to match the image to stitch rightly.

The rest of the paper is organized as follows. In Section II image matching technique is discussed. In section III blending and stitching techniques are discussed. In section IV, the implementation of the proposed work is done and Finally in Section V the results are discussed. The conclusions are made in section VI.

IMAGE MATCHING

Image matching is a technique in image stitching is one of two key technologies, and image matching is image fusion of the base, but the image matching algorithm of the calculation amount is generally very large, so the image stitching technology development depends largely on the image quasi-technological innovation.

The Image Matching based on corner detection:

Image stitching of the key is to find out exactly adjacent two images overlap the location, and then determine the transformation relations between two images, namely image matching. Image matching algorithm can be roughly divided into feature-based image matching and region-based image matching two broad categories. Region-based image matching uses regions of the pixel correlation and others to match. The feature-based image matching uses image of the obvious feature to estimate the transformation relations between the images, rather than uses the image all the information. These obvious features such as image feature points (corner points or key points), contour, and some invariant moment and so on. Correspondences between some features present in both images, in order to determine the geometric transformation necessary to provide their alignment –Image Registration. One of the images is then deformed according to the computed transformation, Image Warping, thereby changing the spatial relationship between pixels coordinates. That is, image coordinates $[x,y]^T$ are mapped into a new set of coordinates $[x' ,y']^T$.

The definition of the correspondences can be done manually or by an automatic process. In the latter, one can use a similarity measure like normalized cross-correlation or correlation coefficient to match the two images. However, due to time restrictions and to the better precision naturally achieved with manual definition, this turned out to be our option for this operation. This is where we come to the importance of corner detection as a pre-processing stage for image mosaicing. When defining the referred correspondences between the features present in both images, we could of course just look at the pictures and trust in our quick glance to determine the positions of those features. What we do instead is to first compute the exact positions of a particular class of features – the corners – after what we are able to perform a safer definition of the correspondences. Manual definition of those correspondences between the two images was supported by a function which prompted the user for alternately clicking on corresponding items on each picture, thereby providing an user-friendly interface. However, in order to achieve good results on the mosaicing operation we thought it was of extreme relevance that the referred function sought the nearest corner (which was previously detected) to the pixel pointed by the user click. This way we protected the computation of the necessary geometric transformation against the nice but not perfect user' s eyes, against his shaky hands and all other kinds of inaccuracies of this process. The reason why we emphasize so much this matter is thinking that, in an operation which requires such a high level accuracy like image mosaicing does, every single point of the process where we can improve that accuracy is precious. And this is certainly one of those.

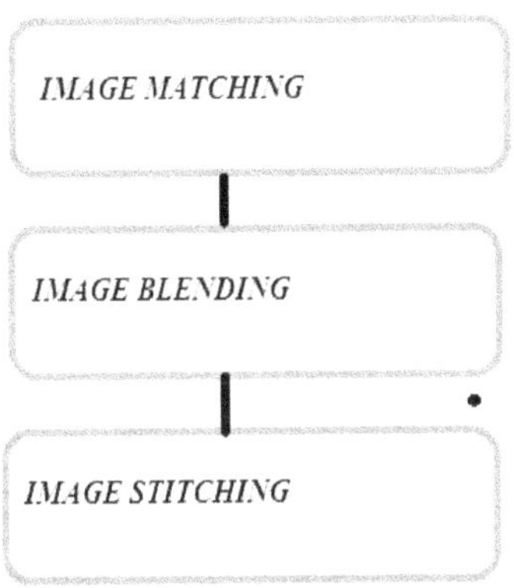

Fig 1

Once corners in each picture have been associated we now have the necessary elements to compute the geometric transformation matrix, which is given by the following relation.

$$\begin{bmatrix} x' \\ y' \\ w' \end{bmatrix} = \begin{bmatrix} m_{11} & m_{12} & m_{13} \\ m_{21} & m_{22} & m_{23} \\ m_{31} & m_{32} & m_{33} \end{bmatrix} \begin{bmatrix} x \\ y \\ w \end{bmatrix}$$

---- (1)

Pixel coordinates on the original image are given by $[x/w, y/w]^T$ and are mapped into $[x'/w', y'/w']^T$ after image warping. In order to enable reverse mapping one can, without loss of generality, force w' to 1. The coefficients of the geometric transformation matrix m_{ij} are the solution of the following system of equations

$$\begin{bmatrix} x_1 & y_1 & 1 & 0 & 0 & 0 & -x_1 u_1 & -y_1 u_1 \\ x_2 & y_2 & 1 & 0 & 0 & 0 & -x_2 u_2 & -y_2 u_2 \\ x_3 & y_3 & 1 & 0 & 0 & 0 & -x_3 u_3 & -y_3 u_3 \\ x_4 & y_4 & 1 & 0 & 0 & 0 & -x_4 u_4 & -y_4 u_4 \\ 0 & 0 & 0 & x_1 & y_1 & 1 & -x_1 v_1 & -y_1 v_1 \\ 0 & 0 & 0 & x_2 & y_2 & 1 & -x_2 v_2 & -y_2 v_2 \\ 0 & 0 & 0 & x_3 & y_3 & 1 & -x_3 v_3 & -y_3 v_3 \\ 0 & 0 & 0 & x_4 & y_4 & 1 & -x_4 v_4 & -y_4 v_4 \end{bmatrix} \begin{bmatrix} m_{11} \\ m_{12} \\ m_{13} \\ m_{21} \\ m_{22} \\ m_{23} \\ m_{31} \\ m_{32} \end{bmatrix} = \begin{bmatrix} u_1 \\ u_2 \\ u_3 \\ u_4 \\ v_1 \\ v_2 \\ v_3 \\ v_4 \end{bmatrix}$$

----(2)

where point $[xi, yi]^T$ is mapped into the corresponding point in the other image $[ui, vi]^T$. It was already said that the minimum number of correspondences to make the perspective transformation possible is four. Nevertheless, the more associations we have between corners on each image, the more accurate the perspective transformation will be. This way, when more than four correspondences are defined, solution in the least squares sense to the overdetermined system of equations is computed, thereby improving the quality and accuracy of the perspective transformation. Bilinear interpolation was used to determine pixel intensities after image warping.

IMAGE BLENDING

The last part of a mosaicing operation is image blending, which consists of bringing the two pictures together into one *Image blending* involves executing the adjustments figured out in the calibration stage, combined with remapping of the images to an output projection. Colors are adjusted between images to compensate for exposure differences. If applicable, high dynamic range merging is done along with motion compensation. Images are blended together and seam line adjustment is done to minimize the visibility of seams between images.

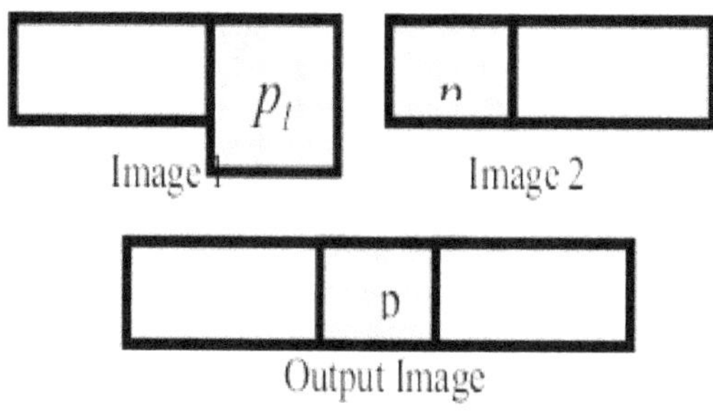

Fig. 2

Our paper describes average weighed method which is simple and fast for blending, which can be described as in figure 2.

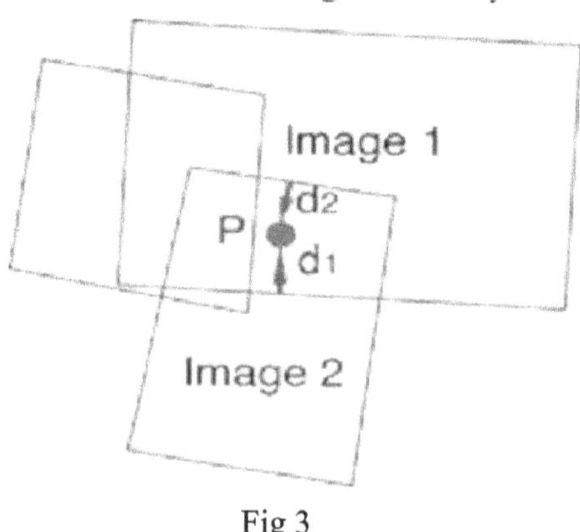

Fig 3

The blending method we chose was the hat function, which can be considered as a particular case of bilinear blending since it calculates a weight for every pixel in each image according to the following relation

$$w(x,y) = \left(1 - \left|1 - \frac{2x}{width}\right|\right) \cdot \left(1 - \left|1 - \frac{2y}{height}\right|\right)$$
---(3)

The resulting intensity of the pixels of the mosaic (Iblend)is then given by the relation below, where xand yrepresent the initial coordinates, before the warping operation, and i is the image index.

$$I_{blend}(x,y) = \frac{\sum_i w_i(x,y) \cdot I_i(x,y)}{\sum_i w_i(x,y)}$$
------(4)

Ii(x,y)is referred to the intensity value of the ith image pixel $[x, y]^T$

In the average weighted blending, the values of features in overlap region are equal to the weighted average values of matching images, which can be shown as following:

$$p = \frac{d_l}{d_l + d_r} p_l + \frac{d_r}{d_l + d_r} p_r$$
------(5)

where d_l(d1) is the distance between the pixel in overlap region to the border of the left matching image, and dr (d2)is the distance between the pixel in overlap to the border of the right matching image. In the following, we will study one particular type of warp namely; the Euclidean warp also called the Euclidean similarity transform. This type of warp involves four parameters: p = [s ® tx ty]^T

Let x = [x y]^T denote a position in the original 2D image, I, and let x0= [x0 y0]^{T}denote the corresponding position in the warped image, I0, (both in homogeneous coor-dinates). Looking at one pixel, equation (1) can now be written as a simple linear transformation, where T denotes the transformation matrix: x0= Tx, which denotes scale, rotation and translation respectively

$$\begin{bmatrix} x' \\ y' \\ 1 \end{bmatrix} = \begin{bmatrix} s\cos\alpha & s\sin\alpha & t_x \\ -s\sin\alpha & s\cos\alpha & t_y \\ 0 & 0 & 1 \end{bmatrix} \cdot \begin{bmatrix} x \\ y \\ 1 \end{bmatrix}$$
.—(6)

Observe that instead of warping a single point we could warp a whole image of n points by replacing x with:

$$\mathbf{X} = \begin{bmatrix} x_1 & \cdots & x_n \\ y_1 & \cdots & y_n \\ 1 & \cdots & 1 \end{bmatrix}$$
----(7)

$$\mathbf{X}' = \mathbf{T}\mathbf{X}$$
-----(8)

Due the discrete nature of raster images, one is in no way ensured that each input pixel exactly maps to an out pixel. Consequently warping is mostly performed backwards – i.e. from the output image to the input image. Since T is square and has full rank we can easily compute the inverse transformation as:

$$\mathbf{X} = \mathbf{T}^{-1}\mathbf{X}'$$
--- (9)

In mosaicing, the transformation between images is often not known beforehand. In this example, two images are merged and we will estimate the transformation by letting the user give points of correspondence (also called landmarks or fiducial markers etc.) in each of the images. In order to recover the transformation we rearrange the warping equation (2) so that the warping parameters is the vector t in:

$$x' = Zt \implies$$

$$\begin{bmatrix} x' \\ y' \\ 1 \end{bmatrix} = \begin{bmatrix} x & y & 1 & 0 & 0 \\ y & -x & 0 & 1 & 0 \\ 0 & 0 & 0 & 0 & 1 \end{bmatrix} \cdot \begin{bmatrix} s\cos\alpha \\ s\sin\alpha \\ t_x \\ t_y \\ 1 \end{bmatrix} \quad ---(10)$$

IMPLEMENTATION
1. Choose one image as the reference frame.
2. Estimate matching between each of the remaining images and the reference image.

To estimate matching between two images use the following procedure:
   a. Detect local features in each image.
   b. Blend Warp each image into the reference frame and composite warped images into a single mosaic images

RESULTS AND DISCUSSION

The videos obtained from the surveillance of UAV Dhaksha are considered for analysis. The videos are transmitted with the transmission frequency of 1.2 GHz from UAV DAKSHA with altitude ranging from 5 to 80 meters. The surveillance is made at Ponneri – andaarkuppam. Few samples were considered for processing and the results are obtained.
* *UAV – DAKSHA is an Unmanned Aerial Vehicle designed and developed at Avionics Division of Madras Institute of technology by Dr. K.Senthil Kumar and with the scholars under his Guidance and directions..*

LOCATION 1:

Fig 4 : No of images stitched - 6

Fig 5 : No of images stitched - 7

Fig 6 : No of images stitched - 5

Fig 7 : No of images stitched - 5

Fig 8 : No of images stitched - 5

Fig 9 : No of images stitched - 7

Fig 10 : No of images stitched - 7

LOCATION 2

Fig 11 : No of images stitched - 7   Fig 12 : No of images stitched - 8

CONCLUSION

Thus the novel approach for stitching images to obtain panoramic view is presented. The algorithm works by extracting the corners of the two images to be stitched. To get the scene image of wide view field, firstly block matching based on corner detection is applied to estimate the motion vector field, and then the parameters of transformation model can be calculated with the backward algorithm to implement image sequence stitching. Because this method is not sensitive to ordering, orientation, scaling and illumination, its stitching precision is much higher than many other methods.

REFERENCES

[1]. H.Y. Shum and R. Szelishi: Panoramic Image Mosaics, Technical Report, Microsoft Research (2008).

[2]. G. Kang: Virtual Environment Touring based on image, master thesis, Nanjing Institute of Technology (2007).

[3]. R. Hartley□A. Zisserman: Multiple View Geometry in Computer Vision□AnHui University Press (2002).

[4]. H.Y. Shum and R. Szelishi: Construction of Panoramic Image Mosaics with Global and local Alignment, International Journal of Computer Vision 36(2) (2000), 101-130.

[5]. Y. Li, L. Ma: A Fast and Robust Image Stitching Algorithm, in proceeding of the 6th World Congress on Intelligent Control and Automation (2008).

[6]. Hui Zhao, Matching Technology based on SIFT Features, Shan Dong University, Institute of Information Science and Engineering, (2006).

[7]. D. Lowe. Distinctive image features from scale-invariant keypoints. Int. Journal of Computer Vision, 60(2):91–110, (2004).

[8] Forsyth, David A, et al, "Computer Vision – A Modern Approach," Prentice-Hall of India, 2006.

[9] R. Szeliski, "Image Alignment and Stitching, A Tutorial," Technical Report MSR-TR-2004-92, Microsoft Research, Microsoft Corporation, 2004.

UAV BASED OBJECT DETECTION USING MULTI-SENSOR IMAGE FUSION

Thillainayagi R[1] and Senthil Kumar K[2]

[1]Research scholar, [2]Associate Professor
Division of Avionics, Department of Aerospace Engineering, Madras Institute of Technology,
Anna University , Chennai 600044, India

Author for correspondence thillaimit@gmail.com

Abstract

With the recent developments in the field of UAV sensing technologies multi-sensor image fusion plays a vital role in the object detection. Image fusion is a process that combines visual data from dissimilar sensors to obtain a single composite image preserving the information of the sources. Here the thermal and visual images are fused together to serve the various requisites from the image. The data is acquired from the image sources. After the image registration, discrete wavelet transform is applied to fuse the images.The simulation results show that the fused image will maintain the discrimination between targets and the environment. Then the segmentation algorithms Otsu and Kmeans is applied to the fused images to segment the object alone. This paper evaluates the effectiveness of the two methods with a set of fused images.

Keywords: Image fusion, IR image, airborne image , dwt, segmentation, Otsu, Kmeans, UAV.

Introduction

Unmanned aerial vehicles (UAVs) are aircrafts which have the capability of flight without an onboard pilot. UAV can be remotely controlled, semi-autonomous, autonomous, or have a combination of these capabilities. UAV can execute given missions in a whole lot of domains [1]. In order to complete these various missions, UAV firstly needs to be equipped with sensor payloads to acquire images of the mission area and realize environment perception. The sensor payloads include infrared sensor, visible light sensor and so on. Infrared and visible image fusion from the same scene is the basis of target detection and recognition for UAV. However, the images shot by airborne sensors are dynamic, which increases more difficulties for visible and infrared image fusion. In order to acquire the situation assessment, it is very important to extract the target information. In fact, information of texture and colour in visible images are very abundant, while the target information, especially an artificial target, is more outstanding in infrared images. According to this, we can divide the image regions based on target regions, which can utilize the target information more effectively.

Image fusion

Image fusion is a process of combining multiple images to form a single image by utilizing certain features from each image. The successful fusion of images acquired from different modalities or instruments is of great importance in many applications such as image analysis and computer vision, concealed weapon detection, and autonomous landing guidance. Image fusion can be performed at four levels of the information representation, which are signal, pixel, feature, and symbolic levels. Multi-scale transforms are widely used for analyzing the information content of images for image fusion[2]. Several multiscale transforms have become very popular. These include the Laplacian pyramid transform [3], the contrast pyramid transform [4], the gradient pyramid transform [5], and the discrete wavelet transform (DWT) Other fusion methods for the infrared and visible images have been compared in [6]-[8], in which the fusion method based on discrete wavelet transform (DWT) performed well. In this paper, we only use DWT as a meaning to research the fusion concerning target detection and recognition, for DWT has less computational complexity [9].

Discrete Wavelet Transform

A signal analysis method similar to image pyramids is the discrete wavelet transform. The main difference is that while image pyramids lead to an over complete set of transform coefficients, the wavelet transform results in a non redundant image representation. The discrete two dimensional wavelet transform is computed by the recursive application of low pass and high pass filters in each direction of the input image (i.e. rows and columns) followed by sub sampling. These basis functions or baby wavelets are obtained from a single prototype wavelet called the mother wavelet, by dilations or contractions (scaling) and translations (shifts). They have advantages over traditional Fourier methods in analyzing physical situations where the signal contains discontinuities and sharp spikes. Image fusion process is achieved by multiresolution decomposition at level 2. The multiwavelet decomposition coefficients of the input images are appropriately merged and a new fixed image is obtained by reconstructing the fused multiwavelet coefficients.

Image segmentation

Segmentation refers to the process of partitioning a digital image into multiple regions (sets of pixels). The goal of segmentation is to simplify and change the representation of an image into something that is more meaningful and easier to analyze. Airborne image segmentation is a challenging task due to the various characteristics of the images, which leads to the complexity of segmentation. Threshold segmentation is wildly used in many fields because of its simplicity and efficiency. Its basic objective is to classify the pixels of a given image into two classes: those pertaining to an object and those pertaining to the background[10]. For the image with clear objects in the background, the bi-level thredsholding method can easily divide the object from the background. But to segment complex images, a multilevel threshold method required. The multilevel threshold segments the pixels into several distinct groups in which the pixels of the same group have gray levels within a specific range. However, when the thresholding method is extended to multi-level thresholding, the computation time grows exponentially with the number of thresholds.

2D Otsu segmentation

The two dimensional Otsu algorithm is given as follows. Suppose an image pixel size is M×N, gray-scale of the image ranges from 0 to L-1. The neighborhood average gray g (m, n) of the coordinate definition (m, n) pixel point is as follows:

$$g(m,n) = \frac{1}{k \times k} \sum_{i=-\frac{(k-1)}{2}}^{\frac{k-1}{2}} \sum_{j=-\frac{(k-1)}{2}}^{\frac{(k-1)}{2}} f(m+i,n+j) \tag{1}$$

Calculating the average neighbourhood gray of each pixel point, a gray binary group (i, j) may form. We use Cij to represent the occurrence frequency of (i, j). Then the probability Pij of vector (i, j) may be determined by the formula:

$$P_{ij} = \frac{C_{ij}}{M \times N} \tag{2}$$

Here, $0 \leq I, j < L$, and $\sum_{i=0}^{L-1} \sum_{j=0}^{L-1} Pij = 1$ Assuming the existence of two classes $C0$ and $C1$ in Two dimensional form, the histogram represents their respective goals and background, and with two different probability density distribution function. If making use of two-dimensional histogram threshold vector (s,t) to segment the image (of which $0 \leq s, t < L$), then the probability of two classes are respectively:

The probability of background occurrence is:

$$t_r(\sigma_B(S,T)) = \max_{0 \leq s,t \prec L} \{t_r(\sigma_B(S,T))\} \tag{3}$$

The probability of object occurrence is: $\omega_0 = P(C_0) = \sum_{i=0\,j+1}^{L-1} \sum_{j=t+1}^{L-1} P_{ij} = \omega_1(s,t) \tag{4}$

The definition of dispersion matrix:

$$\sigma_B = \omega_0[(\mu_0 - \mu_t)(\mu_0 - \mu_t)T] + \omega_1[(\mu_1 - \mu_t)(\mu_1 - \mu_t)T] \tag{5}$$

When the track of the above-mentioned dispersion matrix gets the maximum, the corresponding threshold of segmentation is the optimal threshold (S, T), namely:

$$t_r(\sigma_B(S,T)) = \max_{0 \leq s,t \prec L} \{t_r(\sigma_B(S,T))\} \tag{6}$$

We know that 2-D thermal images with noise segmented by Otsu way may get better results compared to one dimensional threshold segmentation methods. However, the computation cost gets huge, which is because the determination of the optimal threshold need to travel all the s and t, of which $0 \leq s, t < L$. That is to say, the more gray scale value of images is, the longer choice time of the threshold is.

K means segmentation

The k-means method aims to minimize the sum of squared distances between all points and the cluster centre. This procedure consists of the following steps, as described by Tou and Gonzalez [11,12].

1. Choose K initial cluster centres $z_1(1), z_2(1),,,..., z_k(1)$.

2. At the k-th iterative step, distribute the samples {x} among the K clusters using the relation,

$$x \in c_j(k)\, if \, \|x - z_j(k)\| < \|x - z_i(k)\| \tag{7}$$

for all i = 1, 2, ..., K; i, j; where $c_j(k)$ denotes the set of samples whose cluster centreis $z_j(k)$.

3. Compute the new cluster centres $z_j(k+1)$, j =1, 2, ..., K such that the sum of the squared distances from all points in $c_j(k)$ to the new cluster centre is minimized. The measure which minimizes this is simply the sample mean of $c_j(k)$. Therefore, the new cluster centre is given by

$$z_j(k+1) = \frac{1}{N_j} \sum_{x \in c_j(k)} x, j = 1, 2 \ldots k. \tag{8}$$

where N_j is the number of samples in $C_j(k)$

4. If $z_j(k+1) = z_j(k)$ for j = 1, 2, ..., K then the algorithm has converged and the procedure is terminated.
Otherwise go to Step 2.

It is evident in this description that the final clustering will depend on the initial cluster centres chosen and on the value of K.

Simulation results

Image fusion based on wavelet transform and segmentation results based on Otsu and K means are discussed. Two images (Visual and thermal) are taken as shown in Fig. 1 and Fig. 2. The visual image gives a realistic human sensing view. The thermal image identifies the object with the temperature difference coming into the picture with objects possessing different emissivity values. This paper presents two techniques of image segmentation, Otsu thresholding and Kmeans and, they are tested for fused images and their corresponding segmentation results using the two methods, are shown in Fig. 1 to Fig.10.

Fig.1 Visual image

Fig.2 Thermal image

Fig.3 Fused image

Fig.4 Otsu segmented image

Fig.5 K means segmented image

Fig.6 Visual image

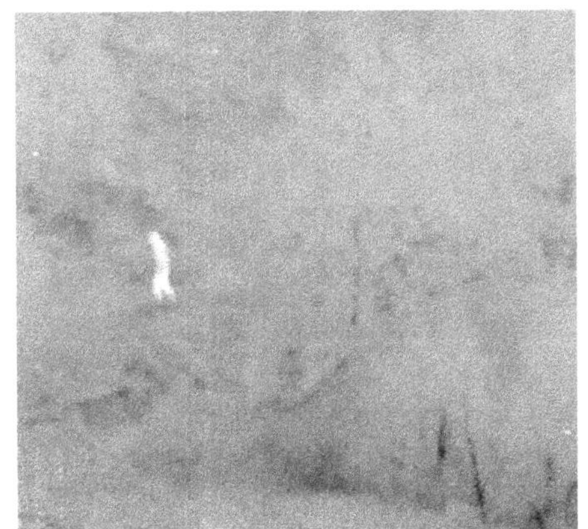

Fig.7 Thermal image

Fig.8 Fused image

Fig.9 Otsu segmented image   Fig .10 Kmeans segmented image

Conclusion

The airborne images obtained from an UAV are analysed in ground control station. By using the thermal images, all weather and night operation are possible. Visual and thermal image fusion is done and the fused image is given for target detection. The effectiveness of the proposed segmentation algorithms are evaluated for fused images, as seen in Fig. 4, Fig.5, Fig.9 and Fig.10. The two algorithms give good segmented images, where the objects of the fused images are almost fully segmented, and Kmeans segmentation is more suitable than Otsu for grey scale with a little noise as seen in the images of Fig.5 and Fig.10.

References

1. R.R. Pitre, X.R. Li, R. Delbalzo, "UAV Route Planning for Joint Search and Track Missions-An Information-Value Approach", *IEEE Transactions onAerospace and Electronic Systems*, vol. 48, no. 3, pp. 2551-2565, 2012.
2. G. Piella, "A General Framework for Multi-resolution Image Fusion: from Pixels to Regions", *Information Fusion*, vol. 4, no. 4, pp. 259-280, 2003.
3. Burt, P.J., Adelson, E.: The Laplacian pyramid as a compact image code. IEEE Trans. Communications 31(4), 532–540 (1983)
4. Toet, A.: Image fusion by a ratio of low-pass pyramid. Pattern Recognition Letters 9(4), 245–253 (1989)660 F. Yao and A. Sekmen
5. Burt, P.J.: A gradient pyramid basis for pattern-selective image fusion, Society for Information Display. Digest of Technical Papers, 467–470 (1992)
6. J.Y. Liu, Q. Wang, Y. Shen, "Comparisons of Several Pixel-Level Image Fusion Schemes for Infrared and Visible Light Images", in *Proceedings of the IEEE Instrumentation and Measurement Technology Conference, (IMTC 2005)*, vol. 3, pp. 2024-2027, May 2005.
7. X.Q. Zhang, Q.L. Chen, T. Men, "Comparison of Fusion Methods for the Infrared and Color Visible Images", in *Proceedings of 2nd IEEE International Conference on Computer Science and Information Technology, (ICCSIT 2009)*, pp. 421-424, 2009.

8.   J. Saeedi, K. Faez, "Infrared and visible image fusion using fuzzy logic and population-based optimization", *Applied Soft Computing*, vol. 12, no. 3, no.1041-1054, 2012.

9.   C. Cattani, A. Kudreyko. "On the Discrete Harmonic Wavelet Transform", *Mathematical Problems in Engineering*, vol. 2008, AID 687318, 7 pages. doi: 10.1155/2008/687318, 2008.

10. OTSU N.A threshold selection method from gray level histogram. *IEEE Transactions on Systems, Man and Cybernetics,* **9**(1),62-66 (1979).

11. J.T. Tou and R.C. Gonzalez, Pattern recognition and principles, Massachusetts: Addison-Wesley, 1974

12. **S Ray and R H Turi**: *Determination of number of clusters in K-means clustering and application in colour image segmentation*, Proceedings of the 4th International Conference on Advances in Pattern Recognition and Digital Techniques ISBN: 81-7319-347-9, pp 137-143,(1999).

ESTIMATION OF FLIGHT PARAMETERS AND TRAJECTORY OF A MAV USING ACOUSTIC SENSORS

A. Saravanakumar and K. Senthilkumar

Division of Avionics, Department of Aerospace Engineering,
Madras Institute of Technology Campus,
Anna University, Chennai-600044, India.
e-mail: saravanakumar_a@yahoo.com

ABSTRACT

An aircraft generates an acoustic impulse that propagates outwards from the source. The position of the source and hence the trajectory can be estimated by measuring the relative time of arrival of the impulse at a number of spatially distributed sensors. The time difference for the acoustic wave front to arrive at two spatially separated sensors is estimated by cross correlating the digitized outputs of the sensors. The time delay estimate is used to calculate the source bearing and the position of the source is found using triangulation technique using the bearings from two widely separated receiving nodes. The flight parameter of the aircraft is obtained by autocorrelation method using acoustic multipath delays. The signal emitted by an UAV arrives at a stationary sensor located above a flat ground via a direct path and a ground-reflected path. The difference in the times of arrival of the direct path and ground-reflected path signal components is known as the multipath delay. A model is developed to predict the variation with time of the multipath delay and based on this model; autocorrelation method is formulated to estimate the speed and altitude of the aircraft.

Keywords: Autocorrelation, Cross correlation, Multipath delay, Trajectory, Sensor

INTRODUCTION

Acoustic techniques have been widely used in variety of fields. An aircraft which is an acoustic source produces sound that can be received by placing sensors on the ground that captures the sound waves propagated through the air. Acoustic sensors are used widely because of their passiveness, affordability, robustness, and compatibility. Ground sensors are often deployed in remote areas for surveillance and early warning purposes. Due to the high levels of acoustic energy radiated by the propulsion systems of aircraft and by the engines of vehicles, it is possible to detect these sources using passive acoustic sensors mounted close to the ground. Here described a technique that utilizes the Lloyd's mirror effect of the radiated sound to estimate the speed, altitude of a low-flying aircraft by using only a single sensor. The localization of acoustic source is usually done with sensor array in which the output of each sensor is a scalar corresponding to the acoustic pressure. For linear trajectory, we consider a different approach for solving this problem using Acoustic Vector Sensor whose output is a vector corresponding to the acoustic pressure and acoustic particle velocity. The main advantage of these vector sensors over traditional scalar sensors is that they make use of more available acoustic information, hence they should outperform sensor array in accuracy of localization. An array of sensors has been used to find the non linear trajectory as it will be a convenient way to find the trajectory of an aircraft.

ACOUSTIC SENSORS

Acoustic wave sensors are so named because their detection mechanism is a mechanical, or acoustic wave. As the acoustic wave propagates through or on the surface of the material, any changes to the characteristics of the propagation path affect the velocity and/or amplitude of the wave. Changes in velocity can be monitored by measuring the frequency or phase characteristics of the sensor and can then be correlated to the corresponding physical quantity being measured. Acoustic sensors are low cost, passive provides non-line-of-sight situational awareness and target acquisition.

Flight parameter estimation methods for an airborne acoustic source can be divided into two categories, depending on whether the narrow- band lines or the broadband component of the received signal spectrum is processed to estimate the flight parameters.

ACOUSTIC VECTOR SENSORS

AVSs comprise three Microflowns sensors and a co-located pressure microphone. The Microflowns measure the three components of the particle velocity vector, which point towards the acoustic source. Acoustic vector sensors have recently come to play an increasingly significant role in defense and security applications.

Any sound field can be described completely by knowing the following dimensions:

• Scalar sound pressure
• Acoustic particle velocity vector.

ACOUSTICAL LLOYD'S MIRROR EFFECT

The parameters of a low-flying aircraft can be estimated using ground-based passive acoustic sensors. For broad band acoustic sources in motion, multipath propagation results in a pattern of interference fringes, known as the Lloyd's mirror effect in the time-frequency distribution of the sensor output. A model has been developed for the acoustical Lloyd's mirror effect observed when an UAV aircraft (or other airborne source of broadband sound) travels with uniform linear motion over an acoustic sensor located above a hard ground.

EXPERIMENTAL PROCEDURE

MULTIPATH PROPAGATION

In a multipath propagation environment, two or more attenuated and delayed replicas of the same radiated signal are received at a sensor. Based on the model of the Lloyd's mirror effect, a method has been formulated to estimate the flight parameters of the source. In that method, the time-frequency distribution of the sensor output is treated as an image. This image is preprocessed to enhance the fringe pattern and then the flight parameters are extracted from the resultant image by optimizing a cost function. The temporal variation of the multipath delay (time difference between the arrival of the ground-reflected path signal and the direct path signal) and then minimize the sum of the squared deviations of the noisy multipath delay estimates from their predicted values over a sufficiently long period of time. The multipath delay is estimated by auto correlating the sensor output over a short time interval. The performances of the proposed flight parameter estimation methods are evaluated using real data recorded from the sensor.

TIME-DELAY MODEL

Consider an airborne acoustic source (aircraft) moving in a straight line at constant subsonic speed v and constant altitude h_t over a hard ground. An acoustic sensor is located at a height h_r above the ground. The source is at the closest point of approach (CPA) to the sensor at time τ_c, with the ground range at the CPA being d_c. However the aircraft which we considered flies at varying velocity and at different altitudes. But for study purpose we consider that it is flying at constant velocity, constant altitude and follow a straight line.

The source emits a broadband random acoustic signal, which arrives at the sensor via a direct path and a ground-reflected path. A model for the temporal variation of the multipath delay is derived below using a quasi-stationary approach. To calculate the multipath delay at a given time t, the source is assumed to be fixed at the position at an earlier time τ ($< t$) which accounts for the sound propagation delay from the source to the sensor. For this model the height of the sensor from the ground surface is about 1 metre. The source-sensor geometry used for the calculation of multipath delay time is depicted below.

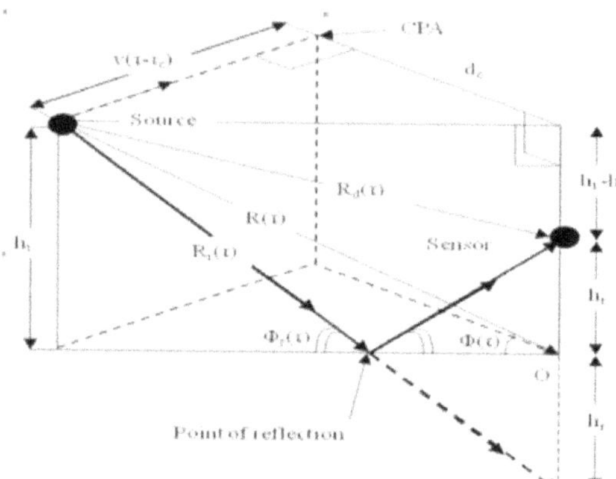

Fig. 1 Source-sensor geometry for calculation of multipath delay at time t, where $\tau = t-R(\tau)=c$

The delay time for the desired model is obtained as

$$D(t) \cong \frac{2(c_r^2 - v_t^2)/c_r^2}{\sqrt{\gamma^2(c_r^2 + v_t^2) + c_r^2 v_t^2(\tau - \tau_c)^2} - v_r v_t(\tau - \tau_c)}$$

where, the values of v_r, v_t, γ, c_r are given by

$$v_r = v/h_r$$

$$v_t = v/h_t$$

$$\gamma = \sqrt{(1 + (d_c/h_r)^2}$$

$$c_r = c/h_r$$

Note that D(t) is a function of $\{v_r, v_t, \tau_c, \gamma\}$ or equivalently the flight parameters $\{v, h_t, \tau_c, d_c\}$.

NON-LINEAR LEAST SQUARES METHOD

The main flight parameters such as $\{v_r, v_t, \tau_c, \gamma\}$ or equivalently $\{v, h_t, t_c, \gamma\}$ estimated using a non-linear least square method. That is by minimizing the sum of squared deviations of delay estimates from their predicted values. Defining the parameter vector as

$$z = [v_r, v_t, \tau_c, \gamma]^T$$

The estimate of z which is $\hat{z} = [\hat{v}_r, \hat{v}_t, \hat{\tau}_c, \hat{\gamma}]^T$ is obtained by minimizing the cost function

$$P(z) = \sum_{k=1}^{K} [\hat{D}(t_k) - D(t_k, z)]^2$$

where,

$\hat{D}(t_k)$ is the multipath delay time at t_k and

$D(t_k, z)$ is the corresponding predicted value using the model for $1 \leq k \leq K$.

Given the sensor height h_r, the speed, altitude and CPA ground range of the source is estimated as

$$\hat{v} = h_r \hat{v}_r$$

$$\hat{h}_t = \hat{v}/\hat{v}_t$$

$$\hat{d}_c = \left| \hat{h}_t \sqrt{(\hat{\gamma}^2 - 1)} \right|$$

The multipath delay at time t_k is estimated using autocorrelation function method.

MODEL FOR LINEAR TRAJECTORY

The relationship between the target trajectory and the single sensor can be simply described by two parameters. One is the slant range from the sensor to the CPA (closest point to approach), denoted as r. The other is x_0, which represents the initial position coordinate to a new axis that is built along the trajectory line and makes the CPA as its origin.

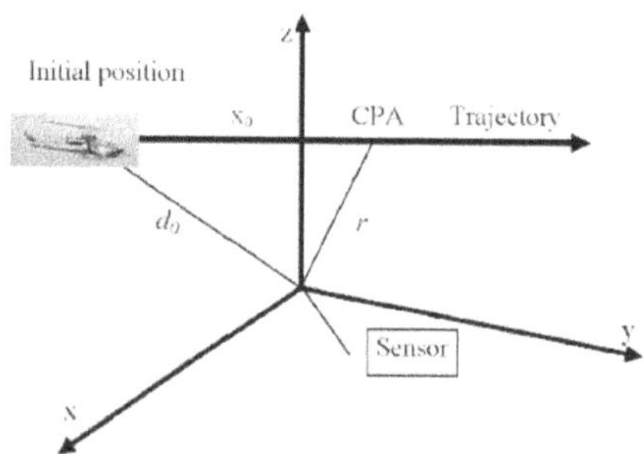

Fig 2 Parameter model for a single sensor

TRIANGULATION METHOD FOR NON LINEAR TRAJECTORY

By knowing the sensor separation distances and the isospeed of sound propagation in the medium, a three-element linear array can be used to estimate the range and bearing of the source by measuring the time delay between the center sensor and each of the other two sensors.

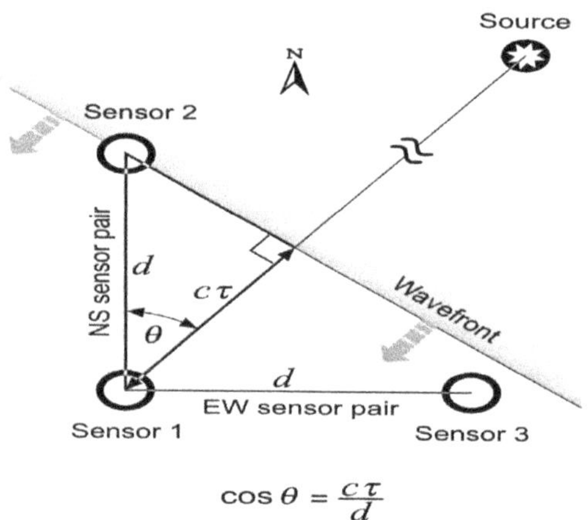

$$\cos \theta = \frac{c\tau}{d}$$

Fig. 3 Triangulation method for finding the position

The above figure shows the principle of triangulation using two nodes. The source and the two nodes, labeled 1 and 2, are located on the $x - y$ plane at coordinates (xs,ys), $(x1,y1)$,and $(x2,y2)$, respectively, with the y axis pointing towards the north. The bearing lines from the two nodes intersect to determine a unique source location. Assuming line-of-sight propagation, the source position is given by

$$x_s = (y_s - y_1)\tan \theta_1 + x_1,$$
$$y_s = (x_2 - x_1 + y_1 \tan \theta_1 - y_2 \tan \theta_2)/(\tan \theta_1 - \tan \theta_2).$$

The source bearing can be estimated using either pair of sensors or both. An estimate of the source bearing with respect to the NS sensor pair axis is given by

$$\theta_{NS} = \cos^{-1}(c\tau_{12}/d)$$

where c is the speed of sound propagation in air and t 12 is the time delay between sensors 1 and 2 and same wise calculated for EW pair. Since both pairs are orthogonal, it could also be found as

$$\theta_{NS} = \tan^{-1}(\tau_{13}/\tau_{12}).$$

AUTO CORRELATION

It is the degree of similarity between a given time series and a lagged version of itself over successive time intervals. It is the same as calculating the correlation between two different time series, except that the same time series is used twice - once in its original form and once lagged one or more time periods. The term can also be referred to as "lagged correlation" or "serial correlation".

DATA ANALYSIS TO ESTIMATE MULTIPATH TIME DELAY

The data from the sensor is processed in overlapping blocks each containing 80000 samples with 50% overlapping between two consecutive data blocks. The autocorrelation of each data block is implemented in the frequency domain using the Fast Fourier Transform (FFT). The location of the peak of the autocorrelation function in the positive time lag axis gives the multipath delay estimate. The multipath delay is calculated for every block and the maximum multipath delay can be obtained by plotting multipath delay for 200 seconds. Error delay time between the estimated and the predicted multipath delay is calculated and it is minimized using a cost function.

CALCULATION OF DIRECTION

Filtering has been done on the pressure signal and velocity vectors and multiplied to obtain the intensity vectors in each axis. By assuming a cylindrical coordinate (r, θ, ϕ) with AVS as origin

$$\theta = \cos^{-1} I_Z/\sqrt{(I_X^2 + I_Y^2 + I_Z^2)}$$

$$\phi = \cos^{-1} I_Z/\sqrt{(I_X^2 + I_Y^2)}$$

Where I_X, I_Y, I_Z are the intensity vectors along X, Y, Z axis respectively. To find the initial direction average of ten samples in each direction has taken and applied the above formula. For getting the direction of CPA time taken from initial position to CPA has calculated using aircraft velocity (v) and distance from CPA to initial position (x_0) both of them obtained through iteration of aircraft parameters.

RESULTS AND DISCUSSION

The total duration of the flight is about 690 seconds i.e., around 11 and ½ minutes. The data has pressure component and three velocity components but the main interest is on pressure component. This is sampled at a frequency of 20 kHz and the number of samples is around 1 crore and 37 lakh. The duration of flight is truncated for about 200 seconds such that this period of duration gives the exact flight period.

 Total duration of flight=689.04695 seconds
 Sampled frequency=20000 kHz
 Total number of samples=13780939
 Truncated duration =200 seconds
 Number of samples=4000000

INITIAL ESTIMATION OF THE PARAMETERS

The initial estimate of $\hat{z}_1 = [\theta_r^0, \theta_\tau^0, \hat{t}_c^0, p^0]^T$ is obtained using the following procedure.

1) Finding the time $\hat{t}_0^0$ at which $\hat{D}^{-1}(t) \equiv 1/\hat{D}(t)$ is the minimum.

2) Computing $\hat{p}^0 = 2\hat{D}^{-1}(\hat{t}_0^{\cdot})/c_r$.

3) Calculating $\theta_r^0 = -c_r \dfrac{\hat{D}_+^{-1} + \hat{D}_-^{-1}}{\hat{D}_+^{-1} - \hat{D}_-^{-1}}$

 and $\theta_\tau^0 = -\left(\dfrac{1}{c_\gamma}\right)\dfrac{\hat{D}_+^{-1}\hat{D}_-^{-1}}{\hat{D}_+^{-1} - \hat{D}_-^{-1}}$

where $\hat{D}_+^{-1}$ and $\hat{D}_-^{-1}$ are the respective gradients of two straight lines that provide best fit to the first few and last few data points of $\hat{D}^{-1}(t)$.

4) Calculating $\hat{t}_c^0 - \hat{t}_0^0 - \dfrac{p^0 \theta_r^0}{c_\gamma \theta_\tau^0}$

CALCULATED PARAMETERS FOR FINDING THE TRAJECTORY

From the recorded pressure signal the following parameters are calculated.
Initial Position from CPA = 14m
Closest Point of Approach = 8.2m
Source frequency = 1692.7 Hz

Time to reach CPA from initial position
t_{CPA}= 5.6 seconds
Initial range to aircraft,d_0= 16.1892m

INITIAL POSITION

In Cylindrical coordinate
$(d_0 ,\theta, \phi) = (16.1892m, 123.1547^0, 32.6635^0)$

In Cartesian coordinate
$(x_1,y_1,z_1) = (11.4020m \quad 7.3134m \quad -8.8656m)$

CLOSEST POINT OF APPROACH

In Cylindrical coordinate
$(r, \theta, \phi) = (8.1807m, 129.3857^0, 44.9625^0)$

In Cartesian coordinate
$(x_2,y_2,z_2) = (5.7617m \quad 3.6956m \quad -4.4800m)$

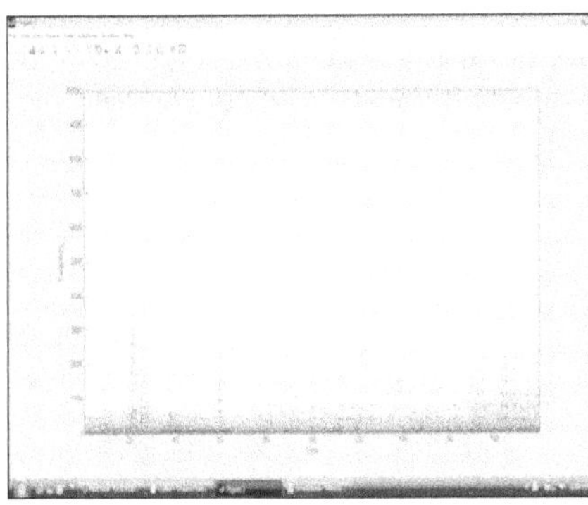

Fig.4 Spectrogram

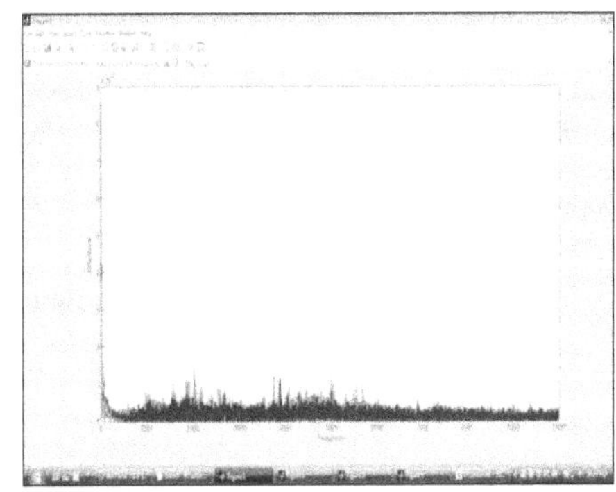

Fig. 5 Fast Fourier transform of a block

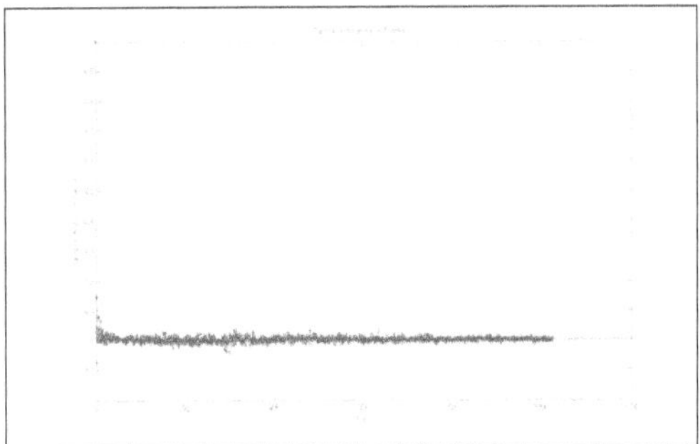

Fig . 6 Autocorrelation of a block

The autocorrelation of a block is shown above. The delayed version a sample has a maximum autocorrelation with the original sample. The time lag between the original and the delayed sample gives the delayed time. So the maximum autocorrelation of all blocks is taken to give the multipath delay time. The depicted figure shows the lag and the autocorrelation values for which a sample having maximum correlation gives the delay value.

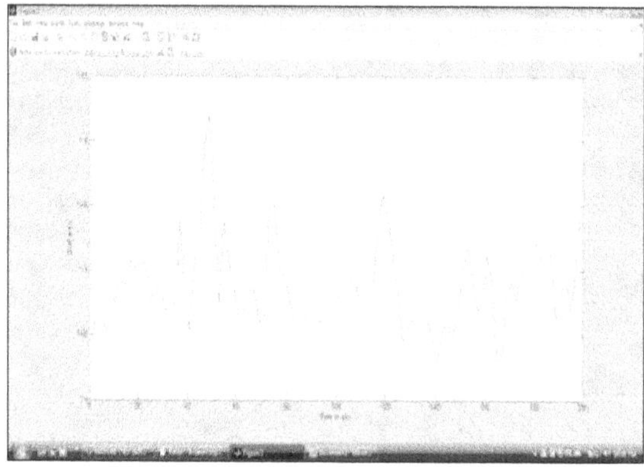

Fig . 7 Multipath delay vs time

The multipath delay is calculated for different values of time. The maximum multipath delay is found to be 1.1ms which occurs at 22nd second. This delay value is used to calculate the delay inverse from which the initial estimation of the parameters is found.

ESTIMATED VALUES OF PARAMETERS

Table. 1. Estimated values of parameters

Type of method	Velocity(m/s)	Altitude(m)
Autocorrelation	91.7299	338.3117

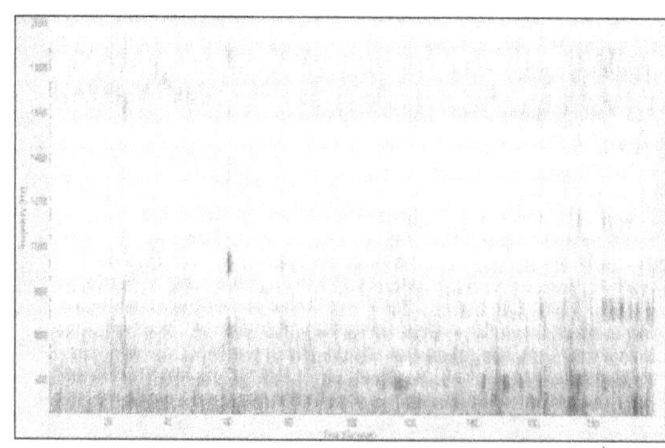

Fig. 8 Spectrogram of Intensity vector in X direction

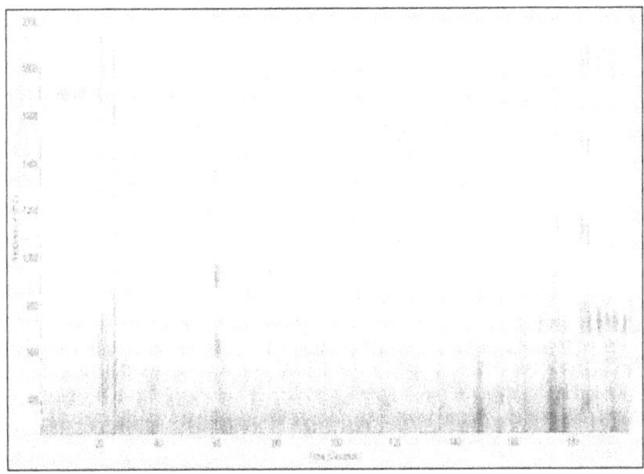

Fig . 9 Spectrogram of Intensity vector in Y direction

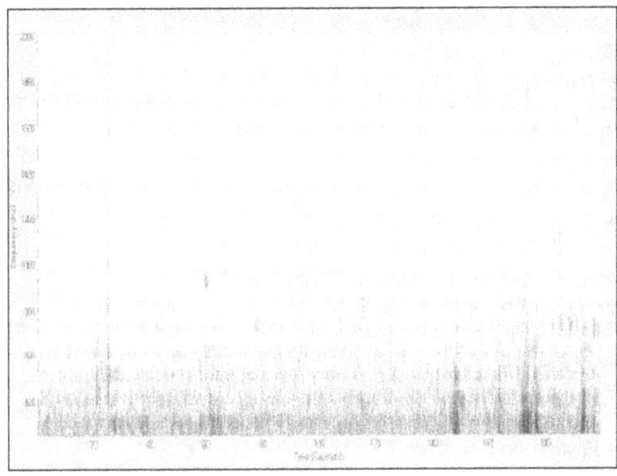

Fig . 10 Spectrogram of Intensity vector in Z direction

FINAL TRAJECTORY

Based on the Cartesian coordinates obtained for initial position and CPA, trajectory has been plotted.

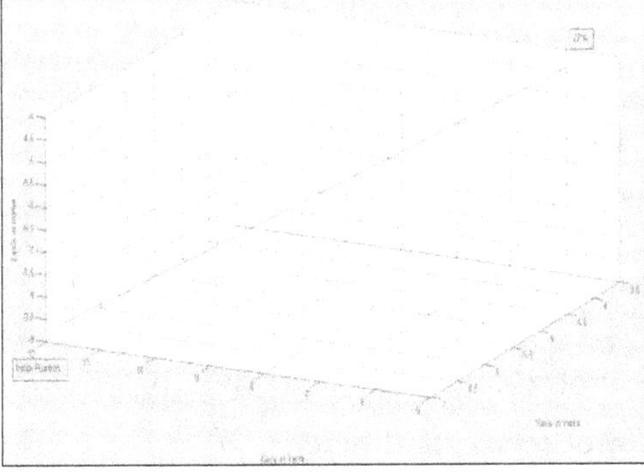

Fig. 11 Linear trajectory of the aircraft

CONCLUSION AND FUTURE WORKS

In this work, flight parameters such as velocity, altitude and straight line trajectory of an aircraft are estimated using a single Acoustic Vector Sensor on the assumption that aircraft following a straight line trajectory. This method exploits the both the scalar as well as vector properties of sound signal for trajectory estimation. NLS algorithm is used to calculate the four flight parameters with the pressure signal obtained from a single AVS. The pressure signal along with velocity vectors are filtered and multiplied to get the intensity vectors. By making use of intensity vectors direction has been found out. Non linear trajectory of the source is found by triangulation method and it is in progress.

REFERENCES

[1]   Andrea Grosso and JeroenVerbeek. "The PU Mini Probe –A Multipurpose Acoustics Sensor", Microflown Technologies, Zevenaar, The Netherlands. Dynamic testing reference issue, Sound & Vibration, pp 17-21,(2010).

[2]   Brian G. Ferguson and KamW.Lo, Turboprop and rotary wing aircraft flight parameter estimation using both narrow-band and broadband passive acoustic signal-processing methods. Acoustical Society of America,(2009).

[3]   Dai Hongyan Zou,Hongxing, Aircraft motion parameter estimation via multipath time-delay using a single ground based passive acoustic sensor. Journal of electronics,(2007).

[4]   TaoWang, QunWan, Motive Parameters Estimation Using Narrow-band Passive Acoustical Measurements. IEEE Transactions of Aerospace and electronic systems,(2007).

[5]   Kam W.Lo, Brian G.Ferguson, YujinGa,Lain Maguer, Aircraft Flight parameter Estimation using Acoustic Multipath delay. IEEE Transactions of Aerospace and electronic systems Vol.39,No.1,(2003).

[6]   Locating far-field impulsive sound sources in air by triangulation BrianG. Ferguson, Lionel G Criswick, Kam W.Lo Acoustical Society of America,(2002).

[7]   KamW. Lo, B.G. Ferguson, Broadband Passive Acoustic Technique for Target Motion Parameter Estimation, IEEE Transactions of Aerospace and electronic systems,(2000).

[8]   KAM W. LO,B. G. Ferguson, Broadband Passive Acoustic Technique for Target Motion Parameter Estimation, IEEE Transactions of Aerospace and electronic systems,(2000).

[9]   K.W. Lo, B.G. Ferguson, Passive estimation of aircraft motion parameters using destructive interference between direct and ground reflected sound waves. Information, Decision and Control Proceedings,(1999).

[10]   K. W. Lo, S. W. Peny, B. G. Ferguson. An image processing approach for aircraft flight parameter estimation using the acoustical Lloyd's mirror effect. Proceedings of the Fifth International Symposium on Signal Processing and Its Applications, vol.2,(1999).

[11]   Arye Nehorai and Eytan Paldi, Acoustic Vector-sensor array processing, IEEE Transaction on Signal processing, Vol. 42, No.9, pp. 2481-2491,(1994).

[12]   Friedlander, Accuracy of source localization using multipath delays. IEEE Transactions of Aerospace and electronic systems,(1988).

DETECTION OF LOW FLYING MAV BASED ON ITS ACOUSTIC SIGNATURE

K.M. Sathis Kumar, A.Saravanakumar and K.Senthilkumar

Division of Avionics, Department of Aerospace Engineering,

Madras Institute of Technology Campus,

Anna University. Chennai-600044. India.

Email:k.m.sathiskumar@gmail.com, saravanakumar79@gmail.com

ABSTRACT

Detection of UAV/Aircraft by radar technology in low altitude is difficult due to its low radar cross section. A novel method of recognition and detection of UAV (Unmanned aerial vehicle)/aircraft when flying in low altitude using low cost ground based acoustic sensor is proposed here. The proposed work investigates to discover and distinguish the UAV/Aircraft by computing energy spectral density (ESD) of its acoustic signature. This task will be accomplished by formulating an algorithm in MATLAB which detect and identify an enemy aircraft/UAV flying in low altitude out of 350 Aircraft/UAV acoustic signatures. These 350 aircraft/UAV acoustic signatures are recorded in real time environment as well as downloaded from the Internet for experimental purpose. The algorithm is examined against each of the 350 acoustic signatures one by one based on Euclidean distance and cross correlation, which affords an accuracy of 99%.These research can be done with help of simple acoustic detectors (wav recorder) established in a desirable location such as country's Line Of Control (LOC). Such a system would be able to collect the threat information and pass it to nearby airbase station.

Keywords: Energy Spectral Density (ESD); Euclidean distance; Cross correlation; Fast Fourier transform (FFT).

INTRODUCTION

Detection and recognition of small UAV/Aircraft flying in low altitude is an important problem. Due to their small size or if the vehicle is flying in very low altitude because of limited radar cross section area they are often difficult to detect with radar technology. This effort analyzes the spying and recognition of UAV/ aircraft using ground based acoustic array of sensor. These ground based acoustic array does not depend on the size of the aircraft for detection, but rather the sound of the engine or propeller, and can therefore serve as an effective acoustic signature of detecting unmanned aerial vehicles /aircraft. A wide variety of real-world targets emit distinct acoustic signatures that not only distinguish them from one another but also provide spectral separation from background clutter. For example, acoustic signature of ground and air vehicle differs from each other. While acoustic signature for each target differs, it made us a easy job to detect and identify the target efficiently. So, this research mainly focuses on ground based acoustic array of sensor which can independently detect and recognize an enemy aircraft/UAV flying in low altitude.

MOTIVATION FOR PROPOSED WORK

In the mundane world threat by any other country has been increased day-by-day, because of the technological development, detection and identification of aerial vehicle by existing radar technology is difficult in the case of low altitude. How the biological earring systems of human being able to identify or distinguish one human voice from other background noise with help of neural intelligence. They might be able to know exact human voice by hearing the same sound one or more time. This effective identification of particular human voice is achieved by storing the sound in memory and used in respective time. This same logic is applied to practical detection and identification system based on some characteristic feature of every sound signal like ESD, power spectral density.

EXPERIMENTAL SETUP

Experimental setup was implemented with simple ground based acoustic detector (wav recorder), which is connected to the computer. Initially the experiment was conducted with single acoustic sensor is located 1.5m above the ground. This same experiment can be conducted with deferent height like (2m,2.5m),then the experiment was conducted with four acoustic sensors are in a same height. Here single sensor data is enough to detect and identify UAV/aircraft but we tried to localize the aircraft with help of four sensor or remaining three sensor can use as redundancy purpose. For experiment we took four UAV with different type of propeller, electric motor (engine), each one produces distinguish acoustic signature. The acoustic signature of four UAV collected with the range of 100m from the acoustic detector.

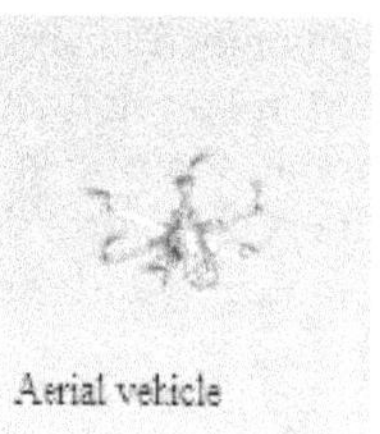

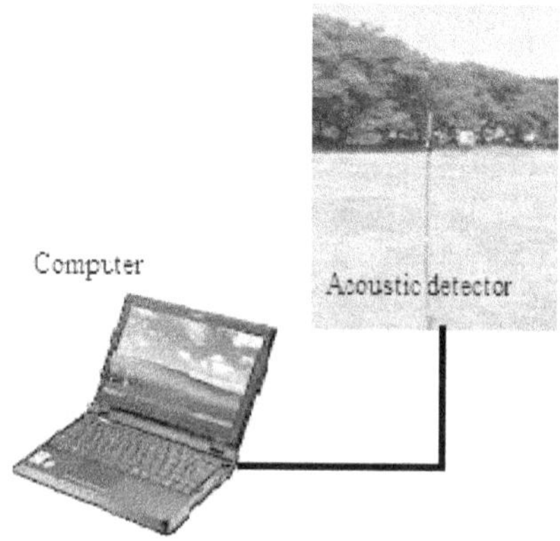

Figure 1. Testing setup of whole detection system

ALGORITHM IMPLEMENTED

Data base is created with acoustic signature of various types of fighter and UAV which are downloaded from the internet as well as recorded in real time environment. Database consisting of signal x_1, x_2, x_3 ...... x_n, normalization process imposed on all the signal of database because the acoustic signature of same aircraft with strength or weak signal can lead to false results. To avoid those faulty results each and every signal of data base and unknown signal is normalized. After completing normalization process energy spectral density of every database signal and unknown signal has been computed with help of Fourier transform. If input signal x has a length N, and then the FFT of signal with length N is

$$X(k) = \sum_{n=1}^{N} x(n) \exp\left(-j2\pi(k-1)\right)(n) \; ; \; 1 \le k \le N \qquad (1)$$

Energy spectral density is computed by taking squared magnitude of FFT coefficients. If X(k) is FFT of signal with length N, then energy spectral density of signal is;

$$1 \le k \le N \qquad (2)$$

Similarity between the energy spectral densities of database signal with energy spectral density of unknown signal is found by calculating Euclidean distance and cross correlation between the two signals are the famous method for 3dimentional geometry, science, economics, Image identification. If Database signal X=(x1,x2,x3........xn) and unknown signal Y=(y1,y2,y3.......yn) then Euclidean distance is estimated as

$$ED = \sqrt{(x_1 - y_1)^2 + (x_2 - y_2)^2 + \cdots (x_n - y_n)^2} \qquad (3)$$

Cross correlation is useful function to find out the similarity between the two signals. Let denote X1*X2 is cross correlation of two signals x1 and x2 then

$$X1(n)*X2(n) = \sum_{i=1}^{n} (yt - xt)^2 \qquad (4)$$

TEST RESULTS

These are the test results obtained after executing the proposed algorithm in the MATLAB. Figure3 shows the energy spectral density of database signal and unknown signal. Table 1 shows Euclidean distance and cross correlation between the various signals with unknown signal.Figure4 shows spectrogram of aircraft versus beep signal.

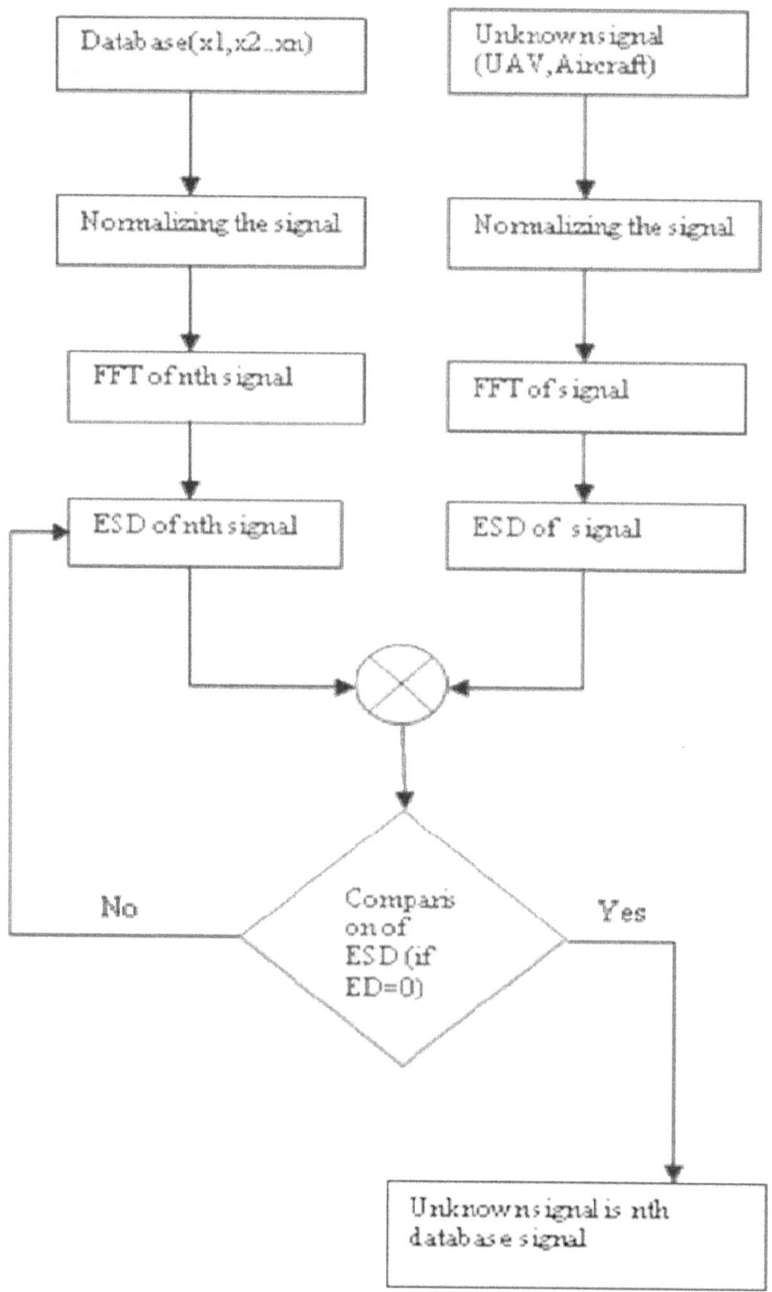

Figure 2. Architecture of implemented Algorithm

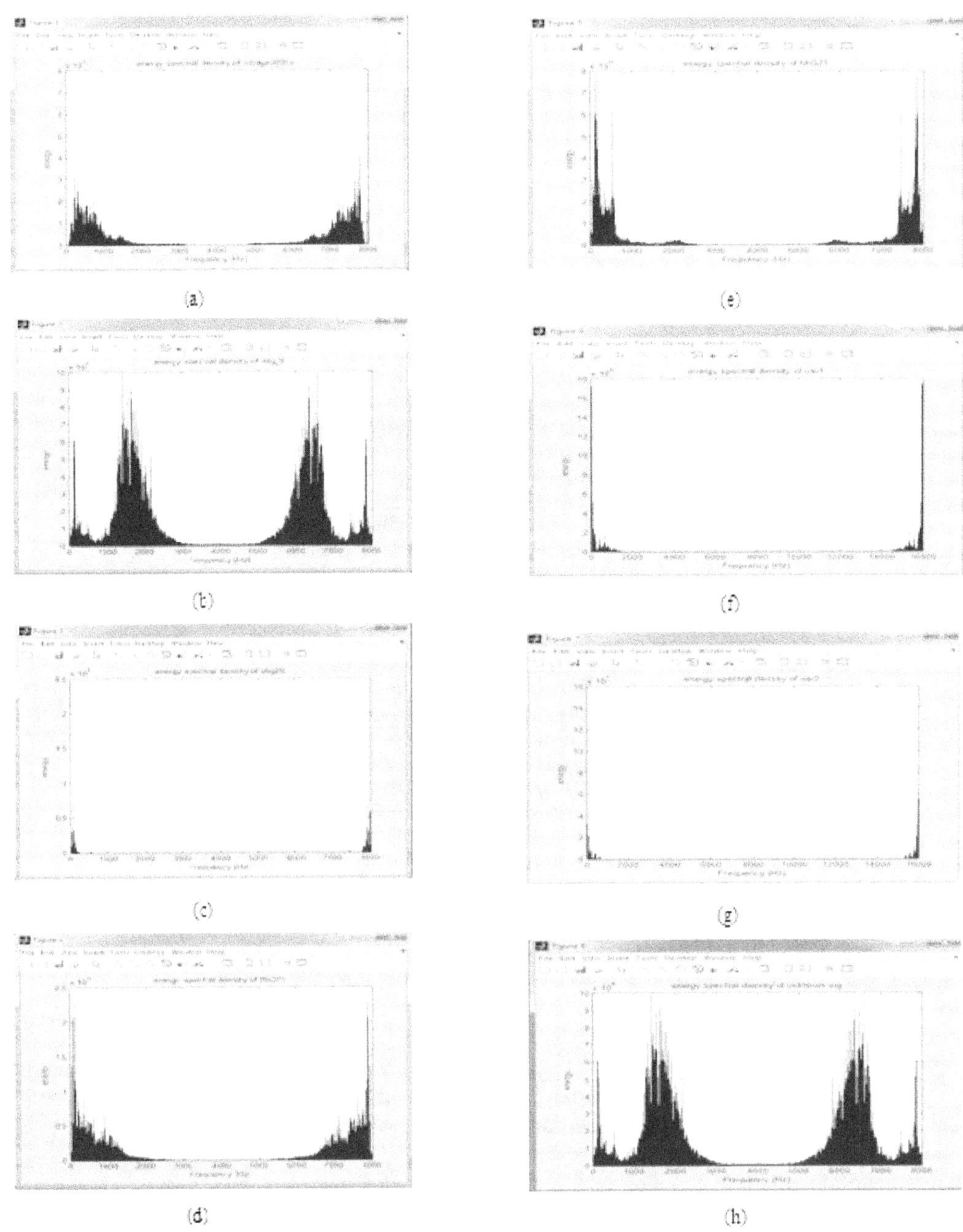

Figure 3. Energy spectral density of signal (a).Mirage2000s (b).Mig19(c).Mig29 (d).Mig25 (e).Mig21(f).UAV1(g).UAV2(h).Unknown signal

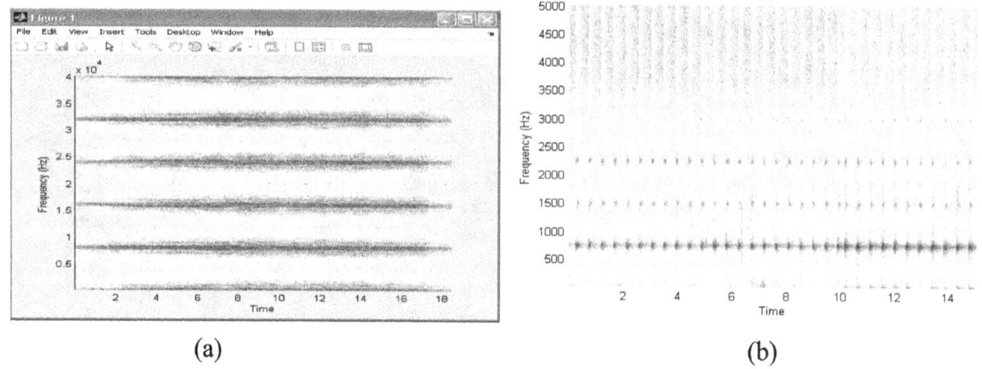

(a)   (b)

Figure 4.Spectrogram of Signal (a).Aircraft (b).Beep sound

TABLE 1 Comparision of Euclidean Distance and Cross Correlation for Decision Making

Database signal	Cross correlation between signals	Euclidean distance between signals
Mirage 2000s	CR1= 0.05961	ED1 =8.4041e+006
Mig19	CR2=1.00000	ED2=0
Mig29	CR3= 0.02631	ED3=6.4813e+006
Mig25	CR4=0.01101	ED4=3.7814e+007
Mig21	CR5= 0.00192	ED5= 3.1639e+006
UAV1	CR6= 0.01279	ED6=1.00741e+09
UAV2	CR7= 0.00658	ED7=2.00562e+08

RESULTS AND DISCUSSION

We have taken seven sound files like mirage2000s, mig19, mig29, mig25, mig21, UAV1, UAV2 for testing the implemented algorithm and from the seven signal one signal taken as a unknown signal. Now we have taken energy spectral density as a one characteristic parameter for each sound file. We can see energy spectral density for each signal varies as shown in figure3 but energy spectral density of unknown signal exactly matches with one of the seven sound signal i.e. from the result we can say unknown signal precisely matches with Mig19.Similarity between the signal is calculated with help of Euclidean distance and cross correlation method. From the result of table, we observed that Euclidean distance corresponding to Mig19 shows very minimum value (i.e. zero) and maximum of cross correlation(i.e. one) which shows unknown signal is belongs to the Mig19 in this case.Figure4 helps to distinguish aircraft versus false alarm signal(bird, car, human being),spectrogram shows different harmonic pattern for aircraft and beep sound. Human speech can be identified with help of MFCC[14](Mel frequency cepstral coefficients).

CONCLUSION

In future utilization of unmanned aerial vehicle (UAV) in every country will increase where size of UAV will be so minimum so it can be easily used for covert operation where radar technology will not be more effective. So this low altitude UAV/aircraft acoustic signature detection system can work efficaciously. The conventional radar emits electromagnetic radiation can sense by anti-radiation missile whereas acoustic signature detection system consist of passive acoustic sensor which cannot be detect easily. This proposed algorithm can be extended with help of cross correlation method. The proper designing of these acoustic detection systems can also be extensively used in ground vehicle identification while violating traffic rules.

REFERENCES

[1]   T. Pham and L .Sim , "Acoustic detection and tracking of small, low- flying threat aircraft ," in23rd Army Science Conference, n o. JP - 25, December 2002.

[2]   T. Pham and L .Sim , "Acoustic detection and tracking of small, low- flying threat aircraft ," in23rd Army Science Conference, no. JP - 25, December 2002.

[3]   T. Pham, N. Srour "TTCP AG-6: Acoustic Detection and Tracking of UAVs", Proc. Of SPIE Vol. 5417, 24-30.

[4]   Clayton Stewart, Victor Larson "Detection and Classification of Acoustic Signals from Fixed-Wing Aircraft",   CH3051-0/91/0000-0025 ' 1991 IEEE.

[5]   Measures   of   distance   between   samples: Euclidean.   Available: http://www.econ.upf.edu/~michael/stanford/maeb4.pdf.

[6]   Aircraft Sounds. Available: http://www.aviationtrivia.info

[7]   Springer "Handbook of Acoustics", Rossing Editor.

[8]   P. Stoica and R.M oses, Spectral analysis of signals. Pearson Prentice Hall, 2005.

[9]   Mario E. Munich "Bayesian subspace methods for acoustic signature recognition of vehicles".

[10] Li Liu. "Ground vehicle acoustic signal processing based on biological hearing models". Master's thesis, University of Maryland, College, Park, 1999.

[11] H.C. Choe, R.E. Karlsen, T. Meitzler, G.R. Gerhart, and   D. Gorsich. Wavelet-based ground vehicle recognition using acoustic signals.Proc. of the SPIE , 2762:434–445, 1996.

[12] G. P. Mazarakis and J. N. Avaritsiotis. "Vehicle classification in sensor networks using time-domain signal processing and neural networks". Microprocess. Microsyst.31(6) 381–392, 2007

[13] Russell Braunling, Randy M . Jensen, and Michael A . Gallo, "Acoustic target detection, tracking, classification, and location in a multiple target environment," in Peace and Wartime Applications and Technical Issues for Unattended Ground Sensors, 1997, vol.3081,pp.57–66.

[14] Almudena Lindoso, Luis Entrena, Judith Liu-Jimneez, Enrique San Milan, "Increasing security with correlation based finger printing." 41st annual IEEE International carnahan Conference on security technology, 8-11 October 2007.

[15] Zheng F., Zhang, G., Song, Z., "Comparison of different implementations of MFCC", J. Computer Science & Technology, 16(6):582-589, Sept. 2001.

MOVING OBJECT SEGMENTATION USING ADAPTIVE BACKGROUND SEGMENTATION TECHNIQUE FOR UAV VIDEOS

Nuzrath Hameedha A[1],   **Athilingam R[2],**   **Senthil Kumar K[3]**

[1] Student , [2]Research Scholar , [3] Associate Prof,
Division of Avionics, Department of Aerospace Engineering,
Madras Institute of Technology, Anna University, Chennai 600044, India
nuzrath5@gmail.com

ABSTRACT

This paper proposes an algorithm that detects moving objects robustly from videos that are obtained from UAV (Unmanned Air Vehicle). Existing algorithms face a common problem, when it comes to Videos from UAV. The problem is timely varying camera position, because of the movement of its platform. While segmenting object of interest, the relative motion of camera and object should be considered. The proposed algorithm effectively detects the object of interest from the video considered. High Definition Input videos are obtained from Dhaksha (UAV) and processed by using proposed algorithm using MATLAB.The experimental results show that the proposed method works effectively in dynamic environment.
Keywords: UAV, High Definition.

INTRODUCTION

The capability of extracting moving objects from video sequence captured is a typical first step in aerial surveillance applications .Background Subtraction is a widely used approach for detecting moving objects in videos . Moving object detection provides a classification of the pixels in the video sequence into either foreground (moving objects) or background. A common approach used to achieve such classification is background removal, sometimes referred to as background subtraction, where each video frame is compared against a reference or background model, pixels that deviate significantly from the background are considered to be moving objects .Background modelling, also referred to as background maintenance, is at the heart of any background removal algorithm .A set of principles to which background modelling modules should adhere. The module performing background modelling should not attempt to extract the semantics of foreground objects on its own, since it is not an end by itself, larger systems use it as a component. One can evaluate background modelling by how closely it comes to finding all foreground, when a foreground object *first* appears in the scene, while simultaneously ignoring all others. Backgrounds are not necessarily defined by absence of motion, e.g. waving trees. No unimodal distribution of pixel values can adequately capture such a background, because these models implicitly assume that the background is static. An appropriate pixel-level stationarity criterion should be defined. Pixels that satisfy this criterion are declared background and ignored. The background model must adapt to both sudden and gradual changes in the background.

RELATED WORKS

The moving object detection system proposed by Lucia maddelena et al[8] is expected to solve the issues such as scenes containing moving backgrounds and gradual illumination variations, shadows cast by moving objects. Besides from intrinsic usefulness of segmenting video streams, the algorithm for moving object detection focuses on recognition, classification and activity analysis since only moving pixels were considered were considered by R.T Collins et al .S. T. Syed Shazali et al.[3] in their work entitled as "Motion Detection Using Periodic Background Estimation Subtraction Method" portrayed a method in which the background image is estimated at every 0.8 second when the sum of absolute difference (SAD) is less than the motion threshold. Kaiqi Huanga et al.[1] proposed an algorithm effective for outdoor night surveillance using contrast analysis. The contrast in local change over time is used to detect potential moving objects. Then motion prediction and spatial nearest neighbour data association are used to suppress false alarms.Anuva Chowdhury et al. [2] considered color information for frame averaging in the paper "A Background Subtraction Method using Color Information in the Frame Averaging Process". Shigang wan et al.[4] calculated the four order statistics of five frames difference are used to automatically separate the motion region from background, and to extract binary mask of moving object, and then to further use morphologic method to fill the gained binary mask. Songyin Fu et al.[5] computed the statistical parameters for background model according to the way in which pixel changes and then extracted the object from background mask.

BACKGROUND REMOVAL

Background removal algorithms normally passes through four major steps, which are (1) pre-processing (simple image processing tasks that change the raw input video into a format that can be processed by subsequent steps), (2) background modelling (also known as background maintenance), (3) foreground detection (also known as background subtraction) and (4) data validation (also referred to as post-processing, used to eliminate those pixels that do not correspond to actual moving objects). Although the terms background subtraction and background modelling are often used interchangeably, they are separate and distinct processes. Background modelling refers to the process of creating, and subsequently maintaining, a model of the appearance of the background in the field of view of a camera. Background subtraction refers to the process in which an image frame is compared to the background model in order to determine whether individual pixels are part of the background or the foreground. So it is also referred to as foreground detection.

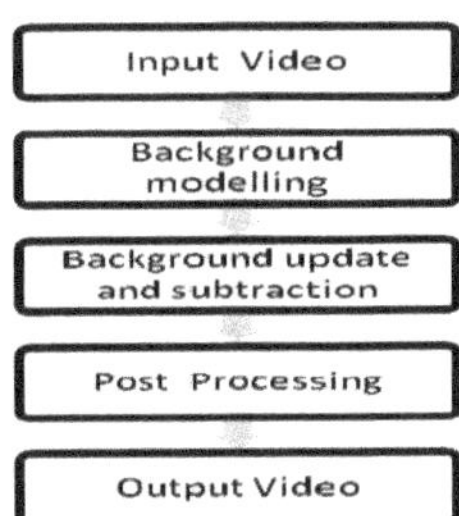

Figure1. Background Removal Methodology

PROPOSED METHOD
BACKGROUND MODEL

Running average is used to model the background , because of its computational simplicity and low memory requirements. The model is based on ideally fitting a Gaussian probability density function (pdf) on the last n pixel's values. In order to avoid fitting the pdf from scratch at each new frame time, t, a running (or on-line cumulative) average is computed instead as:

$$\mu_t = \alpha I_t + \left(1 - \alpha\right) \mu_{t-1}$$

where I, is the pixel's current value and ,u, the previous average; a is an empirical weight often chosen as a trade-off between stability and quick update.

ADAPTIVE THRESHOLDING

The goal of thresholding an image is to classify pixels as either "dark" or "light".Image thresholding segments a digital image based on a certain characteristic of the pixels (for example, intensity value). The goal is to create a binary representation of the image, classifying each pixel into one of two categories, such as "dark" or "light".Under thresholding and Over thresholding will pose problems during foreground detection.

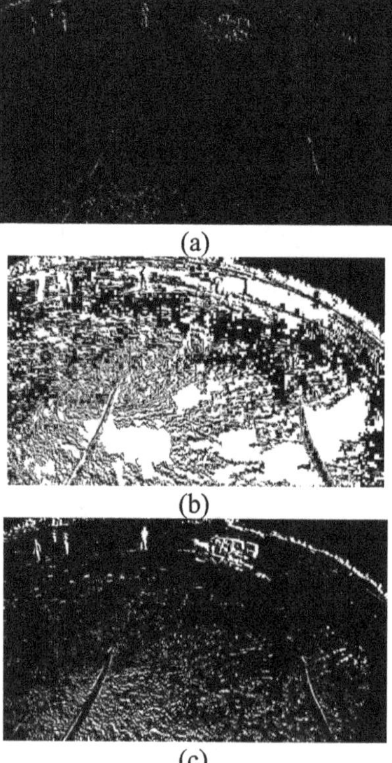

(a)

(b)

(c)

Figure 2. Thresholding

Figure 2 shows results different kinds of thresholding considered .Figure 2(a) shows under thresholding(very low threshold value) and (b) represents over thresholding(very high threshold value).Figure 2(c) represents proposed adaptive thresholding method. Here Mean of the particular frame considered is used as threshold for segmenting Foreground objects.

FOREGROUND DETECTION

The foreground objects are being detected from the video based on the difference between current frame and background model.The foreground detection is effected by thresholding the difference image.The algorithm is stated as follows.

Initially it is assumed that the background model Bi is the first frame of the input video

$$B_i = F_i , i=1 \tag{1}$$

The background model B_{i+1} is obtained by the difference between two consequent frames for i=1,2…,n with the learning rate α.

$$B_{i+1} = \alpha F_i + (1-\alpha) B_i, \tag{2}$$

where α ranges from 0 to 1.

The background update Gi is obtained by subtracting the subsequent frames Fi with the background model Bi

$$F_i - B_i = G_i \tag{3}$$

The resultant binary image R_i is obtained based on the value of threshold T.

$$R_i = \begin{cases} 1, & G_i > T \\ 0, & G_i < T \end{cases} \tag{4}$$

Here the threshold is not fixed. It is made adaptive by calculating the mean of all the pixels in each frame.

$$T = \text{mean } (Gi) = \frac{1}{n^2} \sum_{i,j} G(i,j) \tag{5}$$

where, i, j are the rows and columns in a frame

POST PROCESSING

Data validation is the process of improving the candidate foreground mask based on information obtained from outside the background model. Data validation phase is sometimes referred to as the Post-processing phase of the foreground mask (pixels). There are two kinds of misclassifications that may occur in segmentation results. False positives occur when background regions are incorrectly labelled as foreground. Conversely, false negatives occur when foreground regions are classified as background. Data validation aims to reduce the number of such misclassifications without an appreciable degradation in classification speed. Here median filters are used to remove small groups of pixels that differ from their neighbours' labels (salt and pepper noise) or morphological operations to smooth object boundaries. Post-processing can be applied either to the binary image representing the foreground map Ft(x, y) resulted from the foreground detection phase only, or to both the binary image and the original frame. Here 3×3 median filter is used.

PERFORMANCE ANALYSIS

Performance evaluation allows the appropriate selection of segmentation algorithms as well as adjusts their parameters for optimal performance. The current practice for evaluation involves a representative group of human viewers which is subjective, time consuming and expensive process. Two types of measurements can be targeted when performing video segmentation evaluation; individual object segmentation evaluation when one of the objects identified by the segmentation algorithm is independently evaluated in terms of its segmentation quality, which is valuable when objects are independently manipulated, e.g. for reusing in different contexts, and overall segmentation evaluation when the complete set of objects identified by the segmentation algorithm is globally evaluated in terms of its segmentation quality.

The process of performance evaluation consists of following steps

- Ground truth is generated ,manually.(as the process of establishing the "correct answer" for what exactly the algorithm is expected to produce)

- Resultant image is compared with Ground truth and Pixel based Performance metrics are calculated (TP, TN, FP, FN).

- From the values of TP,TN,FP,FN, Information retrieval measurements (Recall and Precision) are calculated

True Positives (TP)
The number of foreground pixels correctly detected;

False Positives (FP)
The number of background pixels incorrectly detected as foreground (also known as false alarms);
True Negatives (TN)
The number of background pixels correctly detected;
False Negatives (FN)
The number of foreground pixels incorrectly detected as background (also known as misses).
Recall
The ratio of the number of foreground pixels correctly identified by the algorithm to the number of foreground pixels in ground truth.

$$RECALL = \frac{TP}{TP+FN} \tag{6}$$

Precision
The ratio of the number of foreground pixels correctly identified by the algorithm to the number of foreground pixels detected by the algorithm.

$$PRECISION = \frac{TP}{TP+FP} \tag{7}$$

Experimental Results

The proposed background subtraction algorithm is successfully tested on videos obtained from Dhaksha (an efficient scalable UAV from MIT, Chennai). The videos are transmitted with the transmission frequency of 1.2 GHz from UAV DHAKSHA to GCS, with altitude ranging from 10 to 50 meters. The properties of the videos are listed in the following Table No: 1.

Table No: 1 Input Video Properties

Properties	Video Used
Frame rate	60/second
Size	1280 × 720
Objects	Static and Moving objects

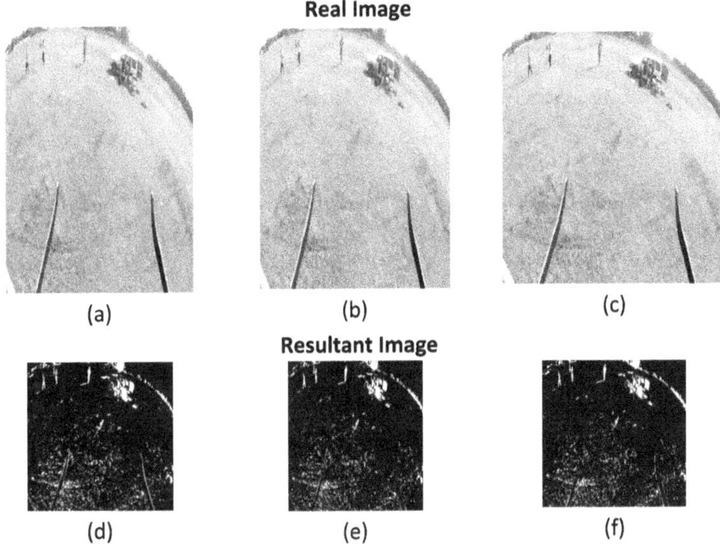

Figure 3. Results –Adaptive Background Subtraction

Sample frames of the video considered is as shown in Figure 3.In Figure 3 a,b,c represents the input frame and d,e,f represents the resultant frame. Recall and Precision values are found to be 0.5256 and 0.1526, from which we infer that the proposed algorithm is working efficiently.

CONCLUSION

In this paper, an effective background subtraction method is proposed for UAV surveillance application. The core of the algorithm includes (1) establishing a simple and effective background model and (2) using adaptive thresholding algorithm. The algorithm can extract moving objects in real time. Experimental results show that the proposed algorithm is promising.

REFERENCES

[1] Kaiqi Huanga, LiangshengWanga, Tieniu Tana, Steve Maybankb,2008 "A real-time object detecting and tracking system for outdoor night surveillance", Pattern Recognition 41.

[2] Anuva Chowdhury, Sang-jin Cho,Ui-Pil Chong , 2011, "A Background Subtraction Method using Color Information in the Frame Averaging Process" ,The 6th International Forum on strategic Technology , 1275-1279

[3] S. T. Syed Shazali, W. L. Cheong, S. Mohamaddan, A. M. N. Abg Kamaruddin, A. Yassin, K. Case, ,2011 "Motion Detection Using Periodic Background Estimation Subtraction Method" , *International Conference on IT in Asia (CITA)*.

[4] Shigang wang , Xuejun Wang, Hexian Chen , 2008 , " Video Object segmentation based on frame differences and its implementation on DSP" ,The Institution of Engineering and Technology, 618-621.

[5] Songyin , Gangyi Jiang, Meu Yu2010, , "An Effective Background Subtraction Method Based on Pixel Change Classification", International Conference on Electrical and Control Engineering , 4634-4637.

[6] Xiaoyu Wu, Lei Yang, Cheng Yang, 2009 "Automatic Real-time Video Background Segmentation System", IEEE.

[7] Yongquan Xia, Shaohui Ning, Han Shen,2010, "Moving Targets Detection Algorithm Based on Background Subtraction and Frames Subtraction" ,International Conference on Industrial Mechatronics and Automation, 122-125.

[8] Lucia maddelena ,Alfredo petrisino, ,July 2011 " A novel organizing approach to background subtraction for visual surveillance applications", IEEE transactions on image processing-Vol 17, No 1.

ON-BOARD TELEMETRY SYSTEM FOR MOVING OBJECTS

Umakanth Jasthi,
umakanth.jasti@yahoo.com

Bhaskar Gorle,
bhaskar_gorle@yahoo.co.in

and

Vegesna Ravivarma
vegesna.ravivarma@gmail.com

ABSTRACT

The advent of UAV has changed the methods of transportation drastically in both military and civilian applications. This is due to the fact that every day, the world demands more efficient, faster and more accurate ways of transporting people and goods to their destinations. Therefore, the on-board instrumentation suite in the aircraft determines much about the usefulness to Navigate and guide the UAV system from a remote monitoring station.

For an UAV to be controlled or autonomous beyond the visual range, sufficient on-board instrumentation is needed to accurately monitor the available alttitude and position information. In general, alttitude and position information can be obtained from a series of sensors equipped on UAVs, such as vision cameras, GPS, gyros, accelerometers, magnetometers and baro-altimeter etc. The On-board Telemetry system is a Bidirectional Communication system to communicate between Ground Stations and UAV. The Telemetry system mainly consists of Telemetry Transmitter and Tele-Command Receiver. The ground station receives the navigation information, demodulates, process the digital data and control / command the UAV guiding towards the target.

INTRODUCTION

Telemetry is the process by which an object's characteristics are measured (such as velocity, pressures, temperatures etc of an aircraft/missile), and the results transmitted to a distant station where they are displayed, recorded, and analyzed. The transmission media may be air and space for satellite applications

The field of telemetry developed from the need to transmit information from one location to another, primarily because of the inability to monitor the signal source directly. The primary objective of telemetry is to transmit accurate information between distant or remote locations.

A telemetry system, also known as a telemetering system, is an electrical/electronic system used for measuring a source quantity and transmitting the measured source quantity to a distant receiving station where the source measurements can be displayed and/or recorded.

In today's telemetry applications, which support large numbers of measurands, it is too costly and impractical to use separate transmission channels for each measured quantity. The telemetry process involves grouping measurements (such as pressure, speed, and temperature) into a format that can be transmitted as a single data stream. Once received, the data stream is separated into the original measurement's components for analysis.

Telemetry lets you stay in a safe (or convenient) location while monitoring what's taking place in an unsafe (or inconvenient) location. Aircraft development, for example, is a major application for telemetry systems. During initial flight testing, an aircraft performs a variety of test maneuvers. The critical flight data from a maneuver is transmitted to flight test engineers at a ground station where results are viewed in real time or analyzed within seconds of the maneuver. Real-time monitoring allows the "safety officer" to make instant decisions on whether to proceed with or terminate a test. With real-time analysis, the flight test engineer can request a maneuver be repeated, the next maneuver be performed, or test plan alternatives be substituted. Real-time data is also captured to storage media, such as disk and tape, for analysis and archiving.

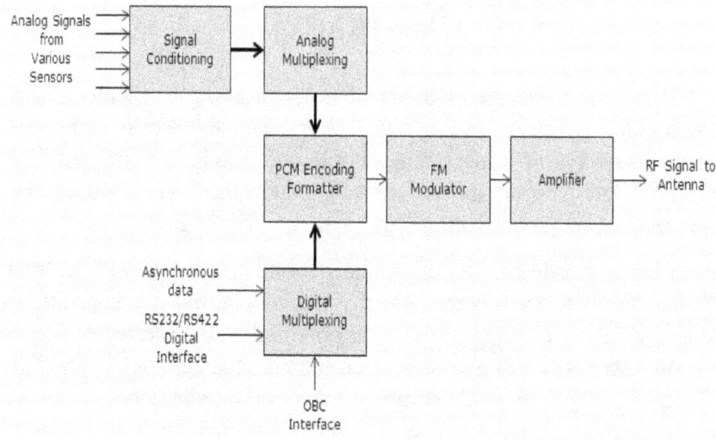

Block Schematic of On-Board Telemetry Transmission System

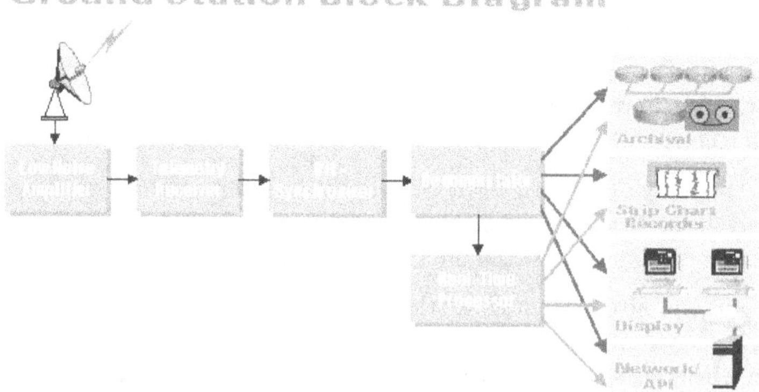

Block Schematic of Ground Telemetry Receive System

OVERVIEW

Today's telemetry systems are built from commercial-off-the-shelf (COTS) products. But while they all have many common elements, they are each uniquely configured to meet specific application requirements.

A telemetry system is often viewed as two components, the Airborne System and the Ground System. In actuality, either or both may be in the air or on the ground.

Data acquisition begins when sensors (aka, transducers) measure the amount of a physical attribute and transform the measurement to an engineering unit value. Some sensors produce a voltage directly (thermocouples for temperature or piezoelectric strain gages for acceleration), while others require excitation (resistive strain gages, potentiometers for rotation, etc.). Sensors attached to signal conditioners provide power for the sensors to operate or modify signals for compatibility with the next stage of acquisition. Since maintaining a separate path for each source is cumbersome and costly, a multiplexer (Commutator) is employed. It serially measures each of the analog voltages and outputs a single stream of pulses, each with a voltage relative to the respective measured channel. The rigorous merging of data into a single stream is called Time Division Multiplexing or TDM.

The scheme where the pulse height of the TDM stream is proportional to the measured value is called Pulse Amplitude Modulation (PAM). A unique set of synchronization pulses is added to identify the original measurands and their value. PAM has many limitations, including accuracy, constraints on the number of measurands supported, and the poor ability to integrate digital data.

Pulse Code Modulation (PCM) is today's preferred telemetry format for the same reasons that PAM is inadequate. Accuracy is high, with resolution limited only by the analog to digital converter (ADC), and thousands of measurands can be acquired along with digital data from multiple sources, including the contents of the computer's memory and data buses. In a PCM-based system, the original PAM multiplexer's analog output is digitized to a parallel format. The Output Formatter along with synchronization data for measurand identification merges this, plus other sources of digital data. The Output Formatter serializes the composite parallel data stream to a binary string of pulses (1's and 0's) for transmission on copper wire, fiber cable, or "the ether." All components from after the sensor to the formatter comprise the encoder (see figure below}. Other, often remote encoders are used to multiplex additional sensor data into the main encoder's output. Not only does this expand the number of measurands to thousands per stream, but it also eliminates the weight of cables required for each sensor.

The output of the main encoder is filtered and transmitted via radio transmitter and antenna, coax cable, telephone line, tape recorder, etc. Filtering rounds or smoothes the square data pulses to reduce frequency content and thus the required transmitter bandwidth. At the Ground Station, the received data stream is amplified. Since the transmission path often distorts the already rounded signal, a bit synchronizer reconstructs it to the original serial square wave train. Then, a decommutator or decom recognizes the synchronization pattern and returns the serial digital stream to parallel data. The decom also separates the PCM stream into its original measurands (also known as prime parameters) and data.

The computer (in the Visual Test System) or the telemetry front end selects prime parameters for real-time processing; archiving to disk or tape; display; output to strip chart recorders and annunciators; or distribution to other computing resources according to the test plan.

In general, a telemetry (telemetering) system can be divided into two major functional. The transmitting subsystem includes a source section (transducers), a source-signal processing section (including signal conditioning and multiplexing), and a transmitting section. The receiving subsystem includes major components such as receiving antennas, preamplifiers, receivers, magnetic recorders, demodulators, data synchronizers, and data processors and displays. Since each of these components can significantly affect the quality of the information to be recovered at the receiving subsystem output, the functions and signal characteristics of each component should be clearly defined. Therefore, a detailed understanding of each component's performance characteristics is critical to ensuring that quality is maintained as the information passes through all stages of the telemetry system

The basic function of a telemetry transmitting system is to convert one or more information sources into a predetermined format and to transmit the information to a receiving facility. The information might include the temperature at a certain location in a missile, the vibration at another point, the position of a fin, the voltage provided by a battery, the digital outputs from a computer, or almost anything else that one may want to monitor.

The transducer outputs usually need to be modified into the desired form for sampling and/or multiplexing. This signal conditioning can include amplification, level shifting, filtering, buffering, rectification, logarithmic amplification, and companding.

The conditioned information signals are then combined with other signals. This process is called multiplexing. The two most common types of multiplexing are frequency division multiplexing (FDM) and time division multiplexing (TDM). The FDM is frequently used in simple systems where only a few channels of data are needed. It is also used in combination with TDM, especially for wideband vibration data. The conditioned information signals are applied to different, compatible subcarrier oscillators (SCOs). These SCO outputs are then added together to form one composite signal which then modulates the transmitter.

Pulse code modulation (PCM) is the most popular form of TDM system. The TDM systems sample the information signals at pre-determined times. Each signal has its own time slots for transmission. The sampling rate is typically 3 to 10 times the maximum expected signal frequency. The required sampling rate is a function of the accuracy required, the roll-off rate of the signal and noise spectrums beyond the maximum expected signal frequency, and the type and roll-off rate of the anti-aliasing and signal reconstruction filters.

The telemetry transmitter converts the multiplexed signal into a modulated radio frequency (RF) signal. The output power is usually limited by the available supply current or the possibility of interference with other systems or signals. The transmitting antenna radiates the RF energy to the outside world.

Whatever the quantities monitored at the data source (whether electrical or physical), the cost to transmit each quantity through a separate channel would be prohibitive. Think of the equipment and cables or frequency spectrum required to monitor and transmit several hundred or thousands of measurands! One way to conserve resources is to share time or frequency spectrum with techniques such as Time-Division Multiplexing (TDM) and Frequency Multiplexing (FM), respectively.

Today, the most popular form of telemetry multiplexing (originally called commutation, as in an electric motor's commutator) is TDM. Here, each channel is serially sampled for an instant by the multiplexer

A complete scan by the multiplexer (one revolution of the commutator) produces a frame of the stream of words containing the value of each measurand. Every scan produces the same sequence of words. Only the value of a measurand is captured, not its address (name). If only the measurand's data is captured, there is no way to distinguish the owner of one value from the next. Thus, a unique word called the frame sync is added at the end of each frame to serve as a reference for the process of decommutating the stream's data (i.e., extracting it into individual measurand values).

PCM COMMUTATOR (PCM ENCODER)

In a simple commutator, each data word is sampled once per revolution at a rate compatible with the measurands with the fastest changing data. Since the rate of change of a measurand's value varies tremendously, the sampling frequency rate must accommodate it. As an example, to characterize vibration requires many more samples per second (thousands) than temperature (fractions).

According to the Nyquist Theorem, you must sample data at twice the maximum frequency component for the signal to be acquired. Sampling rates of 5 times the maximum frequency component are typical. A low pass filter is used to eliminate any frequencies that you cannot accurately digitize to prevent aliasing.

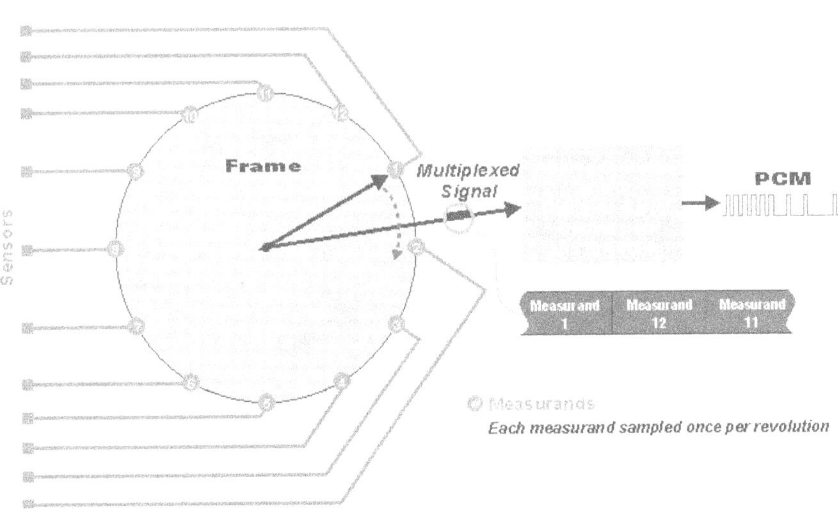

Block Schematic of Commutator

If we were to take a worst-case approach to sampling all measurands at the highest rate, we could expect much waste in carrier frequency spectrum and power. Sampling rates should therefore vary with respect to frequency content and be somewhat independent of other measurands with different periodic acquisition rates. Highly sampled measurands are super-commutated with multiple occurrences of the measurand in each frame.

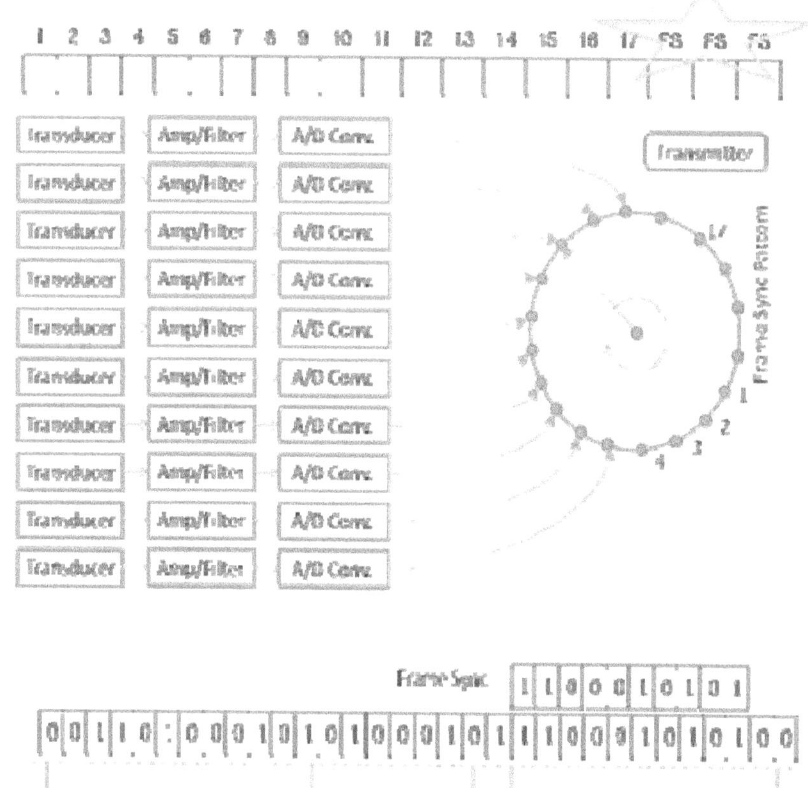

Block Schematic of Commutator with Frame Synchronizer

Data Format

Identifying the end of each minor frame period is the synchronization (sync) word, which is a unique sequence of 1's and 0's. The pattern is generally a pseudo-random sequence that is unlikely to occur randomly in the acquired data and usually occupies two words (or more) in the minor frame. The IRIG-106 Standard lists recommended patterns for lengths 16 through 33 bits. The first three bits transmitted in a frame sync pattern are always a "1," regardless of LSB or MSB alignment.

The length of the frame sync is longer than usual data words to reduce the probability of actual data matching it.

The frame sync should also be commensurate with the number of words in the minor frame (typically, it occupies 1 to 5 percent of the total minor frame). An identical pattern is repeated for every minor frame on the assumption that random data will not consistently match the defined pattern. The decommutator can then be programmed to lock onto this pattern to begin regenerating the original commutated measurands.

TELEMETRY TRANSMITTER

Transmitters are used in telemetry systems for a variety of applications. They are utilized in stationary and mobile vehicle applications (including missiles and satellites) to relay data via digital or analog methods to a ground station, airborne station or relay site. Data can include discrete or analog performance data, Time-Space-Position-Information (TSPI), video, radar, GPS, onboard computer data, etc.

Telemetry transmitters are generally frequency-modulated (Tier 0). The transmitters generate a signal whose power output does not change with or without modulation. In some instances, phase-modulated transmitters are used, but this is less common. However, future high-bit-rate systems should make use of Feher's quadrature phase-shift keying - Revision B (FQPSK-B) or other bandwidth-efficient modulation techniques (Tier 1 and Tier 2).

Telemetry transmitters are available in various types designed for specific applications. Transmitters designed for range applications have typically been Frequency modulated (FM) with analog or digital modulation inputs. However, phase modulated (PM) transmitters are also in use.

An FM transmitter modulates data onto a continuous carrier. The data is conveyed in the deviation of the carrier frequency from nominal.

A PM transmitter modulates its data onto a continuous carrier. The data is conveyed in the deviation of the carrier phase from ainitial reference phase. A PM transmitter can be smaller and less complex than an FM transmitter because the modulator has no direct connection to the oscillator.

TRANSMIT ANTENNAS

The most common types of telemetry antennas utilized for aircraft and missile applications are the blade, slot, and several variations of conformal antennas. The type of test vehicle, flight dynamics and/or the "available real estate" will dictate the type of antenna to be utilized.

Blade Antenna are commonly found on aircraft. Blade antennas are simple and relatively inexpensive compared to other antenna styles. Blades can be purchased for any of the telemetry bands and can also be purchased as multi-band antennas, i.e. lower and upper S-band in one antenna. Blade antennas have an omni-directional pattern in the azimuth plane and generally have more limited coverage in the elevation plane.

Slot Antenna is the common form of antenna found on missiles, drones, and occasionally on aircraft. The antenna is a flat conformal antenna capable of being attached to the surface of the test vehicle. The antenna is sometimes made up of several sections that are fed through a power divider network of some kind in order to create the desired radiation pattern.

REFERENCES

1. Document 120-08 Telemetry (Tm) Systems Radio Frequency (RF) Handbook March 2008 Prepared by Telemetry Group RF systems committee Published by Secretariat Range Commanders Council U.S. Army White Sands Missile Range, New Mexico.

   https://wsmrc2vger.wsmr.army.mil/rcc/

2. Airborne Separation Video System, June 2000 Prepared by Optical Systems Group Range Commanders Council U.S. Army White Sands Missile Range, New Mexico.

3. Flight Safety System (FSS) For Unmanned Aerial Vehicle (UAV) Operation Prepared by Range Safety Group (RSG) by Range Commanders Council U.S. Army White Sands Missile Range, New Mexico.

4. R.K. Barnhart, S.B. Hottman, D.M. Marshall, E. Shappee (Eds.), Introduction to Unmanned Aircraft Systems. CRC Press, 2011.

5. Enhanced Flight Termination System Study - Phase I-IV Prepared by Flight Termination Standing Committee

6. Telemetry Standards, IRIG Standard 106-11 (Part 1), Appendix A, Chapter 1, Chapter 2, Chapter 3, Chapter 4, Chapter 8, Chapter 10. June 2011

7. Musteric, Steven, and Berdugo, Albert, "The Advanced Subminiature Telemetry System (ASMT): A Wireless, Non- Intrusive, Network Based, Instrumentation System", 2007 Test Instrumentation Workshop, Ridgecrest, CA, May 2007.

8. Berdugo, Albert, Grossman, Hy, and Schofield, Nicole, "Wireless Sensor System for Airborne Applications", International Telemeterin Conference, San Diego, CA, Oct 2006

9. Telemetry Tutorial by L-3 Communications Telemetry West, 9020 Balboa Avenue, San Diego, CA www.L-3Com.com/TW

DEVELOPMENT OF A MEMS-BASED ACOUSTIC ENERGY HARVESTER FOR MICRO AIR VEHICLES

M.Sreenivasulu,
ECE Dept, St.Johns College of Engg &Tech,Yemmiganur,Kurnool Dist, A.P, India.

V.U.Shree,
Professor & HOD of ECE , JBIT, Hyderabad. A.P. India.

P.C.S.Reddy
Professor & Coordinator, Dept. of ECE, JNTUH, Hyderabad. A.P, India.
Email:Valasani_usha@yahoo.com, srinuvas42@gmail.com

ABSTRACT

This paper focus on to build energy-harvesting system for micro-air vehicles using MEMS technology. Here electromechanical acoustic resonator is used for reclamation of energy. As in all micro-air vehicles, power is one of the major concerns. There are many methods by which micro-air vehicles might be powered. One of the methods is to draw power from a battery; another method is to harvest energy from the environment. Batteries have the advantage of simplicity, while energy harvesting systems have much longer life and lower mass per total energy delivered. In addition, the total mass of circuitry, MEMS devices, and batteries may severely limit flight duration. Therefore, the best option is for energy-harvesting method.   In MEMS based acoustic energy harvester, resonators are fabricated using MEMS processing techniques. Each resonator have diaphragm. The diaphragm transducers an acoustical pressure fluctuation into a mechanical deformation, while the material transducers that mechanical deformation into an electrical signal in form of charge or voltage.

This paper describes a method for estimating wind field for micro air vehicles. The primary motivation is enabling energy harvesting. Simulation Results of Helmholtz resonator shows the energy harvesting using estimated winds for all turbulence conditions. Plot shows energy change per distance own for each case and control effort; and plot showing the improvement in fight performance over the baseline (constant speed) controller.

KEYWORDS: Electromechanical acoustic resonator, reclamation of energy, energy-harvesting system, MEMS, Micro-Air Vehicles (MAV).

INTRODUCTION

The energy harvesting systems based on the conversion of acoustic signals in to electricalsignal is commonly being used for low power devices due to the reduction in power consumption of electronic devices in the modern world. These applications will be generally used where long term power is unavailable and where cabling and electrical communication are problematic. The acoustic waves are available in the form of longitudinal, transverse, bending, hydrostatic or shear waves of frequencies ranging from few Hertz's to some 10's of KHZ. The input Acoustics or vibration signal can be generated by machines, humans or by nature.

The main objective of this paper is to develop an electromechanical acoustic resonator for storing the acoustic energy. Here, the MEMS device converts the acoustic energy (Vibrational Energy) in to mechanical energy and further it converts in to electrical energy that can be used for micro-air vehicles. The main advantage of this device is we can avoid the wiring. There are many methods by which micro-air vehicles might be powered. One of the methods is to draw power from the battery and another method is to store the energy from the environment by means of the energy harvesting method. The main disadvantage of the battery is its life time is very small and the cost of the device is also very high, for this reason we choose the device which gives more life and cost of the device should also be less. In the proposed system because of the total mass of the circuitry, MEMS devices and batteries will definitely limit the flight duration. Therefore the best option is going for the energy-harvesting method.

In the previous investigations some energy sources such as solar power utilize the optical energy to achieve the greater power densities on the order of 15,000 $\mu W/cm^2$ in the direct sunlight. The main drawback of this is that direct sunlight is not always available in most of the countries. On the other hand the vibrational energy offers power densities up to 250 $\mu W/cm^2$. T his power is very less when we compare with the solar power density but it is useful in the places where direct sunlight is not available but high vibrational energy due to the high winds are available.

Acoustic energy similar to that of vibrational energy, offers power densities on the order of 1 $\mu W/cm^2$ for a 100dB acoustic Signal [1], or approximately 9, 64,000 $\mu W/cm^2$ at 160 Db signal. But, in most of the cases we find the signals to be less than 160 dB, there are some applications where such high signals are present. The important point we have to note here is that acoustic signals do not require the presence of sunlight as that for vibrational energy.

In the MEMS based acoustic energy harvester, resonators are fabricated using MEMS processing techniques. Each resonator has a diaphragm. The diaphragm of the element coverts the acoustical vibrations in to mechanical energy, while the material converts the transformed mechanical energy in to electrical energy in the form of charge or voltage.

THEORY

PIEZOELECTRIC SOUND PRESSURE HARVESTERS

The most common and used piezo materials are ZnO (zinc oxide), AIN (Aluminum nitride), PZT (Lead zirconate titanate).The material is opted depending on the various factors like availability of the material, compatibility and etc. The use of the piezoelectric elements for the transducing of the energy is based on the piezoelectric effect. The most common piezo-electric transducer which is responsible for converting the acoustic pressure in to electrical energy is the Helmholtz resonator which consists of piezoelectric composite black plates shown in the fig1.

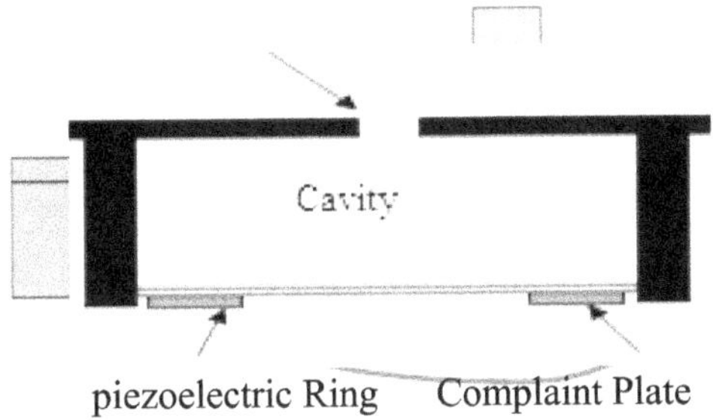

Fig.1: Basic Schematic of the Acoustic energy Harvester

A basic close-up schematic of the acoustic energy harvester is shown in the fig.1.This device consists of a Helmholtz resonator possessing a piezoelectric composite black plate. A Helmholtz resonator is a type of acoustic resonator consists of a cavity connected to the outside environment through a small neck. When exited by an acoustic input signal, a single resonance is seen, where by acoustic pressure inside the cavity is amplified to a level much greater than incident acoustic signal [2,3].

The Helmholtz resonator is similar to an LCR resonant circuit as shown in the fig.3.Both systems have a single degree-of-freedom, with a single resonant peak, where by the amplitude of the forcing functions greatly amplified. In the LCR circuit, the voltage is equivalent to the acoustic pressure. Both the systems operate through the oscillation of energy between generalized potential and kinetic forms. In electrical systems this storage occurs via capacitors and inductors respectively. The large acoustic cavity pressure created by the resonance can be exploited for energy reclamation by converting the acoustic energy in to electrical energy. This conversion is performed by the composite plate. First, acoustical to mechanical transduction is performed via the complaint diaphragm, followed by mechanical to electrical transduction, due to the piezoelectric response of the composite, where by a mechanical strain creates an electrical voltage. The voltage that is created through the electromechanical transduction can be harvested by the energy harvester circuit, which is a necessary part of the energy harvesting process [4].

ACOUSTIC ENERGY HARVESTING USING HELMHOLTZ[4]

RESONATOR

Lumped Element Model (LEM) is applied to the Helmholtz resonator for better understanding. The diagram of the Helmholtz resonator with its equivalent mass-spring-damper system is shown in the fig.3 below and its equivalent LCR circuit is also shown in the fig2. It consists of a cavity connected to the outside environment through a neck as shown in the fig.1.It gets excited by an acoustic input signal .Here 'V ' is the volume of the Cavity, $S=\pi a^2$ is the cross section area of the neck and 'l' indicates the length and 'a' is its radius,P1 is the pressure of the surrounding environment,P2 is the pressure of the cavity.

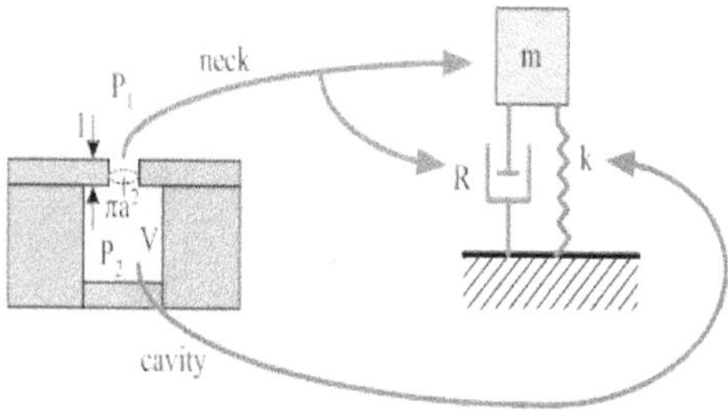

Fig.2: Helmholtz resonator and its equivalent mass-spring-damper system[4].

EQUIVALENT CIRCUIT

The equivalent ciruit of a Helmholtz resonator is presented in the fig3.shown below.

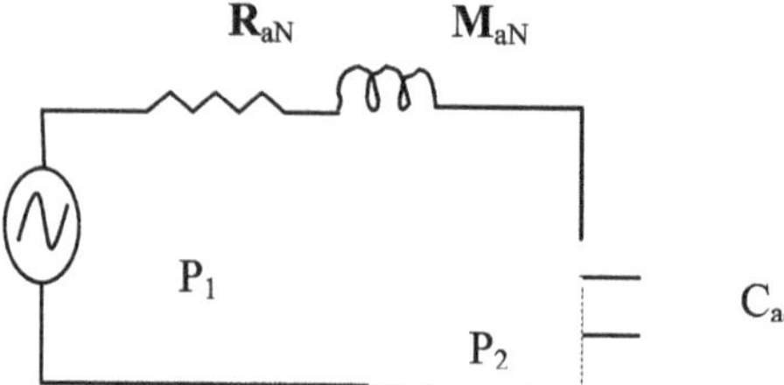

Fig.3: Equivalent circuit representation of Helmholtz Resonator

The frequency response function P_2/P_1, represents the pressure amplification of the resonator. It is the ratio of the cavity pressure amplification of the resonator. It is the ratio of the cavity pressure to the incident pressure and is given by

$$\frac{P_2}{P_1} = \frac{\frac{1}{sC_{ac}}}{R_{aN} + sM_{aN} + \frac{1}{sC_{ac}}}$$

Where s=jω

The single resonant frequency peak when the sum of the reactances is zero for the circuit shown in fig.3 is given by

$$fres = \frac{1}{2\Pi\sqrt{M_{aN}C_{aC}}} \text{ [Hz]}$$

Also the equivalent circuit of piezo-composite plate including the radiation impedance is shown below.

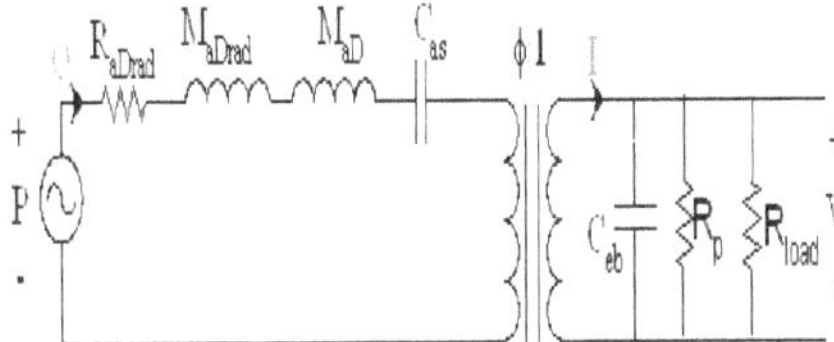

Fig.4: equivalent circuit of piezo-composite plate

ACOUSTIC ENERGY HARVESTER DYNAMIC BEHAVIOR [4]

ACOUSTICAL INPUT BEHAVIOR

When the piezoelectric diaphragm is mounted inside the Helmholtz resonator, the equivalent circuit for the diaphragm (fig.4) is combined with the fig.3 and the formed circuit's acoustical input impedance is given by

$$Z_{in} = R_{aN} + sM_{aN} + \cfrac{\dfrac{1}{sC_{aC}}\left(s(M_{aD}+M_{aDrad}) + \dfrac{1}{sC_{aD}} + R_{aDrad} + \dfrac{\phi R_{load}}{1+R_{load}sC_{eb}}\right)}{\dfrac{1}{sC_{aC}} + s(M_{aD}+M_{aDrad}) + \dfrac{1}{sC_{aD}} + R_{aDrad} + \dfrac{\phi R_{load}}{1+R_{load}sC_{eb}}}$$

Note that the equivalent circuit for the diaphragm is in parallel with the cavity compliance, C_{aC}. The compression of air in the cavity and deflection in the diaphragm is happening due to the pressure in the cavity because the cavity and the diaphragm experience the same input pressure. In this case, we are assuming that $R_{load} \ll R_p$ and therefore the parallel combination can be written as $R_{load} // R_p \equiv R_{load}$.

From the equation above it is seen that the total inout impedance is the impedance of the Helmholtz resonator neck in series with parallel combination of Helmholtz resonator cavity impedance and the Piezoelectric composite diaphragm impedance considering the resistive load.

Acoustic Output Behavior

The Electrical output impedance can be found in the similar manner and is given by

$$Z_{out} = R_{load} // \frac{1}{sC_{eb}} // \left[\frac{1}{\phi^2}\left((s(M_{aD}+M_{aDrad}) + \frac{1}{sC_{aD}} + R_{aDrad} + \frac{1}{sC_{aC}} // (R_{aN}+M_{aN})\right)\right]$$

Expanding the above equation for the output impedance, we get

$$Z_{out} = \cfrac{(R_{load})\dfrac{1}{sC_{eb}}\left[\dfrac{1}{\phi^2}\left(s(M_{aD}+M_{aDrad}) + \dfrac{1}{sC_{aD}} + R_{aDrad} + \dfrac{\dfrac{1}{sC_{aC}}(R_{aN}+M_{aN})}{\dfrac{1}{sC_{aC}}(R_{aN}+M_{aN})}\right)\right]}{\left(R_{load}\dfrac{1}{sC_{eb}} + \left(R_{load} + \dfrac{1}{sC_{eb}}\right)\dfrac{1}{\phi^2}\left[s(M_{aD}+M_{aDrad}) + \dfrac{1}{sC_{aD}} + R_{aDrad} + \dfrac{\dfrac{1}{sC_{aC}}(R_{aN}+M_{aN})}{\dfrac{1}{sC_{aC}}(R_{aN}+M_{aN})}\right]\right)}$$

The output voltage also can be found from the equivalent circuit to be

$$V = \frac{\left(p\phi - \dfrac{\dfrac{1}{sC_{eb}}R_{load}}{\dfrac{1}{sC_{eb}}+R_{load}}\right)}{\left[\left(\dfrac{R_{aN}+sM_{aN}}{\dfrac{1}{sC_{aC}}}+1\right)\left(s(M_{aD}+M_{aDrad})+\dfrac{1}{sC_{aD}}+R_{aDrad}+\dfrac{\phi R_{load}}{1+R_{load}C_{eb}}\right)+R_{aN}+sM_{aN}\right]}$$

The output voltage is displayed graphically and discussed in the next section.

RESULTS AND DISCUSSIONS

The input and output simulation results are displayed graphically.

Input Behavior

The plot of the acoustical input impedance versus frequency is shown in the above fig.5 below for a piezoelectric composite diaphragm.

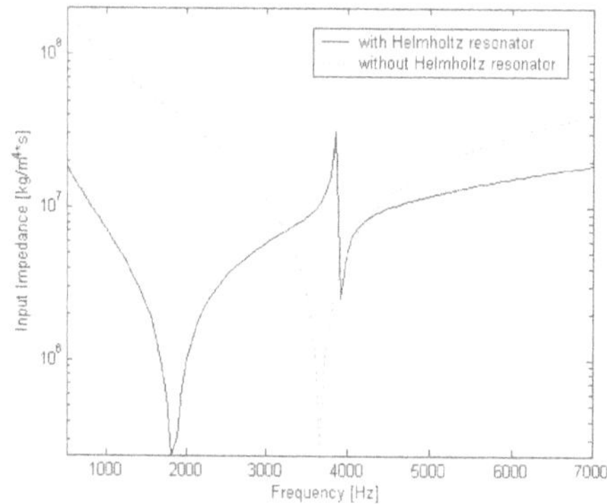

Fig.5: Magnitude of the Acoustic impedance for piezoelectric composite diaphragm with Helmholtz resonator

The above results are obtained by giving the following parameters to the Helmholtz resonator. Lenth of the neck=3.21mm,and aradius R_a=2.45mm and Volume of the Cavity V_{cav}=2015mm^3 in addition to a diaphragm with a thickness of t_{si}=3μm and a outer and inner radius of R_2=1.98 mm and R_1=1.92mm,respectively.From this plot it found that a piezoelectric composite diaphragm has a single resonance frequency at 3.7kHz.when combined with the Helmholtz resonator the two minima are seen,the lower resonance and upper resonance at 1.8k and 3.9 respectively.

OUTPUT BEHAVIOR

The output voltage is shown graphically in the figure.6 for piezoelectric composite diaphragm with the combination of Helmholtz resonator.

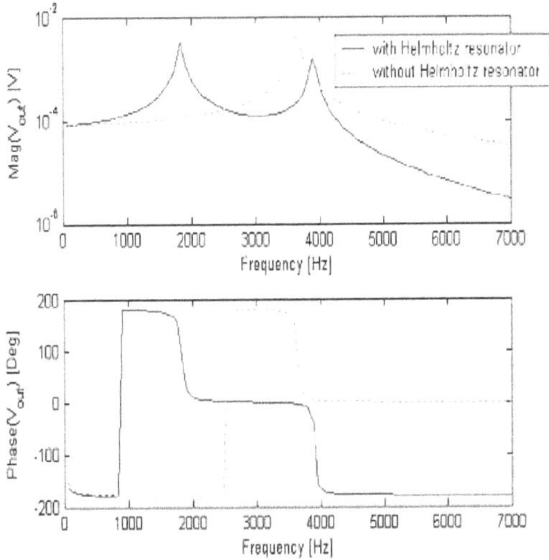

Fig.6: Magnitude and phase of the output voltage for the piezoelectric composite diaphragm with Helmholtz resonator

For both the cases, the magnitude has the same shape to the output impedance. In the output impedance curve, the second resonant peak is higher than the first whereas for the output voltage curve shown in the figure, the first resonant peak is higher than the second.
The electrical power delivered to the resistive load, R_{load}, can be found from

$$R_e\{\pi_{out}\}=V^2/R_{load}$$

The overall conversion efficiency, Γ, can then be found as the ratio of output electrical power to the input acoustical power and is given by

$$Re\{\Gamma\}=Re\{\pi_{out}\}/Re\{\pi_{in}\}$$

CONCLUSIONS

The acoustic energy harvesting technique that was developed in this dissertation provides a unique and innovative method for extracting energy from an otherwise inaccessible source. For the development of acoustic energy harvesters, piezoelectric composite diaphragms were designed based on analytical models of circular piezo-composite structures. Lumped element modeling was then used to further understand the dynamics of the system. The modeling served as a basis around which the geometry and dimensions of numerous diaphragms were determined. The lumped element model developed in this dissertation serves as a design basis for other types of piezoelectric composite structures as well as energy harvesting devices employing other transduction methods. This energy harvesting method is directly applicable to for micro-air vehicle.

ACKNOWLEDGMENTS

This is to acknowledge to my guide Dr.V.Usha,HOD ECE & Dean , JBIT, Hyderabad and to my co-supervisor Dr.P.Chandra Sekhar Reddy, Professor & coordination, JNTUH, Kukat Pally, Hyderabad R.R. District. AP and my family members for supporting me to make this paper.

REFERENCES

[1]   S. Roundy, P.K. Wright, and J. Rabaey, "A Study of Low Level Vibrations as a Power Source for Wireless Sensor Nodes." Computer Communications. 26: p. 1131-1144, 2003..

[2]   M. Rossi, Acoustics and Electroacoustics, Artech House: Norwood, MA, p. 245-308.1988

[3]   D.T. Blackstock, Fundamentals of Physical Acoustics, John Wiley & Sons, Inc.:New York, p. 153-156. 2000.

[4]   Horovitz, St., Development of a MEMS- based Acoustic Energy Harvester, University of Florida, PhD Thesis, 2005.

[5]   O.Brand and G.K.Fedder.,"**Advanced Micro & Nanosystems, Volume 2**", Published by WILEY-VCH Verlag GmbH & Co. KGaA, Weinheim, 2005.

[6]   Stephen Brian Horowitz., "**Development of a MEMS-based Acoustic Energy harvester**"., Ph.D Thesis University of Florida, 2005.

[7]   James J.Allen.,"**Micro Electro Mechanical Systems**" Published by CRC Press Taylor & Francis Group, LLC, 2005.

[8]   Gabriel M.Rebeiz., "**RF MEMS Theory, Design, and Technology**", A JOHN WILEY & SONS PUBLICATION, 2003.

[9]   A. Gross, B. Huang, G. Hwang, C. Lawrence, N. Ghafouri, S. Lee, H. Kim, C. Uher, M. Kaviany, and K. Najafi, "A multi-stage in-plane micro-thermoelectric cooler," in *Proc. 21st Int. Conf.MEMS*, Tuscon,2008,pp. 840–843.

[9]   J. Frechette, R. Maboudian, and C. Carraro, "Effect of temperature on inuse stiction of cantilever beams coated with perfluorinated alkysiloxane monolayers," *J. Microelectromech. Syst.*, vol. 15, no. 4, pp. 737–744,Aug. 2006.

[10]   B. Morgan and R. Ghodssi, "Vertically-shaped tunable MEMS, resonators," *J. Microelectromech Syst.*, vol. 17, no. 1, pp. 85–92, Feb. 2008

A COMPREHENSIVE SENSOR SUITE FOR MINI AERIAL VEHICLE FLIGHT TESTS

Prateek Jolly

Research Assistant, email: jollyboy1@iitb.ac.in

Vaibhav V Unhelkar

Research Assistant, email: v.unhelkar@iitb.ac.in

Hemendra Arya

Associate Professor, email: arya@aero.iitb.ac.in

Department of Aerospace Engineering, Indian Institute of Technology Bombay, Mumbai – 400076

ABSTRACT

Flight   tests, which are primarily used for model creation and validation, are an important part of the aircraft design and development. For larger aircraft, system identification methods based on flight tests have been studied in detail and are well established. However, flight tests of mini unmanned aerial vehicles (MAV) pose new challenges, chiefly because   the small, cheap and lightweight sensors used in MAV instrumentation are not as accurate as those used on larger aircraft. In this paper, we present a comprehensive sensor suite for mini aerial vehicles with specific focus on construction, calibration and implemention of the sensors. The sensor suite is capable of measuring the important flight variables thrust, angle of attack, side slip, airspeed, attitude, inertial rates, altitude, position, velocity, acceleration and actuator deflections - and can be used for flight tests. The paper concludes with a flight data based sys- tem identification methodology, along with simulated results, to arrive at the aerodynamic models of mini aerial vehicles.

INTRODUCTION

Mini unmanned aerial vehicles are remotely pi- loted; sub meter span aircraft that are being widely used for aerial reconnaissance and sensing. Their small size gives them a distinct advantage over their larger counterparts, in that they are easier to trans- port, deploy, and do not require special landing and takeoff facilities. Their low noise electric motors, small size and low velocities also ensure that they are not easy to detect; but, at the same time, they have short endurance and can carry limited payload. Hence, there is a need to optimize the design of a MAV to achieve the desired performance. To extract the optimal performance from these aerial vehicles, their accurate modelling is necessary.

Much technical literature is available to help arrive at the dynamic model of aircraft based on empirical (thumb-rule based) and/or analytical re- lations. Further, computational techniques, such as those   of Computational Fluid Dynamics, are also used while modelling an aircraft.

However, both the analytical and computational techniques suffer from limitations while modelling mini aerial vehicles; the highly unconventional designs of MAVs often pre- vent use of empirical results, and the computational techniques require domain expertise and may be time consuming. To this end, *system identification based on flight tests* can be used to model the aircraft, and predict various forces and moments that affect the aircraft's performance.

For larger aircrafts, methods based on flight tests have been studied in detailed and are well estab-lished ([1], [2], [3]). On similar lines, flight data can be used to model the smaller class of aerial vehicles. However, their short design cycles, reduced cost and noisy on-board sensors, pose new challenges for us- ing flight test based methods for MAVs [4]. Various research studies have been reported in the literature for system identification of UAVs and MAVs, with focus on different aspects of the problem ([4], [5], [6], [7]). For instance, wind tunnel tests have been carried out, albeit without any flight test validation, to determine the lift, drag and moments acting on a MAV [6]. Specifically, Ref. [7] outlines a sensor suite that measures airspeed, altitude, attitude and position. The autopilot based on this sensor suite uses manually tuned PID controller to navigate the aircraft, since additional wind tunnel based tests are required for parameter estimation in absence of α, β, and thrust measurements. Hence, there is a need of a comprehensive sensor suite tailored for MAVs, which can be used to gather important flight variables, namely, thrust, angle of attack, side slip, airspeed, attitude, inertial rates, altitude, position, velocity, acceleration and actuator deflections. Fur- thermore, accuracy and applicability of flight test based system identification methods for MAVs also need to be investigated.

In this paper, we describe the design, implementation and calibration of a novel thrust measurement system, and α and β measurement vanes for MAVs. Details of additional commercially available sensors Inertial Measurement Unit, GPS receiver, altitude sensor, airspeed sensor - required to measure other required flight variables have also been provided. The paper concludes with a flight data based system identification methodology, along with simulated results, to arrive at the aerodynamic models of mini aerial vehicles.

SENSOR SUITE FOR MAVS

The parameters affecting the aircraft can be clas- sified into the following three groups,

1) *Inertia related:* These include variables such as the mass, moment of inertias, and position of center of gravity of the aircraft. These terms contribute to the forces and moments caused due to the Earth's gravity.

2) *Propulsion:* These include variables such as the angular velocity of the propeller, and the associated thrust.

3) *Aerodynamic:* These include the lift and drag associated with the aircraft, along with the aerodynamic moments produced by the aircraft structure as well as its control surfaces.

The inertia parameters can be determined *a priori*, however, determination of thrust and aerodynamic forces requires in flight measurements. To com- pletely characterize these forces and to determine a model of the aircraft one needs to measure *thrust, angle of attack, side slip, airspeed, attitude, inertial rates, altitude, position, velocity, acceleration and actuator deflections.* This can be achieved with the comprehensive sensor suite, which is briefly described in Table I.

The following sections describe the design, calibration and implementation of the individual sensors.

TABLE I

Calibration Curve of the Flex Force Sensor

Measured Quantity	Measurement System	Sensor Element	Dimension (cm)	Weight (g)
Thrust	Compression type motor mount	Piezo-Electric Flex Force Sensor	6x6x1	30
α, β	Vanes	Low Friction Potentiometer	15x10x3	10 x 2
Attitude	Inertial Measurement Unit	Accelerometer and Gyros	3x3x1	7
Actuator Deflection	PWM Tapping	Micro-Controller	-	-
Position, Heading and Ground Speed	Global Positioning System	GPS Receiver (5Hz)	3x3x0.5	6
Altitude	Absolute Pressure	Piezo-Electric Pressure Sensor	3x1x1	2
Airspeed	Differential Pressure	Piezo-Electric Pressure Sensor with Pitot Tube	1x1x1	2

THRUST

Conventionally, in flight thrust has been estimated using a motor-propeller model, which is obtained via wind tunnel tests [5]. The problem with such a setup is that it usually ignores the effects of fuselage and wing blockage, which tend to reduce the applied thrust. To overcome this limitation, a novel thrust measurement device (Fig. 1) has been developed, which uses a highly sensitive piezometric flexi-force resistor [8] as its sensor element.

Fig. 1. Thrust Measurement Sensor

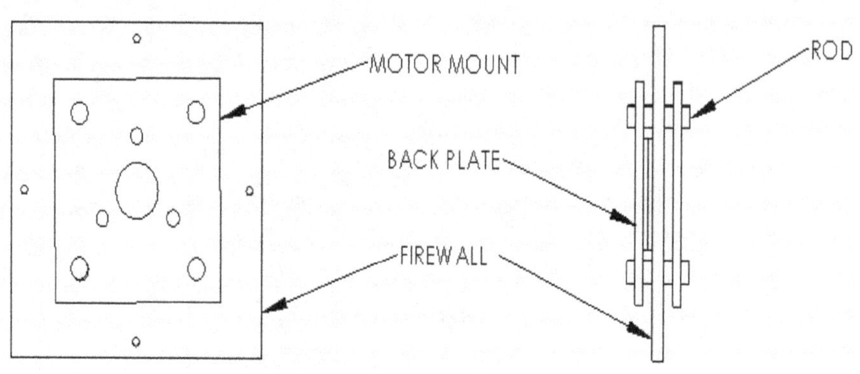

Fig. 2. Thrust Measurement Sensor

The flex force element behaves as a pressure sensor whose resistance varies with compressive loading. The force produced by the propeller, which is tensile in nature, is transferred to the fuselage through the motor mount. A mechanism to convert this tensile force into a compressive force is devel- oped (Fig. 2). The sensor is sandwiched between two plates. A foam backing is used to distribute the pressure evenly. The force produced by the motor is transferred to the sensor via the connecting rod and the backplate. The change in force per unit area is sensed and is reflected as a change in the resistance between the leads of the sensor.

The plates are constructed of 4 mm aluminium sheets to keep the weight to a minimum. The inside of the bearings are smeared with low density grease for smooth operation. The sensor is pre-compressed by 125 grams, to measure the drag produced by the freewheeling propeller can be measured. The calibration curve for the sensor is shown in Fig. 3.

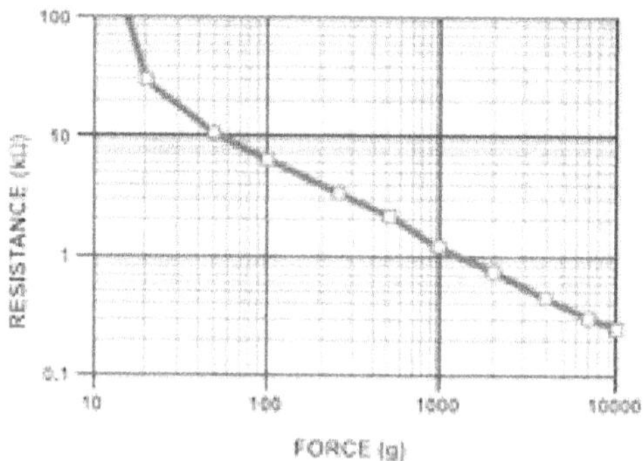

Fig. 3. Calibration Curve of the Flex Force Sensor

Two low pass filters are connected in series to the output of the sensor to dampen the high frequency noise generated by mechanical vibrations of the motor. Fig. 4 shows the effect of the filter on the measured raw data.

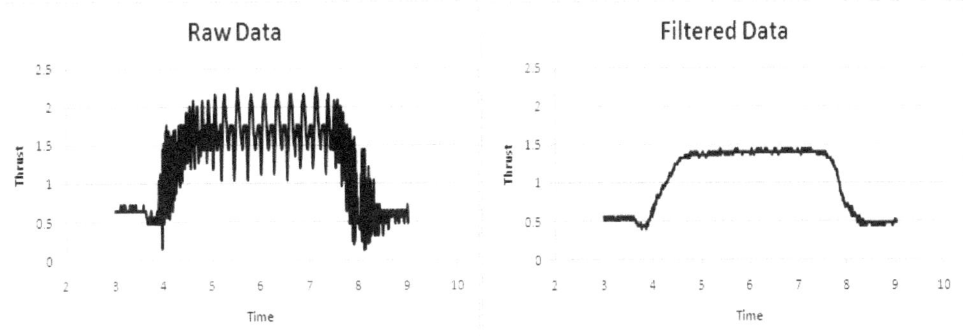

Fig. 4.   Signal Improvement using Low Pass Filters

B. Aerodynamic Angles, α and β

The orientation of the aircraft with respect to the free stream velocity is of high importance,   as it directly affects the resultant aerodynamic force experienced by the aircraft. Two wind vanes are used to measure the aircraft's angle of attack (α) and side slip (β), respectively (Fig. 5).

Alpha and Beta Vanes are small wind vanes that are mounted on the wingtips. They point into the wind and thus the angle between the wing chord and the wind vane measures alpha and beta. The advantage of using wind vanes is that they are very easy to calibrate and do not have any limitation on their operating range. Their downside is that they are heavier, more obtrusive and the inertia of the vanes and the fiction inherent in the measuring instrument implies a   higher dead zone than the multi-hole probes [9].

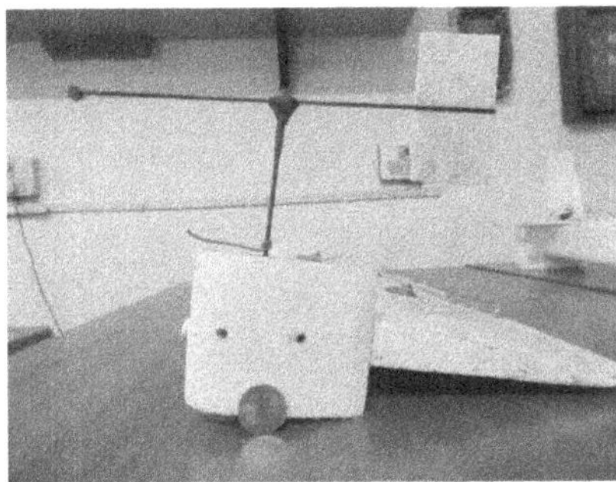

Fig. 5.   Alpha and Beta Vanes

The vanes manufactured   for use   on the mini UAV are constructed of 1mm strips of carbon fiber as the boom, and 2mm balsa as the vane. This light construction ensures a low moment of inertia. Further, two β vane mounts have been used, on either side of the wing, to maintain symmetry. Low friction ball bearings are used to reduce the dead zone. A UNIPOT PS22C [10] high sensitivity low friction, 5 kilo-ohm potentiometer is used as the sensor element. The potentiometer, being of 22mm diameter is mounted in an airfoil shaped mount to reduce drag and support the 5mm diameter carbon fiber cantilevered rod. Fig. 6 shows the calibration curves for the α vane.

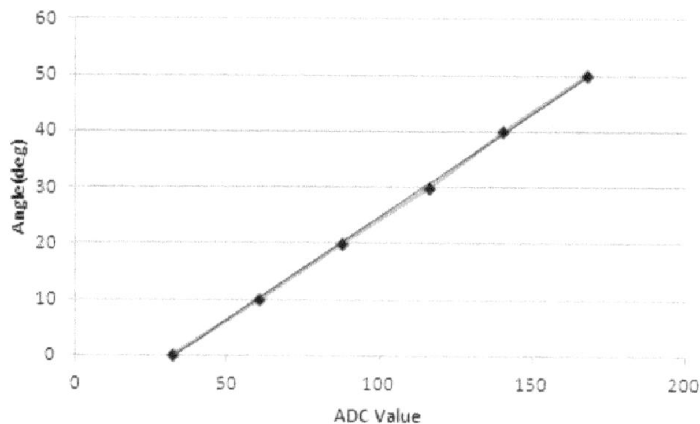

Fig. 6.   Alpha Vane Calibration

C. Attitude

The attitude of the aircraft is defined as the orien- tation of the aircraft with respect to the horizon. The attitude is measured using an inertial measurement unit developed by IdeaForge [11], Fig. 7. The IMU uses MEMS based accelerometers to measure linear accelerations, and gyroscopes to measure angular accelerations. The output of the IMU is a processed and filtered signal, containing attitude, angular rates, and specific accelerations of the aircraft.

Fig. 7.   IdeaForge IMU

$$\delta Vt = kV2 \qquad (1b)$$

D. Position, Heading and Ground Speed

A GPS receiver (Fig. 8) is used to obtain estimates of velocity, heading and position of the aircraft at a 5Hz update rate. The accuracy of velocity and heading measurements is 0.1 m/s and 0.5 deg, respectively.

Fig. 8.   GPS Reciever

E. Air Speed

Airspeed is measured using a MPXV7002DP[12] differential pressure sensor (Fig. 9). One nozzle of the sensor is connected to a thin steel pitot tube, through a flexible rubber tube. The pitot tube is positioned so as to be unaffected by both the airflow over the wing and the prop wash. The sensor has a range of 4 kPa, and an analog output varying from 0.5 to 4.5 volts when operated at its maximum rated voltage of 5V.

Fig. 9. Airspeed Sensor

The sensor is calibrated using a U-tube Manome- ter and the Eq. 1. The sensor senses the change in pressure, which is directly proportional to the voltage variation (δVt). Fig. 10 shows the calibration curves for the airspeed sensor.

$(h\rho g)water = (1/2\rho V 2)air$ (1a)

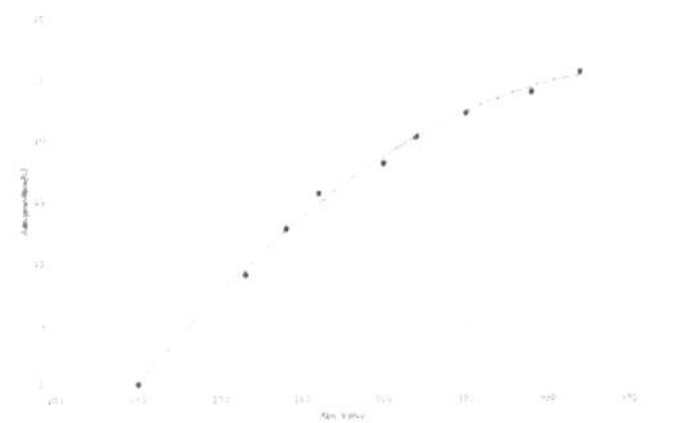

Fig. 10. Airspeed Sensor Calibration

F. Altitude

The altitude is measure using a MPX5100AP[13] absolute pressure sensor. The change in voltage across the pressure sensor is amplified using a differ- ence operational amplifier to increase the readability of the sensor. The sensor is calibrated by using a U- tube manometer, and converting readings in head of water to head of air. Fig. 11 shows the calibration curve of the altitude sensor.

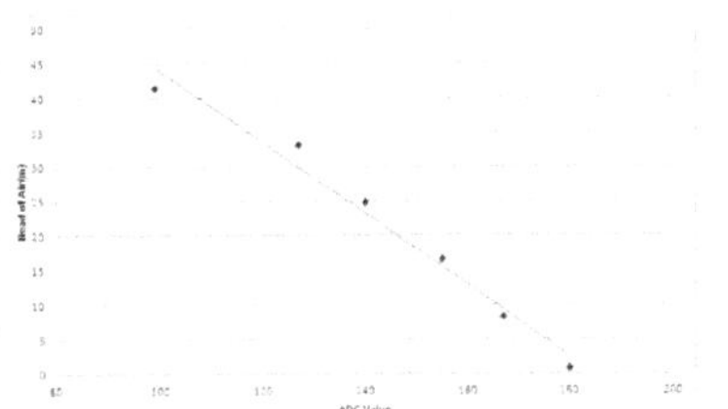

Fig. 11. Altitude Sensor Calibration Curve

G. Actuator Deflections

The control surfaces are actuated by micro ser- vos, which operate using a pulse width modulated (PWM) signal at 5V. The servo motors are con- nected through mechanical linkages to the control surfaces, so as to convert their rotary motion into linear motion, and back to rotary motion. The deflections of the control surfaces are thus direct functions of the PWM value. These PWM signals are recorded and suitably processed to measure the control surface deflections.

H. Data Storage

Fig. 12.   Autopilot and Data Aquisition Board

In order to measure, record and process the mea- surements from various sensors described above, an on-board data acquisition system is required on. The data acquisition system is integrated with the MAV's autopilot system onto a single custom board (hence forth referred to as the autopilot board, Fig. 12). The board uses a MAC 7112 microcontroller[14] as its processing unit and a 2 GB micro-SD data card as the memory storage device for the data acquisition system. The airspeed and altitude sensors are directly mounted on the board. The microcon- troller has 16 multiplexed analog to digital converter pins which are connected to each of our measuring instruments. The DAC records data at a frequency of 50Hz. The boards weighs 15 grams along with the airspeed and altitude sensors mounted on it. The data recorded using this comprehensive sensor suite, system identification of the aircraft can be performed.

SYSTEM IDENTIFICATION

Having described a complete sensor suite for flight tests of mini aerial vehicles, we explain and demonstrate through simulation an offline, time- domain based technique for system identification of aircrafts using the measured and recorded flight data. Knowledge of accurate mathematical models of an aircraft can help drastically improve its design, capability and operating efficiency. For instance, a good model can help the aircraft designer to predict, via simulation, the achievable performance prior to finalizing the design or mission.

The system identification presented in this section are limited to fixed-wing aircrafts, and assume no or minimal wind disturbances during the flight tests. Further, we have reduced the problem of system identification to that of *parameter estimation* of aerodynamic models, and present the analysis only for longitudinal motion of the aircraft. Detailed description of system identification methods for aircrafts can be found in [1], [2]. Ref. [4] provides an overview of additional challenges faced in system identification of small aerial vehicles.

A. Methodology

System identification can be carried out either in time or frequency domain. Here, we explore a time- domain based approach. Further, these time-domain based approaches are broadly classified as equation error, output error, and filter error [1]; these differ in the assumptions considered, their complexity and applicability. In the approach presented, we utilize output error method (which is based on maximum likelihood estimation) for estimation of systematic sensor errors, and equation error (which is based on regression analysis) for estimation of the aero- dynamic parameters.

By assuming a global model for the aerodynamic forces, the problem of system identification can be reduced to that of parameter estimation. The knowledge of aerodynamic model structure, though not imperative, if available greatly aids the param- eter estimation process. The steps of the system identification are now briefly described.

1) Design of Flight Test: Gathering flight data in a proper fashion is critical to success of the param- eter estimation process. To estimate an aerodynamic model which is applicable for the entire flight en- velope from a single flight test, the flight trajectory has to be predesigned. Chapter 2 of Ref. [1] provides a detailed description of the flight maneuvers to be used for system identification. MAV/UAV flight tests have an advantage while designing them, in that the constraint of pilot safety is absent. The tests should be carried out in an environment with minimum wind/gust disturbances.

In the current approach of parameter estimation, measurements of α, β, *attitude, thrust, accelera- tions, angular rates, airspeed, actuator deflections* are required, all of which are available using the sensor suite described in Sec. II. As far as possible, raw data should be recorded so that no information is lost due to onboard processing. The sensor error characteristics should be determined *a priori*, to assist for better data processing. The knowledge of aircraft mass (m), geometry and moment of inertias is assumed to be known for the duration of flight test.

2) Data Compatibility Check: Once the flight data is recorded, it is checked for consistency and accuracy prior to its use for parameter es- timation. This is done using aircraft kinematic equations, which require the IMU measurements (ax, ay, az, p, q, r) as an input and provide the atti- tudes (φ, θ, ψ), wind angles (α, β) and air speed (V) as the output. This reconstructed data is compared with the measured data to check for data compatibil- ity. In case the data is not compatible, sensor errors should be removed, as described next.

3) Removal of Sensor Errors: The sensor errors can be broadly classified as systematic (such as, bias, scale factors) and stochastic (such as, elec- tronic noise). Since the measured data is being pro- cessed offline, digital smoothing is used to remove the random sensor noise. Use of digital smoothing, allows removal of sensor noise without any filter delay.

The smoothed data is used along with the aircraft kinematic (data compatibility) equations to provide an estimate of the systematic sensor errors. An output error formulation is used which estimates the sensor errors based on maximum likelihood estima- tion (Chapter 6 of Ref. [2]). The measurements are then corrected using the estimates of sensor bias and scale factors.

4) Parameter Estimation: A regression based pa- rameter estimation algorithm is used along with the corrected sensor data to arrive at the aerodynamic model. The dependent variable is the aerodynamic coefficient (such as CL, CD, or Cm) to be mod- eled. Using their definitions and measured variables, values of the dependent variables are obtained. For instance, lift is obtained based on accelerometer, thrust (T) and α measurements, which is used for obtaining the value of in flight CL (Eq. 2).

$$L = (max - T)\sin(\alpha) - (maz)\cos(\alpha) \quad (2a)$$

$$CL = 0.5\rho V 2 S \quad (2b)$$

The regressors are measured, independent variables which are selected based on the physical understand- ing of the aerodynamic models. For instance, α, q, and δe may be used as one set of possible re- gressors while estimating CL model. The parameter estimation algorithm provides an algebraic model of dependent variable as a function of the regressors.

5) Model Validation: In order to verify the model thus determined, flights with different trajectories should be flown, and the calculated CL should be compared with the model predicted CL. For a cor- rectly predicted model, the residual, between mea- sured and predicted values, should satisfy a Gausian distribution with zero mean . The standard deviation quantifies the accuracy of the model, and additional flight tests should be carried out in case the observed deviation is not within acceptable limits. Next, we observe the performance of the parameter estimation methodology described above, through a six degree of freedom MAV simulation.

B. Simulation Details

MAV flight simulations have been carried out in order to verify the parameter estimation method described above. An in-house code for MAV flight simulation has been used to generate the true tra- jectory and the corresponding measurements. Com- mensurate measurement errors, as shown in Table II have been included to simulate the measurements as available during a flight. The parameter estimation codes have been adapted from the SIDPAC software package [15], and modified for our use.

Table 2 sensor errors for flight simulation

Quantity	Bias	Random	Scale

In order to simulate the aircraft motion, true aerodynamic models have to be specified. The true aerodynamic models influencing the longitudinal motion are described in Eq. 3. Note that these mod- els are not available with the parameter estimation algorithm, and are used only for flight simulation. Though, the CL and Cm models used here are linear, the parameter estimation algorithm is generic in nature and is equally applicable for nonlinear aerodynamic models.

$$CL = CL0 + CL\alpha\,\alpha + CL\delta\,\delta \quad (3a)$$

$$CD = CD + kC2 \quad (3b)$$

$$Cm = Cm0 + Cm\alpha\,\alpha + Cm\delta\,\delta \quad (3c)$$

As mentioned earlier, the choice of flight manou- ver is critical for successful parameter estimation. The aircraft is trimmed at the start of the flight, and then a pulsed elevator input is provided to excite the longitudinal dynamic mode. Fig. 13 describes the elevator (δe) and angle of attack (α) profile for the flight test.

RESULTS

A data compatibility check is first performed using the noisy measurements obtained from the simulated data. Data compatibility results of Fig. 14, indicate that there is a need for sensor error estimation. First, random errors are eliminated using digital smoothing, and this data is used to estimate sensor bias and scale factor errors. Table III shows the obtained estimates of systematic sensor errors. Note that only measurements affecting the longitu- dinal motion are considered.

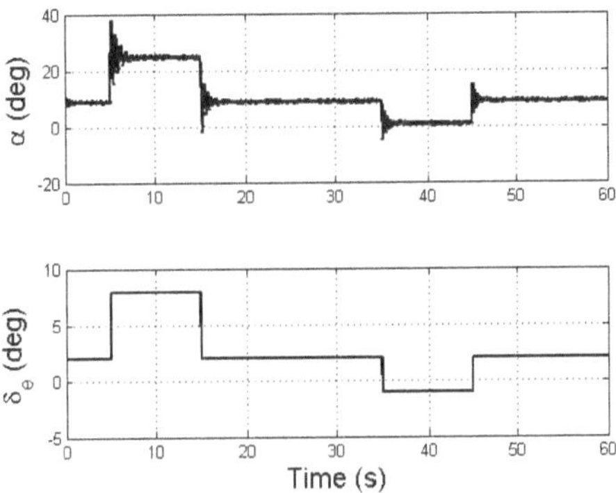

Fig. 13.   Elevator and α Profile

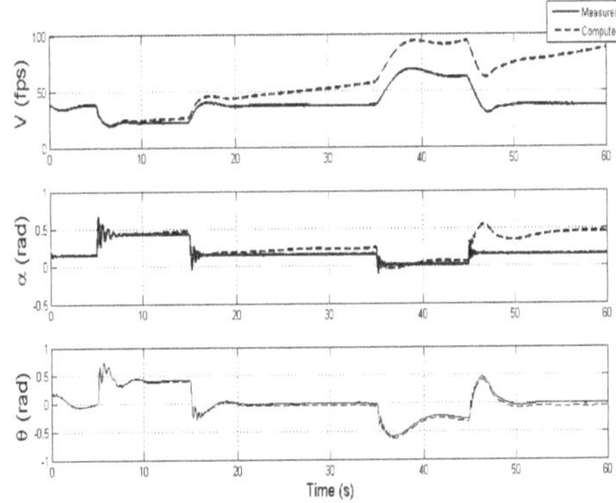

Fig. 14.   Data Compatibility Analysis

Using the corrected sensor parameters, the aero- dynamic   model parameters are being estimated via regression analysis.   Table IV indicates the perfor- mance of the parameter estimation algorithm, by comparing the true and estimated aerodynamic co- efficients.

TABLE III

SENSOR ERROR ESTIMATION

SENSOR ERROR ESTIMATION

Parameter	True	Estimated	Estimation

Both the estimates of systematic sensor errors as well as that of aerodynamic stability derivatives are satisfactory. However, the algorithm shows signifi- cant error in the estimation of control derivatives. Additional flight maneuvers can be carried out to increase the accuracy of these estimates.

Table 4 Aerodynamic Parameter Estimation

Parameter	TRUE	Estimated	Error
$CL0$	+0.1784	+0.1641	+0.0143
$CL\alpha$	+2.453	+2.5325	-0.0795
$CL\delta$	+0.7405	+1.0437	**-0.3032**
$CD0$	+0.0871	+0.0942	-0.0071
k	+0.4	+0.3752	+0.0248
$Cm0$	+0.0385	+0.0386	-0.0001
$Cm\alpha$	-0.5998	-0.5751	-0.0247
$Cm\delta$	+1.5894	+1.4301	**+0.1593**

CONCLUSION AND FUTURE WORK

A comprehensive sensor suite for flight tests of mini unmanned aerial vehicles has been de- signed and presented. This sensor suite is capable of measuring all the required variables for system identification of MAVs. An elementary   algorithm for estimation of aerodynamic coefficients has also been analyzed, and its simulated results have been presented. The algorithm is observed to be working satisfactorily for simulated data.

Future work would include improvements in the sensor suite, in terms of optimizing its weight and size, so as to minimize the effect of sensors on MAV performance. Specifically, the α and β vanes constuction is to be refined to make them more sleek, small and light. Smaller airspeed and altitude measurement   sensors are now commercially avail- able and will further help in miniaturizing and light- ening the autopilot board. The thrust measurement system is to be modified for simpler construction and assembly. Further, detailed error characteristics of the sensors will also be determined so as to assist in design of on-board estimation and control algorithms as well as system identification. The cur- rent system identification algorithm will be validated using real flight data, and required improvements would be made. Alternative system identification algorithms which work even in absence of α, β measurements could also be investigated.

ACKNOWLEDGEMENTS

The authors would like to thank AR&DB for their continued backing of our research. The authors would like to thank the members of the Dynamics and Control Group, Department of Aerospace Engineering at IIT Bombay. Special thanks to Prasanna Shevare for his continued help and support.

REFERENCES

[1]   R. V. Jategaonkar, *Flight vehicle system identification: a time domain methodology*. American Institute of Aeronau- tics and Astronautics, 2006.

[2]   E. A. A. M. Vladislav Autor Klein, *Aircraft System Iden- tification: Theory And Practice*. Amer Inst of Aeronautics & Astronautics, 2006.

[3]   R. D. Kimberlin, *Flight Testing of Fixed-Wing Aircraft*. AIAA Education Series, 1940.

[4]   M. B. T. Colin R. Theodore and J. D. Colbourne, "Rapid frequency-domain modeling methods for unmanned aerial vehicle flight control applications," *Journal of Aircraft*, vol. 41, 2004.

[5]   J. N. Ostler, "Flight testing small, electric powered un- manned aerial vehicles," Master's thesis, Brigham Young University, 2006.

[6]   M. Luke, "Predicting drag polars for micro air vehicles," Master's thesis, Brigham Young University, 2003.

[7]   D. L. Jung Soon Jang, "Automation of small uavs using a low cost mems sensor and embedded computing platform," October 15, 2006.

[8]   "Flex force sensor." http://www.tekscan.com/flexible-force-sensors.

[9]   "Nasa multi-hole pitot probes." http://www.grc.nasa.gov/ WWW/k-

12/airplane/tunp5h.html.

[10] "Unipot ps22c potentiometer." http://www.indiamart.com/ uniautomation/potentiometers.html.

[11] "Idea forge." http://www.ideaforge.co.in/web/home.

[12] "Freescale semiconductor, mpxv7002 series." http://www. freescale.com/files/sensors/doc/data sheet/MPXV7002.pdf.

[13] "Freescale semiconductor, mpx5100 series." http://cache. freescale.com/files/sensors/doc/data sheet/MPX5100.pdf.

[14] "Mac7100 microcontroller family." http://www.freescale. com/webapp/sps/site/prod summary.jsp?code=MAC7121.

[15] E. A. A. Morelli, "System identification programs for aircraft," *AIAA Atmospheric Flight Mechanics Conference,*2002.

EVALUATION OF ATTITUDE ESTIMATION ALGORITHMS USING HARDWARE - IN – LOOP – SIMULATOR

Y. Satya Swaroop

Fifth Year Dual Degree Student, email: y.satyaswaroop@gmail.com

Aseem V Borkar

Research Assistant, email: aseem.v.borkar@gmail.com

Hemendra Arya

Associate Professor, email: arya@aero.iitb.ac.in

Department of Aerospace Engineering, Indian Institute of Technology Bombay,
Powai, Mumbai – 400076

ABSTRACT

In recent years, applications of MAVs have tremendously increased in fields of surveillance, reconnaissance etc. Effective control system design is an essential feature for making these systems autonomous. Use of low cost sensors in MAVs result in uncertain measurements due to presence of noise and bias which make this control ineffective. This paper presents a comparative study of few of these algorithms on the basis of accuracy and computational requirement. Extended Kalman Filter, Complementary Filters and usage of Vector Observation Techniques are considered for the current work. These algorithms make use of measurements from the rate gyros and the accelerometers to estimate the attitude (i.e., the orientation) of the MAV. However, just using accelerometers measurements, the heading of the MAV cannot be estimated. Use of magnetometer or GPS might prove useful.

Offline simulations were performed on C++ to test these algorithms. The IMU measurements, which are the key inputs for these algorithms are generated using the flight dynamics equations. These algorithms are further tested in real time by employing Hardware In Loop Simulator (HILS). The On Board Computer (OBC) used for the current application is an ARM 7 processor. Sensor data was generated in real time and the performance of these algorithms is tested.

INTRODUCTION

The attitude (also known as orientation) of an MAV is defined by 3 quantities, the roll ($\Box$), the pitch (θ) and the heading or the yaw (φ). There are many ways to determine these quantities. One such way is to numerically integrate the angular rates of the MAV as measured by the rate gyros. But, this gives rise to large errors due to the presence of inherent noise and bias. It was proved that the use of accelerometers along with these rate gyros yields better results. In this paper, we discuss about few algorithms for estimating the attitude of the MAV and compare them on the basis of accuracy and computational requirement. These algorithms include complementary Filters, Kalman Filter and Vector Observation Technique.

These algorithms are first simulated on C++ offline where the sensor measurements are generated using flight dynamics equations. Sensor noise and bias characteristics are assumed to be known. Once the algorithms are tested, they are implemented on Hardware In Loop Simulator for testing in real time. The following section gives a brief introduction to these algorithms. The simulation results and the results from HILS setup are presented in further sections.

ATTITUDE ESTIMATION ALGORITHMS

In this section, a brief idea about the following algorithms is given. All these algorithms make use of measurements from accelerometers and rate gyros. Extended Kalman Filter demands an additional input, velocity of the aircraft. Moreover, all these algorithms need an initial guess of the attitude for further estimations in time.

1. Complementary Filters ([1])
    a. Direct Complementary Filter
    b. Passive Complementary Filter
2. Extended Kalman Filter ([2])
3. Vector Observation Technique ([3])

1. **Complimentary Filters:** This algorithm is taken from [1]. The filter acquires noisy data from the rate gyros and the accelerometers, fuses them to determine the attitude of the MAV. Consider the MAV has its orientation defined by the 3 angles $-\phi$, θ and φ. Let the transformation matrix from the body fixed frame to the inertial frame is denoted by **R**. From the rate gyros, the angular velocities of the aircraft in body frame are given by p, q and r. Then, the time derivative of the rotational matrix is given by the equation 1. Here, Ω is the angular velocity vector given $[p, q, r]^{T}$.

$$\dot{R} = R\Omega_{\times} \tag{1}$$

As mentioned earlier, integrating the above equation for finding the attitude of the MAV will lead to large errors due to the presence of noise and bias in the rate gyros. Thus, accelerometer measurements are brought into use. Let the accelerometer measurements be a_x, a_y and a_z. For a level flight, the accelerometer measures only the component of the forces balancing the gravity. Thus, it can be said that:

$$a = \begin{bmatrix} a_x & a_y & a_z \end{bmatrix}^{T}$$

$$v_a = \frac{a}{|a|}$$

$$v_a = R^{T} v_a$$

However, in the above equation, there is no knowledge about the **R** matrix and needs to be estimated. Using the accelerometer measurements, an initial estimate of **R** matrix is obtained as given by the equation 2.

$$R_y = \arg \min |e_1 - Rv_a|^2 \tag{2}$$

Once R_y is evaluated, the following set of equations is used to estimate the attitude of the MAV:

$$\dot{R} = \left(R_y \Omega + k_p \hat{R}\omega \right)_\times \hat{R} \tag{3}$$

$$\omega = \text{vex}\left(\frac{1}{2}\left(\hat{R}R_y - R_y^T\hat{R} \right) \right)$$

$$\hat{R}(t_k) = \hat{R}(t_{k-1}) + \dot{\hat{R}}(t_{k-1})\Delta t$$

$$\hat{\phi} = \tan^{-1}\left(\frac{\hat{R}(3,2)}{\hat{R}(3,3)} \right)$$

$$\hat{\theta} = \sin^{-1}\left(-\hat{R}(3,1) \right)$$

For the current purpose, only accelerometers were used and hence the heading of the yaw (φ) cannot be estimated. In the above equations, Δt represents the time step used for numerical integration. The equation 3 corresponds to that of the direct complementary filter. This equation gets modified in passive complementary filter as follows:

$$\dot{\hat{R}} = \left(\hat{R}\Omega + k_p \hat{R}\omega \right)_\times \hat{R}$$

$$= \hat{R}\left(\Omega + k_p \omega \right)_\times$$

EXTENDED KALMAN FILTER

Since the aircraft dynamics is governed by a set of nonlinear dynamics equations, extended version of Kalman Filter is employed. This algorithm is taken from [3]. Even in the case of EKF, since accelerometer and rate gyro measurements are used, the heading angle cannot be estimated. In this case, the roll and the pitch angles are the states that are to be estimated. The state matrix $\mathbf{x}$ and the measurement matrix $\mathbf{z}$ are given as follows:

$$\mathbf{x} = \begin{bmatrix} \phi & \theta \end{bmatrix}^T$$

$$\mathbf{z} = \begin{bmatrix} a_x & a_y & a_z \end{bmatrix}^T$$

From the equations of rigid body kinematics,

$$\mathbf{f} = \begin{bmatrix} p + \tan\theta\,(q\sin\phi + r\cos\phi) \\ q\cos\phi + r\sin\phi \end{bmatrix}$$

The linearized state update matrix, $\mathbf{A}$ is nothing but the Jacobian matrix of $\mathbf{f}$. The sensor output model is given by equation 5. The linearized sensor matrix, $\mathbf{C}$ is the Jacobian of $\mathbf{h}$. All the values of these Jacobians are evaluated at the estimated angles at the previous instant.

$$\mathbf{h} = \begin{bmatrix} a_x \\ a_y \\ a_z \end{bmatrix}$$

$$= \begin{bmatrix} \sin\theta\left(1 + \dfrac{q\,V_{air}}{g}\right) \\[2ex] -\cos\theta\sin\phi + \dfrac{V_{air}\,(r\cos\phi - p\sin\phi)}{g} \\[2ex] -\cos\theta\sin\phi - \dfrac{V_{air}\,q\cos\phi}{g} \end{bmatrix}$$

(5)

The following set of equations is used to find the estimates of the pitch and the roll angles.

$$\dot{x} = f(x, u) + w(t)$$

$$z_k = h(x(t_k), u) + w(t)$$

$$P = A(\hat{x}, u)P + PA(\hat{x}, u)^T + Q$$

$$L = P^- C^T (R + P^- C(\hat{x}, u)^T)^{-1}$$

$$P = \left(I - LC(\hat{x}, u)\right)^{-1} P^-$$

$$\hat{x} = \hat{x}^- + L(z - h(\hat{x}^-, u))$$

VECTOR OBSERVATION TECHNIQUE

The current algorithm is taken from [3]. This algorithm involves quaternion based approach. In this method, the gravity vector direction and the earth's magnetic field vectors are observed so as to minimize the error in attitude estimation. From the rate gyros, when expressed in quaternion form, the quaternion **q** representing the attitude of the MAV is determined by using the equation 6.

$$\omega = [0 \quad p \quad q \quad r]^T$$

$$\dot{q} = \frac{1}{2}\hat{q} \times \omega$$

$$q_\omega = \hat{q}_{k-1} + \dot{q}\Delta t$$

(6)

Just like as in complementary filters, integrating the above equation results in large errors due to the presence of noise and bias in the rate gyros. Hence, the accelerometer measurements are used to observe the gravity vector direction and this error in attitude estimation is minimized. An estimate of the attitude in quaternion form, denoted by q_T is obtained using the equations [19] to [35] as mentioned in the reference [3]. The quaternion estimate of the attitude is finally obtained by giving a relative weightage to that estimates obtained from the rate gyros and that from accelerometers using the equation 7.

$$\hat{q} = r q_T + (1 - r) q_o \tag{7}$$

HARDWARE IN LOOP SIMULATOR ARCHITECTURE

The HILS System used for running the flight simulation of the MAV for verifying the implementation of the state estimation algorithms discussed above, on the On-board Computer (OBC) of the MAV has been developed for simulating cooperative missions involving up to eight MAVs in real time. The block diagram of the complete HILS system is shown in the figure 1.The Host PC is a computer on which the complete flight simulation blocks for all the eight aircraft is made as two Simulink block diagrams with some embedded MATLAB functions. Each of these Simulink block diagrams simulates four MAVs and these block diagram are compiled and built into C programs and Header files. The host PC then loads the compiled flight simulation codes of each block diagram onto the two target PCs. Both these target PCs run on the xPC-Target Real Time operating system (also provided by MATLAB), and when the execution command is given by the host PC, the simulation is run on these two computers.

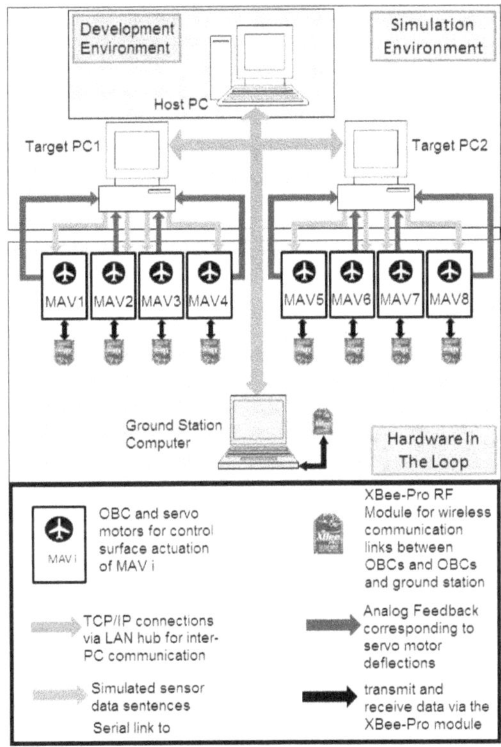

Figure1:Block diagram of the HILS system for real time simulation of cooperative missions of eight MAVs

From the flight simulation states the targets PCs generate the necessary sensor data for the OBC present in the loop. This includes pressure sensor data which are analog voltages generated using a DAC card on each target PC, GPS data which is in the form of serial NMEA format sentences to mimic the EM-406A GPS receiver, IMU data which is in form of serial sentences to mimic the Microstrain 3DM-GX2 IMU. The target PC also has an on-board ADC card which converts analog feedback from the servo motors of each MAV to appropriate control surface deflection values which are given as inputs to the flight simulation of the corresponding MAV.The hardware in the loop which provides feedback consists of the OBC containing the Freescale™ MAC 7121 microcontroller [7] and its peripheral a detailed block diagram of the OBC and its interfacing specifications with its peripherals is as shown in figure 2. The OBC is also equipped with anXBee Pro RF module which it uses to communicate with other MAVs during cooperative mission and with the ground station laptop which is used for monitoring flight parameters of the MAV and tuning controller gains during autonomous flight.

HARDWARE RESULTS

All the algorithms were tested in the Hardware In Loop Simulator. The figure 3 shows the trajectory of the MAV during the simulation. All the sensor data was generated in real time using the flight dynamics equations on this system. As mentioned earlier, these sensor measurements were corrupted using pseudo Gaussian noise of a finite standard deviation. A standard deviation of 0.5 deg/s in rate gyros and 10 milli g in accelerometers were used as noise in the sensors.

The aircraft was commanded to follow a few way points and perform a loop until interrupted. A total of 180 seconds simulation was performed in real time and the results obtained presented in this section. During the entire course of simulation, only the pitch and the roll were estimated the yaw or the heading was assumed to be known from the GPS heading angle. The figures 4 to 7 show the true value of the roll and pitch angles and their estimates for each algorithm. Figure 8 shows the error in estimation of these angles for each algorithm. The statistical data about the errors for each algorithm is presented in the table 1.

Figure 3: Trajectory of the MAV

The Kalman Filter needs an additional state that is to be introduced in order to estimate the bias in the rate gyros. Since no such approach was presented in [2], in all the simulations, no bias in gyros was considered. However, the other two algorithms have special approaches to estimate this bias in gyros. However, it was observed that the complimentary filters could not estimate the roll angles when they are larger in magnitude (even in the absence of gyro bias).

Table 1: Error in Estimation

Algorithm	$\bar{x}_\phi$	σ_ϕ	$\bar{x}_\theta$	σ_θ
DCF	-4.11^0	7.28^0	1.51^0	5.54^0
PCF	-3.72^0	6.89^0	1.08^0	6.21^0
EKF	-2.84^0	4.53^0	0.55^0	1.72^0
VO	-1.65^0	4.17^0	0.55^0	3.22^0

COMPARATIVE STUDY

*Accuracy Comparison:*From the simulation results and the Table 1, it can be clearly seen that the DCF and the PCF did not work well when the turn rates of the MAV are high. They had an unacceptable range of errors in both pitch and roll. However, EKF and the vector observation techniques proved to be useful. From the accuracy point of view, the vector observation seemed to be outperforming the Kalman Filter.

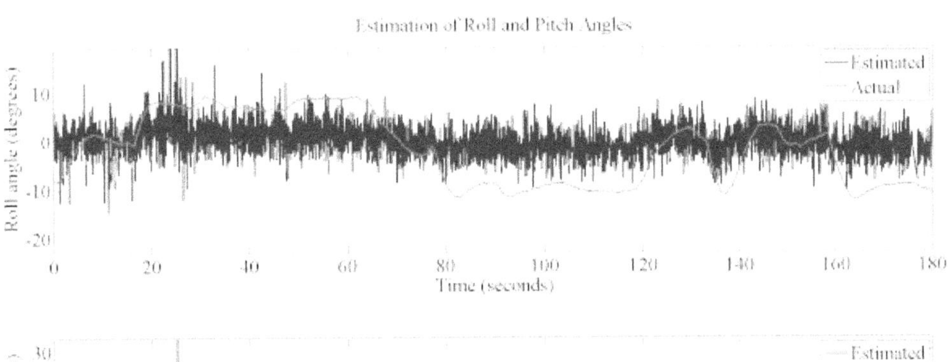

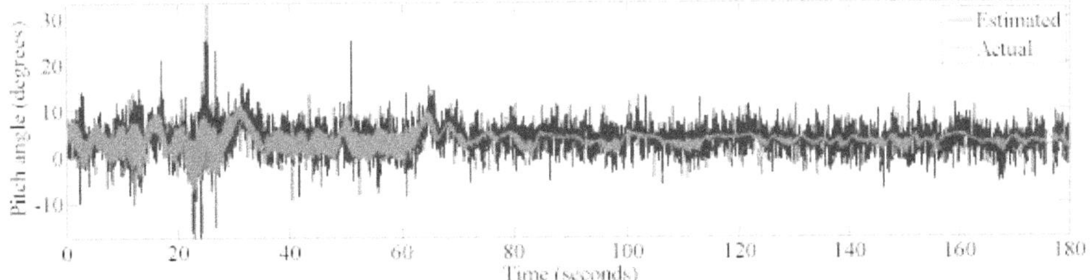

Figure 4: Direct Complimentary Filter

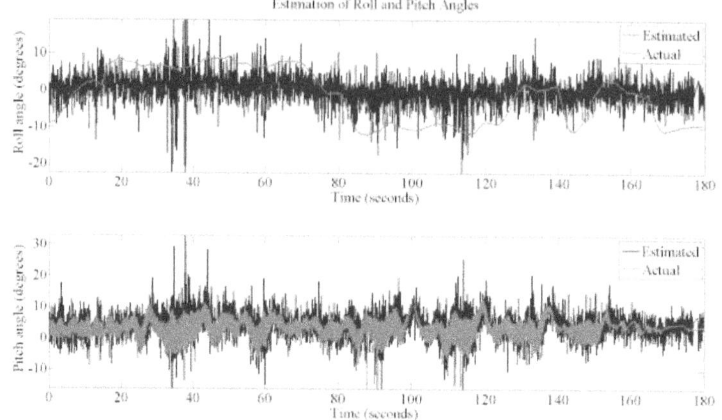

Figure 5: Passive Complimentary Filter

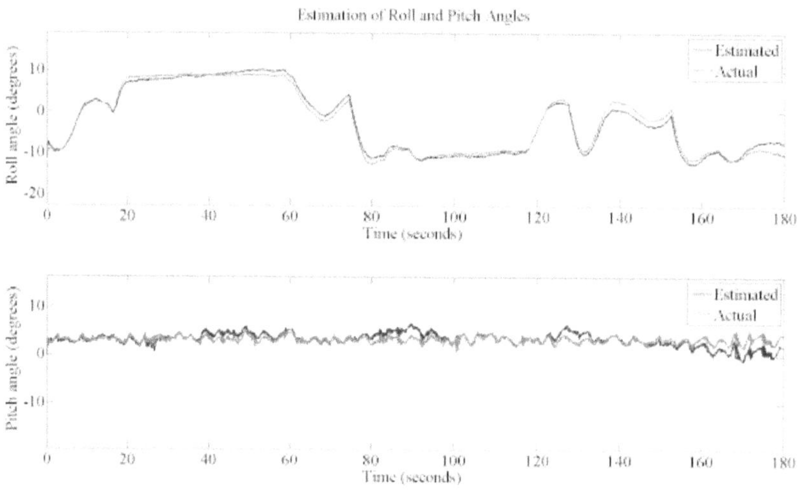

Figure 6: Extended Kalman Filter

Computational Load: All the algorithms were analyzed for computational requirement in terms of 6 aspects. They are:

- No. of trigonometric operations
- No. of matrix inversions
- No. of matrix multiplications
- No. of quaternion multiplications (only for VO)
- No. of evaluations of square root

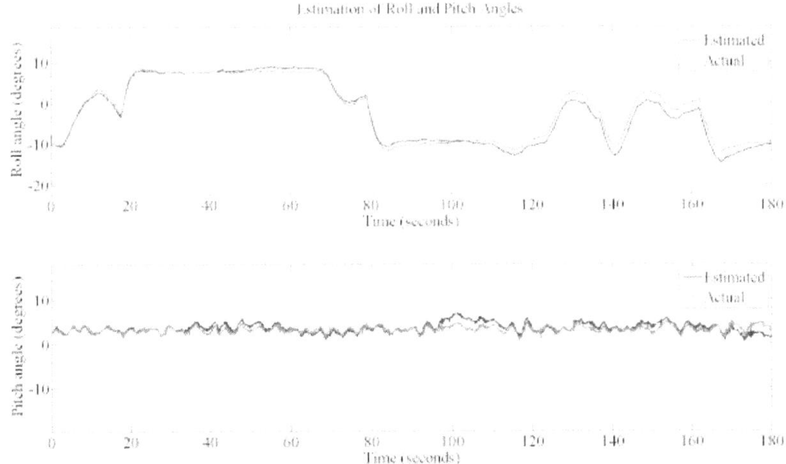

Figure 7: Vector Observation Technique

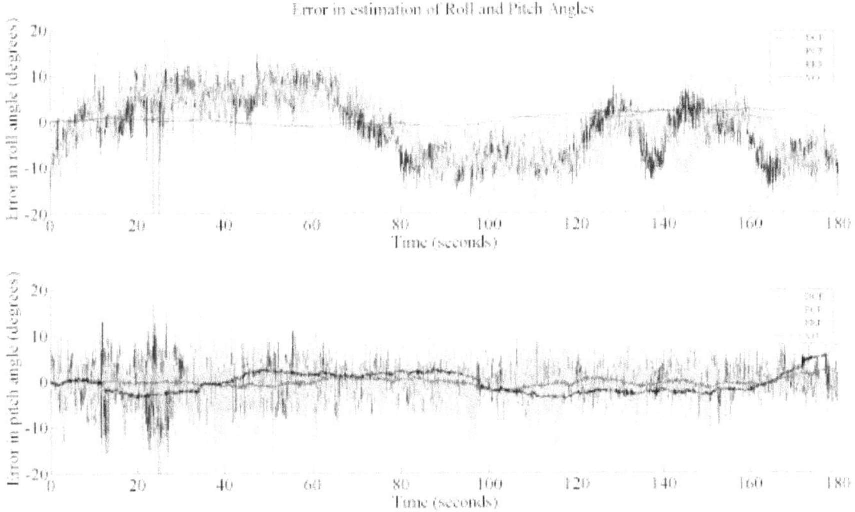

Figure 8: Error in estimation of roll and pitch

The above list shows the decreasing order of computational requirement. All the algorithms involve evaluation of sine and cosine of Euler angles which makes it 6 trigonometric operations. Vector observation technique involves 6 more trigonometric operations. In case of Kalman Filter, there are only matrix additions, multiplications and inversions. On an average, EKF involves 6 2x2 matrix additions, 10 multiplications and 1 inversion operation. Use of complimentary filters involve 5 matrix inversions, 3 inverse trigonometric operations, 3 3x3 matrix multiplications and one square root operation. Vector Observation involves 1 quaternion multiplication, 1 matrix multiplication, 3 inverse trigonometric operations and one matrix addition.

CONCLUSIONS

The HILS generates the sensor data for every 20 milli seconds. However, due to the computational load on the processor (both due to the algorithm and due to other modules for HILS), the algorithms could be run at half the sampling frequency, i.e., for every 40 milli seconds. This resulted in larger numerical errors emerging from Euler's 1 step numerical integration scheme. However, since all the algorithms were run for every 40 milli seconds, this comparative study should be valid.

From computational requirement, it can be said that the vector observation technique proves to be much better than the other 3 algorithms. The Kalman Filter may prove to be better in this regard. In terms of accuracy, the vector observation technique slightly out performs the Kalman Filter. From this, one can conclude that use of Vector Observation Technique proves to be relatively more useful for attitude estimation.

REFERENCES

[1]   Robert Mahony, Tarek Hamel, and Jean Michel Pflimlin, *Nonlinear Complementary Filters on the Special Orthogonal Group*, IEEE Transactions on Automatic Control, Vol.53, No. 5, June 2008, pg 1203 - 1218.

[2]   *Improved State Estimation for Miniature Air Vehicles*, Brigham Young University Graduate Committee Approval of a thesis submitted by Andrew Mark Eldredge.

[3]   Sebastian O.H. Madgwick, *An efficient orientation filter for the inertial and inertial/magnetic sensor arrays*, pg 1-32

[4]   Jianquin Mao, *Optimal orthogonalization of strapdown matrix by using singular value decomposition*, Comp. \&Maths. withAppls. Vol 12 A, No. 3, pp 353-362, 1986

[5]   D. Krishnan, A. V. Borkar, P. Shevare, and H. Arya, *Hardware in loop simulator for cooperative missions*, in Second International Conference on Advances in Control and Optimization of Dynamical Systems, IISc Bengaluru, February 2011.

[6]   D. Krishnan, A. V. Borkar, and H. Arya, *An elegant hardware in loop simulator for cooperative missions of MAVs*, in AIAA Infotech@Aerospace 2012, Garden Grove, California, June 2012.

[7]   http://www.freescale.com/webapp/sps/site/ prod_summary.jsp?code=MAC7121

COMPARATIVE ANALYSIS OF CLASSIFICATION METHODS IMPLEMENTED IN CONTENT BASED IMAGE RETRIEVAL SYSTEMS

P. Nalini

Assistant Professor, Dept. of ECE, Mahatma Gandhi Institute of Technology,
Hyderabad, Andhra Pradesh.

B.L Malleswari

Professor, Dept. of ECE, G. Narayanamma Institute of Technology,
Hyderabad, Andhra Pradesh.

ABSTRACT

The ever increasing amount of multimedia data creates a need for new sophisticated methods to retrieve the information. Content Based Image Retrieval (CBIR) is becoming an important research area in image processing to get faster and efficient retrieval. This paper provides comparison analysis of currently used methods applied in the field of CBIR with their merits and demerits. We used a method of texture based classification by finding texture properties with image co occurrence matrix which has continuous mapping properties. Feature extraction methods in frequency and spatial domain were discussed with and without preprocessing techniques like edge detection and image segmentation were compared and discussed based on their algorithmic performance measures like precision, recall, retrieval speed and efficiency. Various feature extraction techniques are compared using a similarity measure implemented with Euclidean Distance. By applying different level decompositions, image content is captured.

INTRODUCTION

The principle of Content Based Image retrieval is to organize digital image achieves by their visual content. The most common form of CBIR is image search by its visual features. (Greg Pass et al., 1996) Almost all the CBIR systems designed so far widely use features like color, shape and textures and spatial all together or few of these. Paper (B. S. Manjunath et al., 2001) describes a method for image retrieval purely based on color and texture.

The commercial image search engines available as on date are QBIC, PicSom, FIRE, VisualSeek, and AltaVista etc. Region Based Image Retrieval is an extension for CBIR. (Roger Weber et al., 2003).

In this paper apart from visual features like color and texture a new feature extraction algorithm called edge histogram is introduced. Edge conveys essential information so to a picture and therefore can be applied to image retrieval. It is observed by Nandagopalan (2008), the edge histogram captures the spatial distribution of edges (B. S. Manjunath et al., 2001), (Alberto Amato et al., 2003). Our model expects inputs as Query By Example and any combinations of features can be used for retrieval.

The main focus of this paper is to compare performance analysis in terms of precision and recall rate for image retrieval techniques using various global visual features and dividing the image into segments and extracting the features locally. All low level features are determined globally and locally. These are mean, standard deviation and median of RGB channels of color histograms. The two three dimensional color spaces available are HSV and RGB (Arnold W.M. Smeulders et al., 2000). Then the texture features like energy and entropy are retrieved for the full image. The same features are determined after dividing the image into four segments. The same features have been determined for each segment of image and a greedy method is introduced in similarity search.

The rest of the paper is organized as follows. Since there was a lot of work done in this area a comprehensive survey of CBIR is dealt in section II. Section III describes the overview of the proposed CBIR model used in this paper. Sections IV and V focuses on extraction of color and texture features for a global image. In Section VI Image segmentation and extraction of features locally were discussed. In section VIII similarity comparison, Section IX is to provide data tables and types for the feature vector are shown Section X is to provide particulars of experiments conducted followed by results and discussions.

RELATED WORK

Image retrieval work can be traced back to the late 1970's. In the early 1990's as a result of advances in the Internet and new digital Image sensor technologies, the volume of digital images produced by scientific, educational, medical, industrial and other applications available to users increased dramatically. The difficulties faced by text based retrieval became more and more severe. The efficient management of the rapidly expanding visual information became an urgent problem.

Local feature based methods proved good results compared with the global feature based methods. (C. R. Shyu, et. al 1998).For a successful CBIR, the indexing scheme is efficient for searching image database. Recent retrieval systems have incorporated users relevance feedback to modify the retrieval processing order to generate perceptually and semantically more meaningful retrieval results. It is described by Nandagopalan (2008) that the works shown in paper (B. S. Manjunath et al., 2001) was mixture of color and texture features. and edge density for MPEG - 7 standards and in paper (Minyoung Eom, et al., 2005) the edge histogram was used. Papers (Son Lam Phung et al., 2007) and (S. L. Phung et al., 2007) discusses similar kind of approaches based on edge density for detecting people in images. Content Based Image Retrieval with color texture and shape features were discussed in paper (P. S. Hiremath et al., 2007). Image features along with segmentation discussed in paper (Mustafa Ozden et al., 2002. Considerable amount of work had already done in medical image retrieval. For these types of images texture is the highly considerable feature. (John Montagnat, et al., 2007), (Thomas M. Lehmann, et al 2004). To make image retrieval faster several indexing features were designed. The most popular ones are fuzzy based relationship tree, containment tree, Graph based etc.

Proposed CBIR Model

The proposed CBIR model of this paper is shown in figure 1. The test images are kept in a database known as image database. After preprocessing such as enhancement, images are segmented based on method described in paper (Mustafa Ozden et al., 2002) In our model first we obtained a single feature vector by obtaining the color histogram and texture features for all the images of image data base. A feature vector is determined for the given query image. By using Euclidean distance as the similarity measure all closely related images are retrieved and recall rate and precision are determined for different classes of images.

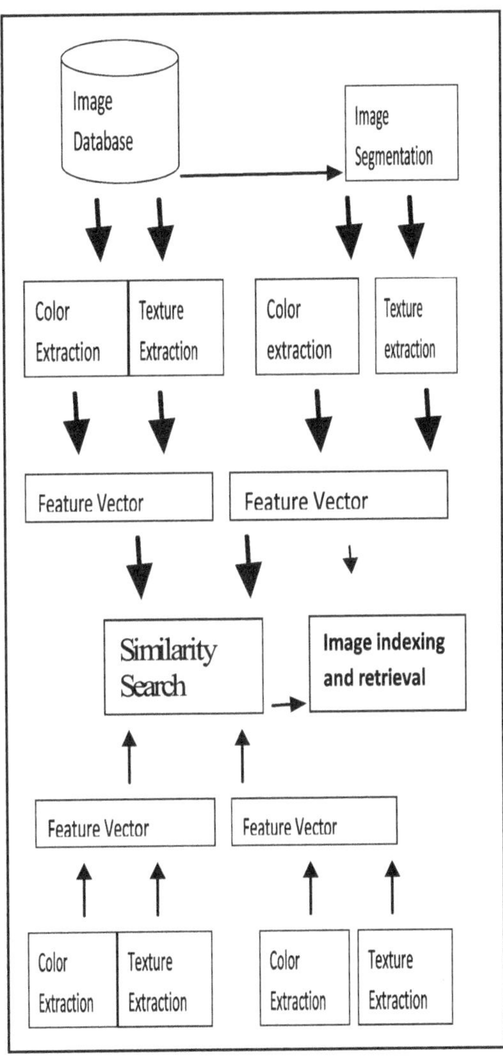

Fig 1: Proposed CBIR Frame Work

Secondly we segmented the images and only the dominant segments are considered for feature extraction namely color histogram features, texture features and image density features. Then a single feature vector is constructed and stored in the feature database, when a query image is submitted by the user, the same work is done like segmentation, feature extraction and creating a feature vector. Using an appropriate threshold, images that are semantically closer are retrieved from the database and displayed as a thumbnail. Then recall rate and precision are calculated for the retrieved images of that particular class.

Finally we compared the results of both cases such as retrieval precision and recall rate without segmentation and after segmentation.

COLOR FEATURE EXTRACTION

Color histogram is the most commonly used feature in Image retrieval systems. The main reason is that image histogram is independent of image size and orientation. Statistically it denotes the joint probability of the gray levels or pixel intensities of all the three color channels. In the first method, color histograms are extracted for the whole image and from which the major statistical features calculated are histogram mean, standard deviation and median for each color channel and totally 3 x 3 = 9 features are obtained for each image. All this are considered as a feature vector. Now the query image color feature vector is compared with database image color feature vector, and the images with in similarity threshold got retrieved.

In other method the image is segmented and color histogram is extracted from each segment and 9 statistical features were calculated for each segment. All this kept as a feature vector segment wise. The query image will also get segmented and feature vector determined for each segment. All the segments may not be considered, but only few segments that are dominant are only considered to speed up the calculation with improved precision. The performance of the color features are tested using test images from Corel image data base.

TEXTURE FEATURE EXTRACTION

There is no precise definition for texture. However, one can define texture as the visual patterns that have properties of homogeneity that do not result from the presence of only a single color or intensity. Texture determination is ideally suited for medical image retrievals (Thomas M. Lehmann, et al 2004). In this work, computation of gray level co-occurrence matrix is done and from which a number of statistical measures are derived.

The autocorrelation function of an image is used to quantify the regularity and the coarseness of a texture. This function is defined for an image I as:

$$\rho(x,y) = \frac{\sum_{u=1}^{N}\sum_{v=1}^{N} I(u,v)\,I(u+x,\,v+y)}{\sum_{u=1}^{N}\sum_{v=1}^{N} I^2(u,v)} \qquad (1)$$

A texture is characterized by a set of values called energy, entropy, and homogeneity. The following formulas are used to calculate the features and are shown in equations 2 to 4 (Dong Yin et al., 2008)

$$\text{Energy} = \sum_i \sum_j P_d^2(i,j) \qquad (2)$$

$$\text{Entropy} = \sum_i \sum_j P_d(i,j) \log P_d(i,j) \qquad (3)$$

$$\text{Homogeneity} = \sum_i \sum_j \frac{P_d(i,j)}{1+|i-j|} \qquad (4)$$

The texture features are determined globally as well as for individual segments of the image after image segmentation. Performance parameters were determined global feature based retrieval and also local feature based retrieval. The performance of the texture features are tested using test images from Corel image database.

SEGMENTATION WITH EDGE DENSITY

The edge histogram is normally used in the area of computer vision primarily in tracking of moving objects (Bohyung Han et al., 2005). Edges convey essential information to a picture, and their accurate detection is of primary importance. The identification of edge inside one image is the first step to recognize geometric shapes within one image (Alberto Amato et al., 2003). The edge histogram descriptor represents the local edge distribution in the image which is obtained by subdividing the image into 4 x 4 sub images. For each of these sub images a histogram is computed.

For each sub image the edge density is calculated using equation (6). Let (x1,y1) is the top left corner and (x2,y2) are bottom right corner of the sub image the edge density *f* is given by

$$f = \frac{1}{a_r} \sum_{x=x1}^{x2} \sum_{y=y1}^{y2} e(x,y) \qquad (6)$$

Where a_r is the region area. All these features put in the feature vector table.

FEATURE VECTORS & FEATURE SPACE

Feature vector is one method to represent an image by finding measurements on a set of features. The *feature vector* is an n-dimensional vector that contains these measurements, where n is the number of features. The measurements may be symbolic, numerical, of both. An example of a symbolic feature is color and of a numerical feature is the area of an object. The feature vector can be used to classify an object, or provide us with condensed higher-level image information. A *feature space* is a mathematical abstraction which is also n-dimensional and is created to allow visualization of feature vectors, and relationships between them. With two- and three-dimensional feature vectors it is modeled as a geometric construct with perpendicular axes and created by plotting each feature measurement along one axis. For n-dimensional feature vectors it is an abstract mathematical construction called a *hyperspace*. The creation of the feature space allows us to define distance and similarity measures, which are used to compare feature vectors and aid in the classification of unknown samples.

SIMILARITY COMPARISON

The feature vector is meant to represent the object and will be used to classify it. The primary method is to either measure the difference between the two, or to measure the similarity. Two vectors that are closely related will have a small difference and a large similarity. For this we used Euclidean distance which is the most common metric for measuring the distance between two vectors, the Euclidean distance is given in equation (5):

$$E_d = \sqrt{\sum_{i=1}^{n}(a_i - b_i)^2} \qquad (5)$$

The main issue in image retrieval systems is the number of dimensions of the feature vector which is normally huge. This can be reduced with the techniques like using Principle Component Analysis (PCA) (Dr. Fuhui Long et al., 2003) It explores exponentially with the increasing of the dimensionality and eventually reduces to sequential searching, to overcome these methods a simple greedy strategy is used in paper (Nandagopalan .S et al., 2008) we used the same method in similarity matching.

The algorithm shown in figure (2) uses greedy strategy to compare the similarity between query image and the database images.

ALGORITHM IMAGE SIMILARITY

// *I[N]* – Image DB with *N* images

// *QI* – Query Image

for each (Image *I* in *I[N]*)

for each (Segment *s* in *SegmentSet*)

if (Euclidean(*QI*[s], *I*[s]) < *threshold*)

// continue to check other segments

else

// no need to check other segments

end.

Fig. 2 Algorithm for Similarity Comparison Based On Greedy Strategy

EXPERIMENTAL RESULTS

A Dell Precision Pentium Core 2 Duo workstation with 2 GB RAM computer is used for conducting the experiments. The main software used was MATLAB 6.0. For image processing work MATLAB 6.0 was used. Graphical User Interface (GUI) window was used to show the retrieved results. Corel Image database with 2000 natural images were used for testing the proposed CBIR system.

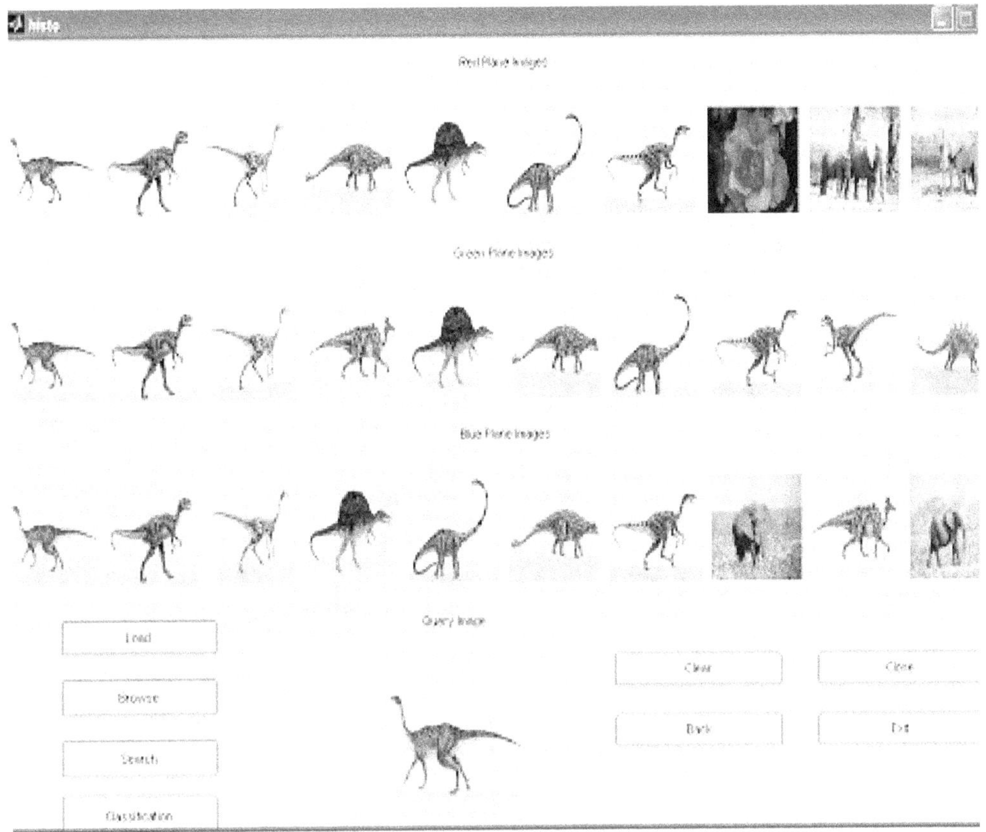

Fig.3: Screenshot of CBIR Model

RETRIEVAL EFFICIENCY

The retrieval efficiency namely precision and recall were calculated using 2000 natural color images (100 in each category) from Corel image database. Figure 3 shows the screen shot of the frame work.

Standard formulas equation (5 & 6) have been used to compute the precision and recall for four query images as shown in figure 4.

$$Precision = \frac{No\ of\ relevant\ images\ retrieved}{total\ number\ of\ images\ retrieved}$$

$$Recall = \frac{No.\ of\ relevant\ images\ retrieved}{Total\ Number\ of\ relevant\ images\ in\ the\ database}$$

By randomly selecting four query images from the Corel Image Database, the system was tested and the results are shown in table 1

Fig.4 Query Images 1 to 4 (top left to bottom right)

Table-1 Precision And Recall Values In %

Query	Color	Texture	EHD	All	Image
1	21.8	50.0	23.6	35.2	(LF)
	28.0	15.0	34.1	60.0	(GF)
2	100.0	75.0	87.0	100.0	(LF)
	98.0	62.0	68.0	78.0	(G F)
3	74.6	20.0	65.0	42.8	(LF)
	59.0	10.0	37.0	90.0	(GF)
4	91.7	75.0	85.6	92.0	(LF)
	24.0	33.0	34.9	28.0	(GF)

GF – Global Features & LF Local Features

CONCLUSION AND FUTURE WORK

This paper proposed two different approaches for the Content Based Image Retrieval system combining the color, texture and edge density features and individual for the whole image and also the segments of the image. We compared the analysis and understood that local feature based retrieval gave good results than the global based search. Users were given options to select the appropriate feature extraction method for good results. The results are quite good for most of the query images and it is possible to further improve by fine tuning the threshold and adding the relevance feedback in the search.

REFERENCES

[1] Alberto Amato, Vincenzo Di Lecce 2003, "Edge Detection Techniques in Image Retrieval: The Semantic Meaning of Edge", 4th EURASIP Conference on Video/Image Processing and Multimedia Communications, 2003, Zagreb, Croatia. pp. 143-148.

[2] Arnold W.M. Smeulders, Marcel Worring, Simone Santini, Amarnath Gupta, and Ramesh Jain, 2000 "Content based image retrieval at the end of the early years."

[3] B. S. Manjunath, Jens-Rainer Ohm, Vinod V. Vasudevan, and AkioYamada 2001, "Color and Texture Descriptors". In: IEEE Transactions on Circuits and Systems for Video Technology, Vol. 11, No. 6, June 2001, pp. 70-715.

[4] Bohyung Han, Changjiang Yang, et al 2005, "Bayesian Filtering and Integral Image for Visual Tracking".

[5] C. R. Shyu, et. al 1998, "Local versus Global Features for Content-Based Image Retrieval", IEEE Workshop on Content-Based Access of Image and Video Libraries,

[6] Dong Yin, Jia Pan, et al 2008, "Medical Image Categorization based on Gaussian Mixture Model", IEEE 2008 International Conference on BioMedical Engineering and Informatics, pp. 128-131.

[7] Dr. Fuhui Long, Dr. Hongjiang Zhang and Prof. David Dagan Feng 2003, "Fundamentals of Content-Based Image Retrieval" - http://research.microsoft.com/asia/dload_files/group/mcomputing/2003 P/ch01_Long_v40-proof.pdf

[8] Greg Pass, Ramin Zabih 1996,, "Histogram refinement for content based image retrieval" WACV '96.

[9] John Montagnat, et al 2007, "Texture-based Medical Image Indexing and Retrieval on Grids", Medical Imaging technology, vol 25 No. 5 Nov 2007. pp. 333-338.

[10] Minyoung Eom, and Yoonsik Choe 2005, "Fast Extraction of Edge Histogram in DCT Domain based on MPEG7", Proceedings of World Academy of Science, Engineering and Technology Volume 9 November 2005 ISSN 1307-6884, pp. 209-212.

[11] Mustafa Ozden and Ediz Polat 2002, "Image Segmentation using Color and Texture features".

[12] Nandagopalan .S, Dr. B. S. Adiga, and N. Deepak 2008, "A Universal Model for Content-Based Image Retrieval" World Academy of Science, Engineering and Technology , June 2008 Pg. No 644 – 647

[13] Nidhi Singhai, and Prof. Shishir K. Shandilya 2010, "A Survey On: Content Based Image Retrieval Systems" International Journal of Computer Applications (0975 – 8887), July 2010, Volume 4 – No.2.

[14] P. S. Hiremath , Jagadeesh Pujari 2007, "Content Based Image Retrieval using Color, Texture and Shape features", 15th International Conference on Advanced Computing and Communications, IEEE Computer Society 2007, pp. 780-784.

[15] Ritendra Datta, Jia Li, and James Z. Wang 2005, "Content-Based Image Retrieval-Approaches and Trends of the New Age", *MIR '05,* November 11-12, Singapore, 2005, Pg. .No 253-262

[16] Roger Weber and Michael Mlivoncic 2003, "Efficient Region-Based Image Retrieval", ACM CIKM '03 November 3-8, 2003, USA

[17] Son Lam Phung and Abdesselam Bouzerdoum 2007, "A New Image Feature for Fast Detection of People in Images", International Journal of 2007 Institute for Scientific Information and Systems Sciences Computing and Information Volume 3, Number 3, pp. 383-391.

[18] S. L. Phung and A. Bouzerdoum 2007, "Detecting People in Images: An Edge Density Approach", IEEE, ICASSP 2007. pp. 1229-1232

[19] Thomas M. Lehmann, et al 2004, "Automatic categorization of medical images for content-based retrieval and data mining", Computerized Medical Imaging and Graphics. Elsevier 2004. pp. 143-155.

DESIGN CONSIDERATIONS OF CCD OPTICAL SIGHTS FOR LONG RANGE OBJECTS IN REAL TIME SUITABLE TO MICRO AIR VEHICLES

R. Sudhakar Rao

Design & Engineering Division, Bharat Dynamics Limited,
Kanchanbagh, Hyderabad – 500 058, A.P., India; E-mail:rsrbdl@yahoo.com

Dr. Matham Chandra Sekher

Design & Engineering Division, Bharat Dynamics Limited,
Kanchanbagh, Hyderabad – 500 058, A.P., India; E-mail: cmatham@gmail.com
and

Prof. Chandra Lingam

Dept. of Physics, College of Engineering, JNTUH, Kukatpally, Hyderabad – 500 085, A.P., India;
E-mail:chandra_lingam@yahoo.com

ABSTRACT

CCD Optical Sight (Fig. 1) was designed to track movement of objects in real time. The concept of introducing a novel CCD Optical Sight on a weapon simulator has been realized. Rigorous field trials at different ranges have proved the robustness and efficacy of the system. Few numbers of these items were fabricated, integrated in to the weapon simulators and delivered to the user. Apart from this specific application, present development enabled to carryout innovative research and development of new devices to capture real time trajectory of a long range weapons (~ 4000 m).

The paper accounts the importance of the selective features of Charge Coupled Device (CCD) sensors and its advantages over film-based imaging. The purpose and features of a CCD Optical Sights for Micro Air Vehicles are accounted. Design considerations of well corrected optical systems and types of configurations suitable to such environments shall be reported. The basic idea is to develop a knowledge based design center to cater to the design and development requirements of long range object detection, recognition and identification and adaptation of a well-corrected optical system matching the imaging properties of CCD sensors, display devices. These details shall be explained with illustrations from the work carried out for a specific task. Few details on the Engineering design; development, description and functioning of the systems for military applications shall be provided. The same philosophy is extended to address the developmental needs of optical systems for micro air vehicles for different applications.

Fig 1. CCD Optical Sight

REFERENCES

[1   EG&G Optoelectronics - Reticon Product Data Book on "Image Sensing & Solid State Cameras", 1994-95.

[2]   R L Lombardo Jr, "Target Acquisition: It's Not Just for Military Imaging", Photonics Spectra, p123, July 1998.

[3]   Rudolf Kingslake, "Lens Design Fundamentals", Chapters 3 – 6, Academic Press, London, 1978.

[4]   R. Sudhakar Rao et al, "Design and development of a Long Range Telescope coupled with CCD Camera for Remote Detection Applications", Jr. of Optics, Vol. 31, No. 3, Pages 145-152, 2002.

[5]   Indian Patent No. 228497 dated 5th Feb 2009.

[6]   R. Sudhakar Rao and Prof. S. Chandra Lingam, "Design and development of CCD Optical Sight for tracking real time objects", Int.Journal of Advancements in Research & Technology, Volume 1, Issue 4, September-2012 (ISSN 2278-7763). tracking real time objects", Int.Journal of Advancements in Research & Technology, Volume 1, Issue 4, September-2012 (ISSN 2278-7763).

INTEGRATION OF IMAGE SENSORS AND IMAGE DATA PROCESSING

Durusoju Hari Prasad, N. Darga Kumar, S. Sreekanth

ABSTRACT

The utilization and of nature for sustainable development is based on material cycling and recycling, capture and utilization of resources through dispersed processes. When humans exploit the nature for their own one-time benefits with scant respect for ecological balance, the system will tend to show the sign of destabilization and deterioration. Thus our towns and cities are now endangered with complex environmental problems due to lack of sustainable planned growth. This is detrimental both to urban as well rural environments. Ground water is are component of nature, which has to be preserved. The central theme of the present study is the impact of Land use / Land cover on ground water quality.

PROBLEM

Rapid urbanization and industrialization has led to unchecked proliferation of hazardous industries in and around Hyderabad, unchecked and unregulated distribution and management of urban resources especially green areas, water bodies and the combination of above aspects with large scale urban sprawl have all made Hyderabad an unmanaged polluting city.

One of the major problems 'Water Pollution' is referred to the addition to water an excess of material that is harmful to humans, animals, or desirable aquatic life, or otherwise causes significant departures from the normal activities of various living communities, in or near bodies of water.

The term water pollution refers to any type of aquatic contamination between two extremes:

1) A highly enriched over productive biotic community such as river or lake with nutrients from sewer or fertilizer.

2) Water body poisoned by toxic chemicals, which eliminate living organizing exclude all forms of life.

GROWTH AND DEVELOPMENT

According to 1991 census Hyderabad is the fifth largest metropolis of India with a population of 4,344,437. The population of Hyderabad has increased from 0.448 millions in 1901 to 0.502 millions in 1911 but came back down to 0.406 million in 1921. It again went up to 0.447 million in 19321, 0.739 million in 1961 (an increase of 65.3%), 1.28 million in 1951 (52.5% increase), 1,429 million increase in 1961 (10.71% increase) and 1.796 million in 1971 (43.8% increase). Between 1971 and 1981 the population went up to 2.759 million, and the rate of increase in the Urban Agglomeration (U.A) was 42.65%. Between 1981 and 1991 the population went up to 4.34 million and the rate of growth the highest so far is 67.04%.

OBJECTIVE

The main objective of the study is utilization of integrated methodology, which derives, a correlation between land use/land cover and water quality parameters.

1) Assessment of impact of Land use / Land cover on ground water quality in the present study (Zone VII).

2) To analyze the estimated ground water quality in the study area for its suitability for drinking purpose.

3) Determination of water quality index by statistical approach.

4) To highlight the areas under environmental stress.

5) To recommend suitable measures for best management mitigation practices.

METHODOLOGY

The methodology or the work schedule gives the following principles steps involved, which are to be adhered, to achieve the set of objection.

1) Collection of toposheet from survey of India (1:50000 scale)

2) Collection of data from satellite imagery' (fused data of IRS-IC PAN and IRS-ID LISS-III Imagery) obtained from NRSA, Balanagar

3) Delineation of the study area (Based on MCH division) and conversion of raster data into vector data.

4) Preparation of thematic maps with the help of secondary data and primary data.

5) Study of the land use features and their influence on the zone.

6) Analyzing the quality of water and to determine water quality index by rating the parameters.

7) Correlating the land use impact with the water quality.

8) Resulting with the previous data for the comparative study and to find the changes.

9) Recommendations at micro level.

Flow Chart Showing Methodology for the Present Study

<u>METHODOLOGY</u>

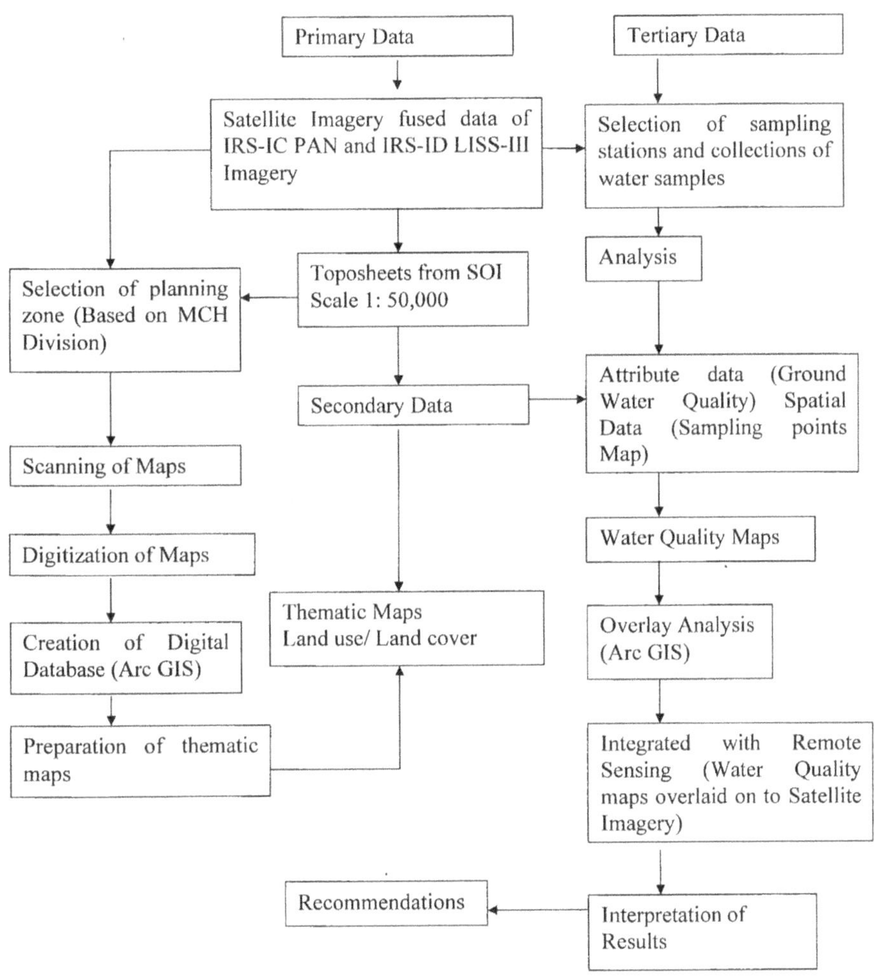

SOURCES OF DATA PRODUCTS

S.No.	Thematic Layers	Source of Acquisition
1	Landuse/Landcover (Level II Classification)	Satellite Data from NRSA (IRS-1DLISS-III + PAN)
2	Map showing Drainage Network	Survey of India Toposheet
3	Map showing transportation network	Survey of India Toposheet
4	Base Map	Survey of India Toposheet

LIMITATION

1) Study is limited to single zone.

2) The selection of environmental parameters and land use features are confined to   data availability and time constraint.

3) Recommendations are not suggested at macro level.

DESCRIPTION OF THE STUDY AREA

INTRODUCTION

The city of Hyderabad is located on undulating ground lends itself to a remarkable terrain of beauty enhanced by its rocks, lakes and landscapes. The total municipal area is divided into 11 planning zones. Some of them are fully developed, like 1, 3, 5, 7, 9 divisions. An increase in the population civic amenities water quality of the area is affected adversely. Most of the lakes and rivers are forced to serve as drainage channels, which in turn are affecting ground water quality of the adjacent localities in the study area.

A single zone from the planning division is selected for the present study (zone VII) and important locations are listed.   This study area falls in Survey of India toposheet 56 k/7, 56k/11 (source SOI). This Zone is adjoined by 3, 6, 9 zones of MCH division.

DELINEATION OF STUDY AREA

As the land use, quality of environment are diversified and are not uniform throughout an Urban area, to evaluate the quality of environment the whole area has to be divided into units. The delineation of study area can be under the following considerations.

i) Considering the planning divisions demarcated by Municipal Corporation of Hyderabad,

ii) Considering the areas, which are declared as Institute by Environmental Monitoring Agencies.

ii) Delineation of study area based on demarcations by natural barriers.

VISUAL INTERPRETATION

The study of land use, land cover has been a focus of interest. Since the early days of aerial photography with the availability of new remote sensing techniques using aircrafts and spacecraft as platforms with a capacity for operating outside the visible part of the electromagnetic spectrums. The limitations of photo interpretation have now been changed to broad spectrum of image interpretation. Success in image interpretation varies with the training and experience of the interpreter, the nature of the objects or phenomena being interpreted and the quality of the image being utilized.

The various aspects of image interpretation are listed below in simpler form:

- Detection: Picking out an object or element from photo or image through interpretation techniques.

- Recognition and Identification: It is a process of classification or trying to distinguish an object by its characteristics or patterns, which are familiar on the image.

- Analysis: It is resolving or separating a set of object or features having similar set of characterizes.

- Classification: It is a process of identification of grouping of objective or features resolved in and analysis.

- Deduction: It is a process where references are drawn about the objects based on direct or indirect evidence of the information or phenomenon under study. Deductions may be firmly confirmed by ground checks to avoid misspecification

- Idealization: It is a process of drawing ideal or standard representation from what is actually identified and interpreted from the image or map.

Basic Elements of Visual Interpretation

The exact characteristics useful for any specific work and the manner in which they are considered depend on the field of application.

- **Shape:** Refers to the general form, configuration, or outline of individual objects. Size and Shape are interrelated.

- **Size:** Size of objects on image must be considered in the context of the image scale.

- **Pattern:** Relates to the spatial arrangements of objects. The repetition of certain general forms or relationships in characteristics of many objects, both natural and constructed,

- **Tone (or hue):** Refers to the relative brightness or color of objects on an image. Different surface objects reflect and emit certain amounts of radiant energy. The true color or false color imagery increase the interpretability by providing a subtle tonal contract between them.

- **Texture:** Is the frequency of tonal change on an image. Texture is produced by an aggregation of unit features that may be too small to be discerned individually on the image, such as tree leaves and leaf shadows. It is a product of their individual shape, size, pattern, shadow and tone. It determines the overall visual "smoothness" or "coarseness" of image features. It is dependant on scale.

- **Site / Location:** Refers to topographic or geographic location and is a particularly important aid in the identification of vegetation types. It also provides the due for identifying objects and understanding their genius.

- **Shadow:** They are cast due to sun's illumination size, and shape of the objective or sensor vectoring angle. The shape and profile of shadows help in aspect of image in interpretation.

- **Association:** Refers to the occurrence of certain features in relation to them and neighboring features.

- **Resolution:** Depends on many factors, but it always places a practical limit on interpretation because some objects are too small or have too little contrast with their surroundings to be clearly seen on the image.

It is of two types: Spatial - Direct and distinguish the smallest objective on the Ground

      Spectral - Refers to picture elementary the image of smallest Area resolvable or identifiable on ground

- **Aspect:** It refers to the direction in which a mountain / hill slope faces particularly with reference to possible amounts of sunshine and shadow. Aspects here marked effect on the sitting of vegetation, settlements and cultivation.

GEOGRAPHIC INFORMATION SYSTEM (GIS)

GIS is an acronym for Geographic Information Systems. In detail GIS is decision support computer based system for collecting, storing presenting and analyzing spatial information.

An information system, a set of process, executed on raw data, to produce information, which will be useful in decision making. GIS is a general-purpose technology for handling geographic data in digital form, and satisfying the following specific needs, among others.

- The ability to preprocess data from large stores into a form suitable for analysis including operation such as reformatting, change of projection, resampling and generalization.

- Direct support for analysis and modeling such that, form of analysis calibrations of models, forecasting and prediction all handled through instructions to the GIS.

- Posts processing of results, including such operations are reformatting tabulations, report generation and mapping.

GIS is a convergence of technological fields and traditional disciplines. GIS has been called an "Enabling Technology" because of the potential it offense for the wide variety of disciplines which must deal with spatial. Many related fields of study provide techniques, which make up GIS many of these related fields emphasis data collection and together by emphasizing integrations, modeling and analysis. This GIS often claims to be the science of spatial information.

Geographical Info System (Geographic) features:

The GIS or geographic features are combined from map futures

- Point

- Polygon

- Line

And attribute (descriptive information)

Overview of GIS for Environmental Problem Solving

In all of these operations, the typical GIS user now expects to be able to define requirements and interact with the system through a "user friendly", intuitive interface that makes me of such contemporary concepts as graphic icons and desktop metaphors.

- GIS applications now span a wide range, from sophisticated analysis and modeling of spatial data to simple inventory and management

- GIS has a broadly based community of interest, drawn together by a common concern for the computerized handling of geographic data.

- It includes established disciplines like surveying, RS, Geocoding and cartography which see GIS as another valuable digital technology with capabilities that increase those of GPS, Image processing, digital cartography etc.

- In some sense GIS is the common ground between all these, the broad technology that attempts to integrate data from a no. of acquisition system, and provide it to   the user with appropriate analytical tools.

- Geographical reality is enormously complex and it can be represented in digital form in a rich variety of ways.   More ever the set of GIS factions is long and growing, as user are found for a greater and greater variety of forms of spatial analysis.

GIS ADVANTAGES IN BRIEF

1) Data consistency can be maintained

2) Data redundancy can be reduced.

3) Multisource data can be extended and integrated.

4) Data update can be easily undertaken and flexible in operation and easy retrieval.

INTEGRATION OF REMOTE SENSING AND GEOGRAPHIC INFORMATION SYSTEM

Remote sensing data can be readily merged with other sources of geocoded information in GIS. One of the most important benefits of a GIS is the ability to spatially interrelate multiple types of information stemming from a range of sources. This permits the overlapping of several layers of information with the remotely sensed data, and the application of a virtually unlimited number offers of data analysis On the other hand, the data in a GIS might be used to aid in image classification. In the other hand, the land cover data generated by a classification might be used in subsequent guesses and manipulation of the GIS database. Remotely sensed data is almost always processed and stored in master data structures. When working simultaneously with an image processing system and raster geographic information system, it usually easy to move data between them.

The most common application of this is the land cover mapping In addition to aerial photographs, it also provides a synoptic view of the surface features

USE OF REMOTE SENSING AND GIS IN THE PROJECT

Implementation of entire project is designed in different phases:

1. Acquisition and derivation of baseline data products.

2. Creation of baseline digital database on GIS platform.

3. Types of data products.

4. Overlay analysis Techniques.

ACQUISITION AND DERIVATION OF BASELINE DATA PRODUCTS

- Toposheets from survey of India.

  Four toposheets 56 k/6, k/7, k/10, k/11 are acquired from SOI Scale: 1:50000.

- Satellite data from NRSA, Hyderabad. It is fused Imagery of IRS-IC PAN and IRS-ID LISS III Imagery

  The main objective is to derive the relevant data from different source and analysis of primary data to extract the layered information as required in project.

Processing of Satellite data for better Resolution and FCC Moderate Resolution Land Satellites

- **IRS System**

  The republic of India has success fully launched, and operated several moderate resolution satellite systems. The Indian Remote Sensing (1RS) program has began with the launch of IRS IA in 1988. This system was extended with the launch of an identical follow-on system IRS-IB in 1991

  A second generation of IRS Satellite operation began with the launch of IRS-IC and IRS-1D in 1995 and 1997 respectively. These systems are identical in design and earn,- three sensors.

  LISS-III with 23 m. Resolution (70 m in the mid-1R band)

  A panchromatic sensor resolution (5.8m resolution)

  And wide field sensor resolution (188 m resolution)

- **Panchromatic Camera (PAN)**

  The panchromatic camera provides data with a spatial resolution of 5.2-5.8 (at nadir) and a ground swath between 63 Km -70 Km (at nadir). It operates in the $0.50 - 0.75$ microns spectral band. This camera can be steered upto ± 26 deg (storable upto ± 398 Km across the track from nadir), which in turn increases the revisit capability to 3 days for most part of the cycle and 7 days in some extreme cases.

- **Linear Imaging and Self Scanning Sensor (LISS-III)**

  The LISS-III sensor provides multi spectral data collected in four bands of the visible, near infra-red (VNIR) and short wave infra-red (SVVIR) REGIONS. While the spectral resolution and swath in the case of visible (two bands) and NIR (one band) regions are between 21.2 m to 23.5 m and 127 Km-141 Km. respectively, they are between 63.6 m to 70.5 m and 133 Km to 148 Km for the data collected in SWIR region.

- **Creation of Base Line Digital Database on GIS Platform**

  o Scanning a digital image of the map is produced by moving an electronic detector across the map surface. The size of the map area viewed by the detector and scanning should be processed or edited to improve quality and convert the raster to vector after digitization.

  o Digitizing of all scanned maps using automated digitizing process by AutoCAD It is a process of converting the spatial features on a map into a digital format point line and area features that for a map are converted into x, y coordinates. Thus digitizing is proceeded by capturing series of points and lines.

  o Editing of the digitized data, detection, and correction of errors in digital data so that is ready for spatial analysis on GIS platform. GIS software used here is Arc, info developed by (ESRI) Redlands, California, USA. It is a vector-based GIS package, capable of handling both spatial and non-spatial data.

  o Execution of project is done keeping all inputs in view and its implementation.

  o Data storage and retrieval subsystem.

TYPES OF DATA PRODUCTS OBTAINED

Data types with their relevance in digital database are classified based on source of acquisition and creation for the preparation of Base line digital database, three different types of data are defined.

a) Topographical data

b) Thematic data.

c) Collateral data.

a) Topographical Data

Comprises of all the topographical details available on soil 1:50000 toposheet. The important topographical data layers:

- Road Network

- Water bodies

- Drainage

- Settlements

b) Thematic Data

It mainly comprises of land use / land cover map. The steps involved in deriving thematic layered data from satellite data are:

➢ Image Rectification and Registration

o Geo coding / Geo referring.

➢ Image enhancement

➢ Image classification / Analysis

o Ground central points (GCP)

o Supervised classification

o Visual analysis of classified output

➢ Landuse/Landcover map generation

A digital color plot output will be developed using all above steps and VIP (visual interpretation techniques) techniques can be applied to prepare land use / land cover map by incorporating the results of field survey.

Satellite data used is fused imagery of IRS-IC PAN and IRS I-D LISS III Imagery.

➢ Software used EASI/PACE: It is a remote sensing software developed by PCI geomatic, Ontario Canada.   It is designed for remote sensing, image processing, data visualization and GIS support. The functional components are:

c. **Image works:** Image classification and image enhancements.

d. **GCP works:** Georeferencing and Image registration, projections and mosaicking.

e. **PCI Models:** Analysis, data interchange, image correction, image processing

radar analysis PACE packages.

f. **ACE:** Advanced cartographic environment is for cartographic functions.

g. **Collateral Data:** It consists of water quality maps.   These maps show the spatial distribution of ground water quality.   Software used is 3D analyst of ARC/INFO package.

h. **Overlay Analysis Technique:** It is the common technology of structuring the digital baseline data of the real world.

SYNTHESIS OF THE PROBLEM

As urbanization implies specific effects on environment it is often characterized as an Environmental Degradation Phenomena, due to following reasons.

1) It results in concentration of the people

2) It breaks the cycle flows of matter found in natural systems
3) It provides a human environment vastly different from that in which mankind   evolved.

Existing Land use Pattern of Area under MCH Along With Areas Corresponding to That Land use Pattern

LANDUSE	AREA IN sq Kms
Water Bodies	7.053
Land with scrub	10.678
Land without scrub	9.505
Plantation	0.303
Barren sheet area	9.36
Medium residential	36.103
Sparse residential	16.691
Dense residential	77.986
Public related places	1.74
Quarries	0.3
Agricultural land	1.605
Rivers	3.422
Industries	1.213
Parks	2.834

Source: EPTRI

IMPACT OF URBANIZATION ON ENVIRONMENT

The growth of urbanization has a tremendous adverse impact on the environment. The expansion / growth of the city in terms of population invariably accompanies a proportional increase in the built-up area of the city. The general increase in the use of resources and the transition from a cyclic flow to a linear flow of matter is the consequence of urbanization, which occurs dominantly on Agricultural land around half a million hectares of such land is annually taken by or in connection with urbanization.

LAND USE / LAND COVER MAP

Comprehensive information on land use / land cover is the basic prerequisite for land resources evaluation, assessment, utilization and management. Today, with increasing population pressure on land and the resulting changes in the land use pattern and processes, a considerable degree of land transformation and environmental deterioration is being witnessed. Therefore it is important to understand the cause and effect of the changes through scientific studies.

OBJECTIVES OF LAND USE / LAND COVER MAP

The main objectives of land use map are:

1) The land use map will be utilized as a basic data base which provides the information for allocating new land use practices.

2) It will incorporate demographic, economic and environmental impacts which have occurred in an area.

3) Not only the information indicate where intensive development has already taken place and where there is open land suitable for future expansion, but it will also make it possible to determine special areas, such as prime agricultural lands.

4) Land use / Land cover map will serve as a basis for monitoring and use change.

5) The land use map will serve as a base in the integrating overall planning of agricultural and industrial development of the region.

USGS LAND USE / LAND COVER CLASSIFICATION SYSTEM

S.No.	Level - I	Level - 11
1	Urban or built-up land	1.1 Residential 1.2 Commercial and service 1.3 Industrial 1.4 Transportation, communications and utilities 1.5 Indusial and commercial complexes 1.6 Mixed urban or built-up land 1.7 Other urban or built-up land
2	Agricultural land	2.1 Cropland and pasture 2.2 Orchards, groves, vineyards, nurseries and ornamental horticultural areas 2.3 Other agricultural land
3	Rangeland	3.1 Herbaceous rangeland 3.2 Shrub and brush rangeland 3.3 Mixed rangeland
4	Forest land	4.1 Deciduous forest land 4.2 Evergreen forest land 4.3 Mixed forest land
5	Water	5.1 Streams and canals 5.2 Lakes 5.3 Reservoirs 5.4 Bays and estuaries

6	Wetland	6.1 Forested wetland
		6.2 Non forested wetland
7	Barren land	7.1 Dry salt flats
		7.2 Beaches
		7.3 Sandy areas other than beaches
		7.4 Bare exposed rock
		7.5 Stip mines, quarries and gravel pits
		7.6 Transitional areas
		7.7 Mixed barren land
8	Tundra	8.1 Scrub and bush tundra
		8.2 herbaceous tundra
		8.3 Bare ground tundra
		8.4 Wet tundra
		8.5 Mixed tundra
9	Perennial snow or ice	9.1 Perennial snowfields
		9.2 Glaciers

HYDROLOGIC CYCLE

From the time the earth was formed, water has been endlessly circulating. This circulation is known as the hydrologic cycle. Groundwater is part of this continuous cycle as water evaporates, forms clouds, and returns to earth as precipitation.

HYDRLOGIC CYCLE

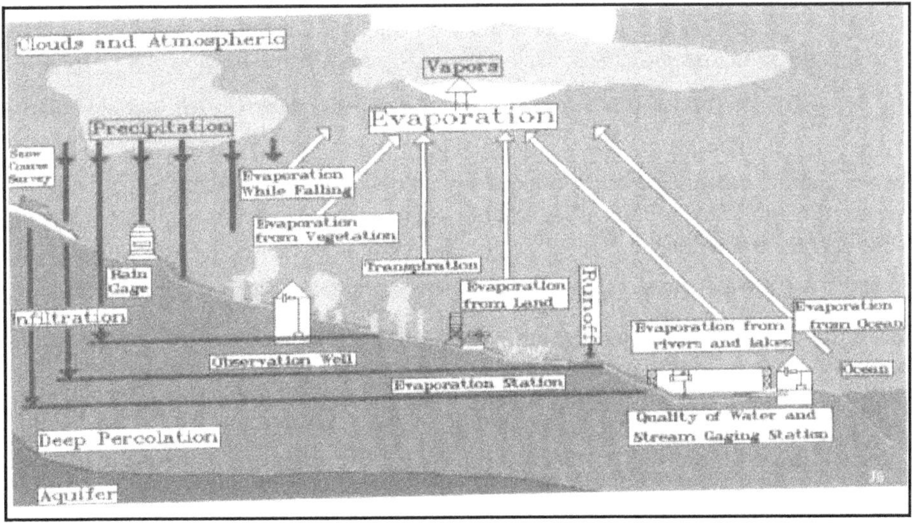

THE PROCESS

Surface water is evaporated from the earth by the energy of the sun. the water vapour forms clouds in the sky. Depending on the temperature and weather conditions, the water vapour condenses and falls to the earth as different types of precipitation. Some precipitation runs from high areas to low areas on the earth's surface. This is known as surface runoff. Other precipitation seeps into the ground and is stored as groundwater.

When rain falls to the ground, the water does not stop moving. Some of it flows along the surface into streams or lakes, some of it is used by plants, some evaporates and returns to the atmosphere, and some sinks into the ground.

Ground water is evaporate, from clouds, and return to the earth to begin the cycle over again.

METHODOLOGY OF SAMPLING

   1) Sampling procedures - (water sampling)

First: To know what data is to be used and how it is useful

The type of investigation, purpose of study and anticipated variations are other points to be considered.

The first stage of planning of the sampling program is the selection of most suitable site to provide the required data.

   2) Site selection

The objective of water quality monitoring systems

1)   To assess the impact of activities by man upon the quality of water and its suitability for required uses.

2)   To determine the quality of water in its natural state this might be available to meet the future needs.

3)   To keep under observations the sources and path way of specified hazardous substances.

The selection of sampling sites is decided by various uses of the water and by their locations, relative magnitude and importance. The chance of accidental pollution is also an important key factor and should be considered.

LOCATION OF GROUND WATER SAMPLING POINTS

For the selection of site for underground water, it will be necessary to obtain hydrological information and the information about the aquifer (should describe its hydrological situation) regarding the location, depth, and the area of the aquifer and its geological, mineralogical characteristics.

Background information will be needed on existing and potential influence on water quality.

TYPE OF SAMPLING

Random Sampling: Sample collected at a place irrespective of time randomly in the study area.

SAMPLING CONTAINERS

1) Bottles are soaked in 10% Hcl for 24 hr and cleaned and rinsed with distilled   water

2) For metal analysis rinse the container with 20% HNO^3 followed by distilled

SAMPLE COLLECTION

1) The container should be rinsed out two or three times with the sample to be examined before filling.

2) The stopper should not be laid down but kept in hand before reinserted.

3) In taking a sample from a tap or pump, the nozzle should be examined to see that it appears clean.

4) If not it should be cleaned and water should be allowed to run waste before filling the bottle.

5) It is better to take the sample first in the morning before anyone has drawn for   other purposes.

GROUND WATER CONSTITUENTS

- ➤ **Dissolved Constituents:** These include Silica, Iron, Calcium, Magnesium, Manganese, Sodium, Potassium, Carbonate and Bicarbonate ions, Sulfates, Chlorides, Fluoride, Nitrates and Boron.

- ➤ **Trace Constituents:** Certain metallic constituents such as Arsenic, Barium, Beryllium, Bismuth, Cadmium, Chromium, Cobalt, Copper, Lead, Nickel, Selenium, Strontium, Tin, Titanium and Zinc and radioactive elements like Radium, Thorium and Uranium are found in ground waters in trace amounts generally not exceeding img/1.

Effects of water quality parameters for drinking purposes of their effects

S.No.	Parameters	Desirable limit	Permissible limit	Probable effects
1	Colour (Hazen unit)	5	25	Makes water aesthetically undesirable
2	Odour	Essentially free from objectionable Odour		Makes water aesthetically undesirable
3	Taste	Agreeable		Makes water aesthetically undesirable

S.No.	Parameters	Desirable limit	Permissible limit	Probable effects
4	Turbidity (NTU)	5	10	High turbidity indicates contamination pollution
5	PH	6.5	8.5	Indicative of acidic or alkaline waters, affects taste, corrosivity and the water supply system.
6	Hardness CaCO^3 (mg/1)	300	600	Affects water supply system (Scaling), Excessive soap consumption, and calcification of arteries. There is no conclusive proof but it may cause urinary concretions, diseases of kidney or bladder and stomach disorder.
7	Chloride (mg/1)	250	1000	May be injurious to some people suffering from diseases of heart or kidneys. Taste, Indigestion, corrosion and palatability are affected.
8	Total dissolved solids (mg/1) (TDS)	500	2000	Palatability decrees and may cause gastro intestinal irritation in human, may have laxative effect particularly upon transits and corrosion, may damage water system.
9	Calcium (Ca) (mg/1)	75	200	Causes encrustation in water supply system. While insufficiency causes a severe type of rickets, excess causes concretions in the body such as kidney or bladder stones and irritation in urinary passages. (Essential for nervous and muscular system, cardiac functions and in coagulation of blood)

S.No.	Parameters	Desirable limit	Permissible limit	Probable effects
10	Magnesium (Mg) (mg/1)	30	100	Its salts are cathartics and diuretic. High cone May have laxative effect particularly on new users. Magnesium deficiency is associated with structural and functional changes. It is essential as an activator of many enzyme systems.
11	Sulphate (SO$_4$) (mg/1)	200	400	Causes gastro intestinal irritation along with Mg or Na, can have a cathartic effect on users, concentration more than 750 mg/1 may have laxative effect along with Magnesium.
12	Nitrate (NO$_3$) mg/1	45	100	Cause infant methaemoglobinaemia (blue babies) at very high concentration, causes gastric cancer and affects adversely central nervous system and cardiovascular system.
16	Fluoride (F) mg/1	1.0	1.50	Reduce dental carries, very high concentration may cause crippling skeletal fluorosis.
17	Alkalinity mg/1 CaCO$_3$	200	600	Impart distinctly unpleasant taste may be deleterious to human being in presence of high pH, harness and total dissolved solids.
18	Phosphate (PO$_4$) mg/1	200	400	High concentration may cause vomiting and diarrhea, stimulate secondary • hyperthyroidism and bone loss

S.No.	Parameters	Desirable limit	Permissible limit	Probable effects
19	odium (Na)	200	400	Harmful to persons suffering from, cardiac, renal and circulatory diseases.
20	Potassium (K) mg/l	200	400	An essential nutritional element but / its excess amounts is cathartic)

pH

pH is defined as the negative logarithm of hydrogen ion concentration present in water and is an indicator of the acidity or alkalinity of water) pH is determined with the help of a pH meter. The permissible value of pH recommended for public water supplies is in between 6.5 to 8.5. The presence of mineral acids like sulphuric acid, iron, cadmium, aluminum (effluents) makes the water acidic i.e., pH= 0-7 and the presence of carbonates and bicarbonates of calcium, magnesium, potassium etc make the water alkaline i.e., pH=7-14 the pH of water samples in the present area pH varied between 6.77 (Kamalanagar DW) to 8.28 (Abids) most of the samples collected in study area showed neutral to slightly alkaline pH.

EFFECTS

- Corrosion of water mains is the main problem associated with acidic waters.

- Higher value of pH encourages the scale formation in water heating systems and also reduces the germicidal potential of chlorine.

- High pH induces the formation of tri halomethanes which are causing cancer in

  human beings.

Total Dissolved Solids: Dissolved minerals, gases and organic constituents may produce aesthetically displeasing colour, taste and odour. The mam source of TDS in ground water is the seepage of industrial and domestic wastes. The permissible value recommended for TDS is 500 mg/1 as per Indian standard. The TDS concentration in the

Present study area ranged from 285 - 1180 mg/1. 56% of water samples collected showed high concentrations of TDS in areas like Amberpet, Kamalanagar, Bazarghat, Bathkammakunta, Chikkadpally, Imlibun, Goshamahal etc High TDS concentrations in Bazarghat Kamalanagar (DW) etc., may be due to seepage of domestic wastes and insufficient rainfall. Samples collected nearby Musi River and outlet of Hussain Sagar at Chikkadpally showed high TDS concentration.

Source: The impact of Residential area (Sewage disposal) Location of Slums along with Musi.

EFFECTS

- Water with high solid content often has a laxative effect.
- Use of water with high amount of dissolved solids may lead to scaling in boilers, corrosion and degrade quality of product.
- The presence of high concentration of TDS depletes dissolved oxygen and causes displeasing dour, taste and colour. High concentrations of 3000 mg/l may also produce distress to livestock.

Hardness: Hardness of water is a measure of its capacity to form precipitates with soap and scales with certain anions present in the water. Temporary hardness also called carbonate hardness caused by carbonates and bicarboriafes of calcium and magnesium is removed by boiling or adding some to water. Permanent hardness or non-carbonate hardness is due to presence of sulfates, chlorides and nitrates of calcium and magnesium. It can be removed only by Ion exchange or Zeolite process.

The desirable limit of Hardness in water is 300 mg/1. Areas like Golnaka, Amberpet, Nampally etc. Showed hardness concentration greater than 300 mg/1, which can be considered as very hard water with maximum value of 710 mg/1 in highly residential area Golnaka.

< 50 mg/1	-	Soft Water
50-150 mg/1	-	Moderately Hard
150-300 mg/1	-	Hard
> 300 mg/1	-	Very hard

EFFECTS

- The precipitate formed by soap and hardness adheres to surface of tubs, sinks and utensils and stains clothes and dishes.
- The precipitate if remains in the pores, skin may feel rough and uncomfortable and may lose its texture.
- Increases cardio vascular diseases.
- Absolute soft waters are tasteless and effect the human cardiovascular system and cause heart attacks.
- Not suitable for production of ice, soft drink and textiles.

Fluoride: Fluoride of water is important in determining the suitability of water for drinking purpose use. Water in contact with natural deposits of fluoride such as fluorspar, calcium fluoride, cryolite and water contaminated with industrial effluents are found to contain excess fluorides. Fluorides can be categorized as:

| < 1 | Good |
| 1-1.5 | Very Good |

| 1.5-2 | Tolerable |
| >2 | Polluted |

Fluorides are determined using ion selective electrode method.

The fluoride concentration in the present study areas ranges from-0.45 mg/1 - 5.83 mg/1 in areas like king Koti, Kachiguda, Patelnagar and with a maximum value of 5.83 mg/1 in dense residential area Golnaka. However for the studies a required to establish the linkages.

EFFECTS

- Fluoride more than 1.5 mg/1 it leads to discoloration of teeth called "Mottling"[11]. The white patches formed later become yellow and turn brown or black

- Fluoride in excess of 5 mg/1 cause bone flourosis and skeletal abnormalities. Fluoride less than 1 mg/1 cause dental cavities.

Sulfate: Sulfate is a naturally occurring cation (SO_4) found in all kinds of natural waters discharge of industrial waste and domestic sewage in water tends to increase sulfate concentration most of them originate from the oxidation of sulphate ores, the presence of shales etc. Sulphate can be readily leached from zone of weathering by infiltrating waters and surface runoff. The desirable limit of sulphate concentration is 250 mg/1 in waters intended for human consumption. Sulfates are determined spectro photometrically at wavelength of 420 mm. The sulfate concentrations in the ground water samples of study area range from 8.5 - 120 mgl indicating that the concentrations are well within the permissible limits as given by Indian standards.

Effects: Sulfate acts a purgative in adults and causes temporary diarrhoea and disorders of alimentary tract.

WATER QUALITY INDEX COMPUTATION

Water Quality Index can be computed using the method proposed by Tiwari and Mishra, 1995. According to them, Quality rating (q) is calculated as

$$Q_{ni} = (V_{actual} - V_{ideal})/V_{standard} - V_{Ideal}) * 100$$

Where,

Q_{ni} = Quality rating of ith parameter for a total of n water quality parameters

V_{actual} = Value of the water quality parameter obtained from laboratory analysis

V_{actual} = Value of that water quality parameter can be obtained from the standard tables.

V_{ideal} for pH = 7 and for other parameters it is equivalent to zero.

(A.K. Srivastava and O.K. Sinha, IJEP, volume 14, No. 5, May 1994 and Mahuya

Dasgupta Adak etal, IJEP, Volume 8, No. 3, 2001)

To determine the suitability of the water for drinking purposes, an indexing system called

Water Quality Index (WQI) has been developed from this water quality rating which is formulated as,

WQI - Antilog [$\Sigma W^n_{n=1} \log_{10} q_n$]

Where,

$W_n = K / S_n$ and

$K = 1 / \Sigma^n_{n=i} I/Si$

ESTIMATION OF WATER QUALITY INDEX FOR ZONE VII

Table No 7.2 shows water quality parameters, their ICMR/WHO standards and assigned unit weights (A.K. Srivastava and D.K. Sinha, IJEP, Volume 14, No. 5, May 1994).

K is calculated as

$$K = \frac{1}{1/8.5+1/120+1/300+1/1000+1/1.5+1/200+1/250+1/50} = 1.2059$$

by substituting this K value in the equation

$W_n = K/Si$

SUMMARY

The objective of the present study is impact of Landuse/Landcover changes on the environmental problems i.e., Zone VII under Municipal Corporation of Hyderabad, which is located on Musi River. With reference to ground water pollution Unbrilled urbanization and industrialization, unchecked proliferation of obnoxious and hazardous industries in and around Hyderabad. Unchecked, unregulated distribution and management of Urban Resources especially water bodies along with the combination of all above aspects have all made Hyderabad environmental quality a unmanaged polluting city

Ground Water is important source of water supply throughout the world. Its use in irrigation, industries and domestic usage continues to increase where perennial surface water sources are absent. The quality of ground water used for these purposes is more important as the case of quantity. But rapid industrialization and Urban growth alter the ground water quality and makes it unfit for use.

In the present study, RS and GIS have been used for the evaluation of impact of Landuse/Landcover on ground water quality of Zone VII. Two types of data base spatial and attribute are created using GIS Software. Spatial database includes all necessary thematic maps and attribute date includes ground water quality of all areas in the study area. Maps showing spatial distribution of various water quality parameter have been prepared using the curve fitting method of are vies, GIS Software. Form these maps areas under stress are identified WQI has be calculated to communicate water quality of different areas of Zone VII.

IMAGE SENSORS AND IMAGE DATA PROCESSING

M. Santhi Swarup
Acad Advisor, Ex Dy. General Manager (GIS), INTEC,
Swaroop_ms@yahoo.com

J. Venkatesh
Assoc. Prof & Head
Centre for Spatial Information Technology
IST, JNTUH, Hyderabad-500085

ABSTRACT

An advanced image processing method which handles high rates of data from the sensors and generates good quality images is presented in this paper

Unmanned aerial vehicles (UAVs), also known as remotely piloted vehicles (RPAs), outfitted with advanced sensor payloads are actively gathering, and even processing a wealth of intelligence, surveillance, and reconnaissance (ISR) data.

General Electric Co. (GE) is harnessing the power of supercomputer installations and replicating it in a package that is small enough to fit where it has to go and robust enough to survive extremes of temperature, shock, and vibration. Supercomputing and high-performance computing (HPC) is now available in the form of high-performance embedded computing (HPEC). HPEC is fundamental to managing sensor payloads on unmanned vehicles because of the high bandwidth of data, processing complexity, need for very rapid response times, and pressure. Moving processing to the mobile platform, near the sensor, could enable faster response times but requires an HPEC system, but performed on the data in real time without the bottleneck of the low-bandwidth link back to a data center. Automated algorithms perform first-order processing and transmit only select imagery and other metadata over the link. This workflow is much more optimized and gets much closer to real-time response. The key is putting rugged processing elements adjacent to the sensor where there are no bottlenecks.

Data compression technology also holds promise in reducing the size of sensor-acquired data files. Engineers are using compact image compression devices to reduce sensor data to the point where it can fit a restricted bandwidth data link on small platforms. Sensors with wider areas, faster processing, and on-board processing to save bandwidth-that's the future.

KEYWORDS: UAV, Sensors, Data Processing, UGVs.

INTRODUCTION

Unmanned aerial vehicles (UAVs), also known as remotely piloted vehicles (RPAs), outfitted with advanced sensor payloads are actively gathering, and even processing a wealth of intelligence, surveillance, and reconnaissance (ISR) data.

The number of unmanned ground, aerial, and undersea vehicles deployed by aerospace and defense organizations has grown exponentially.

As unmanned aerial vehicle (UAV) developments take a larger role in the aerospace industry, we may ask ourselves "What is the main benefit of a UAV?" An assertion of this paper is that the main benefit of a UAV is to focus on cost effective and innovative technologies in sensors and data processing.

"Cost" may be a "soft" cost such as human risk or a "hard" cost such as reduced environmental control requirements or reduced redundancy requirements.

With this assertion in mind, we will review how displacement sensor selection can reduce both UAV acquisition cost and UAV operational cost. Specifically, we will:

1. review aircraft displacement sensor applications

2. discuss what contributes to sensor lifecycle cost

3. review sensor technologies and costs

DISPLACEMENT SENSORS

These are the sensors that can be used to measure distances and heights. A wide variety of models is available, including Laser Sensors, LED Sensors, Ultrasonic Sensors,

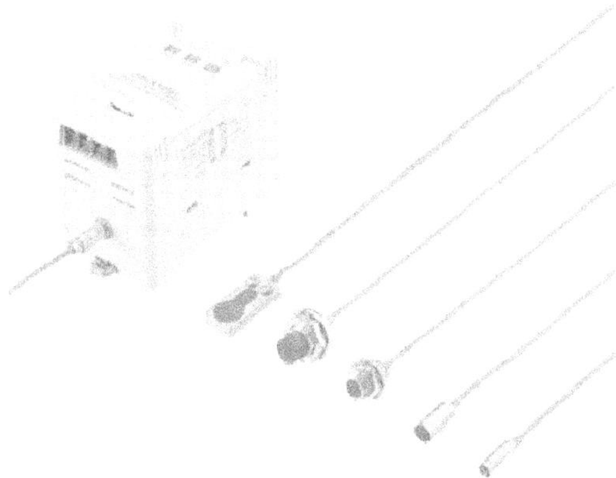

Fig: - 1   Eddy Current Sensors

Contact Sensors, Eddy Current Sensors, and more. We are focused on displacement sensors that monitor movement from one position to another for a specific distance or angle. As such, we are focused on continuous-sensing devices versus discrete-sensing devices such as proximity sensors that signal a critical distance using an on/off output.

TYPES

Displacement sensing can be categorized into 5 categories related to the geometry of motion being sensed (see Figure 2):

- linear

- rotary: motion greater than 360°

- angular: a special case of rotary motion limited to 360° or less

- 2D: flap

- 3D: aero elastic (warping) wing

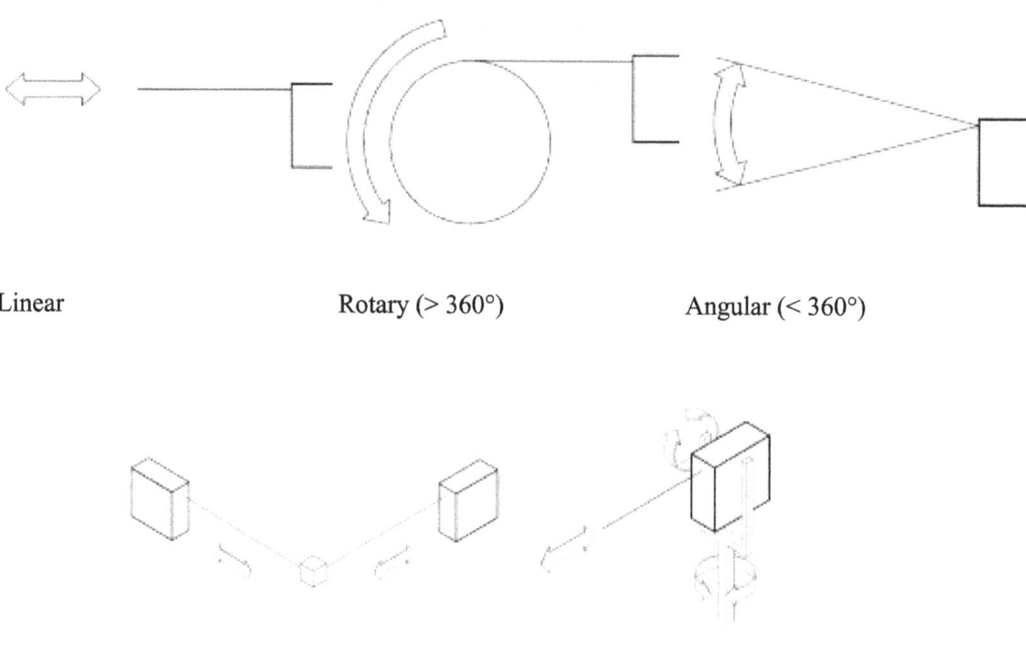

Figure 2 - Displacement sensing categories prevalent on MAVs and UAVs

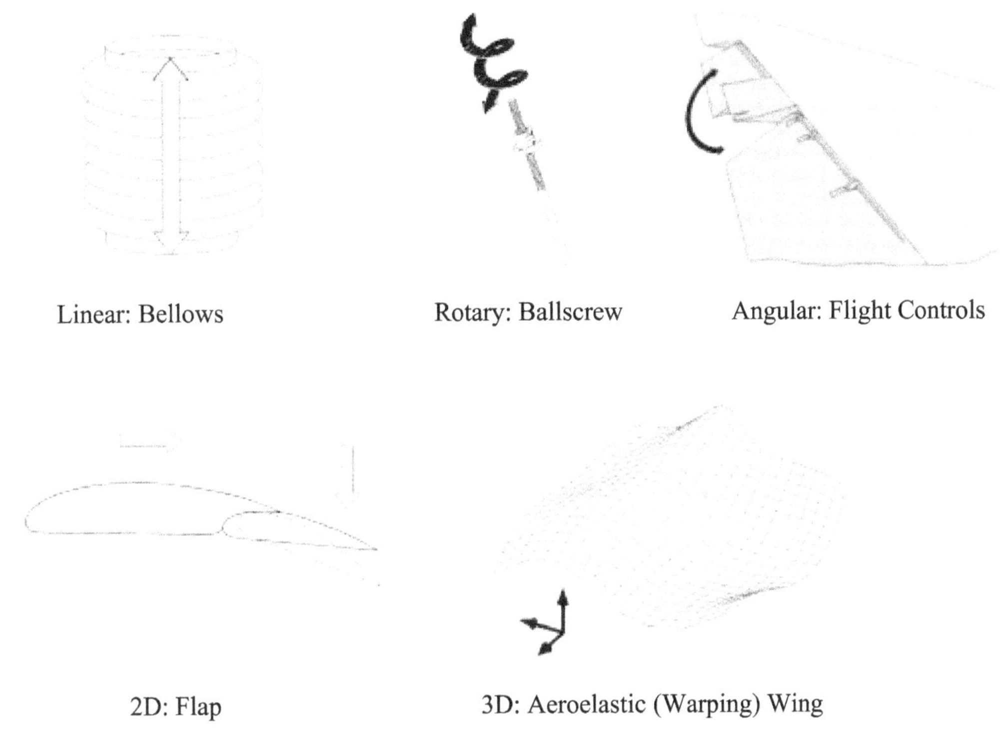

Figure 3&4 - Aircraft displacement sensing examples

Aircraft displacement sensing generally involves linear, rotary, and angular. In situations where two- and three-dimensional sensing is required, two or three displacement sensors are often used.

SENSOR LIFECYCLE COST

The lifetime cost of a sensor involves more than the initial purchase cost. By looking at the total cost of ownership, an optimum purchase decision can be made specific to your application.

If you purchase a car, the initial purchase price may only be 60% of the total lifetime cost of the vehicle. Gas, oil, repairs, insurance, maintenance, taxes, license fees, and other costs can exceed the initial purchase price over a 5- to 10-year typical vehicle lifetime.

If you purchase a PC, the initial purchase price may only be 10% of the total lifetime cost of the computer. Installation, support, training, upgrades, and repairs usually dwarf the initial outlay.

When someone asks you how much did something "cost," you typically state a figure based on what was shown on the quote, invoice, or receipt. In the case of a sensor, this is often only the cost of the sensor and possibly an amount for shipping, taxes, and related transaction costs. This cost accounting may make the boss and the finance department happy. It can also reduce effectiveness and profitability.

You may be thinking, "I'm only buying a simple sensor. What other costs could there be?"

Installation - Does the sensor design require you to make a special mounting plate or is flexible mounting inherent in the product? How long does installation take? Can installation be performed by a lower-skilled employee or must a higher-skilled technician or engineer perform the task?

Cabling, Connectors, And Signal Conditioning - Does the sensor require the purchase of additional electrical cable, electrical connectors, signal conditioning, and related instrumentation?

Reliability - What is the stated lifetime of the product? Does it have an MTBF (mean time before failure) rating? Does the vendor have reliability statistics of the product being used in an environment similar to your own? Unscheduled downtime costs can be huge in factory automation, aviation, and capital-intensive applications.

Scheduled Down Time - Is calibration or scheduled maintenance required? How will this downtime affect your operations? Will alternate sensors need to be installed? Can this work be done during other maintenance periods?

Repairs - Is the product repairable or is it discarded at the end of its lifetime? Are there costs associated with its disposal? Can the repair be performed on site or must the item be returned to the manufacturer?

Calibrations - Can calibrations be performed on site or is factory return required? How often calibrations required and what are the cost?

Usability - Is the signal from the product easy to work with or does it require specialized power supplies, amplifiers, and related equipment that must be procured, installed, learned, and configured?

On-Time Performance - Does the sensor get delivered on time? If you planned for receiving the item in 7 days but the shipment does not show up for 21 days, you will spend valuable time re-scheduling resources and nagging the vendor to get the product to you.

Shipping - It may not seem like much to pay a flat small fee for shipping. But add that flat small fee over spare parts, factory calibrations, repairs, and replacements and the amount can become substantial.

If shipping is based on weight and volume, look at the products you are considering specifying. Are there any size or weight differences? Are there are tariff differences related to the products originating from different countries?

REDUCING DISPLACEMENT SENSOR COST

How can we reduce the lifecycle cost of displacement sensors in UAV applications? The answer to the question depends on the specific UAV application in mind, but here are a ways the total cost of ownership can be lowered.

Easier Signal Conditioning - Many sensors on today's aircraft require expensive and heavy signal conditioning for amplifying, converting, or filtering signals. Selecting sensors that require modest or no signal conditioning will have the synergistic effect of reducing cost, size, and mass.

Reduced, Simpler Power Requirements - Most UAVs use a low-voltage DC onboard power supply. By selecting sensors that use this same power supply, power conversion is eliminated. In addition, by using sensors that use low levels of current, the overall aircraft power requirements can be reduced allowing for lower weight and longer flight durations.

Flexible Installation and Servicing - The cramped confines of a UAV interior can make installation and servicing of components difficult and costly. Using a displacement sensor that can be easily installed and serviced in these confines reduces labor hours and aircraft downtime.

Interfacing to Software and Other Systems - Using displacement sensors with simple electrical interfaces simplifies the task of interfacing with software and other systems. This simplification reduces software development time and reliability as well as development efforts for other systems.

ADVANCES IN SENSOR TECHNOLOGIES

The rapid evolution of electro-optical imagers and infrared cameras is making it possible to employ them for less cost on smaller and lighter systems

Surveillance camera systems are riding an upward spiral. They're improving their image quality, bringing more benefits, and that's sparking more demand for information gathered by unmanned aerial vehicles (UAVs) and unmanned ground vehicles (UGVs).

As this increased interest drives market growth, these unmanned platforms are increasingly carrying a range of sensors, including full-motion cameras and thermal imagers. But still pictures remain the mainstay of data collection. Generally, electro-optic (EO) imagers gather high-resolution data during daylight, while infrared (IR) systems provide nighttime imagery.

However, both technologies are evolving rapidly, providing higher-resolution images while reducing size and power consumption so that cameras can be used on smaller platforms that are often less expensive. Consequently, more of them can be deployed. One of the keys that are driving this spiraling usage is the transition that occurred across America a few years ago: the changeover to high-definition digital technology.

"HD systems have been around for defense applications since 2006, but the military had to make a couple steps before they could gain the benefits of HD, such as going to systems that could deal with digital images," said Robert Kubis, product manager for gimbal systems at FLIR Government Systems Division. "Last year, they reached the turning point where analog was being replaced, so the transition is now moving forward quickly."

The expanding use of EO/IR sensors is prompting many system designers to integrate additional equipment that can enhance the data collected by these imagers. HD cameras have become the norm, and more platforms are adding sensors that make it easier for analysts to determine exactly where the images they're seeing are located.

"We are introducing HD in all our payloads and integrating inertial navigation system/global positioning system (INS/GPS) sensors in many payloads," said Igal Mevorach, marketing director at Israel Aerospace Industries' Tamam Division. "Integrating INS/GPS enables geo-pointing and geo-registration and minimizes target location error."

DATA PROCESSING

All new imaging systems use digital technologies. That eliminates the need for data conversion while also realizing the advantages that occur when system designers can use the advances in semiconductor technology.

One example of the transition to fully digital systems has come as (CMOS) metal-oxide-semiconductor imaging chips used in most consumer cameras have improved. They can now work in lower-light conditions, letting system designers gets rid of image intensifiers, which used analog vacuum-tube technology to improve image quality in low-light environments.

Low-light CMOS supplies a digital video signal that can easily be processed to improve the image. Though these devices are challenging IR for low-light applications, IR components are battling back. Vogelsong noted that the transition to digital is also occurring in IR sensors. Uncooled micro bolometer and cooled mid-wave infrared and long-wave infrared sensors also provide a digital video signal.

IMPROVING IMAGES

Sensor improvements are being matched by advances in the systems that process their input. Semiconductors are able to perform more functions each second, letting engineers devise more techniques to improve the quality of images that are often collected under poor conditions such as low light and dusty atmospheres.

"Today's image processors use low-power, field-programmable gate arrays and digital signal processors (DSPs) that make it possible to embed image processing functions directly behind the sensor focal plane array. These advancements are providing a path to solutions that also support the requirements of portable defense systems. These advances are making huge differences in system requirements. Engineers are leveraging lower power semiconductors and improving software. For example, controls and sensors quickly perform their tasks, and then they go into deep sleep modes so they draw very little power when they're inactive.

"Image quality is often unusable if platform jitter is not removed," Piacentino said. "Additionally battlefield lighting conditions are very extreme, and noise reduction algorithms become critical."

Sophisticated software is also being employed. As microprocessor and DSPs are able to handle more operations per second, systems can better analyze images and make enhancements. The improvements provided by sophisticated electronics continue to advance. For example, algorithms used in General Dynamics' Imaging through Volume Turbulence enhancement technology remove the distortions caused by dust particles and the variations caused by heated air.

The electronics aren't the only components that must change. Mechanical components such as positioning and stabilization motors have to be more precise so the imagers can collect sharp photos.

FUTURE

The sensor technologies discussed in this paper were initially developed 40 to 60 years ago. Tremendous improvements have been made over the years, particularly in the areas of reliability, lifetime, and size. Nevertheless, today's UAV design engineers are specifying the same sensor technologies as their grandfathers did.

What does the future hold for displacement sensor technologies and cost reduction? Will a "wonder sensor" based on MEMS or non-contact technologies come upon the scene? Given the prevalence of incremental improvements in aircraft and technology, a wonder sensor seems unlikely. Yet, if displacement sensor capabilities begin to limit UAV design choices and UAV performance, increased resources will need to be applied if we are to avoid a situation where displacement sensors limitations drive UAV design.

SUMMARY

UAV developments show innovation is alive and well in the aerospace industry. Hybrid designs incorporating rotary- and fixed-wing characteristics, micro UAVs, autonomous control, and a host of other features make UAVs one of the most exciting areas in engineering and technology today.

But, if UAVs are to gain broad acceptance, they must not only demonstrate they can be used safely but that they can also reduce costs compared to the MAV alternative. By doing an analysis of the unique requirements of UAVs and lifetime costs, displacement sensors can be selected that not only reduce the cost of the UAV but that may also improve UAV performance.

REFERENCES

1. Handbook of Unmanned Aerial Vehicles by Valavanis, Kimon P.; Vachtsevanos, George J (New Eds.).
2. Space Age Control, Inc. 38850 20th Street East, Palmdale, CA 93550 USA
3. Advances in Unmanned Aerial Vehicles by by Kimon P. Valavanis (Sep 10, 2007)
4. Zhang, R. and Aktan, E., "Design consideration for sensing systems to ensure data quality", Sensing issues in Civil Structural Health Monitoring, Eded by Ansari, F., Springer, 2005, P281-290
5. Unmanned Aircraft Systems: International Symposium On Unmanned Aerial Vehicles, UAV'08 by Kimon P. Valavanis, Paul Oh and Les A. Piegl (Nov 2, 2011)
6. Predator Drones and Unmanned Aerial Vehicles (UAVs), The New York Times
7. http://www.newscientist.com/blog/technology
8. News Letters, GISCafe Newsletter, Space Imagine, Space Technology etc.
9. http://www.uavfactory.com/news
10. http://www.vectorsite.net/twuav.html
11. Sensors and Control Systems in Manufacturing, Second Edition Book By Greedy

REVERSIBLE DATA HIDING BASED ON HISTOGRAM MODIFICATION OF PIXEL DIFFERENCES

Swarna Kumari K

Asst. Prof, Dept of ECE, Jyothismathi College of Engineering & Technology
Email: swarnakakarla@gmail.com

Prasanna S

Asst. Prof, Dept of EIE, Vignan Institute of Technology & Science
Hyderabad, AP, India.
Email: Prasanna.shesapu@gmail.com

David Solomon Raju Y[3]

[3]Assoc. Prof, Dept of ECE, Holy Mary Institute of Tech & Science, Hyderabad, AP, India.
Email: davidsolomonraju131@gmail.com

ABSTRACT

This Paper presents a reversible data hiding scheme based on histogram modification. Exploit a binary tree structure to solve the problem of communicating pairs of peak points. Distribution of pixel differences is used to achieve large hiding capacity while keeping the distortion low. And also adopt a histogram shifting technique to prevent overflow and underflow. Performance comparisons with other existing schemes are provided to demonstrate the superiority of the proposed scheme. An efficient extension of the histogram modification technique by considering the differences between adjacent pixels rather than simple pixel value is to be presented. One common drawback of virtually all histogram modification techniques is that they must provide a side communication channel for pairs of peak and minimum points. To solve this problem, in this paper introduces a binary tree that predetermines the multiple peak points used to embed messages thus, the only information the sender and recipient must share is the tree level L. In addition, since neighbor pixels are often highly correlated and have spatial redundancy, the differences have a Laplacian-like distribution. This enables   to achieve large hiding capacity while keeping embedding distortion low.

Keywords:

INTRODUCTION

More than 700 years ago, paper watermarks were used in Fabriano, Italy to indicate the paper brand and the mill that produced it. After their invention, watermarks quickly spread over Italy and then over Europe, and although originally used to indicate the paper brand or paper mill, they later served as indication for paper format, quality, and strength and were also used to date and authenticate paper. By the 18th century it began to be used as anti counter fitting measures on money and other documents. They are still widely used as security features in currency today.

Figure 1 watermark in mark and Dollar bank notes

Watermark in Mark and Dollar bank notes. The term watermark was introduced near the end of the 18th century. It was probably given because the marks resemble the effects of water on paper. The purpose of digital watermarks is to provide copyright protection for intellectual property that's in digital format.

Digital watermarking is similar to watermarking physical objects except that the watermarking technique is used for digital content instead of physical objects. In digital watermarking a low-energy signal is imperceptibly embedded in another signal. The low energy signal is called watermark and it depicts some metadata, like security or rights information about the main signal. The main signal in which the watermark is embedded is referred to as cover signal since it covers the watermark. The cover signal is generally a still image, audio clip, video sequence or a text document in digital format.

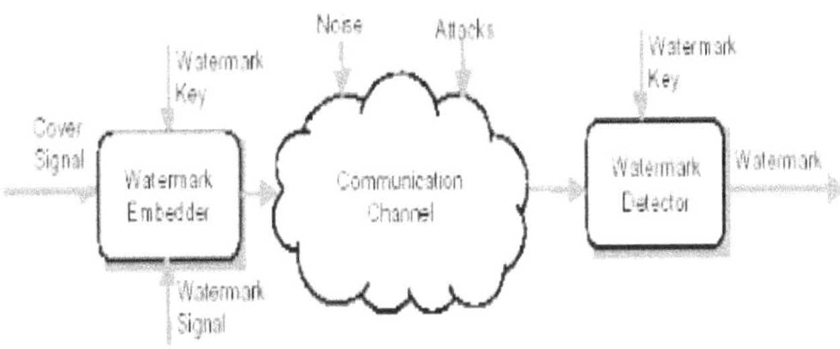

Figure 2 General frameworks for Digital watermarking

STEGANOGRAPHY AND WATERMARKING

Watermarking is not a new technique. It is descendent of a technique known as Steganography, which has been in existence for at least a few hundred years. Steganography is a technique for concealed communication. Here the existence of the message that is communicated is a secret and its presence is known only by parties involved in the communication.

In Steganography a secret message is hidden within another unrelated message and then Communicated to the other party. As opposed to this in Watermarking again one message is hidden in another, but two messages are related to each other in some way. Steganography methods are in general not robust, i.e., the hidden information cannot be recovered after data manipulation. Watermarking, as opposed to Steganography, has the additional notion of robustness against attacks. Even if the existence of the hidden information is known it is difficult ideally impossible for an attacker to destroy the embedded watermark, even if the algorithmic principle of the watermarking method is public.

CRYPTOGRAPHY VS. WATERMARKING

Watermarking is a totally different technique from cryptography. Cryptography only provides security by encryption and decryption. However, encryption cannot help the seller monitor how a legitimate customer handles the content after decryption. So there is no protection after decryption. As shown in the figure 3 in this case Customer can make illegal copies of the digital content. Unlike cryptography, watermarks can protect content even after they are decoded.

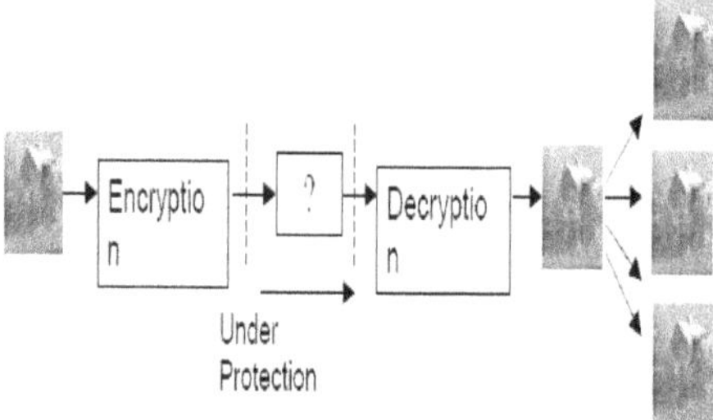

Figure 3 Cryptography block diagram

Addition to protecting content they provide many other applications also, like copyright protection, copy protection, ID card security etc.

Also the concept of breaking the system is different for cryptosystems and watermarking systems. A cryptographic system is broken when the attacker can read the secrete message. But Breaking of a watermarking system has two stages:

- The attacker can Encrypt and Decrypt, under Protection detect that watermarking has been used.
- The attacker is able to read, modify or remove the hidden message.
  The enormous popularity of the World Wide Web in the early 1990's demonstrated the commercial potential of offering multimedia resources through the digital networks. Since commercial interests seek to use the digital networks to offer digital media for profit, they have a strong interest in protecting their ownership rights. Digital watermarking has been proposed as one way to protect such interests. Though much research remains before watermarking systems become robust and widely available, there is much promise that they will contribute significantly to the protection of proprietary interests of electronic media. Collateral technology will also be necessary to automate the process of authentication, non-reputable transmission and validation.

METHODOLOGY

A reversible data hiding scheme based on histogram modification using pairs of peak and zero points. Let P be the value of peak point and Z be the value of zero point. The range of the histogram, P + 1, Z - 1, is shifted to the right-hand side by 1. Once a pixel with value P is encountered, if the message bit is "1," the pixel value is increased by 1. Otherwise, no modification is needed. Data extraction is actually the reverse of the data hiding process. Note that the number of message bits that can be embedded into an image equals the number of pixels associated with the peak point. However, the histogram modification technique does not work well when an image has an equal histogram. While multiple pairs of peak and minimum points can be used for embedding, the pure payload is still a little low.

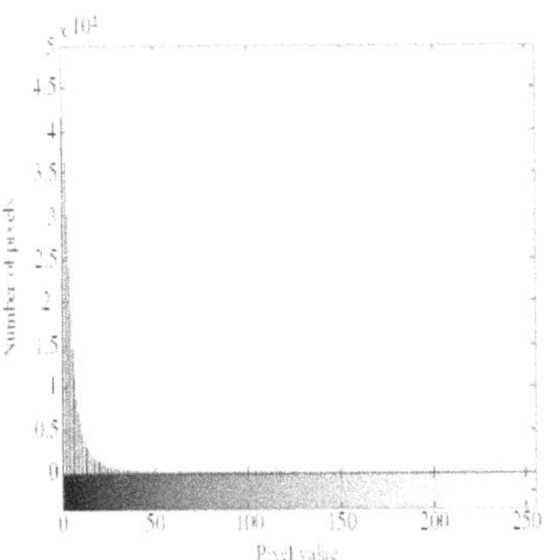

Figure 4 Distribution of pixel differences

Histogram modification technique carries with it an unsolved issue in that multiple pairs of peak and minimum points must be transmitted to the recipient via a side channel to ensure successful restoration.

Thus, present an efficient extension of the histogram modification technique by considering the differences between adjacent pixels instead of simple pixel value. Since image neighbor pixels are strongly correlated, the distribution of pixel difference has a prominent maximum, that is, the difference is expected to be very close to zero, as shown in Figure 4. Can find that the differences have almost a zero-mean and Laplacian like distribution. Distributions of other images also follow this model. Laplacian data can be applied to data hiding schemes to improve their embedding ability. This observation leads us toward designs in which the embedding is done in pixel differences. And also use a tree structure to solve the issue of communicating multiple pairs of peak points to recipients. Having explained r background logic, now outline the principle of the proposed reversible data hiding algorithm.

HISTOGRAM MODIFICATION ON PIXEL DIFFERENCES

For an N-pixel 8-bit grayscale host image H with a pixel value xi, where xi denotes the grayscale value of the i_{th} pixel, $0 < i < N - 1$, xi $\in \mathbf{Z}$, xi $\in 0, 255$.
   1) Scan the image H in an inverse s-order. Calculate the pixel difference di between pixels x_{i-1} and xi by

$$d_i = \begin{cases} x_i, \\ |x_{i-1} - x_i|, & otherwi \end{cases} \qquad (1)$$

   2) Determine the peak point P from the pixel differences.

3) Scan the whole image in the same inverse s-order as in Step 1. If di > P, shift xi by 1   unit

$$y_i = \begin{cases} x_i, \\ x_i + 1, & if \\ x_i - 1, & if \end{cases} \tag{2}$$

Where yi is the watermarked value of pixel i.

4) If $d_i = P$, modify x_i according to the message bit

$$d_i = \begin{cases} x_i + b, \\ x_i - b, 1 \end{cases} \tag{3}$$

Where b is a message bit to be embedded. At the receiving end, the recipient extracts message bits from the watermarked image by scanning the image in the same order as during the embedding. The message bit b can be extracted by

$$b = \begin{cases} 0, \\ 1, \end{cases} \tag{4}$$

Where x_{i-1} denotes the restored value of y_{i-1}. The original pixel value of x_i can be restored by

$$x_i = \begin{cases} y_i + 1, & if\ |y_i - x_{i-1}| > P\ and\ y_i < x_{i-1} \\ y_i - 1, & if\ |y_i - x_{i-1}| > P\ and\ y_i > x_{i-1} \\ y_i, & otherwise \end{cases} \tag{5}$$

Thus, an exact copy of the original host image is obtained. These steps complete the data hiding process in which only one peak point is used. Large hiding capacities can be obtained by repeating the data hiding process. However, recipients may not be able to retrieve both the embedded message and the original host image without knowledge of the peak points of every hiding pass. Thus, here presents a binary tree structure in the following subsection that deals with communication of multiple peak points.

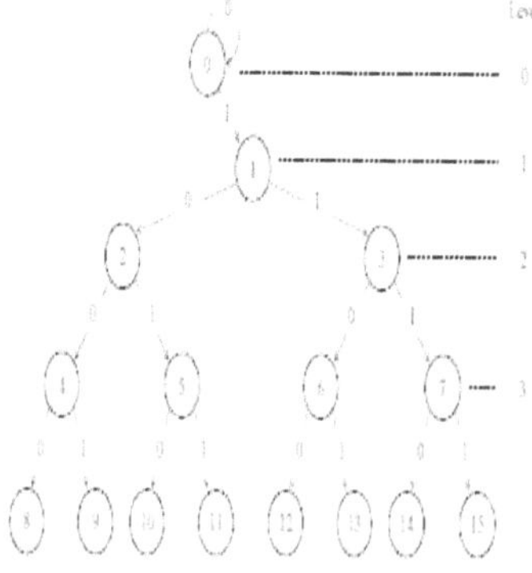

Figure 5 Binary trees for proposed scheme

BINARY TREE STRUCTURE

Figure 5 shows an auxiliary binary tree for solving the issue of communication of multiple peak points. Each element denotes a peak point. Let us assume that the number of peak points used to embed messages is 2^L, where L is the level of the binary tree. Once a pixel difference d_i that satisfies $d_i < 2^L$ is encountered, if the message bit to be embedded is "0," the left child of the node d_i is visited; otherwise, the right child of the node d_i is visited. Higher payloads require the use of higher tree levels, thus quickly increasing the distortion in the image beyond acceptable levels. However, all the recipient needs to share with the sender is the tree level L, because propose an auxiliary binary tree that predetermines multiple peak points used to embed messages. A detailed embedding algorithm with the auxiliary binary tree is given below.

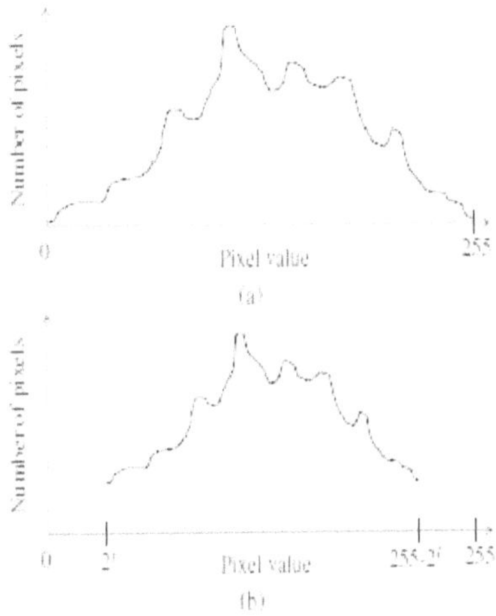

Figure 6 Histogram shifting a) original histogram b) Histogram shifting

PREVENT OVERFLOW AND UNDERFLOW

Modification of a pixel may not be allowed if the pixel is saturated (0 or 255). To prevent overflow and underflow, adopts a histogram shifting technique that narrows the histogram from both sides, as shown in Figure 6. Let us assume that the number of peak points used to embed messages is 2^L, where L is the level of the proposed binary tree structure. Thus, shift the histogram from both sides by 2^L units to prevent overflow and underflow since the pixel xi that satisfies di $> 2^L$ will shift by 2^L units after embedding takes place.

After narrowing the histogram to the range 2^L, $255 - 2^L$, must record the histogram shifting information as overhead bookkeeping information. For this purpose, create a one-bit map as the location map, which is equal in size to the host image. For a pixel having grayscale value in the range 2^L, $255 - 2^L$, assigns a value 0 in the location map; otherwise, we assign a value 1. The location map is losslessly compressed by the run-length coding algorithm, which will yield a large increase in compression ability since pixels out of the range 2^L, $255 - 2^L$ are few and are almost always contiguous. The overhead information will be embedded into the host image together with the embedded message. Note that the maximum modification to a pixel is limited to 2^L according to the proposed tree structure. As a result, shifting the histogram from both sides by 2^L units enables us to avoid the occurrence of overflow and underflow.

EMBEDDING PROCESS

For an N-pixel 8-bit grayscale host image H with a pixel value xi, where xi denotes the grayscale value of the ith pixel, $0 < i < N - 1, x_i \in Z, x_i g\ 0,255$.

1) Determine the level L of the binary tree.
2) Shift the histogram from both sides by 2^L units. Note that the histogram shifting information is recorded as overhead bookkeeping information that will be embedded into the image itself with payload.
3) Scan the image H in an inverse s-order. Calculate the pixel difference di between pixels xi-1 and xi.
4) Scan the whole image in the same inverse s-order. If $di > 2^L$, shift x_i by 2^L units

$$y_i = \begin{cases} x_i + 2^i, i \\ x_i - 2^i, i \end{cases} \qquad (6)$$

Where yi is the watermarked value of pixel i.

5) If $di < 2^L$, modify x_i according to the message bit

$$y_i = \begin{cases} & \end{cases} \qquad (7)$$

Where b is a message bit to be embedded and b {0, 1}. Note that the overhead information is included in the image itself with payload. Thus, the real capacity Cap that is referred to as pure payload is Cap = N_p — |0|, where N_p is the number of pixels that are associated with peak points and | O | is the length of the overhead information.

EXTRACTION PROCESS

This process extracts both overhead information and pay-load from the watermarked image and losslessly recovers the host image. Let L be the level of the proposed binary tree. For an N-pixel 8-bit watermarked image W with a pixel value yi, where y_i denotes the grayscale value of the ith pixel, $0 < i < N — 1$, yi $\in Z$, yi $\in 0, 255$.

1) Scan the watermarked image W in an inverse s-order.
2) If $|y_i x_i - 1| < 2^{L+1}$, extract message bit b by

$$b = \begin{cases} 0 \\ \end{cases} \qquad (8)$$

where $x_i - 1$ denotes the restored value of y;_1.

3) Restore the original value of host pixel x_i by

$$x_i = \begin{cases} y_i + \left\lceil \frac{|y_i - x_{i-1}|}{2} \right\rceil, & \text{if } |y_i - x_{i-1}| \le 2^{L+1} \text{ and } y_i < x_{i-1} \\ y_i - \left\lceil \frac{|y_i - x_{i-1}|}{2} \right\rceil, & \text{if } |y_i - x_{i-1}| \le 2^{L+1} \text{ and } y_i > x_{i-1} \\ y_i + 2^L, & \text{if } |y_i - x_{i-1}| > 2^{L-1} \text{ and } y_i < x_{i-1} \\ y_i - 2^L & \text{if } |y_i - x_{i-1}| > 2^{L-1} \text{ and } y_i > x_{i-1} \\ y_i & \text{otherwise} \end{cases} \quad (9)$$

4) Repeat Step 2 until the embedded message is completely extracted.
5) Extract the overhead information from the extracted message. If a value 1 is assigned in the location i, restore x_i to its original state by shifting it by 2^L units; otherwise, no shifting is required.

EXPERIMENTAL RESULTS

For single peak point

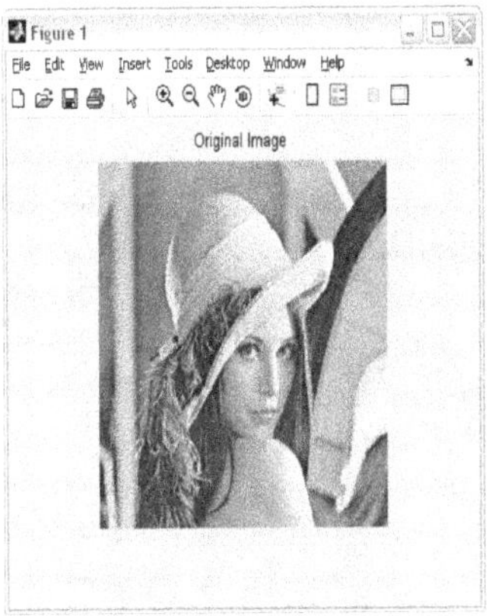

Figure 5.1 original image

This is original image Lena taking as input image to make watermarked image.

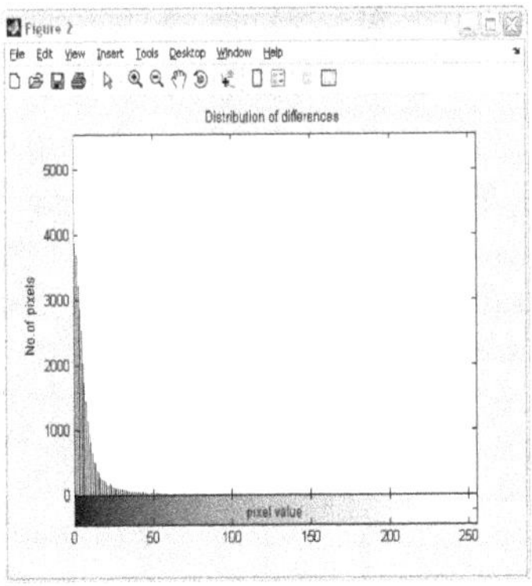

Figure 5.2 Distribution of pixel differences

This figure shows distribution of pixel differences it means after finding difference between the pixels histogram is plotted.

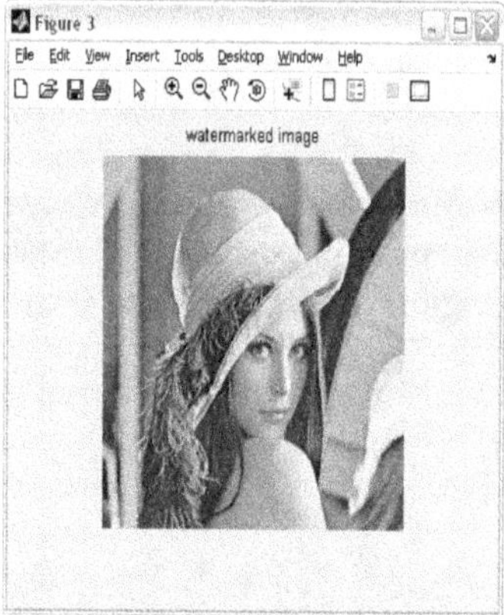

Figure 5.3 watermarked image

This is watermarked image of Lena image. In this after finding the pixel differences data is embedded to the peak point of pixel differences.

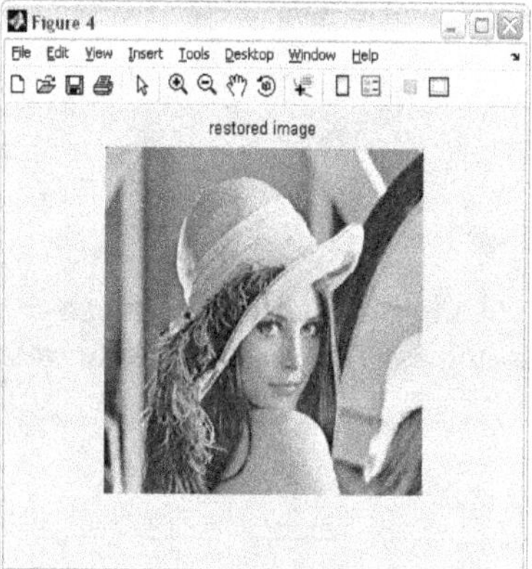

Figure 5.4 Restored image

This image shows restored image it means after extraction of watermarked data and original image. In command window outputs like PSNR, bits per pixel, transmitted bits, received bits and capacitance of bits are displayed.

RESULTS FOR BINARY TREE STRUCTURE FOR LEVEL 1
For image Lena

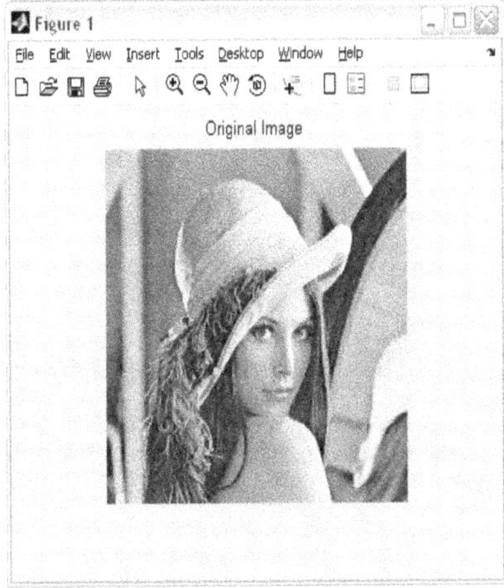

Figure 5.6 original image

This is original image Lena taking as input image to make watermarked image.

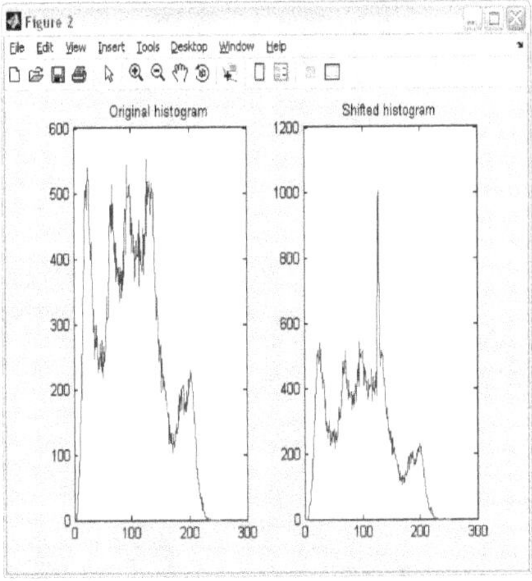

Figure 5.7 original histogram and shifted histogram

Original histogram is shifted to prevent overflow and underflow. This figure shows original histogram and shifted histogram

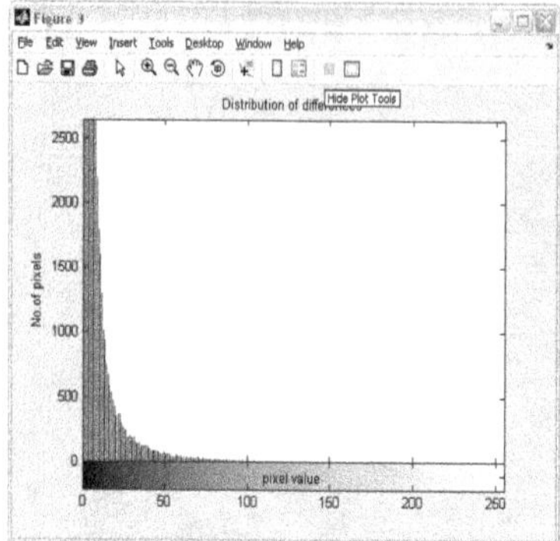

Figure 5.8 Distribution of pixel differences

This figure shows distribution of pixel differences it means after finding difference between the pixels histogram is plotted.

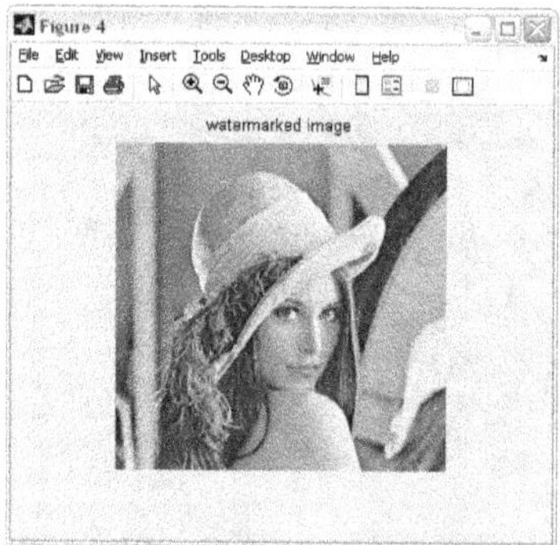

Figure 5.9 watermarked image

This is watermarked image of Lena image. In this after finding the pixel differences data is embedded to the peak point of pixel differences.

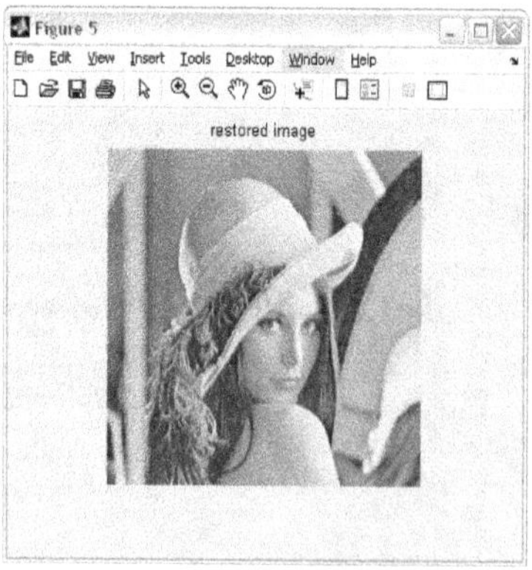

Figure 5.10 Restored image

This image shows restored image it means after extraction of watermarked data and original image. In command window outputs like PSNR, bits per pixel, transmitted bits, received bits and capacitance of bits are displayed.

Similarly for image Mandrill and for image Boat all the above are analyzed and given in the below table in comparison.

Table 1 Hiding capacity and distortion for test images with L=1

Host image(256x256)	Np	Cap(bits)	PSNR(db)	Bit rate(b/pixel)
Lena	7388	7382	30.5037	0.0035
Mandrill	2773	2767	24.8860	0.0055
Boat	8583	8577	31.3989	0.0071

CONCLUSIONS AND FUTURE SCOPE

In this paper, an efficient extension of the histogram modification technique is presented by considering the differences between adjacent pixels rather than simple pixel value. One common drawback of virtually all histogram modification techniques is that they must provide a side communication channel for pairs of peak and minimum points. To solve this problem, introduces a binary tree that predetermines the multiple peak points used to embed messages; thus, the only information the sender and recipient must share is the tree level L. In addition, since neighbor pixels are often highly correlated and have spatial redundancy, the differences have a Laplacian like distribution. This enables us to achieve large hiding capacity while keeping embedding distortion low. The future scope of this paper is to increase hiding capacity and reducing the bit error rate and in this used only images for hiding data in the future supposed to do with the videos also.

REFERENCES

1. M. Wu and B. Lin, "Data hiding in image and video: Part I-fundamental issues and solutions," IEEE Trans. Image Process., vol. 12, no. 6, pp. 685–695, Jun. 2003.

2. M. Wu, H. Yu, and B. Liu, "Data hiding in image and video: Part IIdesigns and applications," IEEE Trans. Image Process., vol. 12, no. 6, pp. 696–705, Jun. 2003.

3. J. Fridrich, M. Goljan, and R. Du, "Invertible authentication," in Proc. SPIE Security Watermarking Multimedia Contents III, San Jose, CA, Jan. 2001, vol. 4314, pp. 197–208.

4. J. Fridrich, M. Goljan, and R. Du, "Lossless data embedding for all image formats," in Proc. SPIE Security Watermarking Multimedia Contents IV, San Jose, CA, Jan. 2002, vol. 4675, pp. 572–583.

5. C. De Vleeschouwer, J. F. Delaigle, and B. Macq, "Circular interpretation of bijective transformations in lossless watermarking for media asset management," IEEE Trans. Multimedia, vol. 5, no. 1, pp. 97–105, Mar. 2003.

6. M. U. Celik, G. Sharma, and A. M. Tekalp, "Lossless watermarking for image authentication: A new framework and an implementation," IEEE Trans. Image Process., vol. 15, no. 4, pp. 1042–1049, Apr. 2006.

7. M. U. Celik, G. Sharma, A. M. Tekalp, and E. Saber, "Lossless generalized-LSB data embedding," IEEE Trans. Image Process., vol. 14, no. 2, pp. 253–266, Feb. 2005.

8. J. Tian, "Reversible data embedding using a difference expansion," IEEE Trans. Circuits Syst. Video Technol., vol. 13, no. 8, pp. 890–896, Aug. 2003.

9. A. M. Alattar, "Reversible watermark using the difference expansion of a generalized integer transform," IEEE Trans. Image Process., vol. 13, no. 8, pp. 1147–1156, Aug. 2004.

10. L. Kamstra and H. J. A. M. Heijmans, "Reversible data embedding into images using wavelet techniques and sorting," IEEE Trans. Image Process., vol. 14, no. 12, pp. 2082–2090, Dec. 2005.

11. D. M. Thodi and J. J. Rodríguez, "Expansion embedding techniques for reversible watermarking," IEEE Trans. Image Process., vol. 16, no. 3, pp. 721–730, Mar. 2007.

12. M. Van der Veen, F. Bruekers, A. Van Leest, and S. Cavin, "High capacity reversible watermarking for audio," in Proc. SPIE Security Watermarking Multimedia Contents V, Santa Clara, CA, Jan. 2003, vol. 5020, pp. 1–11.

13. D. Rui and J. Fridrich, "Lossless authentication of MPEG-2 video," in Proc. IEEE Int. Conf. Image Process., vol. 2. Rochester, NY, 2002,   pp. 893–896.

14. J. Dittmann and O. Benedens, "Invertible authentication for 3-D meshes," in Proc. SPIE Security Watermarking Multimedia Contents V, Santa Clara, CA, Jan. 2003, vol. 5020, pp. 653–664.

15. Y. Hu and B. Jeon, "Reversible visible watermarking and lossless recovery of original images," IEEE Trans. Circuits Syst. Video Technol., vol. 16, no. 11, pp. 1423–1429, Nov. 2006.

Session II
Guidance, Navigation and Control

DESIGN OF AUTOMATIC PID GAIN SCHEDULING FOR AN UNMANNED AERIAL VEHICLE

Kaviyarasu A[1], K.Senthilkumar[2], and G.Umanath[3]
[1]Research scholar, [2]PG student, [3]Associate Professor
Aerospace Engineering, Madras Institute of Technology, Anna University, Chennai
Contact id: isrokavi@gmail.com ksk_mit@rediffmail.com

ABSTRACT

This paper presents the Automatic Gain scheduling of the Aircraft Parameters by using MATLAB. Here, Aircraft does not require an external operator for tuning the gain, while loitering phase. The Proposed tuning method to estimate its required PID gain by means of dynamic response from the system. So, it is adapted to all kind of platforms, and it not required for making the complete mathematical model to calculate the required gain. An Experienced and Skilled Person is needed for tuning an autopilot in manual technique. In Normal PID tuning, the Gains are tuned by trail and error method or by means of making complete mathematical model of the system. It consumes more time to make the complete mathematical model of the system and then find the required gain for the particular system. Aircraft Control Algorithm for the three axis system is designed by using relay test and Ziegler Nichols tuning technique. For continuous changes in the proportional gain system oscillate in random manner. At particular proportional value the system oscillate constant amplitude. While substituting ultimate gain value and corresponding time period for the Ziegler Nichols formula system will be stabilized. Here due to the more oscillation in the tuning value system will be more unstable. To avoid that relay test was introduced for the system loop. It will reduce the unstable condition for the system.

KEYWORDS: Autopilot systems, Gain Scheduling, Unmanned aerial vehicle (UAV),

INTRODUCTION

The gain-scheduling approach is perhaps one of the most popular nonlinear control design approaches which has been widely and successfully applied in fields ranging from aerospace to process control. Despite the wide application of gain-scheduling controllers and a diverse academic literature relating to gain-scheduling extending back nearly thirty years, there is a notable lack of a formal review of the literature. Moreover, whilst much of the classical gain-scheduling theory originates from the 1960s, there has recently been a considerable increase in interest in gain-scheduling in the literature with many new results obtained. An extended review of the gain-scheduling literature therefore seems both timely and appropriate. The scope of this paper includes design procedures relating to automatic gain-scheduling for an unmanned aerial vehicle. Gain scheduling has over the years been widely used for aircraft flight control for two reasons: (1) It allows the designer to use a suite of linear control design tools that are powerful and well understood, and (2) Flight control clearance and certification procedures are usually based on linear methods. However, gain scheduling has had its share of problems as well, mainly due to its ad hoc nature.

UAV

The term UAV is an abbreviation of Unmanned Aerial vehicle, meaning aerial vehicles which operate without a human pilot. UAVs are commonly used in both the military and police forces in situations where the risk of sending a human piloted aircraft is unacceptable, or the situation makes using a manned aircraft impractical.

Currently, UAVs are most often used for the following tasks:

Aerial Reconnaissance – UAVs are often used to get aerial video of a remote location, especially where there would be unacceptable risk to the pilot of a manned aircraft. UAVs can be equipped with high resolution still, video, and even infrared cameras. The information obtained by the UAV can be streamed back to the control center in real time.

Scientific Research – In many cases, scientific research necessitates obtaining data from hazardous or remote locations. A good example is hurricane research, which often involves sending a large manned aircraft into the center of the storm to obtain meteorological data. A UAV can be used to obtain this data, with no risk to a human pilot.

Logistics and Transportation – UAVs can be used to carry and deliver a variety of payloads. Helicopter type UAVs are well suited to this purpose, because payloads can be suspended from the bottom of the airframe, with little aerodynamic penalty.

UAV CONTROL SURFACES

In order for an UAV to be automatically flied, all the flight fixes need to be defined for the autopilot system. In this manner, the autopilot system will be able to control the heading, the altitude, and the speed of the UAV compared to the air speed. The autopilot also needs noticeable capabilities for accepting commands corresponding to maintenance of the roll and pitch angles. To achieve this goal, one needs to use a nested structure for the basic PID controller. The commands relevant to elevator, aileron, rudder and throttle will be controlled by an internal closed-loop PID controller in such away that sufficient stability in roll and pitch axis could be met.

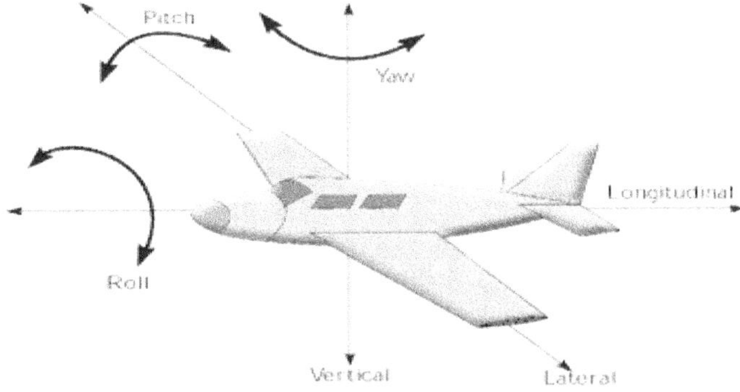

Fig 1. Rotational motion

An airplane has three axis of movement. They are known as the yaw, pitch, and roll. Combinations of these three controls enable an airplane to maneuver. The pitch refers to the movement of the airplane's nose either up or down. The movement is done about a horizontal axis stretching from one wing tip to the other. The elevator controls the pitch. The elevator is also located on the rear of the aircraft on the tail, along with the rudder. Roll is known as the rising or dipping of the airplane's wing. The movement is done about a horizontal axis stretching from the nose to the tail of the aircraft. Ailerons control the roll. The ailerons are located on the trailing edge of both wings. The yaw allows the airplane to move towards the left or right while in flight. The movement is done about a ventricle axis. The yaw is controlled by the rudder. The rudder is located in the rear of the aircraft on the tail.

TUNING TECHNIQUES

The aim of the controller tuning is system should behave fast response and good stability. Unfortunately, for most practical processes being controlled with a PID controller, these two wishes can not be achieved simultaneously. In other words: 1) The faster response, the worse stability, and 2) The better stability, slower responses. For control system it is important that it has good stability than being fast so it should be Acceptable stability. Here three important tuning techniques are discussed. Such as Ziegler Nichols technique, Ultimate gain technique, relay test.

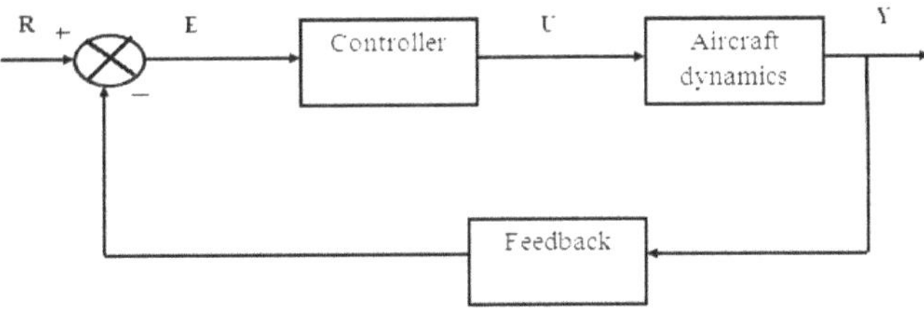

Fig. 2 Block Diagram of Feedback system

Figure 2 shows the block diagram of basic feedback closed loop system. here the output is corrected from the reference value. So that the required output is obtained.

ZIEGLER NICHOLS TECHNIQUE

Ziegler-Nichols tuning rule would serve as the basis for a coming new generation of PID technology. The time-honored Ziegler-Nichols tuning rule as introduced in the 1940s had a large impact in making PID feedback controls acceptable to control engineers. PID was known, but applied only reluctantly because of stability concerns. With the Ziegler-Nichols rule, engineers finally had a practical and systematic way of tuning PID loops for improved performance. Never mind that the rule was based on science fiction. After taking just a few basic measurements of actual system response, the tuning rule confidently recommends the PID gains to use. The Ziegler-Nichols rule is a heuristic PID tuning rule that attempts to produce good values for the three PID gain parameters:

1. Kp - the controller path gain
2. Ti - the controller's integrator time constant
3. Td - the controller's derivative time constant
   Given two measured feedback loop parameters derived from measurements:

1. The period Tu of the oscillation frequency at the stability limit
2. The gain margin Ku for loop stability

ULTIMATE GAIN

Ziegler and Nichols also described a "closed loop" tuning technique that is conducted with the controller in automatic mode (i.e., with feedback), but with the integral and derivative actions shut off. the controller gain is increased until any disturbance causes a sustained oscillation in the process variable. the smallest controller gain that can cause such an oscillation is called the ultimate gain (ku). the period of those oscillations is called the ultimate period (tu). the appropriate tuning parameters can be computed from these two values.

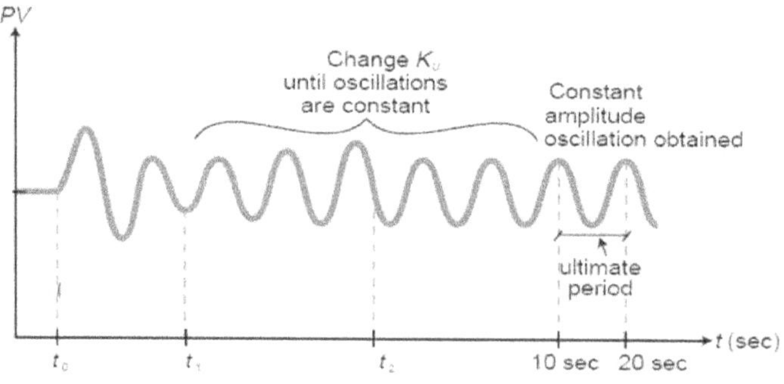

Fig. 3 System oscillation response

From that ultimate gain value and ultimate time period controller type parameters are calculated. Such as Kp, Ki, Kd for PID controller.

Ziegler – Nichols method			
Control type	Kp	Ki	Kd
P only	0.5Ku	-	-
PI	0.45Ku	1.2Kp Tu	-
PID	0.6Ku	2Kp Tu	KpTu 8

Table:1 Formula Ziegler – Nichols method

RELAY TEST

CAVEATS

Ziegler-Nichols methods aren't so straightforward in actual practice. Different PID controllers use different versions of the PID formula, and each must be tuned according to a different set of tuning rules. The rules also change when:
- The controller is not equipped with integral and/or derivative terms
- The process itself is inherently oscillatory
- The process behaves as if it contains its own integral term (as with level control)
- The process dead time is very small or significantly larger than the process time constant.

For many years, the Ziegler-Nichols methods were strictly manual operations executed whenever a new control loop was commissioned. An engineer would run a test, tune the loop according to the results, and then start production with the new loop in automatic mode. It was tedious and repetitive work to commission every loop in the plant this way, and the results weren't always satisfactory. Several iterations were often necessary to generate tuning parameters that produced acceptable closed-loop performance.

Worse still, the closed-loop test deliberately forces the process variable to fluctuate between two extreme values that can not be predicted beforehand. Many control engineers have held their breath while ramping up the controller gain, hoping that the process variable would reach a sustained oscillation without destroying the process.

AUTOMATING THE TUNING PROCESS

Their relay method generates a sustained oscillation of the process variable but with the amplitude of those oscillations restricted to a safe range. With all three PID terms temporarily disabled, the controller uses an on/off relay to apply a step-like control effort to the process. It then holds the control effort constant and waits for the process variable to exceed the set point. At that point, it applies a negative step and waits for the process variable to drop back below the set point. Repeating this procedure each time the process variable passes the set point in either direction forces the process variable to oscillate out of sync with the control effort but at the same frequency. Although the process variable's oscillations aren't strictly sinusoidal, their period turns out to be a close approximation of the ultimate period that Ziegler and Nichols used for their tuning rules. And the amplitude of the process variable's oscillations relative to the amplitude of the control effort's oscillations approximates the process's ultimate gain when multiplied by 4/π. So once the ultimate period and ultimate gain have been determined, tuning the loop becomes a simple matter of plugging those two values into the Ziegler-Nichols tuning rules. To identify the ultimate period TU and ultimate gain PU of the process, the control engineer running the test temporarily disables the PID block and replaces it with an on/off relay that forces the process variable to oscillate.

Where a denotes the amplitude of the controller's square wave and b denotes the amplitude of the process variable's resulting oscillations. By fixing a at a relatively small value, the engineer can also limit the value of b and thereby prevent the wild swings that sometimes plague the original Ziegler-Nichols open-loop test.

The critical gain measured from the Relay test is

$$\text{Critical gain} = \frac{4a}{\pi b}$$

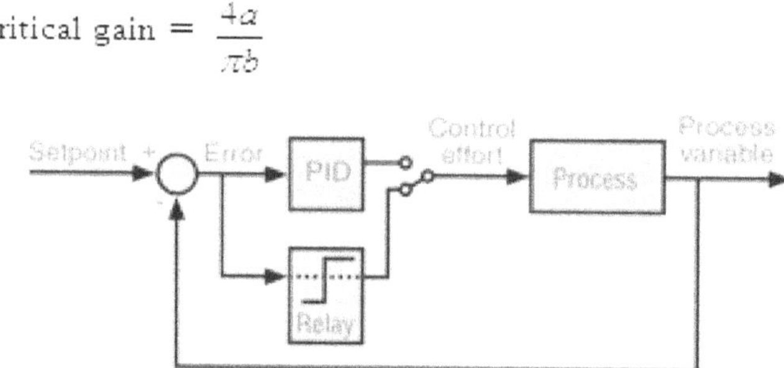

Fig. 4 Block Diagram of Relay test

But unlike the original Ziegler-Nichols closed-loop test, the relay test can be configured to limit the amplitude of the process variable's oscillations by fixing the amplitude of the control effort's oscillations at a user-defined value. This allows the controller to force the process variable to oscillate just enough to distinguish the process's behavior from measurement noise. The process variable needs not swing so wildly as to endanger the process.

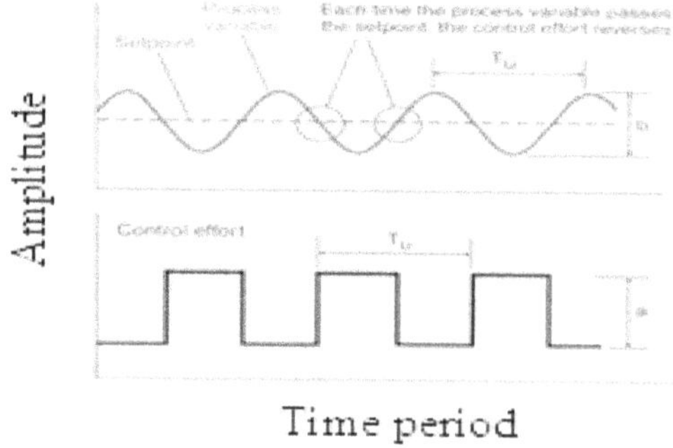

Fig. 5. Relay feedback test response

Better still, the controller can be configured to conduct the relay test and tune the loop without operator intervention. Theoretically, even an operator unfamiliar with the fundamentals of tuning theory can press a button and let the controller conduct its own relay test and select its own tuning parameters accordingly. If the resulting closed-loop behavior proves unacceptable, the operator can simply push the tune button again.

Forcing the closed loop system into sustained oscillations with a proportional-only controller reveals the process's ultimate gain Ku and ultimate period Tu. Unfortunately, doing so can also cause dramatic and sometimes dangerous swings in the control effort and the process variable.

X-PLANE

X-Plane is an engineering tool that can be used to predict the flying qualities of fixed-and rotary-wing aircraft with incredible accuracy. Because X-Plane predicts the performance and handling of almost any aircraft, it is a great tool for pilots to keep up their currency in a simulator that flies like the real plane, for engineers to predict how a new airplane will fly, and for aviation enthusiasts to explore the world of aircraft flight dynamics. X-Plane is used by world-leading defense contractors, air forces, aircraft manufacturers, and even space agencies for applications ranging from flight training to concept design and flight testing.

SYSTEM REQUIREMENTS

Given X-Plane's incredible capabilities and accuracy, it is not possible to run a current release of X-Plane on an exceptionally old computer. A good rule of thumb is that any machine built in the last 18 to 24 months will probably be able to run the simulator acceptably. Computers up to about 36 months old may be fine if they were top-of-the-line machines when manufactured. Even if they weren't, X-Plane may still be able to run, albeit with its rendering options turned down.
X-Plane 10 requires a computer with at least the following specifications:
- A 2 GHz, dual-core processor,
- 2 GB of RAM (physical memory),
- A DirectX 9-capable video card with 128 MB of on-board, dedicated video   RAM (VRAM),
- 10 GB of hard drive space, and
- A DVD-ROM drive.

However, for the best experience, we recommend the following:

- A 3 GHz, multi-core CPU (or, even better, multiple processors),
- 4 GB of system RAM (physical memory), and
- A   DirectX 10-capable (DirectX 11 preferred) video card with 1 GB of on-board, dedicated VRAM, and
- 10 GB of hard drive space, and
- A DVD-ROM drive.

X-Plane, of course, is a flight simulator. A typical flight consists of some, if not all, of the following steps:

- Choosing an aircraft,
- Going to a location (either an airport's runway, a location some distance out from an airport in order to make an approach to the airport, or a random location),
- Setting the weather and time of day, and
- Actually flying.

SIMULATION MODEL

The simulation performed using MATLAB SIMULINK, M-file and the simulation results are obtained. Ziegler Nichols PID tuning and Relay test technique can be implemented for aircraft control axis.

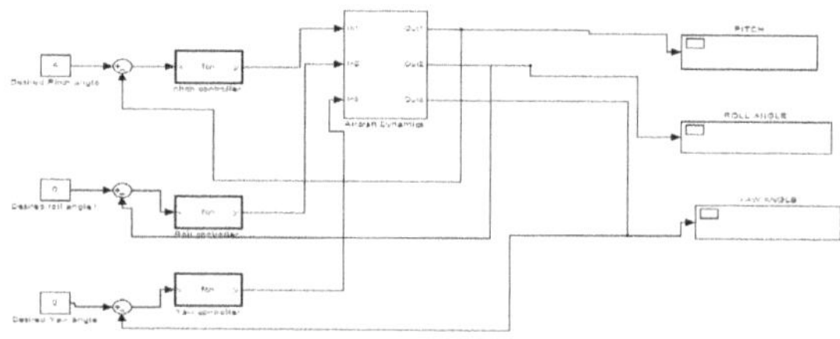

Fig 6 Block diagram representation in X-Plane

While implementing Ziegler Nichols technique for the pitch controlled autopilot response of the system more unstable one. It leads to harmful effect to the aircraft structure. Setting the integral and derivative gain values as zero and increasing the proportional gain value (kp) from zero to until it reaches the ultimate gain (ku). Here the following diagram shows that system response for the different proportional gain values.

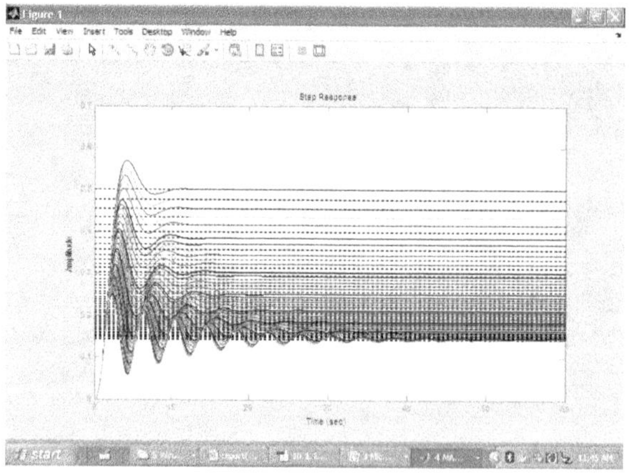

Fig 7 System Response for different proportional gain value

At particular value the output of the control loop oscillates with constant amplitude.

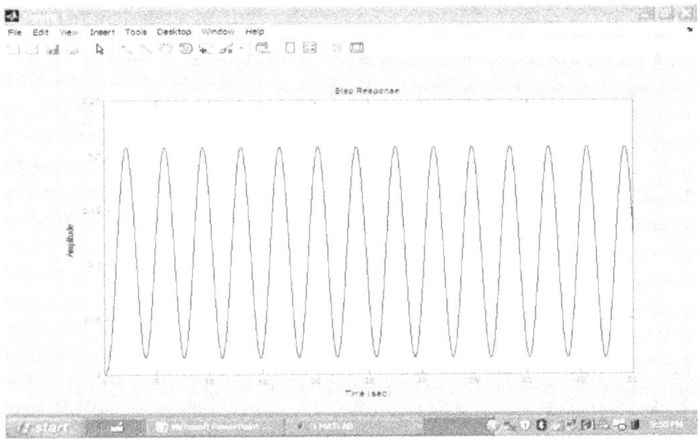

Fig 8 System oscillates with constant amplitude

Calculated ultimate gain value is substituted in Ziegler Nichols formula. Then the system response stabilized output.

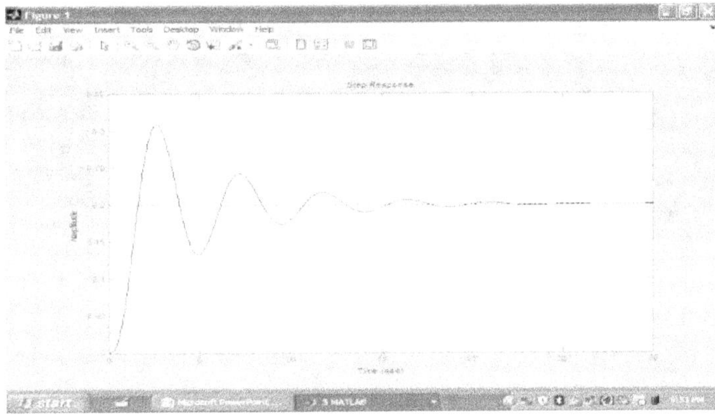

Fig 9 Stabilized system response for Ziegler Nichols method

RELAY TEST

It is a simple test method to measure the critical gain from its feedback using a relay. Here the set point is directly fed to the process using a relay. Due to the result, the system will changes its states (on/off) with low time period and oscillation. These oscillation time period and amplitude gives a key idea to measure the critical amplitude and critical time period.

$$\text{Critical gain} = \frac{4a}{\pi b}$$

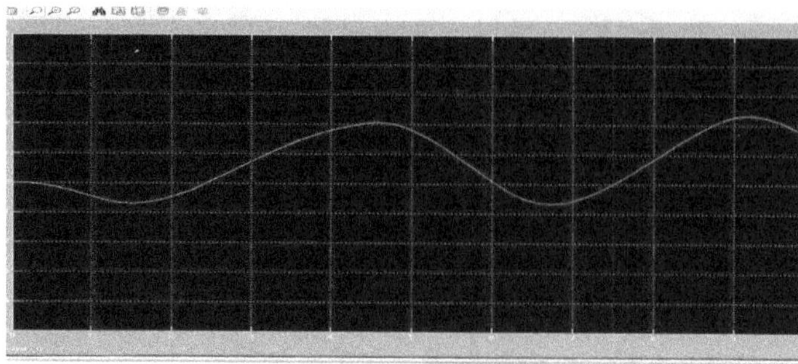

Fig 10 Process variable output

From the process variable response amplitude of the process value (b) can be calculated. And relay system gives the control effort amplitude (a) value. The a,b values are submitted to the Ultimate gain equation we get the desired value. Time period (Tu) also calculated from the relay test.

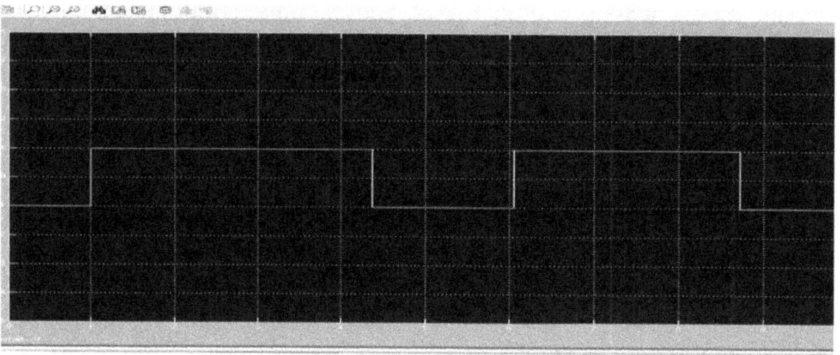

Fig. 11 Control effort output

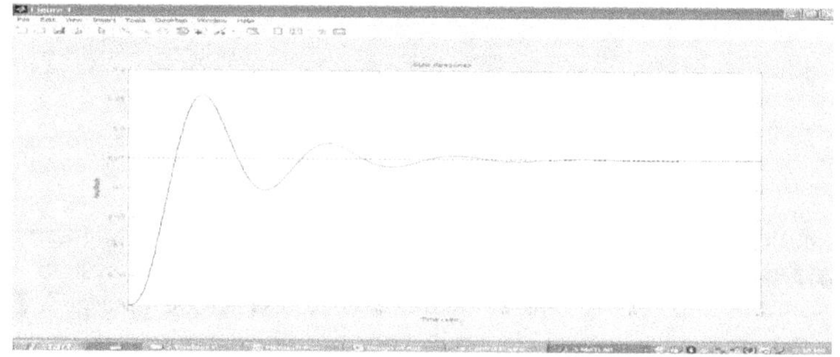

Fig 12 Improved stabilized output after the relay test

REFERENCES

1. Unmanned aerial vehicle (UAV), by Lum Yue Hao Joseph. National at university of Singapore, 2008.

2. UAV Autopilot Controllers test platform using MATLAB/SIMULINK and X-Plane by Lucio R. Ribeiro and Neusa Maria F. Oliveira, Aeronautic Institute of technology, 2007.

3. MATLAB based flight control design scheme for UAV's by Iftikhar H. Makhdoom and Shi-Yin Qin, School of automatic, 2008.

4. Design and simulation of a Basic controller for Unmanned Aerial Vehicle by Abolfazl Sheibani and Mohammad Ali Pourmina, fellow, 2011.

5. Autonomous Flight Control System for Longitudinal Motion of a Helicopter by Atsushi Fujimori, Kyohei Miura, 2007.

6. Auto tuning Autopilots for Micro-ROVs by N.Miskovic, Z.vukic, Department of Control and Computer Engineering, 2009.

7. Autopilots for Small Unmanned Aerial Vehicles by HaiYang Chao, YongCan Cao, and YangQuan Chen, International Journal of Control, Automation, and Systems, 2010.

8. Improved auto tuning using shape factor from relay feedback by T.Thyagarajan and cheng-ching Yu, 2002.

9. Automatic tuning of PID- regulators by H.Rasmussen, Aalborg University, Dept. of Control Engineering, 2002.

10. Path Tracking of UAV using self –Tuning PID controller based on Fuzzy logic by Theerasak Sangyam, Pined Laohapiengsak, 2010.

SIMULATION OF QUADROTOR DYNAMICS

Kaviyarasu A[1]   **Dr.K.Senthilkumar[2]**   **P.Gopinath[3]**

[1]Research scholar, [2]PG student, [3]Associate Professor

Department of Aerospace Engineering, MIT, Anna University, Chennai
Contact id: ksk_mit@rediffmail.com isrokavi@gmail.com

Quad rotor is rotorcraft that has four lift-generating propellers. This project proposes a mathematical model of quad rotor dynamics. A simplified approach is adopted where gyroscopic effect and air friction on frame of machine has been neglected. Thrust force from each motor is calculated and mathematical calculation of Moment of Inertia of quad rotor model is done .Equations of motion is developed for quad rotor model, thereby dynamics of quad rotor model are obtained from it. Simulation of quad rotor dynamics is developed in MATLAB. Optimum control law is developed for the quad rotor model and simulation of quad rotor dynamics is obtained. Simulation of quadrotor dynamics is useful for designing a controller to stabilize the machine.

KEYWORDS: Quad rotor, simulation, Dynamics.

INTRODUCTION

A Quadrotor Helicopter, also called Quadrotor, is a helicopter with four lift-generating propellers mounted on motors. Two of the motors generate thrust by spinning their propellers clockwise and other two counter-clockwise. Control of the machine can be achieved by varying the relative speed of the propellers.

In quad rotor lift is generated by four rotors symmetrically fixed around its centre. The required flight-maneuvers (i.e. yawing, rolling and pitching) and the vertical or lateral flight are realized by independently varying speeds of the four rotors. Moreover not only it is capable of performing Vertical Take-Off and Landing (VTOL), also has a simpler configuration for a compact mechanical design. Therefore the quad rotor vehicle has become an attractive candidate for small scale and medium scale Unmanned Aerial Vehicles (UAVs) for applications such as reconnaissance, search, rescue and surveillance. However the dynamics of this aerial vehicle represents a marginally stable system and an active control system is essential to stabilize it.

The quad rotor is considered an effective alternative to the high cost and complexity of standard rotorcraft.

CONSTRUCTION AND ASSUMPTIONS

A quad rotor simply consists of four dc motors on which propellers are mounted. These motors are arranged at the corners of a + shaped frame, where all the arms make an angle of 90 degrees with one another.

As shown in Figure 3, two of the rotors or propellers spin in one direction and the other in the opposite direction. The motors labeled as MI and M3 spin in the clockwise direction with velocity and other two in the opposite direction. Each spinning propeller generates vertically upward lifting force. All the motion of machine is a consequence of this force.

The mathematical model developed in this text is based on certain basic assumptions as given below:

a) Quad rotor body is rigid
b) Propellers are rigid.
c) There is no air friction on quad rotor body.
d) Free stream air velocity is zero.
e) Drag torque td is proportional to propeller speed with D as drag constant.
f) Design is symmetrical.

FORCES AND MOMENTS ACTING IN QUADROTOR

In order to create an accurate model of the platform, the various forces and moments induced on the craft must be understood and accounted for. The forces and moments induced on the craft are responsible for its movement and overall attitude

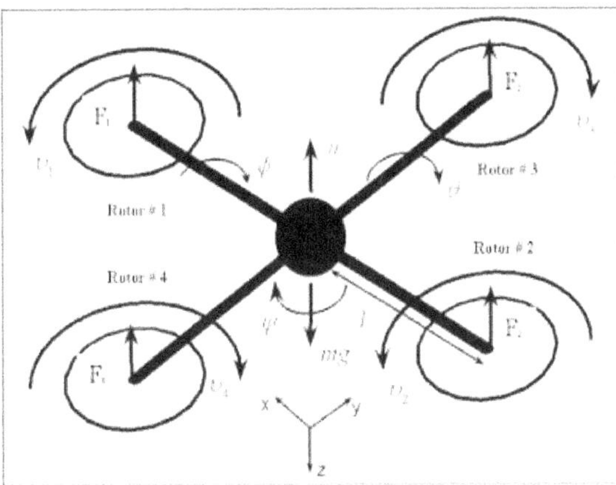

Fig. 1 Forces and moments acting on quad rotor

Each of the forces can be broken into an x, y, and z component. The following Newton-Euler form equation (1) defines the total influence of the net forces and moments on the craft. Using this equation with the individual forces and moments defined for each degree of freedom below, we can determine the full equations of motion for the craft.

$$\begin{bmatrix} mI_{3x3} & 0 \\ 0 & I \end{bmatrix} \begin{bmatrix} \dot{V} \\ \dot{\omega} \end{bmatrix} + \begin{bmatrix} \omega \times mV \\ \omega \times I\omega \end{bmatrix} = \begin{bmatrix} F \\ \tau \end{bmatrix} \qquad (1)$$

The forces and moments are primarily due to gravity and the four propellers. From the reference of the onboard craft coordinate system, the thrusts generated by the motors/propellers are always in the crafts z-direction.

The gravity vector, however, is always in the fixed frame z direction (towards the center of the earth). In this instance, it is important to utilize the rotation matrix from equation (2). We can therefore write the force of gravity as

$$F_g = mg \begin{bmatrix} -\sin\theta \\ \cos\theta \sin\phi \\ \cos\theta \cos\phi \end{bmatrix}_{body} \qquad (2)$$

It is important to remember that this force is taken with respect to the craft coordinate system, affixed to the center of gravity of the quad rotor platform. Along with gravity, the only other forces to be considered are the forces generated by the propeller/motor combos. These forces combined with the force of gravity, allow us to solve equation (1) for the forces acting on the platform, and determine the acceleration of the craft in terms of the craft fixed frame.

$$\begin{bmatrix} \ddot{X} \\ \ddot{Y} \\ \ddot{Z} \end{bmatrix} = -\frac{1}{m} \begin{bmatrix} 0 \\ 0 \\ F_{thrust} \end{bmatrix} + g \begin{bmatrix} -\sin\theta \\ \cos\theta \sin\phi \\ \cos\theta \cos\phi \end{bmatrix}$$

QUADROTOR DYNAMICS

Lifting forces generated by the spinning propeller and the weight, are responsible for all the motion of body, as the external effects such as air friction, wind pressure etc. have been neglected. Linear Acceleration in the X-axis Direction is given by

$$\ddot{X} = \frac{u(1)(\sin\psi \sin\phi + \cos\psi \sin\theta \cos\phi)}{m}$$

Linear Acceleration in the Y-axis Direction is given by

$$\ddot{Y} = \frac{u(1)(\sin\psi \sin\theta \cos\phi + \cos\psi \sin\phi)}{m}$$

Linear Acceleration in the Z-axis Direction is given by

$$\ddot{Z} = \frac{u(1)\cos\theta \cos\phi - k_3 \cdot \dot{Z}}{m} - g$$

Rolling angular Acceleration in the x-axis Direction is given by

$$\ddot{\theta} = (c_2 \times l)/I_x$$

Pitching angular Acceleration in the y-axis Direction is given by

$$\ddot{\theta}=(c_3 \times l)/I_y$$

Yawing angular acceleration in Z-axis Direction is given by

$$\ddot{\psi}=(c_4 \times l)/I_z$$

Where,

$C_1 = F1+F2+F3+F4$
$C_2 = F3-F1$
$C_3 = F4-F2$
$C_4 = F1-F2+F3-F4$

Moment of Inertia

Moment of inertia is a property of a distribution of mass in space that measures its resistance to rotational acceleration about an axis. An object that is rotating at constant angular velocity will remain rotating unless acted upon by an external torque. In this way, the moment of inertia plays the same role in rotational dynamics as mass does in linear dynamics, describing the relationship between angular momentum and angular velocity, torque and angular acceleration.
Calculating the moments of inertia about the various axes is the next step towards accurate modeling of the quad rotor dynamics.

SUPERPOSITION PRINCIPLE

Body is decomposed (either physically or conceptually) into several constituent parts, and then the moment of inertia of the whole body about a given axis is equal to the sum of moments of inertia of each part around the same axis

While the notation (for example, Ixx) denotes the moment of inertia around the x-axis while the platform is rotation around the x-axis (or rolling), we will assume the rolling, pitching and yawing of the platform will not change the moment for any specific axis.

Assuming perfect symmetry between the x- and y-axis, it is safe to assume the moments about each of these axes are numerically equivalent. To simplify the modeling process, all mass components of the platform will be modeled as solid cylinders attached by zero mass and frictionless arms.

The moment of inertia of a cylinder rotating about an axis perpendicular to its body is given by

$$I = \frac{mr^2}{4} + \frac{mh^2}{12}$$

Where m refers to the cylinder mass, r to the cylinder radius, and h to the cylinder height.

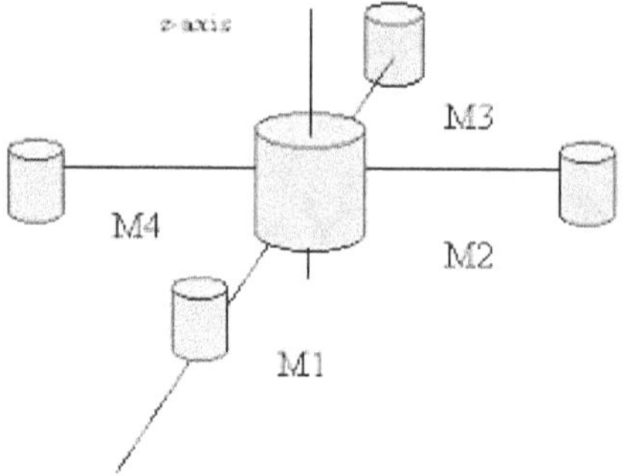

Fig. 2 Moment of inertia about x, y, and z axis

For this implementation, the cylinder includes the motor, motor bracket and landing gear. Taking the x-axis as the first effort, the moment of inertia due to the motors on either side of the axis (motors 2 and 4) is approximated by

$$I_{xx_1} = 2ml^2$$

Again, m refers to the mass of a single cylinder and l refers to the arm length of one side of the craft. The last items of concern to the moment of inertia are the two motors in line with the x-axis (motors 1 and 3) and the central hub where the arms meet.

The equation governing the effect of these objects on the moment of inertia is given by

$$I_{xx_2} = 2\left[\frac{mr^2}{4} + \frac{mh^2}{12}\right] + \frac{m_h r_h^2}{4} + \frac{m_h r_h^2}{12}$$

The first bracketed portion of the equation accounts for motors 1 and 3. The latter portion refers to the central hub, which includes all the electronic speed controllers for the motors, the avionics, sensors, and the batteries and power distribution system.

The overall moment of inertia approximation for the x- and y-axis is given by

$$I_{xx} = \frac{mr^2}{2} + \frac{mh^2}{6} + 2ml^2 + \frac{m_h r_h^2}{4} + \frac{m_h r_h^2}{12}$$

Due to symmetry, above equation applies to both the x- and y-axis. The moment of inertia for the z-axis rotation (yaw) can be attributed to all 4 motor/mount/gear cylinders and the central hub.

The moment of the central hub modeled as a cylinder rotating about an axis through and parallel to its center is given by

$$I_{zz_1} = \frac{m_h r_h^2}{2}$$

For the 4 motors at an arm's length away from the axis of rotation, the total moment of inertia is

$$I_{zz_2} = 4ml^2$$

Therefore, by combining equations two equations, the total equation for the approximation of the moment of inertia about the z-axis is

$$I_{zz} = \frac{m_h r_h^2}{2} + 4ml^2$$

THEORETICAL CALCULATION OF MOMENT OF INERTIA FOR QUAD ROTOR MODEL

a) Motor Specifications

Mass of motor = 50g
Radius of motor = 1.385cm
Height of motor = 3.7cm

b) Central hub specification:

Mass of central hub = 450g
Radius of central hub = 8cm
Height of central hub = 5cm
Length of each Arm = 20cm

c) Moment of Inertia for quad rotor model

Moment of Inertia about x-axis = 49169.9gcm^2
Moment of Inertia about y-axis = 49169.9gcm^2
Moment of Inertia about z-axis = 96000gcm^2

THRUST CALCULATIONELECTRIC MOTOR

Electric powered model aircraft has gained popularity, mainly because the electric motors are more quiet, clean and often easier to start and operate the combustion motors. The motors selected for use on the quadrotor are Hacker KDA20-22L brushless out runner motors. These motors have a stationary internal core and windings with magnets on the outer cylinder. This outer cylinder is the portion that rotates. Since no brushes are involved and the only friction points are at the shaft, these motors are of much higher efficiency when compared to standard brushed motors.

PROPELLER

Propellers are measured by the diameter and pitch. The pitch is the theoretical distance travelled, by the prop, in one revolution. The higher the pitch the higher the air speed and vice-versa.
The propellers selected for quad rotor dynamics are a propellers with a diameter of 10 inches and a pitch of 4.7".

Experimental thrust calculation from motor

An experimental setup is developed for calculating thrust force from motor used in quad rotor model..Componens used for thrust calculation are motor, electronic speed contoller,weighing machine,hindge type setup, transmitter and receiver. Thrust is calculated from weight machine using following relation

$$P_1 * L_1 = P_2 * L_2$$

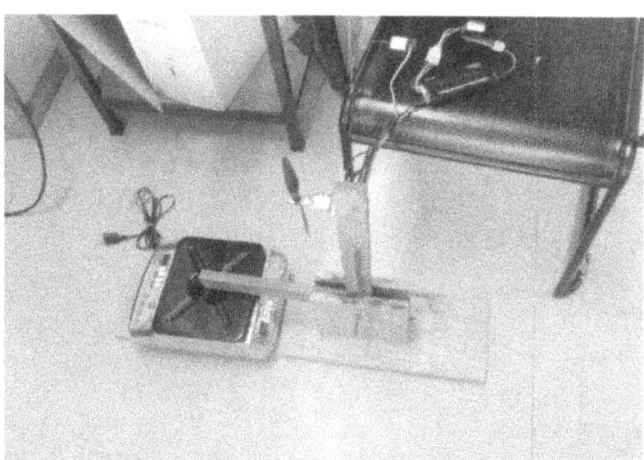

Fig. 3. Experimental setup for calculating thrust

P_1 = Thrust force acting at motor end.
L_1 = Length of arm from hindge point to motor end.
P_2 = Thrust force of motor acting at weighing machine.
L_2 = Length of arm from hindge point to weighing machine end.

Hacker KDA20-22L brushless out runner motors is used with propeller of 10 inches diameter and 4.7 inch pitch.

Measured thrust force from motor is 870Kgs.

QUADROTOR SIMULINK MODEL

In order to accurately simulate the behavior of the ANGEL system, we must build a loop through which an input command to the ANGEL can be applied and the resulting state space vector is updated. Figure 4 shows a block diagram of how this system should be implemented.

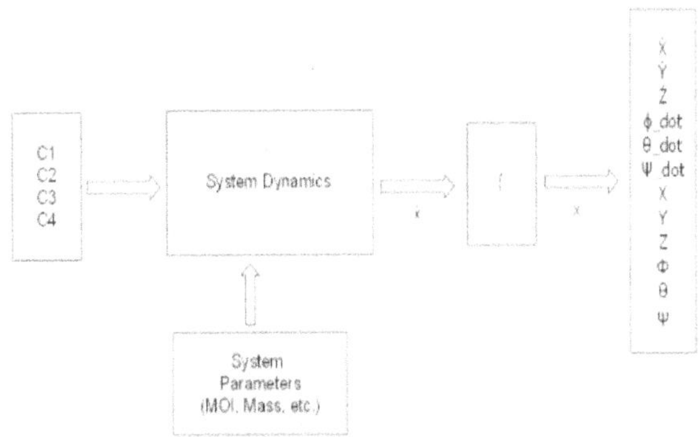

Fig. 4 Block diagram of Simulink model

The input block in the simulation diagram allows us to change the commands (C1, C2, C3, and C4) going to the motors. Thus, disregarding any external disturbances or noise from the physical implementation of the system, we can track how the system will react to changes in the actuator output. This will give us some idea of how the craft will react, and will allow us to design the control system based on desired performance parameters.

IMPLEMENTATION IN MATLAB

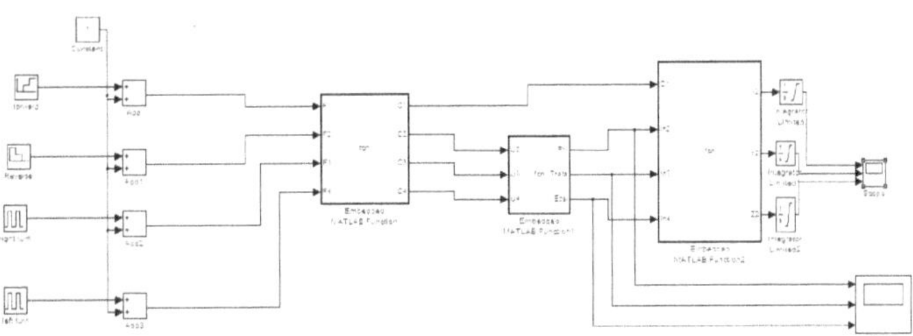

Fig. 5 Simulink models of quadrotor dynamics

The simplified model of the quad-rotor consists of three subsystems.

a) **Resultant Force Diagram**
   The inputs of this block diagram are the resultant forces (F1, F2, F3 and F4 of the motors) and it calculates the sum of forces acting on the quad-rotor due to variation of motors speeds in order to estimate the transitional and rotational motion of the quadrotor.

b) **Euler Angles**
   This block calculates Euler angles (Phi, Theta and Epsi) as a result of variation of forces acting on the quad-rotor.

c) **System Dynamics**

It contains the equations of the system dynamics and it calculates the position of the quad-rotor at any given time.

This is a cut down model and has a clear simple mathematical modeling. Some of the parameters affecting the system were neglected such as ground effect and residual angular speed of the motors.

SIMULATION RESULTS

The most straightforward movements the craft can make are a simple pitch or roll in order to move either forward/back or left/right. To the novice user unfamiliar with the actuator interactions and coupling, the first attempt may involve changing the output speed of only one motor. For example, if a slight forward propagating pitch angle is desired, the first attempt may be to turn on all actuators to gain altitude, provide a negative pulse to the front motor momentarily in order to cause the craft to pitch forwards, travel forwards for a few seconds before providing a positive pulse to the front motor to kick the craft out of forward pitch.

The actuators are powered on at 5s, steadily gaining in altitude. At 10s, a negative pulse is provided to the front motor, causing a drop in speed, which should cause the craft to pitch forward and move in the positive x-direction. At 12s, a positive pulse is applied to presumably bring the craft out of forward pitch and back into steady hover. This however, is not what occurs.

Figure 6 illustrates the input signal described in the preceding paragraph.

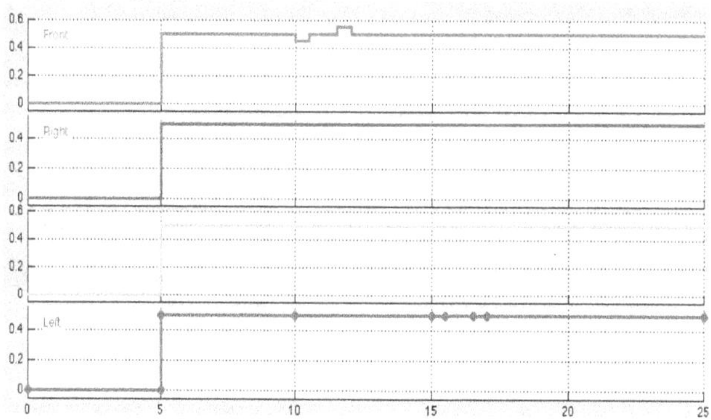

Fig. 6. Motor Input Signal for Pitch Forward attempt

As is evident, the craft does not exhibit the desired behavior. We can analyze what occurs by studying the roll, pitch, and yaw moments as a function of time. Figure 7 shows these values for this particular simulation.

From the figure, we see the desired pitch angle response previously predicted. The actuators all turn on equally at 5s, and there is no deviation in the roll, pitch or yaw angles. At 10s, the effect of the short negative pulse on the front motor is evident in the pitch response curve, followed closely by the short positive pulse to bring the pitch back to a nearly zero offset.

Thus, the pitch acts in accordance to the expectations. The yaw angle, however, does not look correct. There was no intended yaw movement in our signal description, and the presence of this deviation is entirely responsible for the odd trajectory of the craft.

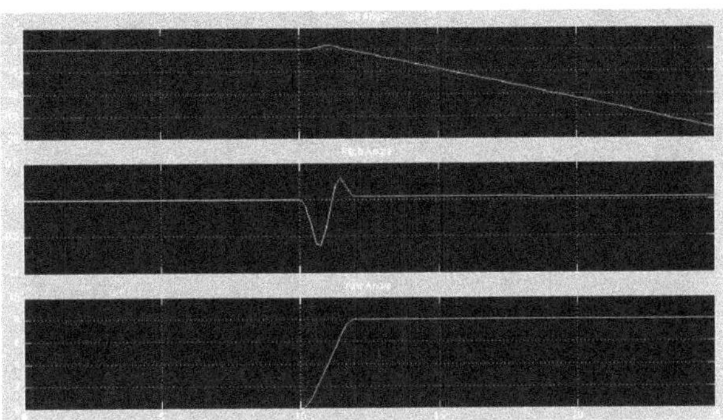

Fig. 7 Roll, pitch, Yaw angles for single actuator simulation

Due to the change in ratio of counter-clockwise propeller speed to clockwise propeller speed, an overall yaw moment was induced. As the front motor speed was decreased, the back motor speed should have increased simultaneously to compensate for the decreased overall clockwise thrust.

The pitching and yawing movements, when combined, changed the thrust vector of the craft. The correct method for implementing a forward movement will rectify the CCW/CW thrust ratio problem that caused the erratic behavior in the first simulation attempt. The new input signals for the actuators are shown in Figure 8

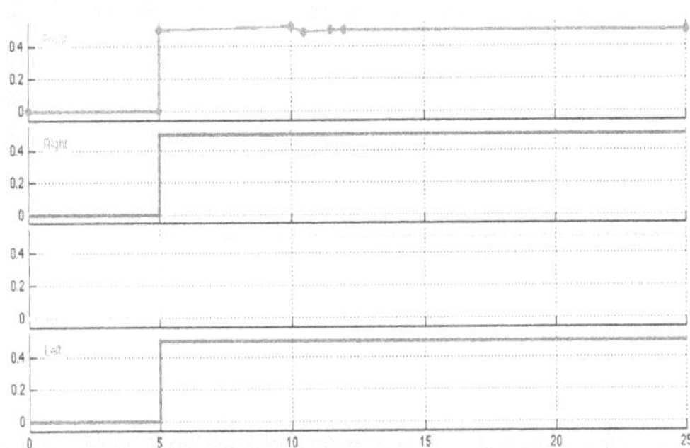

Fig. 8 New Yaw compensation Input signal

These signals provide something much closer to the intended behavior looking at the angle graphs, the first reaction may be that we did not solve anything by changing the input signals. There is no deviation in yaw and roll; there is desired pitch deviation which is described in figure 9

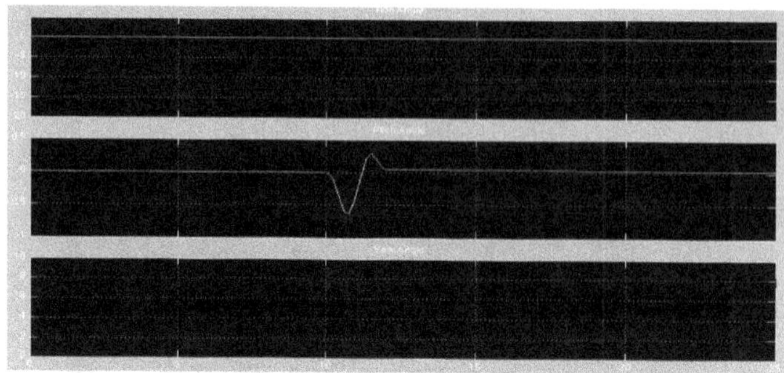

Fig. 9 Roll, Pitch, Yaw graph with updated input signal

The results of this simulation also verify the need for a control system to provide input to the motors .From these two simulations, the complexity of by-hand control of the quad rotor should be clear. Each axis will need an independent control system implementation tuned to the specific characteristics and variables of the axis.

REFERENCES

[1] Arda Ozgur Kivrak, "Design of Control Systems for a Quadrotor   Flight   Vehicle equipped with inertial Sensors", pg 4 to 29, Master's Thesis, Atilim University Turkey, 2006.

[2] P. Ponds, R. Mahony, J. Gresham, P. Corke and J. Roberts,   "Towards   Dynamically Favourable Quad-Rotor Aerial Robots", In   Proc. Of Australasian   Conference on Robotics and Automation, Canberra, Australia 2004.

[3] I. Prinz, "The Mesicopter: A Meso-Scale Flight Vehicle", h a er/.

[4] S. BouabdAllah, P. Murrieri, R. Siegwart, "Towards Autonomous Indoor Micro VTOL", Autonomous Robots 18,171-183, Springer   Inc, 2005.

[5] P. Ponds, R Mahony and P. Corke, "Modeling and Control of Quad-Rotor Robot", In Proc of Australasian Conference on Robotics and   Automation, Sydney, Australia 2005.

[6] R. Resnick, D. Halliday, K. Krane, Physics Volume 1 Fourth   Edition, pg 231 to 240, Published by John Wiley and Sons, Inc.

[7] G. Goodwin, S. Graebe, M. Salgado, Control Systems Design, pg41 to 65, Published by Pearson Education, Inc.

[8] G. M.Hoffmann, H Huang, S L. Waslander, C J. Tomlin, "Quadrotor Helicopter Flight Dynamics and Control: Theory and   Experimentation", AIAA Guidance, Navigation and Control   Conference, 2007.

[9] S. BouadAllah, A. Noth and R. Siegwart, PID versus LQ Control Techniques Applied to an Indoor Micro Quadrotor", In Proc, of the IEEE International Conference on Intelligent Robots and Systems, Sendai Japan, 2004.

SELECTION AND CONTROL ASPECTS OF BRUSHLESS DC MOTORS FOR UNMANNED AERIAL VEHICLES

R. Shanmugasundram

Dept. of Electrical and Electronics Engg., Sri Ramakrishna Engg. College, Coimbatore, India.

K. Muhammad Zakariah

Dept. of Electrical and Electronics Engg., SNS College of Engg. and Technology, Coimbatore, India.

N. Yadaiah

Dept. of Electrical and Electronics Engg., JNTUH College of Engg., Hyderabad, India.
Email: rss_is@yahoo.com[1], kmzakariah@yahoo.co.in[2], svpnarri@yahoo.com[3]

ABSTRACT

Brushless dc (BLDC) motors are now-a-days preferred in many applications such as robotics, space vehicles, electric vehicles, Instrumentation systems and Industrial control applications due to its excellent speed-torque characteristics, wide speed range, compact in size, less in weight, less maintenance, sparkless operation, and easily controllable. However, its potential applications in the emerging areas such as Unmanned Aerial Vehicles (UAVs) need to be explored. Even though brushless dc motors can be used as electric propelling source instead of micro combustion engines, there is a need to study method of selection of BLDC motors, its characteristics and control aspects so as to improve the performance of brushless dc motors. Recent advancements in embedded technology and power semiconductor devices have paved the way to develop intelligent control techniques such as fuzzy logic and artificial network based controllers to improve the performance of BLDC motor drives. This paper discusses the method of selection of brushless dc motors for UAVs and its characteristics and control aspects.

INTRODUCTION

The unmanned Ariel vehicles are now-a-days used in applications such as wildlife tracking, traffic monitoring, border patrols, environmental research, geographical mapping and numerous military applications. In such applications, the reconnaissance should be carried out in the platform of low heat and noise under invisibility to enemy targets. Data collected in the presence of atmospheric and combustion gasses from UAVs powered by internal combustion engine may be comprised. In order to overcome these problems, the only alternative is to use electrical propulsion systems instead of internal combustion engines. In this paper, method of selection of brushless dc motors for electric propulsion systems and its characteristics and control aspects are discussed. The information gathered from various literatures is as follows: Modeling and control aspects of UAVs are discussed in (Castillo et al., 2004, 2005). Design of brushless dc motors and control aspects are discussed in (Gieras et al., 2002; Hanselman et al., 2004). Design and control of electric UAVs are discussed in (Escare et al. 2006; Harmon et al. 2006). Development of fuel cell based UAVs and energy management in the propulsion system is discussed in (Herwerth et al., 2006; Karunarathne et al., 2010). The design of PID, Fuzzy and ANN controllers for BLDC motor drives are discussed in (Rubaai et al., 1997; Shanmugasundram et al., 2012; Tipsuwanporn, 2012). Information about small electric motors and their drives are available in (Emadi, 2005; Yeadon, 2001). Comparison of technical features between electric aircraft and a hybrid electric vehicle is discussed in (Zhang, 2008). This paper is organized as follows: Followed by the introduction in section I, theory of electric power transfer in electric propulsion is discussed in section II, Section III discusses the various power sources used in UAVs, Section III deals with the selection of BLDC motors for UAVs, Section IV discusses the various controllers used to improve the performance of BLDC motor drives and finally Conclusion is presented in section VI.

Electric propulsion systems produce zero gaseous emissions, low heat, and generate less audible noise than an internal combustion system. An electric propulsion system consists of four main components as shown in Fig. 1. They are an electrical power source, electronic speed controller, electric motor and a propeller. The selection of each component for the electric propulsion system should be made such that it is efficient and suit the application requirements. The various power sources used to derive power for modular UAVs are battery, solar power and hydrogen fuel cell. Out of these power sources, fuel cells have higher energy density > 800 Wh/kg.

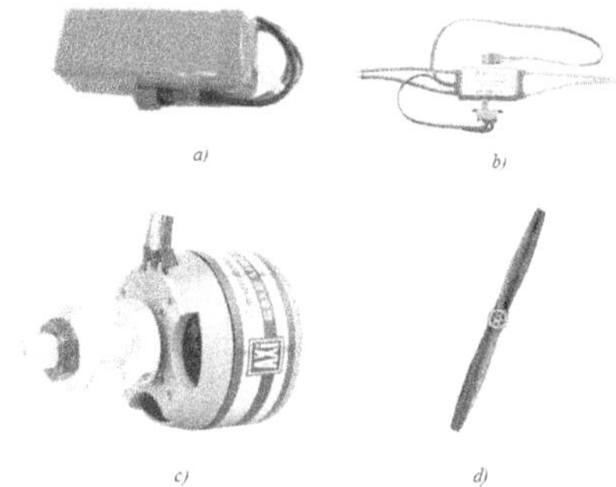

Fig. 1 Electric propulsion system components: a) battery, b) electronic speed controller, c) motor and d) propeller

Theory of Power Transfer in Electric Propulsion Systems

The electric propulsion system converts the electrical energy into the mechanical power in the form of thrust. The power is transferred through four components as shown in Fig. 2.

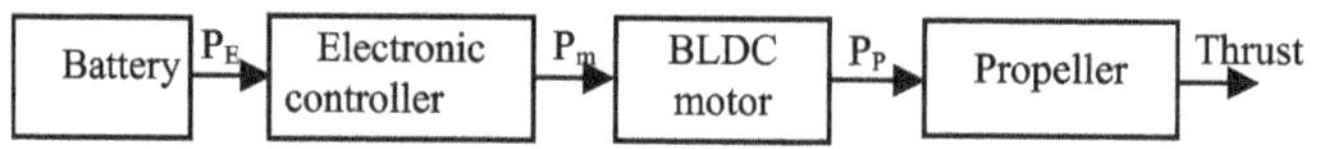

Fig. 2 Power transfer in electric propulsion systems

The electronic controller receives control signals either from an on-board Radio Frequency (RF) receiver or an auto-pilot system. The Electronic controller controls the motor speed according to the percentage duty-cycle of the control signal. Power is transferred from the Electronic controller to the BLDC motor. The output mechanical power of BLDC motor is applied to the propeller which in turn converts the input torque and rotational speed to an aerodynamic thrust.

Power Sources

A. Solar cell

The solar cells are usually made from semiconductor materials such as silicon and they are most expensive. Due to the advancing solar technology the solar cells are light in weight and more efficient. The efficiency of the solar cells has increased over the years as shown in Fig.3 and the current solar cells have the efficiency of 45%. The factors that are considered for the selection of solar cells for UAVs are weight and efficiency. Among these factors, weight is the crucial factor to be considered.

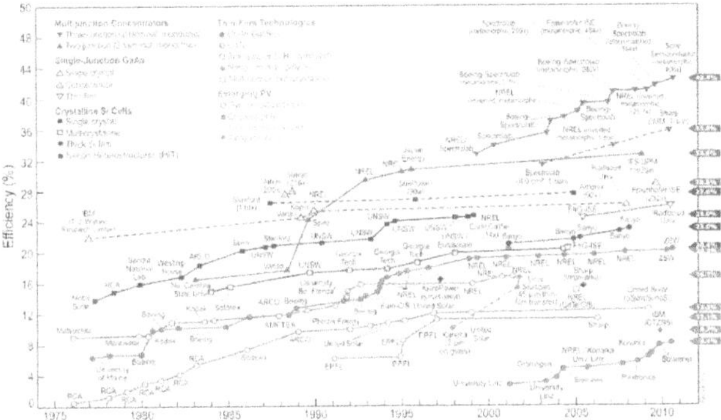

Fig. 3 Efficiency of different types of solar cells

For the purpose of UAVs, Spectrolab NeXt Triple Junction (XTJ) solar cell may be chosen as it has an efficiency of 29.5% and lighter than other solar cells.

B. Rechargeable batteries

The energy collected from the solar cells during day-time may be stored in the batteries and may be utilized for night flight. The various types of batteries are Lead-Acid, Lithium-ion, Nickel-Cadmium, and Lithium-Polymer. The Lithium-Polymer batteries are used in small flights as they are lighter in weight than other batteries. The other rechargeable batteries are two heavy and have low energy density, so they are not preferred for UAVs.

C. Fuel Cell

A fuel cell is one which can generate electrical energy from chemical energy. The fuel cell has higher energy densities and has almost 25% - 50% lighter than batteries. The energy density of H_2-O_2 fuel cell can range between 0.45 to 0.65kWh/kg. Since hydrogen is not readily available in nature, it should be extracted from its compounds and it has explosive characteristics. On the other hand, the solid oxide fuel cell does not require hydrogen to operate. The solid oxide fuel cells are more efficient as the temperature raises and they operate with natural gas as a fuel. Therefore, solid oxide fuel cells may be preferred as power source in UAVs.

Selection of BLDC Motor

Among the various types of motors, brushless dc motors can be chosen for electric propulsion system to convert electrical energy into mechanical energy. The brushless dc motors are preferred due to the following advantages: excellent speed-torque characteristics, sparkles and noiseless operation, less maintenance, high efficiency, high speed range, less in weight and small in size.

The required motor size is a function of the amount of payload weight the aircraft must carry. Selection of a motor then depends on choosing the proper "input watts per kg" rating provided by motor manufacturers. Therefore the first step in the motor selection process was estimating the maximum weight of the UAV. When fully assembled the aircraft weights 0.86kg. Since autopilots and other flight electronics are generally very small and light weight, the total weight was estimated at 0.23kg. The vision payload was estimated to weigh less than 0.68kg. The most significant weight component is the batteries. To provide a flight time of up to 60 minutes it was estimated that a battery weight of 1.77kg was required. This brought the entire system weight to 2.63kg. Using the watts per kg chart provided by E-Flite, a value of 580 watt power input was selected [5]. The closest available motor was an 850 watt BLDC motor. This would keep the performance of the aircraft in the trainer category but provide sufficient power for the flight. This motor would weigh 0.18 kg and require three 4-cell Lithium Polymer (LiPo) battery packs.

Controllers

The controllers are required to improve the acceleration and deceleration characteristics of BLDC motor drive. The drawbacks of the present BLDC motor drives used in the UAVs, are its speed cannot be accurately controlled, poor acceleration and deceleration characteristics. Therefore, the performance of UAVs has to be improved by providing suitable controller with the BLDC motor drive. The various types of controllers are PID, Fuzzy and ANN based controller. These controllers are briefly discussed in the following sections.

A. PID controller

The conventional PID controllers are used in most of the industrial applications. These controllers need exact mathematical model of the overall system to determine the controller gains such as proportional gain, integral time and derivative time. The PID controller based BLDC motor drive can provide better transient and steady-state response as long as the parameters of the system remain unchanged. The general control law for the PID controllers is given by,

$$u(t) = K_P\left(e(t) + (1/T_i)\int e(t)dt + T_d\, de(t)/dt\right) \qquad .....(1)$$

Where, $u(t)$ is the continuous control signal, K_P is the proportional gain, T_i is the integral time constant, T_d is the derivative time constant, and $e(t)$ is the error signal.

The corresponding discrete equation for the control signal [12] can be written as,

$$u(k) = u(k-1) + K_1 * e(k) + K_2 * e(k-1) + K_3 * e(k-2) \qquad .....(2)$$

Where $u(k-1)$ is the previous control output, $e(k-1)$ is the previous error, and $e(k-2)$ is the error preceding $e(k-1)$. The constants K_1, K_2 and K_3 are given by,

$$K_1 = K_P + TK_i / 2 + K_d / T \qquad .....(3)$$

$$K_2 = -K_P - 2K_d / T + TK_i / 2 \qquad .....(4)$$

$$K_3 = K_d / T \qquad .....(5)$$

$$K_i = K_P / T_i \qquad .....(6)$$

$$K_d = K_P T_d \qquad .....(7)$$

$$T = 1 / f \qquad .....(8)$$

Where f is the sampling frequency and T is the sampling rate.

The popular tuning methods such as Ziegler-Nichols method and Genetic algorithm based tuning method can be used to obtain the controller gains of the PID controller. The PID controller can be implemented using inexpensive digital controllers such as microcontrollers or DSP processors.

B. Fuzzy Controller

There has been a significant and growing interest in the application of artificial intelligence (AI) type control techniques such as neural network and fuzzy logic to control the complex, non-linear systems. To design a fuzzy controller, we need to understand the general behaviour of the system, and describe the general behaviour of the system in a linguistic manner by forming IF-THEN rules in the form of statements. The great challenge is to design and implement the fuzzy controller quickly by framing minimum number of rules based on the knowledge of the system. The general fuzzy logic controller consists of four parts as illustrated in Fig.4. They are fuzzification, fuzzy rule-base, fuzzy inference engine and defuzzification.

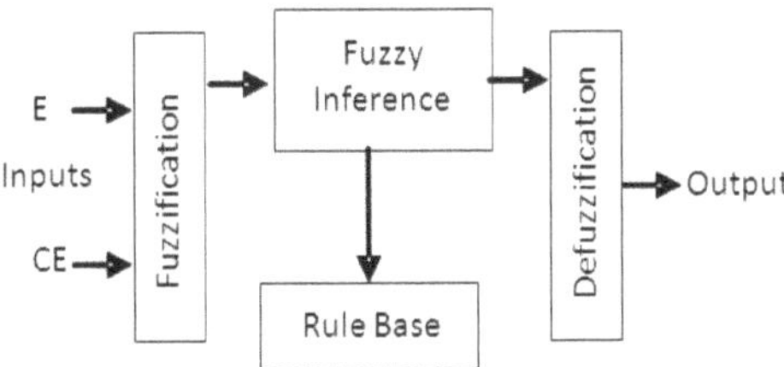

Fig. 4. Block diagram of fuzzy inference system

The design steps are as follows.

Step 1: Define inputs, outputs and universe of discourse

In this step, the inputs and outputs for the fuzzy controller are indentified and their universe of discourse is defined.

Step 2: Defining fuzzy membership functions and rules

In this step, the membership functions are chosen for the input and output variables of the fuzzy controller. Fuzzy membership functions are used as tools to convert crisp values to linguistic terms. In order to define fuzzy membership function, the designer can choose many different shapes based on their preference and experience. The popular shapes are triangular and trapezoidal because these shapes are easy to represent designer's ideas and they require less computation time. Fuzzy controller uses fuzzy rules to make a decision and generate the control action. The rules are in the form of IF-THEN statements. The number of rules to be used to describe the system behavior is entirely based on the designer's experience and the previous knowledge of the system. The performance of the controller can be improved by adjusting the membership function and rules. Finally the fuzzy output is converted into real value output i.e. crisp output by the process called defuzzification. Even though many defuzzification methods are available, the most preferred one is centroid method because this method can easily be implemented and requires less computation time when implemented in digital control systems using microcontrollers or digital signal processors (DSPs). The formula for the centroid defuzzification method is given by,

$$z = \sum_{x=1}^{n} \mu(x)x / \sum_{x=1}^{n} \mu(x) \qquad \ldots (9)$$

where z is the defuzzified value, $\mu(x)$ is the membership value of member x. This crisp value is used to control the average voltage applied across the phase windings, hence the speed of the motor.

C. ANN Controller

Artificial Neural Network has been applied successfully to a wide range of control system applications in recent years. Artificial neural networks have high learning and nonlinear mapping essences and its parallel and distributed structure can provide a nonlinear mapping between inputs and outputs of an electric drive system, without the knowledge of any predetermined model. The structure of the ANN Controller based BLDC drive system is shown in Fig.5.

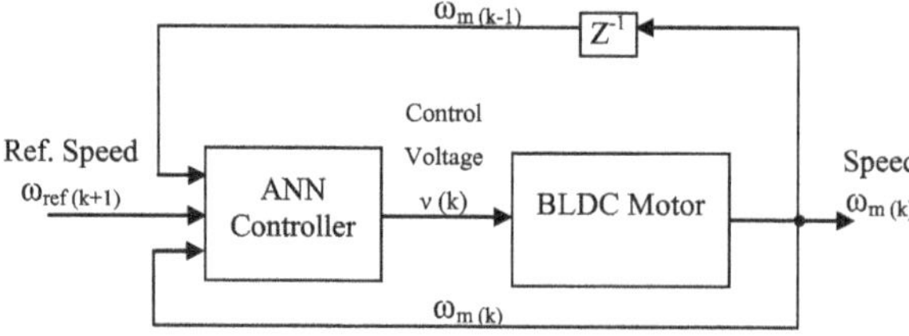

Fig. 5. Control structure of ANN controller based BLDC motor drive

The first and foremost step in implementation of ANN controller is the selection of reference model. Selection of the reference model dynamics is a very important part of the adaptive system design procedure that ultimately defines the desired system behavior. A model that gives the desired transient response specifications is usually taken as the reference model. The control voltage of the BLDC motor drive is a non-linear function of three samples of speed and it is given by,

$$v(k) = f(\omega_{ref}(k+1), \omega_m(k), \omega_m(k-1)) \qquad \dots\dots(10)$$

where $\omega_{ref}(k+1)$ is the reference speed, $\omega_m(k-1)$ is the previous speed, $\omega_m(k)$ is the present speed and $v(k)$ is the control voltage of the BLDC motor. ANN can be trained with three speed inputs $\omega_{ref}(k+1)$, $\omega_m(k)$ and $\omega_m(k-1)$ and the corresponding output control voltage $v(k)$ to build inverse dynamic model of the nonlinear BLDC motor drive system. The ANN is trained to emulate the characteristics of the reference model. The ANN is made to learn this non-linearity between control voltage and consecutive speeds. The trained ANN controller generates control voltage $v(k)$ for the BLDC motor based on the three consecutive speed inputs $\omega_{ref}(k+1)$, $\omega_m(k)$ and $\omega_m(k-1)$ to control the voltage applied to the BLDC motor.

Conclusions

The need for electrical propulsion systems in UAVs is emphasized in this paper. The selection of electrical components of UAVs such as power source, motor and power drive circuit is discussed. Moreover, the need for controllers and the various types of controllers used for BLDC motor drives are briefly discussed. The information provided in this paper will be useful for the design of UAVs with electrical propulsion systems to suit the application requirements.

References

1.   Castillo, P., A. Dzul et and R. Lozano. 2004. "Real-time stabilization and tracking of a four rotor mini rotorcraft", IEEE Transactions on Control Systems Technology, Vol. 12, No. 4, pp.510-516.

2.   Castillo, P., R. Lozano and A. Dzul. 2005. "Modeling and Control of Mini-Flying Machines", Springer-Verlag. London.

3.   Emadi, A. 2005. "Handbook of automotive power electronics and motor drives", Ed. Florida, United States of America: CRC Press.

4.   Escare no, J., S. Salazar-Cruz and R. Lozano. 2006. "Embedded control of a four-rotor UAV", in Proc. American Control Conference, Minnesota, pp.3936-3941.

5.   E-flite motor power selection chart, Horizon Hobbies,

     http://www.horizonhobby.com/ProdInfo/Files/EFLDeterminingPowerRequirement.pdf

6.   Gieras, J. F. and M. Wing 2002. "Permanent Magnet Motor Technology – Design and Applications", Second Edition, ISBN 0-8247-0739-7.

7.   Hanselman, D. C. 1994. "Brushless permanent-magnet motor design", H. B. Crawford, United States of America: McGraw-Hill.

8.   Harmon, F. G., A. A. Frank, and J. Chattot, 2006. "Conceptual Design and Simulation of a Small Hybrid-Electric Unmanned Aerial Vehicle", Journal of Aircraft, Vol. 43, No. 5, Sept.– Oct. 2006, pp. 1490-1498.

9. Herwerth, U., C. Ofoma, S. Wu, Matsuyama and S. Clark. 2006. "Development of a fuel cell powered UAV for environmental research", in Proc. 44th AIAA Aerospace Sciences Meeting and Exhibit, Reno, Nevada, Jan. 2006, pp. 2006–237.

10. Karunarathne, Lakmal, Economou, T. John, Knowles, and Kevin. 2010. "Model based power and energy management system for PEM fuel cell/ Li-Ion battery driven propulsion system", 5th Int. Conf. on Power Electronics, Machines and Drives, pp. 1-6.

11. Rubaai, A., R. Kotaru, and M.D. Kankam. 1997. "A Real-Time Neural Network Based Controller for Brushless DC Motor Drives", IEEE Trans. on Industry Applications Conference, IAS'97, October 5-9, New Orleans, LA, USA. pp. 828-835.

12. Shanmugasundram R., K. M. Zakariah, and N. Yadaiah. 2012. "Modelling, simulation and analysis of controllers for brushless direct current motor drives", Journal of Vibration and Control. pp. 1-15.

13. Tipsuwanporn, V., W. Piyarat, and C. Tarasantisuk. 2002. "Identification and Control of Brushless DC Motors Using On-line Trained Artificial Neural Networks", Proc. of the Power Conversion Conference PCC'2002, pp.1290–1294.

14. Yeadon, W. H. 2001. "Handbook of small electric motors", United States of America: McGraw-Hill.

15. Zhang, H., C. Saudemont, B. Robyns and M. Petit. 2008. "Comparison of Technical Features between a More Electric Aircraft and a Hybrid Electric Vehicle", IEEE Vehicle Power and Propulsion Conference (VPPC), Harbin, China.

CURRENT TRENDS IN STATE ESTIMATION FOR AUTOMOUS MICRO AIR VEHICLES

G.Krishnaiah

Professor of ECE, Joginpally B.R.Engineering College, Hyderabad, INDIA

S.Vathsal

HOD (EEE) & Dean (R&D), JBIET, Moinabad, Hyderabad-75, INDIA

D.N.Rao

Principal & Professor of ECE, Joginpally B.R.Engineering College, Hyderabad, INDIA

ABSTRACT

A Micro air vehicle (MAV) is limited to a maximum dimension of 15cm, gross weight of 100 g, with up to 20gm payload and the Reynolds number must be below 10^6. The size limitation of MAV for international competitions is 100 cm for outdoor mission and 70/80 cm for indoor mission with maximum weight of 1 Kg. In MAV, the dimension of 15cm is very critical because it is considered border line between bird flight and insect flight. Flight control algorithms used for MAVs have been based on radio controller (R/C) or Tele operation Guidance navigation and control GNC has been an important area in MAV research. To cope with size on board sensors are MEMS gyro, MEMS accelerometer magnetometer and GPS.These sensors are needed for state estimation of MAV. In this paper the problem of modelling and selection of estimation algorithms are addressed. Different parameterizations are discussed for attitude. Representation of quaternion pameterization is best suited for getting good navigation accuracy. Due to limitation of GPS at low altitudes and indoors, vision based state estimation for autonomous micro air vehicles is proposed. This techniques poses challenges in diverse fields such as image processing, trajectory planning, control theory and micro hardware design. This paper explores one aspect of the overall problem which is robust real time estimation of craft states from a set of tracked feature points. This vision based approach is realized using an implicit extended Kalman filter. Superior overall results can be achieved using higher –fidelity dynamic models using quaternion errors in the Kalman filter. Further research directions are presented for using unscented Kalman filter, Particle filter etc. which are the current research in the field of non linear filtering.

KEY WORDS: Micro Electro Mechanical System (MEMS), Micro Air Vehicle (MAV), Measurement Covariance Matrix, Process Covariance Matrix, Global positioning system, Pseudolite, attitude estimation, kalman filter

Nomenclature:

P_N	Inertial North Position of MAV
P_E	Inertial East Position of MAV
W_N	Wind from North
W_E	Wind from East

V_{air} Total Airspeed
P Angular Rate about x-axis
Q Angular Rate about y-axis
R Angular Rate about z-axis
$\Box$ Roll Angle
Θ Pitch Angl
ψ Yaw Angle

INTRODUCTON

Micro Air Vehicles are the type of aircrafts which are very small in size and dimensions as compared to others. Some of these vehicles have a wing span smaller than 6 inches and can carry payloads that are measured in grams. Substantial progress has been made in the fabrication, structural design and development of Radio controller (R/C) that use vision –based, horizon-tracking algorithms [1-3].

The task of flying autonomous MAVs in environments confined by buildings and trees with a additional features from ground vehicles and civilians as well as poor weather requires a host of innovations in vision – based flight control.

In developing these airborne entities ,the primary concerns are the size sensors,communication and the computational power[4].In general MAVs are small usually less than 15 cm in length and 100 grams of height, autonomous aircrafts which have a great potential in fact-finding missions in confined spaces ,inside buildings including both civilian search and connaissance missions.
Inertial Navigation (INS) integrated with other Navigation schemes such as Global Position System(GPS) ,Pseudolite   systems have gained significance   due to its enhanced performance [5].in this paper GPS/pseudolite and INS/GPS are integrated for improved accuracy[6].

BACKGROUND

Navigation [7] is very interesting and important area for the automation of any micro air vehicle. Different   types of schemes are in aircraft navigation including Inertial Navigation System(INS),Global Positioning system(GPS) and Pseudolite (PS) systems for providing the best estimation of attitude of the MAVs in terms of height ,longitude ,latitude ,roll, pitch and yaw[8].
The Inertial Navigation System(INS) itself calculates attitude ,velocity and position of the MAV ,but introduces large errors over a period of time due to sensor ,gyro, accelerometer drifts and biases .therefore GPS/Pseudolite is required to aid the INS using an extended kalman Filter[EKF][9] that helps in the estimation of errors in INS and more accurate positioning is updated.

KALMAN FILTER EQUATION

We proceed to derive the Kalman filter equations as the best linear estimator for linear, non-Gaussian model. We slightly generalize the model by allowing correlation between the state noise and measurement noise. Thus, we consider the model

$$x_{k+} = F_k \, x_k + G_k \, w_k \, , \quad k \geq 0$$
$$z_k = H_k \, x_x + v_k \, ,$$

With $[w_k; v_k]$ a zero-mean white noise sequence covariance

DIFFERENT DAT A FUSION SCHEMES

this paper focuses on problem of autonomous ,vision –based flight for MAVs and vision –based state estimation .specifically this paper explores state estimation algorithms that synthesize vision –based measurements ,in the form of tracked feature points provided by the image processing with a vehicle dynamic model using a variety of kalman filter like EKF and Particle Filter.INS/GPS based state estimation is very important for navigation of MAVs.Another important scheme is the integration of GPS and Pseudolite system based state estimation.

DIFFICIENCIES IN EXISTING SCHMES

The INS itself can calculate the position of the micro-air vehicle without the aid of anything from outside, but a large range of errors are introduced by sensors leading to drift in the sensors output. Therefore GPS is required to aid the INS using an extended Kalman Filter(EKF) for improved accuracy. Although GPS is a useful source of navigation data it can subject to dropout in dense urban environments. Therefore vision – based state estimation augmented with GPS is capable of operating with the benefits of GPS updates for extended period of time. The main assumptions made are the Linear System and measurement model corrupted by additive Gaussian noise but in practise we find Non-linear system and measurement models with Non-Gaussian noise distributions.

PROPOSED SCHEMES USING NON-LINEAR FILTERS

Mathematical modelling of INS/GPS based state estimation with quaternion Parameterization. In the Single Seven Stage Discrete Time Extended Kalman Filter [18], the state equations which relate body frame rotations to changes in roll, pitch and heading are nonlinear. Letting the states be roll angle and pitch angle, ϕ and θ, and letting angular rates p, q and r and Airspeed Vair be inputs

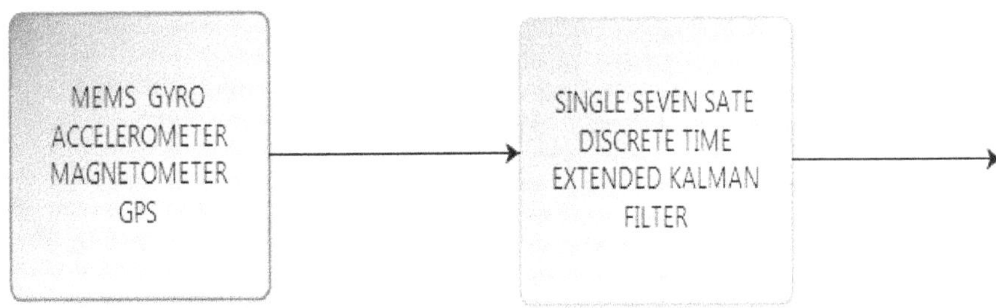

Fig1: Single Seven State Extended Kalman Filter Scheme [19],

Now consider Two Stage Cascaded Extended Kalman Filter Scheme [16].As the heading update equation uses the same inputs as the Pitch and Roll equations, it is not unnatural to lump them together in the same estimation block. The exclusiveness of the heading state lies in the output equations. There is not a sensor output equation which will relate heading to accelerometer readings, which is why it was convenient to split heading estimation into it's own stage as mentioned earlier. One of the merits of heading and attitude at the same time is that magnetometer information may be beneficial in the estimation of pitch and roll, since no maneuver will upset the earth's magnetic field, as they may the accelerometer readings. And, depending on the attitude and heading of the MAV, projecting pitch and roll onto the magnetic field vector may refine the pitch and roll estimates.

In three stage Cascaded Extended Kalman Filters [16] work independently, each imparting the information that it estimates to the stage below. This three stage filter assumes the least coupling.

Vision –based Kalman filter

The estimation of aircraft states is achieved using an implementation of the extended Kalman filter [5]. The standard Kalman filter employs a linear dynamic model to propagate the states and the state covariance matrix. When considering nonlinear models, a common practice is to linearize the equations of motion about a nominal trajectory, or trim condition. The accuracy of such an approximation tends to deteriorate over time, however, as the actual trajectory often deviates substantially from the nominal one. The extended Kalman filter was developed as an alternative approach whereby the equations of motion are linearized about the current state estimate. Therefore, a linearized model is computed at each time step.The resulting linearized state transition matrix is then used for propagating the state covariance matrix. The original nonlinear equations of motion are typically used for state propagation. The standard implementation of the extended Kalman filter uses measurements that are explicit functions of the states. The approach taken in this paper is to use the positions of tracked feature points to update the state estimates, resulting in measurements that are implicit functions of the aircraft states, the tracked feature points, and the camera parameters. Soatto et al. [7] and Soatto and Perona [8] employed two distinct implicit measurement constraints that can be incorporated into the extended Kalman filter: the epipolar constraint and the subspace constraint. The resulting Kalman filtering algorithm has been termed the implicit extended Kalman filter.

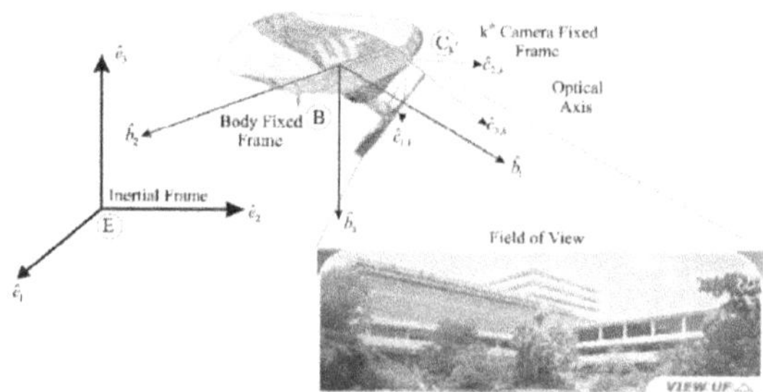

Fig. 2   Kinematics of the vehicle, camera, and field of view.

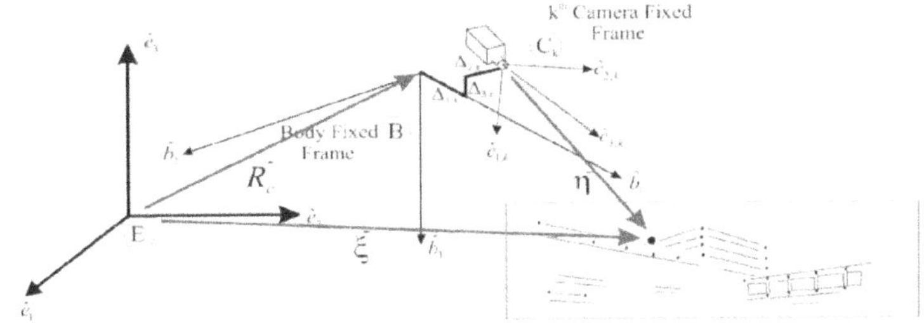

Fig. 3 Aircraft, camera, and feature point geometry.

Integration of GPS and pseudolite systems

When the GPS signal dropouts due to dense environmental areas like huge building structures ,hill areas ,underground cellars ,deep open cut areas (coal mines),and underground water , the so called pseudolite system signals can be used to obtain the state estimation of the MAVs with aid of the Global positioning system. The attitude, velocity and position are estimated accurately as alignment errors/Tracking errors; position and Attitude errors are minimized   as shown in below figure.

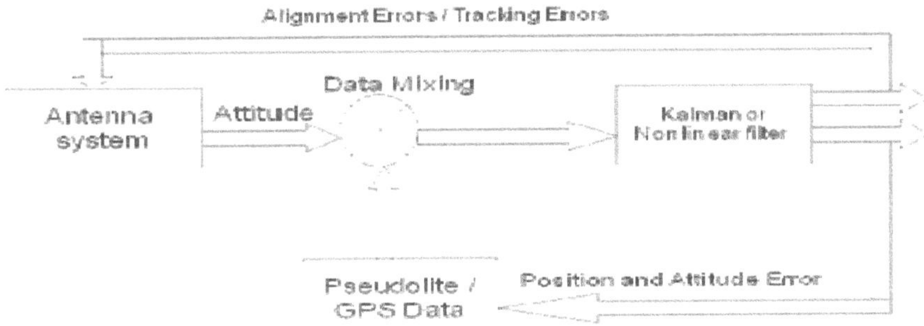

Fig. 4 Integration of GPS and Pseudolite signals for precision attitude estimation of MAVs.

DEVELOPMENTAL PLAN.

The fusion of GPS and visual data for optimal state estimation is a future developmental plan. Integration of MEMS and GPS/Pseudolite systems for better estimation of states of the MAVs is a topic for future research and is not treated in this Paper.

CONCLUSIONS

This paper has studied the performance of a vision –based kalman filter for state estimation for MAVs .the estimation is based on the epipolar constraint by implementing the IEKF.Because the kalman filter incorporates a dynamic model, it will produce state estimates regardless of the number of available feature point.

The MAVs are playing a significant role in military surveillance and reconnaissance and civilian search and rescue .Kalman filter is one of the techniques that can be used for the Navigation of MAV.

REFERENCES

[1]   James M. McMichael (Program Manager Defense Advanced Research Projects Agency) and Col. Michael S. Francis, USAF (Ret.) (Defense Airborne Reconnaissance Office), *"Micro air vehicles - Toward a new dimension in flight"* dated 8/7/97

[2]   INTEGRATION OF GPS/INS/VISION SENSORS TO NAVIGATE UNMANNED AERIAL VEHICLES" by Jinling Wang [a,*], Matthew Garratt [b], Andrew Lambert [c], Jack Jianguo Wang [a], Songlai Han[a], David Sinclair[d] · [a]School of Surveying & Spatial Information Systems, University of New South Wales, NSW2052, Australia(Jinling.Wang@unsw.edu.au). [b]School of Aerospace, Civil and Mechanical Engineering, [c]School of Information Technology and Electrical Engineering, Australia Defence Force Academy, Canberra, [d] QASCO Surveys Pty. Limited, 41 Boundary St. South Brisbane, Qld, 4101, Australia.

[3]   http://www.ornithopter.net/history_e.html.

[4]   James M. McMichael (Program Manager Defense Advanced Research Projects Agency) and Col. Michael S. Francis, USAF (Ret.) (Defense Airborne Reconnaissance Office), "Micro air vehicles - Toward a new dimension in flight" dated 8/7/97

[5]   "Ground control station development for autonomous UAV "by Ye Hong, Jiancheng Fang, and Ye Tao. Key Laboratory of Fundamental Science for National Defense, Novel Inertial Instrument & Navigation System Technology, Beijing, 100191, China

[6]   Schmidt, G.T., \Strapdown Inertial Systems - Theory and Applications,"AGARD Lecture Series, No. 95, 1978.

[7]   Grewal, M.S., Weill, L.R., and Andrews, A.P., Global Positioning Systems, Inertial Navigation, and Integration, John Wiley and Sons, New York, 2001.

[8]   "Integration of MEMS inertial sensor-based GNC of a UAV" by Z. J. Huang and J. C. Fang (huangzhongjun@buaa.edu.cn, fangjiancheng@buaa.edu.cn) from School of Instrumentation & Optoelectronics Engineering Beihang University, Beijing 100083, China. International Journal of Innovative

[9]   Randle, S.J., Horton, M.A., \ Low Cost Navigation Using Micro – Machined Technology," IEEE Intelligen Transportation Systems Conference, 1997.

[10]   Grejner-Brzezinska, D.A., and Wang, J., \Gravity Modelling for High-Accuracy GPS/INS Integration," Navigation, Vol. 45, No. 3, 1998, pp. 209-220.

[11]   Wolf, R., Eissfeller, B., Hein, G.W., \ A Kalman Filter for the Integration of a Low Cost INS and an attitude   GPS," Institute of Geodesy and Navigation, Munich, Germany.

[12]   Gaylor, D., Lightsey, E.G, \ GPS/INS Kalman Filter desing for Spacecraft operating in the proximity of teInternational Space Station," University of Texas - Austin, Austin.

[13] Optimal state estimation Kalman,$^{H}\Box$ and nonlinear approaches" by Dan Simon, Cleveland State.niversity. A John Wiley &Sons, INC., Publication.

[14] Elbert Hendricks, Ole Jannerup, Paul Haase Sorensen, "Linear Systems Control" Deterministic and Stochastic Method, ISBN: 978-3-540-78485-2, Library of Congress Control Number: 208927517, 2008 Springer-Verlag Berlin Heidelberg.

[15] 'Mathematical Modeling of INS/GPS Based Navigation System Using Discrete Time Extended Kalman Filter Schemes for Flapping Micro Air Vehicle' by Sadia Riaz in International Journal of Micro Air Vehicle, March 2011

[16]"State estimation for micro air vehicles" by Randal W. Beard, Department of Electrical and Computer Engineering, Brigham Young University, Provo, Utah. Studies in Computational Intelligence (SCI) 70, 173–199 (2007). Springer-Verlag Berlin Heidelberg 2007

[17] Moore, J.B., Qi, H., \Direct Kalman Filtering Approach for GPS/INS Integration", IEEE Transactions on Aerospace and Electronic Systems , Vol 38, No.2, April 2002.

[18] 'Single Seven State Discrete Time Extended Kalman Filter for Micro Air Vehicle' by Dr. Afzaal M. Malik1 and Sadia Riaz2, World Congress of Engineering (WCE), International Conference of Mechanical Engineering (ICME-2010).

[19] Ettinger, S., Nechyba, M., Ifju, P., and Waszak, M., "Vision Guided Flight Stability and Control for Micro Air Vehicles," IEEE International Conference on Intelligent Robots and Systems, IEEE, Piscataway, NJ, Oct. 2002, Vol. 3, pp. 2134–2140.

[20] Grewal, M., and Andrews, A., Kalman Filtering: Theory and Practice,Prentice–Hall, Upper Saddle River, NJ, 1993.\

[21] Webb, T., Prazenica, R., Kurdila, A., and Lind, R., "Vision-Based State Estimation for Uninhabited Aerial Vehicles," AIAA Paper 2005-5869,2005.

VISION–GUIDED ADVANCED NAVIGATION SYSTEMS FOR MICRO AIR VEHICLES

Srinivas Viswanath V
Research scholar, VIT University, Tamilnadu, India.

ABSTRACT

Ongoing progress and research has been developing towards designing, building and test-flying remotely piloted Micro Air Vehicles (MAVs) and small UAVs. We seek to appreciate this progress in overcoming the aerodynamic obstacles to flight at very small scales with a vision-guided flight navigation and autonomy system, based on a well designed location detection algorithm. In this paper, we first motivate the use of computer vision and Machine vision , some of the artificial intelligence, Wireless connectivity for the navigation system defence measures for MAV development, competing that given current sensor technology, vision may be the only practical approach to the problem. We then describe our statistical vision-based location detection, which has been demonstrated at 25Hz with over 99.9% correct horizon identification. Next, we develop ideological schemes for the detection of extreme MAV attitudes, where no horizon is visible, and for the detection of location estimation errors, due to essential factors such as video transmission noise. Lastly, determine the scope and range of the navigation for security measures for the controller for self-stabilized flight, and report results on vision-based automatic flights of duration exceeding 10-15 minutes. We conclude with an overview of our on-going and future MAV-related research. These estimates are then identified by a nonlinear controller for achieving various navigation tasks such as take-off, landing, hovering, target tracking, etc.

KEY WORDS: Micro Air Vehicles (MAV), Navigation, Artificial Intelligence (AI), Controller

INTRODUCTION

From the early beginning the human always trying to improve the Micro Air vehicles(MAV) which are autonomous and useful in the precision of estimating the states i.e., the design ,position and 3D formulation of surrounding environment. However the design for the portable and low altitude flying machine for aerial surveillance of designing and testing the aircraft whose altitude range are small and in other words in the flying range of small birds which satisfy the aerodynamic laws.

The MAV equipped with video cameras and the transmitters are much enough for data capturing and have high potential for having surveillance and monitoring tasks. Operational MAV is much suitable to carry the tasks which are even very dangerous for human such as chemical radiation for spill monitoring, aerial survey for volcanic explosion, visual monitoring for natural disasters . Defence Research and Development Organisation (DRDO), has successfully designed and developed many versatile Unmanned Air Vehicles (UAV) systems that have been inducted into the Indian Armed Forces. Aeronautical Development Establishment (ADE), Bangalore, the major aeronautical systems research laboratory of the DRDO involved in practically all major aspect of aeronautical research, design, and development relevant to Military aviation—is in the forefront, as the nodal agency, in the development of UAVs for the Services. Now a days the Artificial Intelligence based sensors and GPS navigation plays a key role in Aviation sector. Wide range of commercialisation is possible in coming years. Key issue in the aviation is safe landing and the tracking of the miniature vehicles.

Vision Guided Advanced Navigation System Architecture

The main aim of the architecture is of Horizon detection. The low moments of inertia of MAVs make them vulnerable to rapid angular accelerations; a problem further complicated by the fact that aerodynamic damping of angular rates decreases with a reduction in wingspan. Within a few decades, these systems have evolved from performing a single role/mission to performing multiple missions like surveillance, monitoring, acquiring,

Tracking and destruction of target with the use of advanced technologies. The biggest advantage of UAVs, however, is that there is no risk to human lives. Unmanned platforms are the emerging lethal and non-lethal weapons of choice and have transformed the way the armed forces now prosecute operations. a vehicle is most important since the operation is continuous and more over the wide range of application for the simulation of the artificial intelligence and the navigation based system for the effective condition of the longitude and the latitude selection. The ideal design of MAV is operating very heuristically and simple navigation like a robust one. Clearly we can determine the path which enables the armed forces also very curiously and sincerely for the reporting of tasks time to time. A wide range of analysis can be the navigation data and target resource be scheduled

 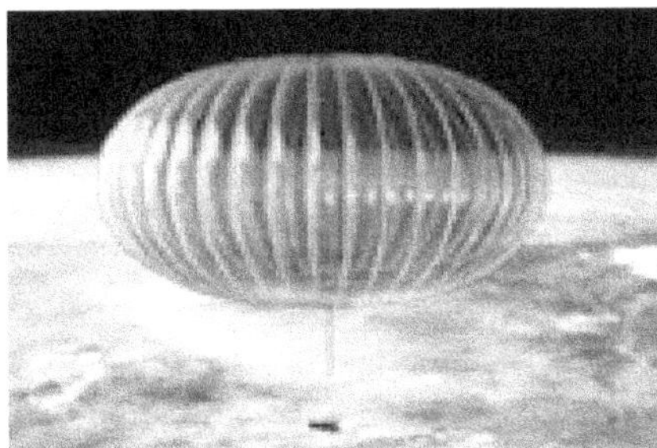

Fig 1.1 (a) and (b) University of Minnesota developed the prediction of parachute opening dynamics

Bio concept for the design of new MAV

Birds, the biological counterpart of mechanical MAVs, can offer some important insights into how one may best be able to overcome these problems. In studying the basic system of the birds we can observe the verticality of the fling aspects for the different variety of birds. Birds generally have large eyes with sharp vision to guide themselves to have vision enhancement of their behavior. The biological concepts always very simple while forceful evidence of the importance of vision in flight, do not, however, in and of themselves warrant a computer-vision has been developed for the

Vision based approach to MAV autonomy. More over the more critical aspects such as the technologies used in rate and acceleration sensors on larger aircraft are not currently available at the MAV scale. It has proven very difficult, if not impossible, to scale these technologies down to meet the very low payload requirements of MAVs. While a number of sensor technologies have came to existence in small enough packages to be used in MAV systems, these small sensors have contributed accuracy for reduced size and weight. Take, for example, MEMS (Micro Electro-Mechanical Systems) rate gyros and accelerometers. MEMs piezoelectric gyros, while only weighing approximately one gram, have drift rates on the order of 100_ per minute and are highly sensitive to changes in temperature.

Algorithm for Horizon Detection

The proposed vision based Navigation system determines the vehicle's relative velocity and 3D position with respect to the initial aircraft location or some target which may be stationary moving. It relies on tracking features appearing in a "target template" initially chosen at the picture center. With the computed template location (xi, yi) in the picture frame, the relative distance (Xi, Yi) between the rotorcraft and the ground objects appearing in the picture template "i" is estimated after compensating the rotation effects using IMU data and recovering the range Z (or height) using an adaptive algorithm. The rotorcraft horizontal motion (X, Y) in the inertial frame is then estimated by accumulating or summing the relative distances (Xi, Yi) for $i = 1...n$ as shown in Figure 2.In this section, we describe the different components of our system which are: 1) picture processing algorithm; 2) rotation effects compensation; 3) flight path integration; 4) range (height) sensing; and 5) velocity and position estimation.

Integrated Digital Flight Control Processor for Micro Air Vehicles:

Integrated digital flight control processor (IDFCP) is a full-advanced compact simplex hardware that caters to the computational and various interface requirements with other sub-systems. The integrated package performs flight control and mission navigation functions, telemetry and ranging functions. The package has provision for GPS, Micro electro mechanical systems (MEMS) sensors, and other real-world interfaces also.

The flight control functionality is performed by a 32-bit PowerPC (MPC 555) operating @40 MHz The telemetry functionality is performed by a 32-bit Power PC (MPC 860) @66 MHz Autonomous handlers have been provided for GPS, heading. The package provides for fault logging through non-volatile random access memory (NVRAM),built-in-self-test (BIT) and power-on self- test (POST) features.

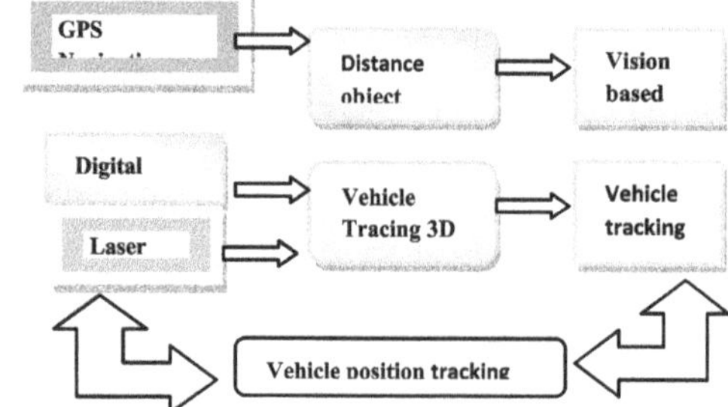

Fig 2.1 Basic idea for Vision Based Navigation

Basic Principle of working Micro Air Vehicle

The simplest design is an MAV that can remain within the line of sight of a small base station that tracks the vehicle, maintains the communications link, and performs navigation calculations. A vehicle that flies behind buildings or hills—beyond the line of sight—must depend on some other approach to communications and needs an independent means of navigation. One figuration that meets these requirements stores data on board with later readout when the vehicle returns to line of sight. Another configuration includes an overhead communications relay. Without a line of sight for navigation, alternative navigation approaches such as dead reckoning, inertial navigation, and the Global Positioning System (GPS) might be tapped, with the latter two depending on the availability of small components.

Requirements for Autonomous Vehicle operation

- Mobility and agility are fundamental to autonomous vehicles
    - negotiating complex terrain environments
    - handle difficult operational conditions (i.e. atmospheric disturbances)
    - making full use of vehicle dynamic capabilities
- Involves interplay between lower-level flight control and higher-level trajectory planning
- Requires technique that integrate local and global scales\

Reducing trajectory operation for easy tracking of MAV

In the beginning, an picture area of 50×50 pixels, which can be considered as a "target template", is initially chosen at the picture center. About 20 features are then selected automatically in that template using the Shi-Tomasi [13] algorithm. These features are then tracked in the successive pictures using the pyramidal Lucas-Kanade algorithm [14]. The outputs of this tracker are the features positions in the picture frame. We have slightly modified that algorithm in order to provide also estimates about the optic flow at each feature location. The position (x_i, y_i) and velocity $(\dot{x}_i, \dot{y}_i)$ of the target template "i" is simply computed by taking the mean of the tracked features positions and velocities (or optic flow). For accurate and robust picture template tracking, we have implemented simple routines that detect and handle features dispersion (features go out of the template) and erroneous feature correspondences which are mainly due to picture noise and large attitude changes. For example, new features are selected in the same template when the variance of features positions exceeds some threshold. As the rotorcraft moves, older features leave the camera FOV and new features enter the FOV. Therefore, a new target template (or new set of features) is selected at the picture center when the current template is about to go out of view.

Traffic Monitoring for MAV in GPS Navigation system:

In aeronautical applications, satellite positioning and timing services have long been an additional means of navigation. They provide supplementary service for many flight phases, in leisure flying as well as commercial air transport. In recent years, scheduled traffic has increased by about 4% per year worldwide, doubling the number of flights within 20 years. Higher navigation accuracy and service integrity are required to allow aircraft separation reduction, allowing for increase of traffic capacity.

All other aviation-related operations, like airport surface movement and guidance control, require precise assistance from air traffic controllers. Airports may have surface radar, but sometimes pilots report taxi movements manually and aircraft are managed using visual aids only. This has lead to severe accidents. Satellite navigation and GALILEO in particular will improve operations safety.

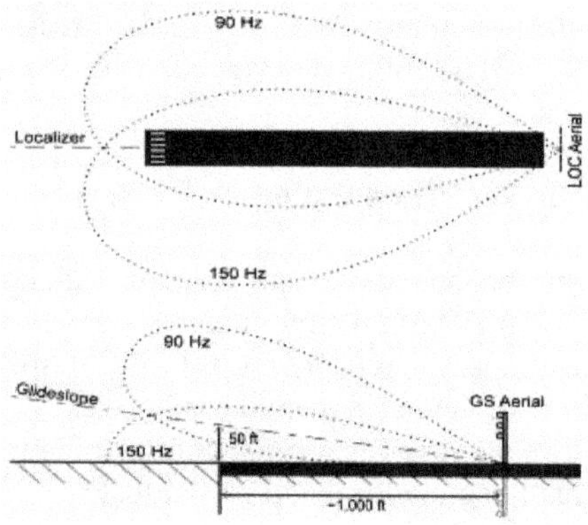

Fig 2.4.1: Instrument Landing system based Navigation Range

Instrument Landing System

An **instrument landing system (ILS)** is a ground-based instrument approach system that provides precision guidance to an aircraft approaching and landing on a runway, using a combination of radio signals and, in many cases, high-intensity lighting arrays to enable a safe landing during instrument meteorological conditions (IMC), such as low ceilings or reduced visibility due to fog, rain, or blowing snow.

Instrument approach procedure charts (or *approach plates*) are published for each ILS approach, providing pilots with the needed information to fly an ILS approach during instrument flight rules (IFR) operations, including the radio frequencies used by the ILS components and the minimum visibility requirements prescribed for the specific approach. The navigation system range for the proposed MAV and aerial antenna capability for receiving the data has been given in the suitable example. Localizer receiver on the aircraft measures difference in depth of modulation – when the difference is 0 the localizer is in line with the center line of the runway.

Localizer antenna located at the end of the runway – 2 signals transmitted on a carrier frequency between 108 MHz and 111.975 MHz .

WAAS Navigation Programs - Wide Area Augmentation System (WAAS)

WAAS is an extremely accurate navigation system developed for civil aviation. Before WAAS, the U.S. National Airspace System (NAS) did not have the potential to provide horizontal and vertical navigation for approach operations for all users at all locations. With WAAS, this capability is a reality. WAAS provides service for all classes of aircraft in all phases of flight - including en route navigation, airport departures, and airport arrivals. This includes vertically-guided landing approaches in instrument meteorological conditions at all qualified locations throughout the NAS.

Fig 3.2 WAAS Navigation for GPS satellite based systems

Integrity with our proposed systems

Integrity of a navigation system includes the ability to provide timely warnings when its signal is providing misleading data that could potentially create hazards. The WAAS specification requires the system detect errors in the GPS or WAAS network and notify users within 6.2 seconds. Certifying that WAAS is safe for instrument flight rules (IFR) (i.e. flying blind) requires proving there is only an extremely small probability that an error exceeding the requirements for accuracy will go undetected. Specifically, the probability is stated as 1×10^{-7}, and is equivalent to no more than 3 seconds of bad data per year. This provides integrity information equivalent to or better than Receiver Autonomous Integrity Monitoring (RAIM).[15]

Methodology

Bio-inspiration:
A new trend in the MAV community is to take inspiration from flying insects or birds to achieve unprecedented flight capabilities. Biological systems are not only interesting to MAV engineers for their use of unsteady aerodynamics with flapping wings; Symposiums bringing together biologists and aerial robots have been held in 2007[3] and some books[4] have recently updated.

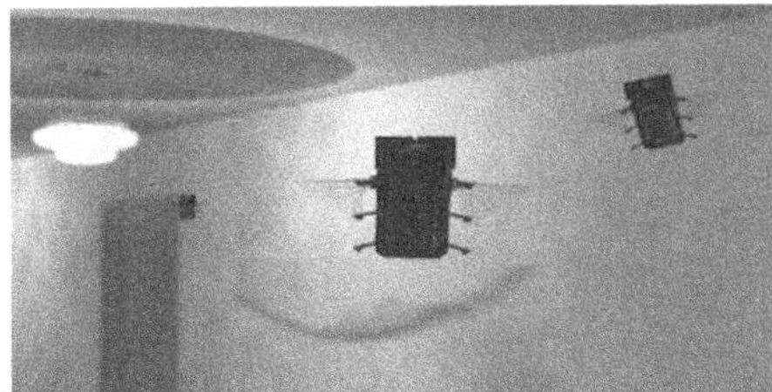

Fig 4.1 Concept based on Bionics

In January 2010, the Tamkang University (TKU) in Taiwan realized autonomous control of the flight altitude of an 8-gram, 20-centimeter wide, flapping-wing MAV. The MEMS Lab in the TKU has been developing MAVs for several years, and since 2007 the Space and Flight Dynamics (SFD) Lab has joined the research team for the development of autonomous flight of MAVs. Instead of traditional sensors and computational devices, which are too heavy for most MAVs, the SFD combined a stereo-vision system with a ground station to control the flight altitude,[6][7] making it the first flapping-wing MAV under 10 grams that realized autonomous flight.

Fig 4.2: Aerial picturistic view of the captured image

Terrestrial Image Acquisition

3D data acquisition and object reconstruction can be performed using stereo image pairs. Stereo photogrammetry or photogrammetric based on a block of overlapped images is the primary approach for 3D mapping and object reconstruction using 2D images. Close-range photogrammetric has also matured to the level where cameras or digital cameras can be used to capture the close-look images of objects, e.g., buildings, and reconstruct them using the very same theory as the aerial photogrammetry. An example of software which could do this is Vexcel FotoG [3] his software has now been replaced by Vexcel GeoSynth[4].

Using laser scans and images taken from ground level and a bird's-eye perspective, Fruh and Zakhor present an approach to automatically create textured 3D city models. This approach involves registering and merging the detailed facade models with a complementary airborne model. The airborne modeling process generates a half-meter resolution model with a bird's-eye view of the entire area, containing terrain profile and building tops. Ground-based modeling process results in a detailed model of the building facades. Using the DSM obtained from airborne laser scans, they localize the acquisition vehicle and register the ground-based facades to the airborne model by means of Monte Carlo localization (MCL). Finally, the two models are merged with different resolutions to obtain a 3D model.

Software used for airborne laser scanning includes OPALS (Orientation and Processing of Airborne Laser Scanning data), used for analyzing the data.

Role of Actuators and Sensors in GPS Navigation:

Actuators

Convert a command to a physical stimulus (e.g., heat, light, sound, pressure, magnetism, or other mechanical motion)

Sensors:
Capture physical stimulus (e.g., heat, light, sound, pressure, magnetism, or other mechanical motion) Typical generate a proportional electrical current A Bionic sensor plays a key role in identifying the Artificial intelligence techniques and wide applications from the nature based experimentation.

Equipment for image capturing

After the data has been collected, the acquired (and sometimes already processed) data from images or sensors needs to be reconstructed. This may be done in the same program or in some cases; the 3D data needs to be exported and imported into another program for further refining, and/or to add additional data. Such additional data could be gps-location data, also, after the reconstruction; the data might be directly implemented into a local (GIS) map or a worldwide map such as Google Earth.

Fig 4.4 : Process of Image Capturing

Principle of Operation

Initialization

Major process for computer integration of the camera. Leveling; internal block Temp adjustment (to –15oC min); gyro compassing (to True or Mag); latitude testing; Total 10 to 45 minutes, depend on temp, gyro type & latitude (takes longer @ higher latitudes to sense rotation & find True North which is essential to the alignment process.Easy for pilot Can use NAV or ALIGN mode ALIGN does a more thorough job but process can be negated if don't switch to NAV before moving Also want to be careful NOT to leave NAV mode in flight because can't initialize in flight. Military can air initialize but civilian units must be stationary, on ground, in a known position

Evaluation of Processed Information

Instrument always knows where True North is and senses all movement, so it always knows what aircraft is and it is easy to derive other information such as Track and Position – as a series of fixes, Ground speed, Drift angle, Cross track error/Cross track distance (Degrees or Miles), Distance to go/ Time to go [mts/sec], and other necessary information Errors & Recommended Actions
Normally has a separate WAAS discussed earlier to provide fully corrected navigation information used to determine location of aircraft. Air data system can also provide instrument with altitude information for systems with only 2 accelerometers.

Guidance and Navigation:

A combination of GPS and inertial sensing is ideal. Geographical information system to provide a map terrain for infrastructure would be great .This combination enables the tracking of unmanned vehicle correctly and quick essential information to the source station.
GPS Advantages for Unmanned Air vehicles

Reduced costs to each individual State while increasing overall benefits to individual States and the entire region. Further economies from reduced maintenance and operation of unnecessary ground-based systems. Improved ground and cockpit situational awareness Increased landing capacity for aircraft navigation and other factors.

Creating the Better navigation, developing a robust system:
The Instrument allows accurate navigation with less than four satellites, helps reject multipath and aids the GPS in re-acquiring signals. The following errors may be occurred the intermittent loss of satellite signals caused by the aircraft moving into or out of a satellite's view, by an aircraft maneuver, or by the weather[some bad conditions], some of the Problems may occur during a turn.

Furthermore the some of the advantages in the aviation by GPS Systems such as include:
- Enhanced safety of flight throughout the region
- Seamless navigation service based on a standardized navigation service and common avionics
- More efficient, optimized, flexible, and user-preferred route structures
- Significant savings from shortened
- Weather updates

Even more Reduced costs to each individual State while increasing overall benefits to individual States and the entire region. Which has commercially many applications and even more the better opportunity for giving an idealistic overview of the predetermined task completion strictly..Further economies from reduced maintenance and operation of unnecessary ground-based systems .A much improvement in the sense of the finding the correct location of longitude and latitude. Improved ground and cockpit situational awareness, which ultimately increased landing capacity for aircraft.

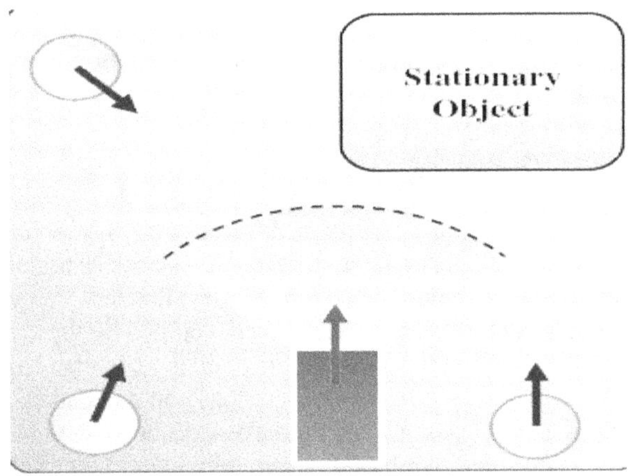

Fig 5.1 Localized pattern

Vision based Navigation Application:

Scan Matching is the most simplest type of the algorithm search and Most low-cost GPS is not robust. The Road intersections have less road information, no lane and no curb. The process require a more robust localization sensor especially when GPS doesn't work.

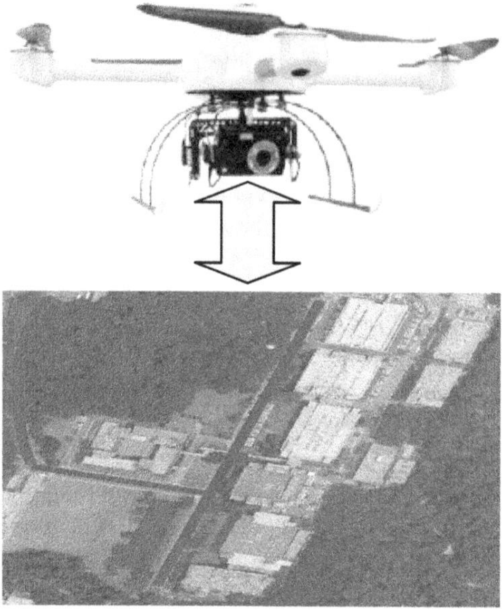

Fig 5.2 : Scan Matching

Bio based sensors designed used especially for the low cost sensors give poor heading angles. Furthermore Scan matching algorithm, Direct State Extraction from Software and directly can do mapping using only a sensor.

Highly suitable for the Defense and Security:

Mechanical assemblies are evaluated for quality at different stages of mechanical integration and various checks are carried out on airframe before clearance for integration. Avionics system integration is a mandatory and vital phase of an aircraft development lifecycle, since it brings out the important issues relating to the interfacing of LRUs developed by the various work centers.The major issue for the tracking based on the visualization enables us to capture and data reference for the future applications.A track is a trace of somewhere that you have actually been (often called a "breadcrumb trail"). The GPS unit (external or internal) periodically sends details of the location which are recorded by the software, either by taking a reading based on a set time interval,

Application Software for vision based Navigation:

Software can be used on a laptop computer with an attached GPS receiver. Most commercial software runs only on Windows and Mac OS X, yet some projects have started to support Linux as well. For most daily use however, a PDA, a dedicated device, or a Smartphone might be handier. Even Wifi controllers are now capable of utilizing the software effectively. Some of the application software's are listed below:

Commercial navigation software with embedded maps :

- DeLorme Street Atlas USA and Topo USA
- Destinator
- C2Logix Route Optimization
- Microsoft Streets and Trips 2009
- Navigon
- Nav N Go (iGO)
- NDrive
- ROUTE 66
- TomTom Navigator
- TomTom Mobile
- TeleType WorldNavigator

Navigation software with maps downloaded from a remote server include the following

- Google Earth (Windows, Mac, Linux)
- Google maps (platform independent)
- Navit (Cross-platform) open source and free
- VZ Navigator (smartphone)The GPS unit (external or internal) periodically sends details of the location which are recorded by the software, either by taking a reading based on a set time interval, based on a set distance, based on a change in direction by more than a certain angle, or a combination of these. Each point is stored together with its date and time.

Identification and credentials

One of the simplest forms of identification is the carrying of credentials. Some nations have an identity card system to aid identification, whilst many, such as Britain, are considering it but face public opposition. If the form of the identity card is "machine-readable", usually using an encoded magnetic stripe or identification number (such as a Social Security number), it corroborates the subject's identifying data.

RFID and Geolocation devices:

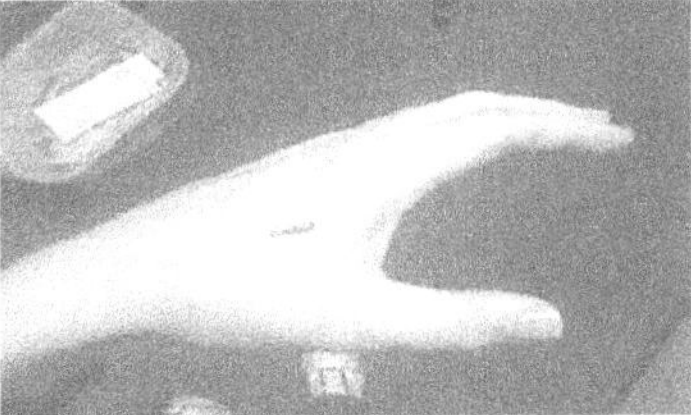

Fig : RFID Scanning

Radio Frequency Identification (RFID) tagging is the use of very small electronic devices (called "RFID tags") which are applied to or incorporated into a product, animal, or person for the purpose of identification and tracking using radio waves. The tags can be read from several meters away. They are extremely inexpensive, costing a few cents per piece, so they can be inserted into many types of everyday products without significantly increasing the price, and can be used to track and identify these objects for a variety of purposes.

Surveillance devices:

Surveillance devices, or "bugs", are hidden electronic devices which are used to capture, record, and/or transmit data to a receiving party such as a law enforcement agency.
Law enforcement and intelligence services in the U.K. and the United States possess technology to remotely activate the microphones in cell phones, by accessing the phone's diagnostic/maintenance features, in order to listen to conversations that take place nearby the person who holds the phone[22][23].

Conclusions and Future Scope:

Integrate the image collection phase with the region decomposition stage to yield an on-line process for simultaneous exploration and localization (SLAM).The design of MAV should include Detect and cope with environmental change. Formulating the commercialization which include Reduced costs to each individual State while increasing overall benefits to individual States and the entire region.

References:

[1]   Office of the Secretary of Defense, Ed., *Unmanned Aerial Vehicles Roadmap 2002-2027*. Washington DC, USA: United States Government 2002.

[2]   L. Chaimowicz, B. Grocholsky, J. F. Keller, V. Kumar, and C. J. Taylor," Experiments in Multirobot Air-Ground Coordination," in *Proceedings of the 2004 International Conference on Robotics and Automation*, New Orleans, LA, April 2004, pp. 4053–4058.

[3]   International Symposium on Flying Insects and Robots, Monte VERITAS, Switzerland, http://fir.epfl.ch

[4]   Zufferey, J.-C. (2008). *Bio-inspired Flying Robots: Experimental Synthesis of Autonomous Indoor Flyers*. EPFL/CRC Press. ISBN 978-1-4200-6684-5. http://book.zuff.info.

[5]   The photogrammetric principles and the data acquisition for the 3D and 2D images.

[6]   Cheng-Lin Chen and Fu-Yuen Hsiao*, *Attitude Acquisition Using Stereo-Vision Methodology*, presented as Paper VIIP 652-108 at the 2009 IASTED Conference, Cambridge, UK, Jul. 13-15, 2009

[7]   Sen.-Huang Lin, Fu-Yuen Hsiao*, and Cheng-Lin Chen, *Trajectory Control of Flapping-wing MAV Using Vision-Based Navigation*, accepted to present at the 2010 American Control Conference, Baltimore, Maryland, USA, Jun. 30 - Jul. 2, 2010

[8]   T. Kanade, O. Amidi, and Q. Ke, "Real-time and 3d vision for autonomous small and micro air vehicles," in *Proc. of the 43rd IEEE Conference on Decision and Control*, Atlantis, Paradise Island,Bahamas, December 2004, pp. 1655–1662.

[9]   F. Kendoul, I. Fantoni, and K. Nonami, "Optic flow-based vision system for autonomous 3D localization and control of small aerial vehicles," *Robotics and Autonomous Systems (Elsevier)*, vol. 57, pp.591–602, 2009.

[10] B. Lucas and T. Kanade, "An iterative image registration technique with an application to stereo vision," in *Proc. DARPA IU Workshop*, 1981, pp. 121–130.

[11] E. N. Johnson, A. J. Calise, Y. Watanabe, J. Ha, and J. C. Neidhoefer, "Real-time vision-based relative aircraft navigation," *Journal of Aerospace Computing, Information, and Communication*, vol. 4, pp. 707–738, April 2007.

[12] K. L. Van Dyke, "The World After SA: Benefits to GPS Integrity," *IEEE 2000: Position Location and Navigation Symposium,* pp. 387-394, 2000.

[13] J. Shi and C. Tomasi, "Good features to track," in *Proceedings of the IEEE Conference on Computer Vision and Pattern Recognition*, ser. Seattle, WA, USA, 1994, pp. 593–600.

[14] B. Lucas and T. Kanade, "An iterative picture registration technique with an application to stereo vision," in *Proc. DARPA IU Workshop*, 1981, pp. 121–130.

[15] "WAAS Navigation syatems", Wikipedia , Integration and computability.

[16] D.L. Polla, P.J. Schiller, and L.F. Francis, "Microelectromechanical Systems Using Piezoelectric Thin Films," *SPIE* **2291,** 1994, pp. 108–124.

[17] T. Niino, S. Egawa, H. Kimura, and T. Higuchi, "Electrostatic Artificial Muscle: Compact, High-Power Linear Actuators with Multiple-Layer Structures," *Proc. IEEE Conf. on Micro Electro Mechanical Systems, Oiso, Japan, 25–28 Jan. 1994,* pp. 130–135.

[18] S.M. Bobbio, M.D. Kellam, B.W. Dudley, S. Goodwin- Johansson, S.K. Jones, J.D. Jacobson, F.M. Tranjan, and T.D. DuBois, "Integrated Force Arrays," *Proc. IEEE Conf. on Micro Electro Mechanical Systems, Fort Lauderdale, Fla., 7–10 Feb. 1993,* pp. 149–154.

[19] *Unmanned Vehicle Handbook 2006.* Bucks, United Kingdom: The Shepard Group, December 2005.

[20] Vladykin, Oleg. "Unmanned Aerial Breakthrough." MIGnews.com.ua Web site, <http://mignews.com.ua/en/articles/213361.html> (Accessed October 2006).

[21] "Science and Technology: High Times; Future of Flight." *The Economist* [London], December 13, 2003 (via ProQuest ID: 499922401).

[22] McCullagh, Declan; Anne Broache (December 1, 2006)."FBI taps cell phone mic as eavesdropping tool". *CNet News*. Retrieved 2009-03-14.

[23] Odell, Mark (August 1, 2005). "Use of mobile helped police keep tabs on suspect". *Financial Times*. Retrieved 2009-03-14.

STUDY OF WINGBEAT FREQUENCY OF A BIO-MIMICKING MICRO AERIAL VEHICLE

N. Chari[1], K.M.R.Achary[2], V.S.K.Reddy[3], A. G. Sarwade[4],

1. Visiting Professor, Malla Reddy College of Engineering and Technology, Secunderabad
2. Professor, Department of Physics, Malla Reddy College of Engineering & Technology, Hyderabad
3. Professor in ECE Malla Reddy College of Engineering & Technology, Hyderabad,
4. Professor, Department of Aeronautical Engineering, Malla Reddy College of Engineering & Technology, Secunderabad,

Abstract: Wing beat frequency is one of the important parameter in biological flight. The biological flight is mainly due to gliding and frequency. The present study relates to flapping frequency of flight. Mechanical Oscillatory theory (Greenwalt 1962), Crawford's theory(1972), Norberg's theory (1990), Pennycuick's theory(1996), Theory based on Newton's laws, Mass flow theory (2007) and Wing beat frequency of dimensional analysis (Dakin's theory, 2010) are well known theories which try to elucidate wing beat frequency of animal fliers. The basic information on all the seven theories has been compared and possible frequency range for bio-mimicking MAV's has been suggested in the present study.

Keywords: Frequency, flapping wing, MAV

1. Introduction

For a biological flier the wing beat can be considered as oscillatory. The vibrations of wings fixed to a point fulcrum or thorax of a flier are similar to the vibration of prongs of a tuning fork. The frequency of a tuning fork is determined by Melde's experiment. In this, the vibrations of the prongs are transferred to the thread with a load, passing over a fixed pulley. Frequency is determined by measuring length of loop formed in the standing wave in the string and by knowing linear density of the thread. However the frequency of any vibrating or oscillatory body depends on mass, displacement and acceleration of the body.

The frequency of wing beat of a flier is an important parameter in aerodynamic and bioenergetic studies of animal flight. This information is very much required while designing MAVs. Many attempts have been made to elucidate the relation between wing beat frequency and body parameters of biological fliers. Following seven theories have been reviewed in detail as these theories have been used in calculating and comparing wing beat frequencies of hovering fliers and also in horizontal condition.

1. Mechanical oscillator theory (Greenewalt, 1962)[1]
2. Crawford's theory (1972)[2]
3. Norberg's theory (1990)[3]
4. Pennycuick's theory (1996)[4]
5. Theory based on Newton's laws[5]
6. Mass Flow theory (1977),(2011)[5]
7. Wing beat frequency on the basis of dimensional analysis (Deakin, 2010)[6]

2. Materials And Methods

For the present investigation, following biological fliers were collected from nature and kept alive in laboratory for 10 to 15 days . 1.M.orientalis-- Bee eater, 2.A.affinis-- House Swift, 3.S.d. decaacto-- Ring dove, 4.C.Livia—Pigeon, 5.H.Speoris --Leaf nosed bat and 6.T.Javanica-- Soap –nut bug.

The wing beat frequencies have been calculated by using the formulae from 7 theories as mentioned in the introduction. A sensitive balance and a measuring scale have been used for measuring flier parameters in M.K.S Units. A stroboscope either from Philip's company or Toshniwal has been used in measuring tethered wing beat frequency. Stroboscopic tachometer was used in the present study.

3. Results and Discussions

The flight parameters such as mass, wing length, wing span, wing area and effective breadth of the wing of the fliers are studied. Six samples have been listed in Tables 1&2.Under each theory formula for frequency (v_h) used has been listed in Table 3. The frequency values as calculated on the basis of these seven theories and the formulae arrived and used in these calculations have been compiled and listed in the above tables. In any theory on wing beat frequency of natural flier constraints and approximations become unavoidable and analytical approach is rather too complex. Therefore empirical formulae are suggested.

1. Mechanical oscillator theory (1962)

A flier whether it may be an insect, a bird or a bat Greenewalt (1962) considered the oscillating wing as a mechanical Harmonic oscillator. To compute its wing beat frequency, the damped harmonic wave equation was initially adopted as mentioned below with usual notations.

$$I\frac{d^2\emptyset}{dt^2} + B\frac{d\emptyset}{dt} + f(r) = F_0 Cos(\omega t) \dots\dots\dots (1)$$

If the oscillations of the wings are free and not damped. then B = 0 and $F_0 = 0$. Hence

$$I\frac{d^2\emptyset}{dt^2} + f(\emptyset) = 0 \dots\dots\dots (2)$$

The Solution in terms angular Frequency (ω) for the undamped oscillation is given by,

$$\omega^2 = \frac{kb_0 r^2}{I}$$

Where k is constant,

b_0 = length of contracting muscle,

r = the effective radius, and

I = Sum of internal and external moments of inertia.

As $v = \dfrac{\omega}{2\pi}$ and wing length "l" measured in millimeters, $b_0 \propto l$, $r \propto l$, also m $\propto l^3$. Sotavalta with his data on insects and Magnan with his data on birds it was shown that

$$vl^{\,n} = \text{a constant } (K) \quad \dots\dots (4)$$

The wing length power (n) lies in between 1 and 1.25.

Finally Greenewalt generalises the frequency in terms of wing length for insects and small birds as,

With n = 1.25,

$$vl^n = constant \dots\dots\dots\dots (5)$$

$$vl^{1.25} = 3540 \dots\dots\dots\dots (6)$$

2. Crawford's theory (1972)

Crawford proposed a relationship for the wing beat frequency of a flier and the relation is given as,

$$v_h = \left(\frac{g}{4\pi\rho}\right)^{1/2} \times \frac{(m_f)^{1/2}}{S_w}$$

Where

S_w = Wing swept area = stroke angle * (wing length)^2,

m_f = mass of the flier

ρ = density of air and

g = acceleration due to gravity

This theory is applicable where the wing swept area and insects are small as in misquote. Similarities between Mass Flow theory and Crawford theory are interesting. The S_d (disc area) concept in Mass Flow theory is replaced by S_w (wing swept area).

3. Norberg's Theory (1990)

This is based on scaling laws and finally hovering frequency (v_h) is calculated on mass consideration alone.

The formula is $v_h = 3.98(m)^{-0.27}$.

This formula is applicable for few birds.

Any formula for calculating v_h as derived finally in terms of mass or wing length alone of the flier has limited applications.

4. Pennycuick's Theory (1996)

Pennycuick's formula is

$$v = g^{1/2}\rho^{-3/8}m^{3/8}b^{-23/24}S^{-1/3}$$

Pennycuick has calculated the frequency in horizontal flight instead of hovering state.

The birds were assumed to be flying at horizontal level with minimum power speed. Effect of air density can be studied for a bird flying at different altitudes.

5. Theory based on Newton's laws

Based on Newton's II Law, for a hovering flier the weight (mg) of a flier is supported by the reaction of the wings, as

$$mg = \left(\frac{d}{dt}\right)(m_{air}v_{air})\ldots\ldots\ldots(1)$$

However, during downstroke, mass of the air pushed down is ρ times the volume of air swept out by the wing, (where ρ is the density of air). This volume swept out by the wing is the wing area (A) times twice the amplitude Z_0 of the wing stroke.

Therefore $m_{air} = 2\rho A Z_0$

Assuming that this air is pushed down and attains the maximum velocity ωZ_0

So we have $\quad dt = 2\pi / \omega$

If we average over one cycle, equation (1) can be written as,

$$mg = (2\rho A Z_0)(\omega Z_0)\left(\frac{\omega}{2\pi}\right)\ldots\ldots\ldots(2)$$

i.e., $mg = \dfrac{\rho A Z_0^2 \omega^2}{\pi}$

For reasonable wing shape, taking the amplitude as large as possible and setting.

$$Z_0 = \sqrt{A}\ in\ eqation\ (2)$$

and gets simplified as

$$\omega^2 = \frac{\pi mg}{\rho A^2}\ldots\ldots(3)$$

Now, $v = K\dfrac{\sqrt{m}}{A}\ldots\ldots\ldots(4)$ where

$$K = \frac{g}{2\pi\rho} = 41 \quad \left(since\ v^2 = \frac{\omega^2}{4\pi^2}\right)$$

The hovering frequency of the flier can be calculated by equation (4) for different fliers.

6. Mass Flow Theory

Puranik et al (1977) have worked out a theory for deriving the wing beat frequency of a flier in hovering state on the basis of mass flow of air. In hovering state of any flier, the system can be said to be in a state of dynamic equilibrium, which is achieved by the flier by generating a reacting force R, just to balance its own weight (m_f. g). The reacting force, R, thus generated is proportional to the disc area, effective breadth, density of air and half of wing beat frequency.

Reacting force, 'R' is produced to balance its own weight and the system comprises the flier and the induced air, which constitute action reaction pair during hovering flight. According to Mass Flow theory, during hovering, the reacting force, R, is in dynamic equilibrium with its rate of mass flow of air.

Here,

$R = W = m_f g$: m_f = mass of flier,

g = acceleration due to gravity.

Therefore $R = m_f g$ …..(1)

Since there is no upward pull and the reacting force argued as,

$$R \alpha S_d, B_{eff}, \rho, v_h/2$$

Therefore $R = \dfrac{kS_d B_{eff}\, \rho v_h}{2}$ ………(2)

Where k = proportionality constant.

From (1) and (2)

$$m_f g = \dfrac{kS_d B_{eff}\, \rho v_h}{2} \text{……(3)}$$

Substituting value of $S_d = \dfrac{\pi L^2}{4}$ in the equation (3), the hovering frequency

$$v_h = K\,\dfrac{m_f}{L^2 B_{eff}} \text{… …(4)}$$

Where $K = \dfrac{8g}{\pi \rho k} = 2086.$ (Since 8, g, $\pi, k\, \&\rho$ are constants.)

The value of K' is determined from the experimental results by drawing a log graph between m_f and $L^2 B_{eff} v_0$. The graph is a straight line, the slope of which gives a value of K. The dimensions of K' are $M^{-1}L^3T^{-1}$.

7. Deakin's Theory (2010)

An expression is derived for the wing beat frequency by Deakin by considering the basic dimensional analysis. Deakin worked on several species of insects having typical mass, wingarea and wing beat frequencies. For theoretical computation of wing beat frequency, he considered those species, which are geometrically similar.

Deakin 2010 developed a formula for wing beat frequency of insects considering the

"Buckingham Pi theorem", $f(\pi_2, \pi_3) = 0$ and by writing the frequency as

$$v_h\ \alpha\ g^{1/2}A^{-1/4}f(\pi_3)\ \alpha\ g^{1/2}A^{-1/4}f(\rho A^{3/2}/m)$$

Function $f(\pi_3)$ *may be expanded by means of Frobenius series as*

$$f(\pi_3) = k\pi_3^{-\alpha}(1 + \alpha_1 \pi_3 + \alpha_2 \pi_3^2 + \cdots)$$

where k and α are dimensionless constants i.e., pure numbers. Power α is assigned for negative sign for convenience and also as π_3 is small, neglecting terms beyond 1, the frequency can be written as

$$v_h = kg^{1/2}A^{-1/4}f(\rho^{-\alpha}A^{-(3/2)\alpha}/m^{-\alpha})$$

$$v_h = kg^{1/2}\rho^{-\alpha}m^{\alpha}A^{-(3/2)\alpha - (1/4)}\text{……(1)}$$

$$v_h = K m^{\alpha}A^{-\beta} \text{………(2)}$$

Where $K = kg^{1/2}\rho^{-\alpha}$ is considered as another constant.

By Solving we get $\alpha = \dfrac{1}{2}$ $\beta = 1.$ and substituting these values in (1)

$$v_h = k(g/\rho)^{1/2}\left(m^{1/2}/A\right)\text{… … …(3)}$$

OR

$$v_h = K\,\dfrac{\sqrt{m}}{A}$$

Where $K = k(g/\rho)^{1/2} = 317$

4. References

1. Greene Walt C.H.(1962): Dimensional relationship for flying animals . Smithsonian Misc Coll. 144(2),1-46.

2. Crawford Frank S.(1972): comments on the physics and physiology of insect fight. Am.J physiology 39,584,

3. Norberg's (1990): Vertebrate flight mechanics, Morphology , Ecology and Evolution Springer – verlag, New York.

4. Pennyquick C.J.1996:Wing beat frequency of birds in steady cruising flight. New data and improved predictions . the journal of experimental Biology, Vol. 199,1613-1618.

5. Puranik P.G., Gopalkrishna G., Adeel Ahmed and Chari N (1977): Wing beat frequency of a flier – Mass Flow Theory. Proc. Ind.Acad.Sci Soc. A., 85, 327-339 and Bio-Physics of bird flight (Bio-aerodynamics) 5th edition.

6. Deakin MAB.2010. Formulae for insect wingbeat frequency. Journal of Insect Science 10:96 available online: insectscience.org/10 96

S.No	Fliers	Body Mass(m_f) (kg)	Wing Length(l) (m)	Wing Span (l) (m)	WING Area (2A) (Sq.m)	Effective Wing Breadth B_{eff}= 2A/2l m	Hovering Frequency (Cps) $v_h = \dfrac{m_f.2086}{L^2 B_{eff}}$
1	M.orientalis - Bee eater	0.01542	1.20	0.272	0.0117	0.05	7.87
2	A.affinis House Swift	0.018	0.1330	0.301	0.0101	0.04	11.3
3	S.d. decaacto Ring dove	0.168	0.256	0.542	0.0579	0.11	10.532
4	C.Livia Pigeon	0.262	0.280	0.671	0.0796	0.14	8.501
5	H.Speoris Leaf nosed bat	0.008	0.127	0.283	0.0114	0.04	4.6
6	T.Javanica Soap –nut bug	0.001	0.022	0.056	0.004	0.02	53.00

The Frequency values calculated have been verified by using a stroboscope where ever necessary

Table 2. Frequency (v_h cps) comparison as Calculated from various theories as follows

S. no	Fliers	Greenwalt's Mechanical Oscillator	Crawfod's Theory	Norberg's Theory	Pennycuick's Theory	Theory Based on Newton's laws	Mass Flow Theory	Deakin's Theory
1	M.orintalis (Bee eater)	14.38,11.38,8.91	0.06	1.86	10.9	8.9	8.0	10.6
2	A.affinis (House swift)	12.7810.51,7.83	0.05	1.80	10.7	11.3	12	13.4
3	S.d decaacto (Ring dove)	6.01,4.82,3.03	0.04	0.99	8	5.96	10.50	7.1
4	C.Livia (Pigeon)	5.42,4.33,3.09	0.045	0.88	7	5.41	9	6.2
5	H.speoris (Leaf nosed bat)	13.40,11.10,8.30	0.037	2.08	8	6	4.60	7.8
6	T.Javanica (soap-nut bug)	102.80,90.90, 75.58	0.43	4.14	48	50.73	53.0	67

Table 3. Theories on wing beat frequencies of hovering fliers.

S.No:	Name of Theory	Frequency
1	Greenewalt's Theory (1960)	$v_h l^{1.25} = 3540$
2	Crawford's Theory(1971)	$v_h = \left(\dfrac{g}{4\pi}\right)^{1/2} \times \dfrac{(m_f)}{S_w}$
3	Norberg's Theory (1990)	$v_h = 3.98(m)^{-0.27}$
4	Pennycuick's Theory (1996)	$v_h = K \dfrac{m^{3/8}}{S^{1/3} b^{23/24}}$ $v_h = m^{3/8} g^{1/2} b^{-23/24} S^{-1/3} \rho^{-3/8}$ or $v_h = (g^{1/2} \rho^{-3/8}) \dfrac{m^{3/8} b^{-23/24}}{S^{-1/3}}$
5	Based on Newton's Laws	$v_h = \dfrac{1}{2A} \sqrt{\dfrac{mg}{\rho\pi}}$ or $v_h = K \dfrac{\sqrt{m}}{A}$ where K = 41
6	Mass Flow Theory (1977)	$v_h = K \dfrac{m_f}{L^2 B_{eff}}$ $Where K = 2086$
7	Deakin;s Theory (2010)	$v_h = K. \dfrac{\sqrt{m}}{A}$ where K=317

* The value of K differs from theory to theory.

Table 4. Comparison of Pennycuick's theory, Newton's theory, Mass Flow theory and Deakin's theory

S.No:		Pennycuick's Formula	Newton's Formula	Mass-Flow Formula	Deakin's Formula
1	Formula	$v_h = \left(\dfrac{g^{1/2}}{\rho^{3/8}}\right) \cdot \dfrac{m^{3/8}}{s^{1/3} b^{23/28}}$	$v_h = K\dfrac{\sqrt{m}}{A}$	$v_h = \dfrac{m_f}{L^2 \times B_{eff}}$	$v_h = k\dfrac{\sqrt{m}}{A}$
2	Parameters: m,g,b,s,ρ and π		m,A,g,π,ρ	R= m_f g (hovering) R α Sd , $B_{eff.}$ ρ, $v_h/2$	
3	Dimensionless numbers α, β, and n				α=1/2 and β=1
4	Values of K		41	2086	317

LOW-COST NAVIGATION SYSTEM FOR COMPUTATIONALLY CONSTRAINED MAVS

Kadur Aditya E-mail : aditya.kadur

Jassar Gulsagar E-mail : gulsagar.jassar

Kashyap Abhyudaya E-mail : abhyudaya.kashyap

M.Santhakumar E-mail : santhakumar} @iiti.ac.in
IIT Indore, PACL Campus, Pithampur, Indore

ABSTRACT

This paper describes a novel navigation system for obstacle avoidance and autonomous navigation in an unknown, GPS-denied environment for a quadrotor helicopter. The proposed system is extremely cost-effective with minimal computation requirements, allowing a highly efficient alternative to camera-vision based and laser-scanner based navigation systems, for application in areas where complete terrain mapping is not required. The system consists of a rotating point sensor (a point laser range finder with a high refresh rate) mounted on a high-resolution precision servo and another sensor fixed at the bottom of the MAV for altitude measurement for low-altitude hover and landing. Modern microcontrollers have enabled the construction of relatively low-cost quadrotor MAVs which rely primarily on inertial data from the IMU along with real-time onboard computation, for flight stability. However, for autonomous navigation, the cost of MAV rises steeply due to added cost of components, such as a camera, along with much higher computation requirements for real-time processing of sensor-data. This paper presents the system architecture and core algorithms for the proposed low-cost navigation system requiring minimal computation. Several algorithms for altitude maintenance andpath traversal with obstacle avoidance in a hypothetical environment are presented and analyzed.

INTRODUCTION

Quadrotors are one of the most popular MAV designs. Owing to their high stability and manoeuvrability, they find use in a wide variety of applications. With their ever-expanding scope, modern applications are increasingly relying on the autonomous flight capability of quadrotors. MAVs capable of autonomous outdoor flight have been developed [2,3]. Speeds of over 5m/s and avoidance of obstacles like wires less than 6mm in diameter have been achieved [4].

In recent years extensive work on indoor autonomous navigation has also been done [5,6].Due to the absence of external positional systems like GPS in these environments, indoor navigation systems must overcome certain challenges like position drift over time. Simultaneous localization and mapping (SLAM) algorithms map the environment around the vehicle while simultaneously using the sensor data to estimate the position of the vehicle in real-time or with a minimal lag. The map and positioning data is then used to determine the optimum route between two specified nodes.

However, most of these systems rely on expensive laser rangefinders for continuous high resolution mapping of indoor the environment in 3D. The main drawback of this approach is the extremely high price of laser rangefinder devices. With prices upwards of $2000 these systems cannot be used as part of low-cost or multi-unit systems . Another popular approach is the use of stereoscopic cameras for video capturing and either conversion to a 3D map using image processing techniques, or by feature extraction for detecting known objects and path-markers. Both thesetechniques offer excellent mapping resolution but at the expense of added cost and added complexity. The processing requirement here are extremely high. Consequently, in most systems, a high-speed data link to the ground station is necessary to relay the sensor data and receive the computed map-data. Thus, implementation of these systems is not feasible in low-cost computationally constrained MAVs .

In this work, we present an obstacle avoidance system which can easily be incorporated into existing quadrotor designs to provide autonomous flight capability. The system uses simplistic point range sensors mounted on a precision servo (DC stepper motor with a digital encoder). We believe the simplistic hardware along with the explained algorithms can be used to autonomously navigate a quadrotor from point A to B in both indoor and outdoor environments. The key difference here is that in our proposed system, only a crude map of the terrain or environment is generated and stored. This effectively minimizes the computational requirement which allows onboard processor to perform the required computations in real-time, without a dedicated link to the ground station.

In section 2, some works related to the system described are discussed. Section 3 gives the system overview from a hardware perspective. The main algorithms are described in section 4 followed by the conclusion.

RELATED WORKS

Autonomous navigation has become one of the most popular areas of MAV research. In the past years, enormous advances have been made in this field with a shift in focus from outdoor to indoor navigation.

A. Bachrach et al [5] have demonstrated an autonomous quadrotor capable of indoor flight. However, this system is based on a laser rangefinder and creates a detailed map of the entire environment. The high price of rangefinder along with the high computation requirements are a limiting factor for implementation in low-cost general designs.

Use of monocular camera sensors for autonomous indoor flight is also a popular approach. G.P Tournier et al [8] used Moire patterns and performed visual servoing to extract the full 6 degree-of-freedom state of the vehicle and used this for vehicle control. In works [9] and [10], visual data was used to track lines and visible edges, respectively, in an office environment. These systems rely on camera data and thus, are computationally intensive. Also, they only work on environments with specific features and thus, aren't well suited for general GPS denied environments.

Use of simple point rangefinders has also been explored and covered in various works. In [7], the authors have used 2 IR sensors and an ultrasonic sensor for simple collision avoidance and altitude control respectively. However, the IR sensors are fixed and can only give a binary danger signal when the distance to a large static wall-like object is below a threshold. This only covers 1 quadrant of the $360°$ space. A more elaborate implementation of a similar setup is described in [1] where 4 range sensors are used to sense obstacles in all directions. Here, again the sensors are fixed and can only work for large extended objects. The systems does not account for moving objects and suffers from several blind spots.

SYSTEM OVERVIEW

For the description of this system, we shall assume a quadrotor MAV design. All onboard computations are done on an arduino-based microcontroller with various digital I/O pins. A 6-DoF IMU (3-axis accelerometer and roll,pitch,yaw gyroscope)is used for measuring relative changes in the vehicle state over time and for providing stable flight. A 3 axis magnetometer is used for direction sensing. For algorithm evaluation, we shall consider that the entire frame of the quadrotor is able to fit within a 50cm sided cube.

In the present work, we shall primarily concern ourselves with the obstacle avoidance and navigation system. All sensors (IMU, proposed system sensors, etc) and actuators (ESC, proposed systemactuators,etc) are assumed to work on compatible TTL logic directly interfaced with the microcontroller.

On the hardware front, the system consists of a single point laser range-finder mounted on a precision servo. The precision servo can be realized using a DC stepper motor with a digital encoder feedback to control the position accurately. The rangefinder can effectively measure the distance of the nearest obstacle in all directions over a fixed angle, in the horizontal plane. We make this angle limited to a fixed value

$$\theta_{max}= - \theta_{min} =30°(1)$$

The hardware parameters of importance here are the beam width and sensor range. Also, the maximum sampling rate of the sensor determines the speed of the system. The sensor is directly connected to the microcontroller (assuming the sensor provides direct TTL digital output).

For evaluating our algorithm performance, we consider a sampling rate of 50 Hz and a maximum sensing range of 5m, which fall well within the reach of commercially available laser point range sensors.

For altitude measurement, an ultrasonic range sensor mounted on the bottom of the quadrotor and facing the ground is used.

ALGORITHM DESCRIPTION

The algorithm for the proposed system is highly simplified in order to keep the processing requirements minimal. All the storage and processing is done on the onboard microcontroller. For the present work, we are only assuming that the input given to the MAV is a general direction of propagation and it is required to move in the specified direction while avoiding obstacles and following an optimized route to minimize deviations from a direct straight-line path.

We shall assume that the MAV is required to maintain a fixed altitude. Thus, all movement is restricted to a horizontal plane.

ALTITUDE CONTROL

The altitude control system is extremely straightforward. The bottom range-sensor continuously samples the range of the nearest obstacle at a fixed frequency say 5 Hz. This is used as feedback to the z-direction motor throttle. The z-direction motion is adjusted so as to counter changes in the sensor input. In order to counter for noise and for intermediate spikes in sesor input (due to small point objects above the ground over which the MAV crosses) the sensor output is first averaged over last 3 samples and then fed as input.

OBSTACLE AVOIDANCE

The servo arm is always kept zeroed at the intended direction of motion, i.e., $\theta=0°$ in the direction of motion.

Now, the servo continuously sweeps from θ_{min} to θ_{max} and back. If the sensor is assumed to operate at a frequency of f=50Hz and the range is r= 5m, then the spatial resolution (minimum size of the obstacle that can be sensed) would be

$$d = r \times \frac{\theta s \times 2\pi}{360} \ m(2)$$

If we want a resolution of say D=5cm ,then we get θ s = 0.57°

$$\text{Thus, } \omega = \theta s \times f = 28.5°/\text{sec} \qquad \qquad .....(3)$$

Also, at this angular velocity, the size of the gap covered (linear distance swept perpendicular to the direction of motion) would be (D), in 2 sec

$$D = 2x \ \sin\theta_{max} \ x \ r = 4.77m \qquad \qquad .....(4)$$

Thus if we only cover an area wide enough for the passage of the quadrotor (D= 1m), we get

$$\theta_{max} = 5.74°$$

Thus, at f= 50Hz and ω=10°/sec, we get from equation (3) :

$$\theta s = 0.2°$$

Thus, from equation (2), we get resolution

$$d = 1.74 \ cm$$

Thus, we can make sense obstacles greater than 1.74cm in cross-sectional size.

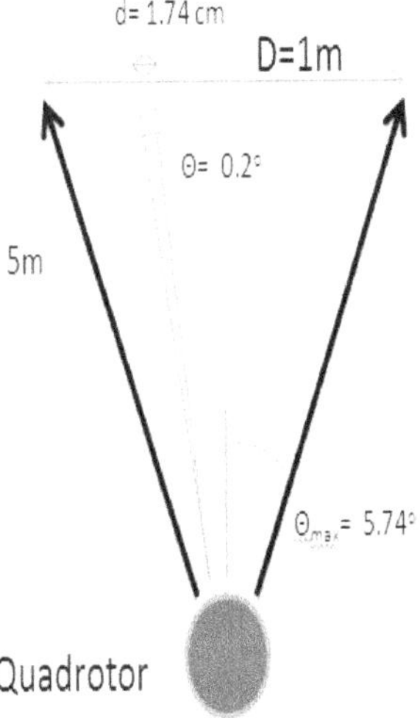

At the chosen specs, the system can make one complete sweep in 1.2 sec.

Thus, the MAV continues to move in the required direction at a maximum speed of say r/5 = 1m/sec .

When an obstacle is sensed the MAV movement stopped . The MAV then follows the path planning algorithm for deciding how to proceed.

PATH PLANNING

The path-planning algorithm essentially aims to minimize deviation from the intended direction of motion.

The present direction of motion is sensed using the magnetometer and accelerometer.

Whenever an obstacle is sensed, the servo sweep (θmax) is doubled to check for alternate paths on both left and right-hand sides. If no such path is found, the MAV makes sideward lateral movement by unit distance. The direction and no. of such movements is stored by the controller. This is then used to calculate the net deviation from path. Now, another sweep is done to check for available paths. This process repeats iteratively until a path is found. The direction of side stepping changed after a threshold (say 5 steps in one direction).

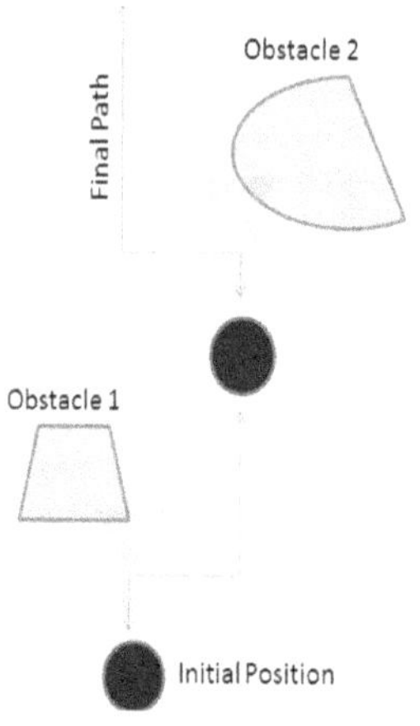

Figure 2

Using the no. and direction of steps, the the controller tries to minimize the net deviation by taking side-steps in the appropriate direction whenever possible.

The primary challenge here is that the side step movement needs to be calibrated for ensuring the same unit movement is achieved using only the onboard IMU. In the absence of external positioning systems, even calibrated measurements tend to drift.

Another issue is that, during a sideward step, we need to ensure that the MAV does not move into an obstacle. Hence, it would be advisable to turn the quadrotor by 90o and take a forward step, while employing the obstacle avoidance system.

CONCLUSION

In this work, we have described a novel navigation system for obstacle avoidance and autonomous navigation in an unknown, GPS-denied environment for a quadrotor helicopter. The proposed system is extremely cost-effective andhigh potential for applicationin areas where complete terrain mapping is not required. The algorithms described are currently in a rudimentary stage. We hope to implement them in an actual quadrotor system and optimize the algorithm using real-world tests. With subsequent improvements, we are hopeful that this system can rival the performance of more complex and resource intensive navigation systems.

REFERENCES

[1] A. Bachrach, R. He, and N. Roy. Autonomous flight inunstructed and unknown indoor environments. In *Proceedings of EMAV*, 2009.

[2] A. Bachrach, A. Garamifard, D. Gurdan, R. He, S. Prentice,J. Stumpf, and N. Roy.Co-ordinated tracking and planning using air and ground vehicles. In *Proc. ISER*, 2008

[3] A. Coates, P. Abbeel, and A.Y. Ng. Learning for control from multiple demonstrations. In *Proc. ICML*, pages 144–151.ACM, 2008.

[4] S. Scherer, S. Singh, L. Chamberlain, and S. Saripalli. Flying Fast and Low Among Obstacles. In *Proc. ICRA*, pages 2023–2029, 2007.

[5] Bachrach, Abraham, Ruijie He, and Nicholas Roy. "Autonomous Flight in Unknown Indoor Environments." International Journal of Micro Air Vehicles 1 (2009): 217-228.

[6] Roberts, Stirling, Zufferey and Floreano. "Quadrotor Using Minimal Sensing For Autonomous Indoor Flight".In proc.EMAV (2007).

[7] Matsue, A., Hirosue, W., Tokutake, H., Sundada, S., Ohkura, A., "Navigation of Small and Lightweight Helicopter", Trans. Japan Society Aeronautical and Space Sciences.Vol.48, NO.161, pp. 177-179, 2005.

[8] G.P. Tournier, M. Valenti, J.P. How, and E. Feron.Estimation and control of a quadrotor vehicle using monocular vision and moir`e patterns.In Proc. of AIAA GNC, Keystone, Colorado, 2006.

[9] N.G. Johnson. Vision-assisted control of a hovering air vehicle in an indoor setting. Master's thesis, BYU, 2008

[10] C. Kemp. Visual Control of a Miniature Quad-Rotor Helicopter.PhD thesis, Churchill College, University of Cambridge, 2006.

VISION BASED INDOOR POSITIONING AND NAVIGATION FOR ROTOR WING UNMANNED AERIAL VEHICLE

Mohamed Rasheed A[1],   Dr K. Senthil Kumar [2],   Magesh Giridharan T[3]

[1]PH.D Scholar, [2] Associate Professor, [3] P.G Student

Division of Avionics, Department of Aerospace Engineering, MADRAS INSTITUTE OF TECHNOLOGY, ANNA UNIVERSITY, Chennai-600044.

ABSTRACT

This project deals with the positioning and navigation of UAV in indoor environment. This project involves the usage of the vision based system. The wide range of applications and configurations of UAVs raises the need for different types of navigation methodologies compared to the conventional INS/GPS system. For instance, the limitation in cost, size and weight of indoor UAVs makes conventional navigation system unsuitable for these vehicles. In addition, the INS/GPS navigation system is impractical for indoor applications because the GPS signal is not reliable in closed territories. This project proposes a new, cost-effective and simple indoor navigation system using four IR LEDS fixed to the body of the UAV. Then, camera and computer vision algorithms are used to capture the bright spots in the image and determine their coordinates. The position of bright dots is used to obtain full information about the position and orientation of the UAV. The proposed navigation system can be classified as a vision based navigation system, yet, it does not depend highly on the quality of the video shots taken from the vision camera and does not require a heavy image processing techniques. An illustrative simulation study is conducted to demonstrate the validity of the proposed navigation system.

Keywords: Unmanned aerial vehicle (UAV), IR LED, tri-rotor, zigbee, GUI, web camera, FOV.

INTRODUCTION

The vision-based control of Unmanned Aerial Vehicles (UAVs) has become a very active field of research in the last decade. Until now, most of the efforts have been concentrated on developing vision-based control methods for autonomous take off, landing, stabilization and navigation, in which the visual information is usually obtained using a known model of a target or the environment. For autonomously performing such a task, one has to be able to robustly extract the object location from images despite difficult constraints: large displacements, occlusions, image noise, illumination and pose changes or image blur. Moreover, controlling such systems is quite different from controlling more conventional outdoor micro aerial vehicles, which can rely on high-precision inertial measurement units, global positioning systems, radars or other conventional distance sensors, and/or visual horizon detection systems. While most of the current researches deal with the attitude estimation and with the control of UAVs, few works propose navigation strategies. Research interest in Unmanned Aerial Vehicles (UAVs) has grown rapidly due to their wide range of applications. Nevertheless, seeking full autonomy is one of the main factors behind the milestone developments in UAV systems. To achieve autonomy, the UAV needs a navigation system that gives information regarding the status of the vehicle and feeds this information to the controller for an appropriate action to be taken.

The common navigation system used for large scale UAVs is the Inertial Navigation System coupled with the Global Positioning System (INS/GPS). The most popular sensor for navigation of UAVs is the GPS receiver. In this case, the navigation task consists generally to reach a series of GPS waypoints. Unfortunately, GPS data are not always available (for instance in indoor environment) or can be inaccurate (for instance in dense urban area where buildings can mask some satellites). For those reasons, it is necessary to use other sensors. However, for small UAVs or indoor applications, there is a need for alternative navigation strategies. This is due to the fact that INS/GPS system is relatively expensive and needs a good access to the GPS signal which is not available in case of indoor areas.The challenge of simple, less costly and efficient indoor navigation schemes has been undertaken by many UAV research. Different techniques such as pressure sensors for altitude, magneto-resistive magnetometer, laser range finder, radar and ultra sound for position estimation and obstacles detection have been investigated. In this regard, employing vision-based systems for autonomous navigation has attained considerable interest due to the fact that vision based systems are lightweight, passive and produce rich information about the motion of the vehicle.The use of camera is very attractive to solve those problems because in place where the GPS is difficult to use such as city centers or even indoors, there are usually a lot of visual features. A navigation system based on vision could thus be a good alternative to GPS. The vision based navigation algorithms were used initially for ground mobile robots and then imported to UAV systems. To usevision based navigation systems, the path of the UAV needsto be known a priori. Images of the surrounding environmentof the UAV's flying path are taken and analysed to identifythe basic features of this path before the flying missioncommences. Then, the real time images taken from on-boardcamera(s) during the flight mission are compared with thevisual memory of the vehicle to identify the known featuresand estimate the motion of the UAV. Different algorithmsand schemes are developed to excel the feature detectionand speed up the matching process. The computer vision systems might be used solely to estimate the motion and orientation of the vehicle by tracking the movement of the captured features in two consecutive shots from the on-board camera. This method is complicated, involves heavy computational burden, affected highly by the quality of the image and the number of features to be analysed, and works efficiently only in specific territories withgood images. To reduce the complexity and computational costs, the image analysis is chosen in to be performed on a ground station where the data is communicated from/to the UAV via a wireless link. This type of implementation reduces the autonomy of the UAV and puts the vehicle at risk in case of wireless communication failure. The vision based information is integrated with other sensors such as GPS/gyroscope into refine and correct the estimation of the UAV's motion. This again requires a good GPS signal and is impractical for indoor applications or urban territories. To sum up, computer vision based systems are still developing and havemany software and hardware difficulties. Some vision-based navigation systems originally developed for ground vehicles have been transposed to the context of UAV navigation. . Vision based control scheme have been also employed to drive the UAV during the autonomous navigation step.The rest of this paper is organized as follows: in Section II, a schematic design for the structure of the proposed system is explained. Section III discusses the mathematical formulation of the navigation scheme and the required steps to obtain the states of the DAV while Section IV is devoted to discuss the implementation process of the proposed strategy. Asimulation study is presented in Section V to demonstrate the proposed navigation scheme.

NAVIGATION SYSTEM DESCRIPTION

This paper introduces a new UAV navigation system for indoor applications. The introduced navigation system consists of four IR LED that are fixed on the body of the UAV. Here the bright spots of four diode in the image are used to estimate the position of the UAV. However, the estimation of the motion of the UAV is made by the maximum likelihood method. The literature lacks such an indoor navigation strategy that is effective and easy to implement. In this work, we tackle the indoor navigation problem by introducing a new navigation strategy that benefits from the computer vision algorithms, but in the same time is cost-effective and easy to implement. The proposed navigation strategy uses the locations of four IR LED bright spots in the image generated by the four IR LED fixed to the body of the vehicle to identify the status of the UAV without further requirement for optimization algorithm, fusing data estimation method like Extended Kalman Filter or additional GPS/IMU sensors. To capture the bright spots in the image analyze their coordinates, a computer vision technique is used. The computer vision algorithm is needed only to identify the positions of the four IR LED bright spots, and therefore, the required image processing is easy and can be in the ground control station. This in turn helps to reduce the computational cost of the algorithm and increase the level of autonomy of the vehicle.

MATHEMATICAL FORMULATION

In this work, the navigation problem is formulated as specifying the position and orientation of the Unmanned Aerial Vehicle from the positions of the four IR LED bright spots a1, a2, a3 and a4 as shown in the figure.

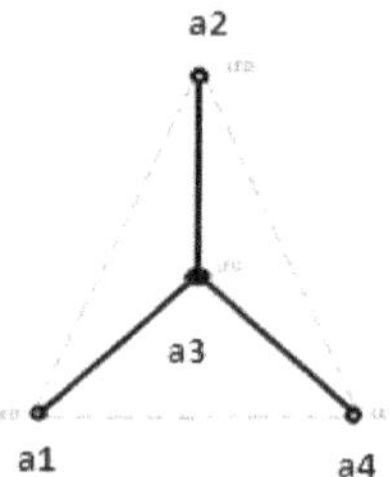

Fig. 3.1 Position of IR LEDs in the vehicle

Therefore, the problem now is to obtain the vector $P = [X_v, Y_v, Z_v, \Phi_v, \theta_v, \Psi_v]$ of the vehicle by knowing the positions of the four IR LED bright spots (X1, Y1), (X2, Y2), (X3, Y3) and (X4, Y4) respectively.

To find the distance between the bright spots, I have used the distance formula,

$$D = \sqrt{(x1 - x2)^2 + (y1 - y2)^2} \qquad .....(1)$$

Where,

(x1, y1) are the x & y co-ordinates of the bright spot 1 in the image.
(x2, y2) are the x & y co-ordinates of the bright spot 2 in the image.

Like that the distance between each point is calculated by the following way.

$$a12 = \sqrt{(x1 - x2)^2 + (y1 - y2)^2} \qquad \text{.....(2)}$$

$$a13 = \sqrt{(x1 - x3)^2 + (y1 - y3)^2} \qquad \text{.....(3)}$$

$$a14 = \sqrt{(x1 - x4)^2 + (y1 - y4)^2} \qquad \text{.....(4)}$$

$$a21 = \sqrt{(x2 - x1)^2 + (y2 - y1)^2} \qquad \text{.....(5)}$$

$$a23 = \sqrt{(x2 - x3)^2 + (y2 - y3)^2} \qquad \text{.....(6)}$$

$$a24 = \sqrt{(x2 - x4)^2 + (y2 - y4)^2} \qquad \text{.....(7)}$$

$$a31 = \sqrt{(x3 - x1)^2 + (y3 - y1)^2} \qquad \text{.....(8)}$$

$$a32 = \sqrt{(x3 - x2)^2 + (y3 - y2)^2} \qquad \text{.....(9)}$$

$$a34 = \sqrt{(x3 - x4)^2 + (y3 - y4)^2} \qquad \text{.....(10)}$$

Then the area of this triangle will be calculated by the Heron's formula.

$$\text{Area of the triangle} = \sqrt{[s(s - a) \times (s - b) \times (s - c)]} \qquad (11)$$

$$\text{Where } s = (a + b + c)/2$$

By calculating the area of that triangle, the vehicle can be hovered in a preselected altitude. Because the area of the triangle will be varying when the altitude varies. Here the camera is placed in the roof so that if MAV move upwards then the area will be increased and vice versa.

IMPLEMENTATION

The previous section shows that the position and orientation of the UAV can be determined when the information about the IR LEDs bright spots is available. In this project, camera is used as the visual sensor. Here an ordinary web camera is converted into an IR camera by the following process.

First the web camera is dismantled and the small lens is pushed out.

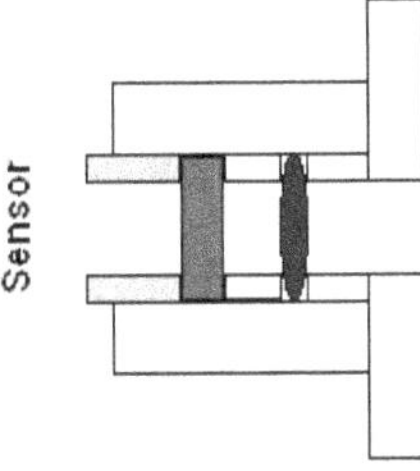

Figure 4.1 IR Filter in Web camera

The red one is the IR filter. Remove the IR filter. Then make a new filter that blocks visible light and only lets IR through it.

Figure 4.2 Filter fitted in to the camera

In order to identify the co-ordinates of the IR LEDs bright spots, different techniques can be used. The Web camera is interfaced with the image acquisition toolbox in MATLAB. And a video object is created by the toolbox. The pixel selected should be remaining the same. Here the pixel taken is 320x240. A camera is used to capture the video of the UAV. And the images will be extracted from the video. Then image is analysed, to determine the IR LEDs bright spots coordinates and then supply the required information to the UAV. This technique needs a communication between the UAV and the ground station to communicate the data of the IR LEDs bright spots coordinates. Moreover, this option can be used in small area applications. This choice increases the independence of the vehicle and makes the system flexible to be used in different environments. Then, a computer vision algorithm can be used to analyse the captured bright spots in the image. Then the distance between each bright spots will be calculated by the Matlab coding. And this distance between the bright spots will change according to the attitude of the vehicle. The distance between the IR LEDs in the vehicle will remain same, but in the image the co-ordinates of the bright spots will be changing according to the movement of the vehicle. This is the concept of the project. Here the video will be taken continuously and frames will be extracted from the video and the distance between the bright spots will be calculated with an interval of 0.01 seconds.By tracking the position of the four LEDs the computer will calculate the attitude, height, velocity, and position of the UAV, by MATLAB coding.

Then a Graphical User Interface is created in MATLAB. In that we will get the status of the vehicle. Here we are not using any RC transmitter to control the vehicle. The wireless transmission is done by using the Zigbee. Here two zigbees are used i.e. one will be in the Vehicle and another will in the GCS. And once we get all the necessary data such as the position and distance between the bright spots, the vehicle will be positioned and by maintaining the area of the three bright spots (area of triangle) the tri copter is hovered in a preselected altitude and it will be in the FOV of the web camera.

SIMULATION AND RESULTS

The Camera system has been made to track the four IR LEDs is shown in the figure below

Figure 5.1 4 IR LEDs are tracked

A single frame is extracted from the video and the co-ordinates and the distance between each bright spots have been calculated. The co-ordinates and distance of the IR LEDs has been calculated for various positions of the vehicle. The important thing is that we need only four bright spots, but in the image we had some noise so it is reduced and that image is shown below

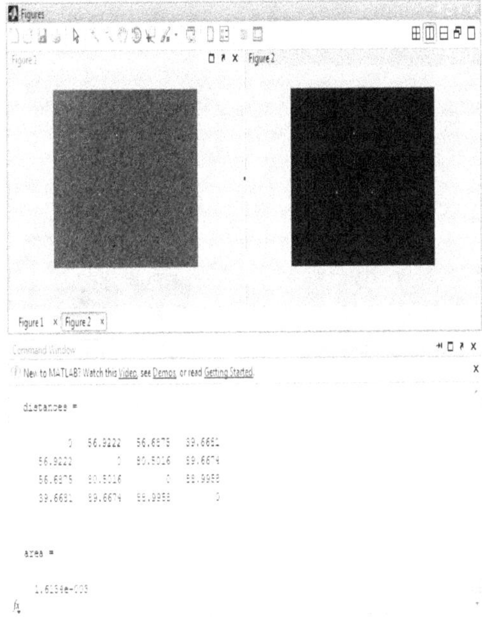

Fig 5.2 co-ordinates and distances of the IR LED

When the position of the vehicle is changed then the co-ordinates and distance between each bright spots is also changing, it shown in the figure below

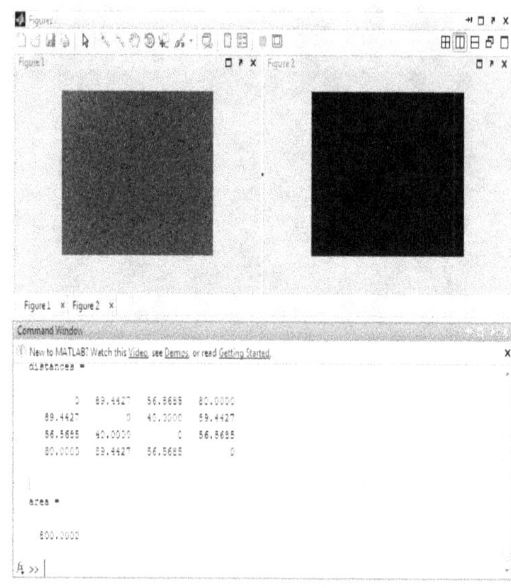

Fig 5.3 co-ordinates and distances of the IR LED when position changed

The co-ordinates and distance between each bright spots when the vehicle is yawed is shown in the figure below

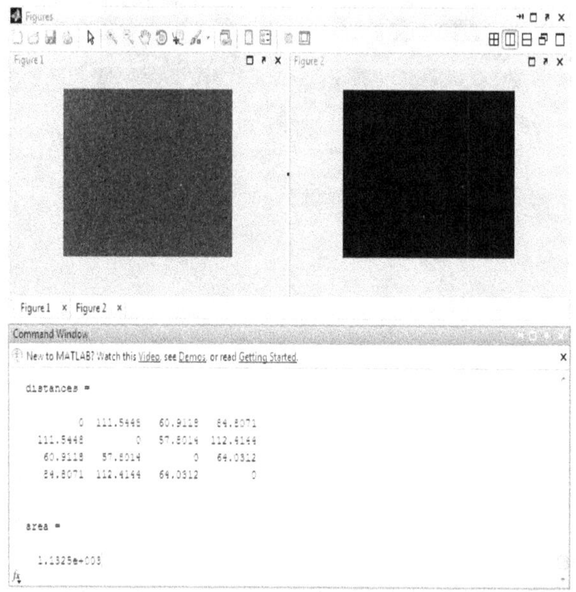

Fig 5.4 co-ordinates and distances of the IR LED when yawed

When the position of the vehicle is changed then the co-ordinates and distance between each bright when the vehicle is pitched up, it shown in the figure below

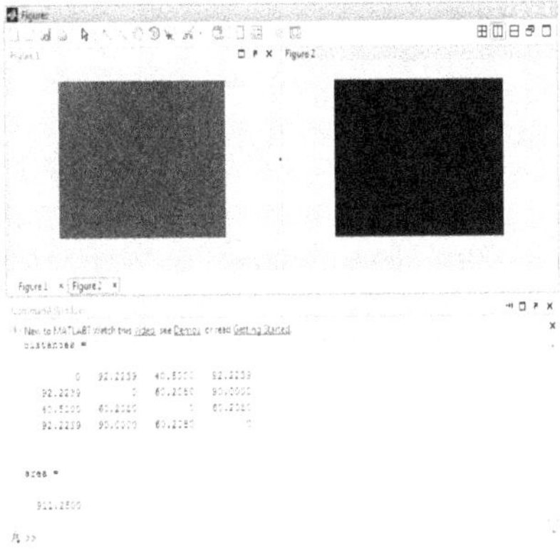

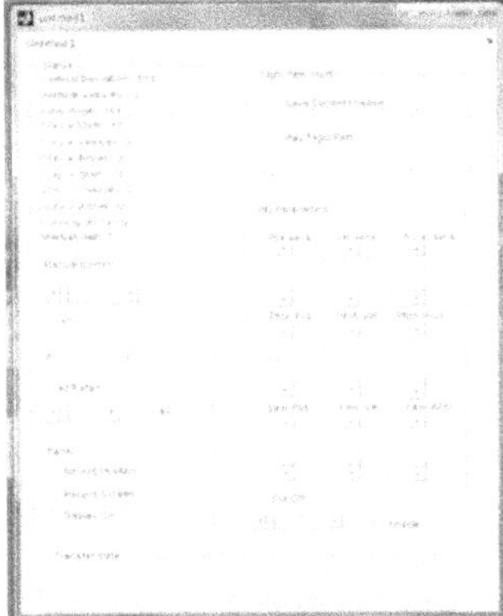

Fig 5.5 co-ordinates and distances of the IR LED when pitched up

Fig 5.6 GUI

A Graphical User Interface has been created in MATLAB using the GUIDE. And in this the status of the vehicle will be displayed.

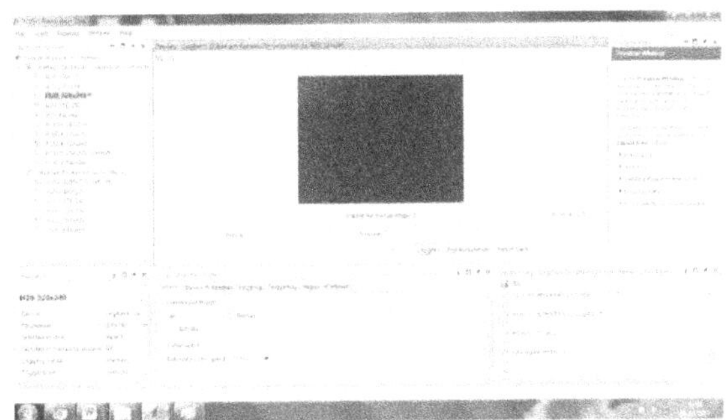

Fig 5.7 Video object created in Image acquisition tool box.

CONCLUSION

"Refinement in man and science is a never ending process" .This project proposes a new, cost-effective and simple navigation system for UAVs in the indoor applications.. The system provides full information about the position and orientation of the UAV using four IR LEDs fixed to the vehicle's body. A computer vision algorithm is needed to identify the dots and determine their positions. A simple simulation study is conducted to show the ability of the proposed navigation system to record the status of the unmanned aerial vehicle.

REFERENCES

1. "Designing Simple Indoor Navigation System for UAVs" (June 20-23, 2011), Mohamed Kara Mohamed, SouravPatra and Alexander Lanzon, the University of Manchester, UK.

2. "Indoor Localization Techniques based on Wireless Sensor Networks" Hyo-Sung Ahn1 and Wonpil Yu2 1Department of Mechatronics, Gwangju Institute of Science and Technology (GIST) 2Electronics and Telecommunications Research Institute (ETRI) Korea.

3. "Dynamic Modelling and Stabilization Techniques for Tri-Rotor Unmanned Aerial Vehicles" Dong-Wan Yoo*, Hyon-Dong Oh*, Dae-Yeon Won* and Min-JeaTahk** Department of Aerospace Engineering, Korea Advanced Institute of Science and Technology, Daejeon 305-701, Korea.

4. "Foot motion tracking using Vision" Tri-NhutDo , Sch. of Elect. Eng., Univ. of Ulsan, Ulsan, South Korea Young-SooSuh.

5. "Design of Wii mote indoor localization technology for Omni-directional vehicle trajectory control" Gu, D. . Cheng-Kung Univ., Tainan, Taiwan.

6. "Real-time awake animal motion tracking system for SPECT imaging", Goddard, J. S. Oak Ridge National Laboratory, TN 37831 USA.

7. "Robust Infrared Vehicle Tracking across Target Pose Change using L1 Regularization" Haibin Ling1, Li Bai2, Erik Blasch3, and Xue Mei4 1Computer and Information Science Department, Temple University, Philadelphia, PA U.S.A.

8. "T. Hamel and R. Mahony, "Attitude estimation on SO(3) based on direct inertial measurements," in IEEE International Conference on Robotics and Automation, ICRA'06, Orlando, Florida, may 2006, pp. 2170–2175. [3] N. Guénard, T. Hamel, and L. Eck, "Control laws for the tele-operation of an unmanned aerial vehicle known as x4-flyer," in IEEE/RSJ Interna-tional Conference on Intelligent Robots and Systems, IROS'06, Beijing,China, oct 2006, pp. 3249–3254.

9. "2D simultaneous localization and mapping for micro aerial vehicles," Angeli, D. Filliat, S. Doncieux, and J.-A. Meyer, in European Micro Aerial Vehicles (EMAV 2006), 2006.

10. "Adaptive planning horizon based on information velocity for vision-based navigation," E. Frew, J. Langelaan, and M. Stachura, in AIAA Guidance, Navigation and Controls Conference, 2007

DESIGN AND DEVELOPMENT OF AN AUTOPILOT AIR VEHICLES

K.Dheeraj Kumar, M.Shreya Reddy, S.Vinay and Anu Sivadass

INTRODUCTION

There are not yet autonomous flying robots capable of manoeuvring in small cluttered environments as insects do. Encouraged by this observation, this paper presents the development of flying robots and control systems going one step toward fully autonomous indoor aerial navigation. The substantial weight and energy constraints imposed by this indoor flying robots preclude the use of powerful processors and active distance sensors.

Micro air vehicles are either fixed-wing aircraft , rotary-wing aircraft (helicopter), or flapping-wing (of which the ornithopter is a subset) designs; with each being used for different purposes. Fixed-wing craft require higher, forward flight speeds to stay airborne, and are therefore able to cover longer distances; however they are unable to effectively manoeuvre inside structures such as buildings. Rotary-wing designs allow the craft to hover and move in any direction, at the cost of requiring closer proximity for launch and recovery. Flapping-wing-powered flight has yet to reach the same level of maturity as fixed-wing and rotary-wing designs. However, flapping-wing designs, if fully realized, would boast a manoeuvrability that is superior to both fixed- and rotary-wing designs due to the extremely high wing loadings achieved via unsteady aerodynamics.

MAVs have been built for hobby purposes, such as aerial robotics contests and aerial photography.

One focus of development involves the batteries used in the MAV's. Batteries that weigh less but produce more power are needed to power MAV's to allow more capabilities while getting smaller and having longer flight times.

Other capabilities under development include:

- Smaller size and lighter – there is active research on "nano air vehicles" the size of insects
- More capabilities – live video and chemical\biological monitoring
- Sensors for tracking enemy troop movements and other activities
- Longer flight times
- Longer range
- Advanced flight control
- Navigation and communications capabilities
- New designs such as flapping wings
- Lower cost
- Fly at higher altitude

GPS

A GPS tracking unit is a device that uses the Global Positioning System to determine the precise location of a vehicle, person, or other asset to which it is attached and to record the position of the asset at regular intervals. The recorded location data can be stored within the tracking unit, or it may be transmitted to a central location data base, or internet-connected computer, using a cellular (GPRS or SMS), radio, or satellite modem embedded in the unit. This allows the asset's location to be displayed against a map backdrop either in real time or when analysing the track later, using GPS tracking software. A GPS tracker essentially contains GPS module to receive the GPS signal and calculate the coordinates. For data loggers it contains large memory to store the coordinates, data pushers additionally contains the GSM/GPRS modem to transmit this information to a central computer either via SMS or via GPRS in form of IP packets.

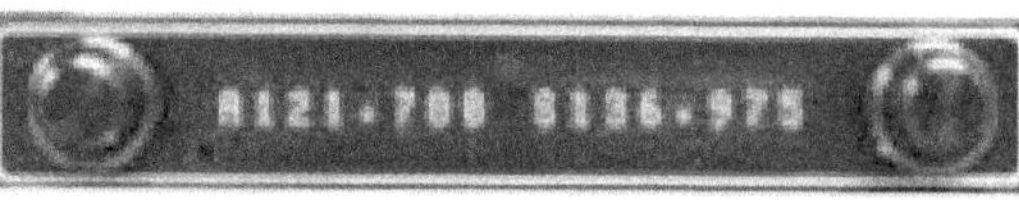
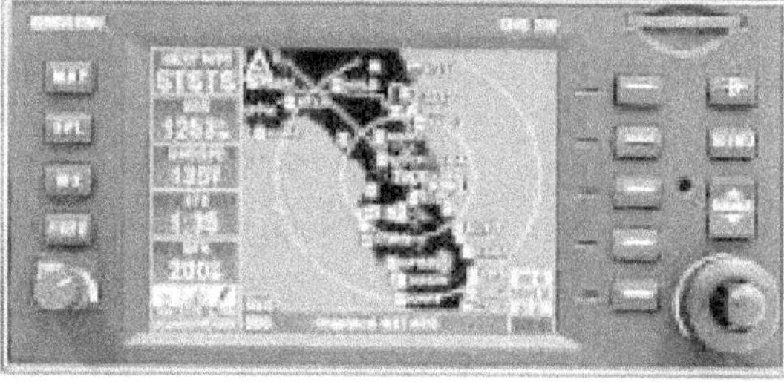

Global Positioning System (GPS) receivers can also determine altitude by trilateration with four or more satellites. In aircraft, altitude determined using autonomous GPS is not precise or accurate enough to supersede the pressure altimeter without using some method of augmentation. In hiking and climbing, it is not uncommon to find that the altitude measured by GPS is off by as much as a thousand meters,[citation needed] if all the available satellites happen to be close to the horizon.

BAROMETRIC ALTITUDE

An altimeter or an altitude meter is an instrument used to measure the altitude of an object above a fixed level. The measurement of altitude is called altimetry.It is an instrument optional in off-road vehicles to aid in navigation.

Altitude can be determined based on the measurement of atmospheric pressure. The greater the altitude the lower the pressure. When a barometer is supplied with a nonlinear calibration so as to indicate altitude, the instrument is called a pressure altimeter or barometric altimeter. A pressure altimeter is the altimeter found in most aircraft, and skydivers use wrist-mounted versions for similar purposes. Hikers and mountain climbers use wrist-mounted or hand-held altimeters, in addition to other navigational tools such as a map, magnetic compass, or GPS receiver.

A barometric altimeter, used along with a topographic map, can help to verify one's location. It is more reliable, and often more accurate, than a GPS receiver for measuring altitude; GPS altimeters may be unavailable, for example, when one is deep in a canyon, or may give wildly inaccurate altitudes when all available satellites are near the horizon. Because barometric pressure changes with the weather, hikers must periodically recalibrate their altimeters when they reach a known altitude, such as a trail junction or peak marked on a topographical map.

The barometric (baro) altimeter, measures altitude based on a model of the atmosphere, in particular, how the pressure and temperature of the air changes with altitude, the higher the altitude, the lower the pressure and the lower the temperture. The standard pressure at sea level is 29.92 inches of mercury and the standard temperature is 15 degrees C or 59 degrees F. If the pressure is other than the standard, the altimeter provides a barometer setting to adjust to the non standard value. But, there is no capability to adjust for a non standard temperature.

In aircraft, an aneroid barometer measures the atmospheric pressure from a static port outside the aircraft. Air pressure decreases with an increase of altitude—approximately 100 hectopascals per 800 meters or one inch of mercury per 1000 feet near sea level.

A radar altimeter measures altitude more directly, using the time taken for a radio signal to reflect from the surface back to the aircraft. The radar altimeter is used to measure height above ground level during landing in commercial and military aircraft. Radar altimeters are also a component of terrain avoidance warning systems, warning the pilot if the aircraft is flying too low, or if there is rising terrain ahead. Radar altimeter technology is also used in terrain-following radar allowing fighter aircraft to fly at very low altitude.

Spaceborne radar altimeters have proven to be superb tools for mapping ocean-surface topography, the hills and valleys of the sea surface. These instruments send a microwave pulse to the ocean's surface and record the time it takes to return. A microwave radiometer corrects any delay that may be caused by water vapor in the atmosphere. Other corrections are also required to account for the influence of electrons in the ionosphere and the dry air mass of the atmosphere.

Combining these data with the precise location of the spacecraft makes it possible to determine sea-surface height to within a few centimetres (about one inch). The strength and shape of the returning signal also provides information on wind speed and the height of ocean waves. These data are used in ocean models to calculate the speed and direction of ocean currents and the amount and location of heat stored in the ocean, which in turn reveals global climate variations.

Attitude Flight Stabilization System of an Autopilot

The autopilot utilizes a 3-axis gyroscope and tri-axial accelerometer to form an accurate drift free attitude stabilization system. The unit also utilizes Global Positioning System (GPS) and a barometric altitude sensing for accurate 3-dimentional positioning of the aircraft. By combining attitude control and positioning, a comprehensive inertial navigation system and autopilot is provided to you in a small compact light-weight package.

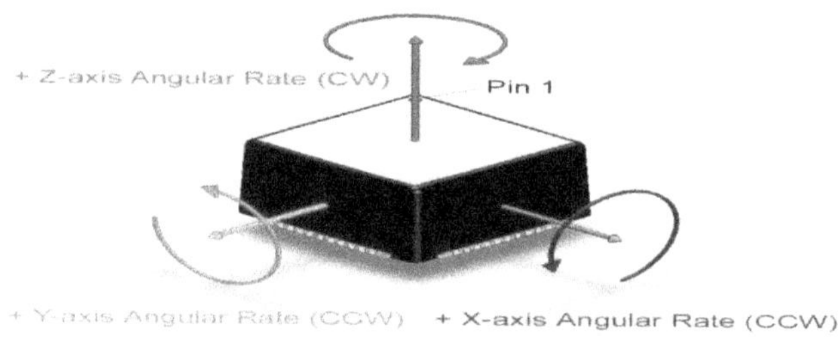

Fig: 3-axis gyroscope

Fig: Tri-axial Accelerometer

The unit calculates the aircraft attitude in 3-dimentions and detects any changes to the model's horizontal position. If attitude change occurs, controlling signals will be sent out to the plane's ailerons, elevator and rudder to counter that change. By continuously doing this, the plane is kept in a state of stabilized equilibrium, resulting in a smooth level flight.

When activated, release the flight control sticks so that they return to the transmitter *neutral* (middle) position. The model will immediately revert to level flight. The unit can be activated or de-activated via a spare channel from your receiver.

GPS, Barometric Sensing and Autopilot

Upon initial boot up, the autopilot will search for GPS positioning signals. When a minimum of 3 satellites have been detected, a fixed position is established. The autopilot will record that position as the return to launch (RTL) point.

The on-board barometric sensor and GPS altitude readings will be combined to establish an accurate relative-altitude of the aircraft.

When the Autopilot Mode is activated via a spare channel, the aircraft will automatically turn and fly back to the take off point (RTL). The aircraft altitude will be constantly maintained. After reaching the home point, the plane will automatically circle with a radius of 120 meters. By using the same autopilot algorithm, the pilot can also activate an auto circling flying pattern at a fixed altitude anywhere he wishes.

DEVICE FUNCTIONS

Attitude Flight Stabilization System

- **Constant stabilized flight in any condition** – the aircraft will automatically level the flight attitude of your aircraft in any weather condition.

- **Emergency Recovery** – The aircraftt will immediately bring the plane back to stabilized level flight.

- **Precision Flying**: It can level and precise flight paths, especially when flying and landing in strong wind.

Auto stabilizer is used in the plane. It has eight pins, of which four are connected to the receiver and other four to the components. It enhances two-way communication.

It automatically corrects flight of the plane when there is any deviation.

GPS and Barometric Sensor

- **Fixed altitude flight** - When this function is activated, the unit will maintain aircraft altitude.

- **Return to Launch (RTL)** – The aircraft will return to the take-off point.

- **Fixed point circling** – by activating this function, the aircraft will automatically circle the selected area at a fixed altitude. Very useful for aerial photography.

We use two GPS modules, of which one is connected to the OSD and the other to the plane to record the position of the plane.

OSD

An on-screen display (abbreviated OSD) is an image superimposed on a screen picture, commonly used by modern television sets, VCRs, and DVD players to display information such as volume, channel, and time.

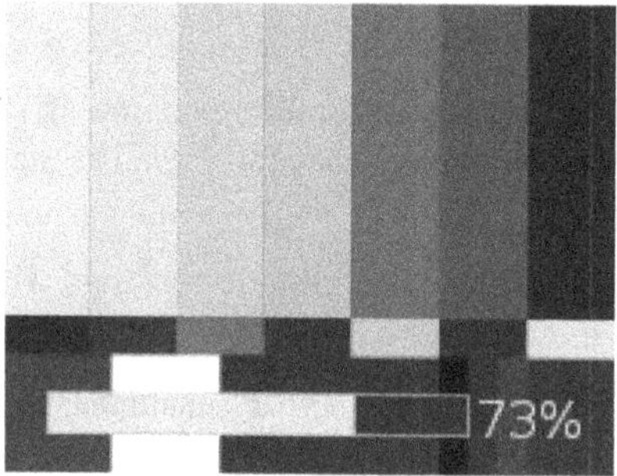

OSD is used in our model plane to display video shoot by the video camera. It displays the surroundings of the plane and also gives details about the degree of stabilization, altitude of the plane and its distance from the home.

Technical Specification and working requirements:

- Working voltage : 4.0~6.0 Volt
- Current draw : 52mA (at 5V)
- Size : 55 x 33 x 20 mm
- Weight (exclude wire)　 : 20g
- Working Temperature : -25°C~ +70°C
- Maximum rate of rotating : ≤ 1200 °/s

Autonomous Flight Modes

Two autopilot modes are incorporated into the autopilot algorithm:

AUTOPILOT MODE 1: Deactivated. The autopilot function is not activated.

AUTOPILOT MODE 2: Auto Return to Launch (RTL). When activated in this mode, the FY-21AP will automatically fly the plane to the take-off point, while maintaining altitude.

Upon reaching the launch area, the unit will automatically fly the plane in a circle at a default circling radius of 120 meters.

AUTOPILOT MODE 3: **Auto circling mode (ACM)**. When activated in this mode, the plane will immediately fly in a circle. The centre of the circle is the point of activation. The default circle radius is 120 meters. The aircraft altitude will be automatically maintained throughout the ACM.

Electrical Connection and Diagram

- **Power supply**
- The aircraft operates between 4 to 6 volts input.

- It is powered via the Receiver connection.

- If your plane is Electric powered, the Receiver power supply is normally from the ESC built-in Battery Elimination Circuit (BEC).

- Alternately, you can supply the Receiver and aircraft via a separate BEC.

- For Gas or Nitro powered planes, you will require a battery to power the Receiver and aircraft.

- Connection between the FY-21AP and Receiver output is via the supplied wire:

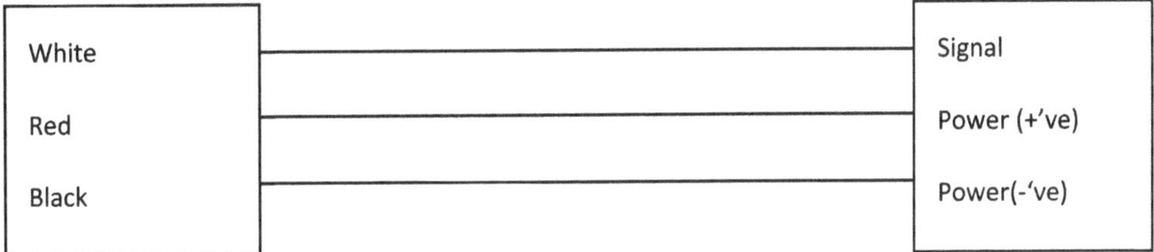

➤ **Connection to RC Receiver**

- FY-21AP requires a minimum of 6-channel RC receiver.
- 3 Receiver channels are used for aileron, elevator and rudder signal output. Connect this 3 receiver output signals to the autopilot with the supplied wires.
- 2 free Receiver channels are required to control the autopilot Flight Modes (3-position switch) and Autopilot Mode (3-position switch).
- Example of 6 channel receiver utilization .

  Channel 1 = Aileron Signal Output

  Channel 2 = Elevator Signal Output

  Channel 3 = Throttle Signal Output

  Channel 4 = Rudder Signal Output,

  Channel 5 = Switch IN1 controlled by radio 3-position switch for Flight Modes.

  Channel 6 = Switch IN2 controlled by radio 3-position switch for Autopilot Modes.

Plane Connection Layout:

1. Normal / Traditional servo layout:

Servo 1	Servo 2	Servo 3
Aileron servo	Elevator servo	Rudder servo

2. Plane without aileron servo:

Servo 1	Servo 2	Servo 3
Rudder servo	Elevator servo	Don't connect

3. Delta-winged plane servo:

Servo 1	Servo 2	Servo 3
Differential servo 1	Differential servo 2	Rudder servo

CONCLUSION

An **autopilot** is a mechanical, electrical, or hydraulic system used to guide a vehicle without assistance from a human being.The autopilot of an aircraft is sometimes referred to as "George", after one of the key contributors to its development.The hardware of an autopilot varies from implementation to implementation, but is generally designed with redundancy and reliability as foremost considerationsSoftware and hardware in an autopilot is tightly controlled, and extensive test procedures are put in place.

MODERN CONTROL LAW DESIGN AND DEVELOPMENT OF DYNAMIC ATTITUDE CONTROL FOR MODERN MULTIROTOR UNMANNED AERIAL VEHICLE

Mohamed Rasheed A[1], Dr K. Senthil Kumar [2], and Magesh Giridharan T [3]

[1] PH.D Scholar, Division of Avionics, Department of Aerospace Engineering, MADRAS INSTITUTE OF TECHNOLOGY, ANNA UNIVERSITY, Chennai-600044.

[2] Associate Professor, Division of Avionics, Department of Aerospace Engineering, MADRAS INSTITUTE OF TECHNOLOGY, ANNA UNIVERSITY, Chennai-600044.

[3] P.G Student, Division of Avionics, Department of Aerospace Engineering, MADRAS INSTITUTE OF TECHNOLOGY, ANNA UNIVERSITY, Chennai-600044.

Abstract:

The main purpose of this paper is to develop the modern control algorithm for modern multi-rotor UAV (more than 6 rotors). The control law is designed in such a way that, even in the failure of any one of the power system, the unmanned aerial vehicle will sustain its flight and completes the mission successfully. This paper mainly concentrates in the design & development of mathematical model and modern control law design for multi-rotor unmanned aerial vehicle, which will make system reliable and more tolerant towards the power failure.

Keywords: UAV, control law, power system, multi-rotor

1. INTRODUCTION

Recently, the significance of the small-size unmanned aerial vehicle (UAV) has grown enormously due to its heightened necessity in missions such as surveillance and reconnaissance in military operations, rescue missions in disaster sites, the acquisition of visual information from steep terrains, etc., where manned or regular-sized aerial vehicles are likely to fail to accomplish the above-mentioned missions even with their full operational capabilities. Concerning the situations mentioned above, the UAV committed to such missions requires the capability of vertical take-off and landing (VTOL), as well as the ability to achieve rapid and stable motion in every direction. To satisfy all of the above requirements, the rotary wing type aerial vehicle is selected as the best choice. Rotary wing aircraft, otherwise known as 'rotorcraft', can be categorized into several types depending on the number of rotors installed in the system. The advantages of multi-rotor UAVs include hovering, VTOL and agile mobility capabilities. In this kind of rotor wing UAVs, the brushless motors are playing a vital role. Intuitively, it can be inferred that increasing the number of rotors will require more power to operate, meaning that a UAV with a lower number of rotors will require less power than one with more rotors. The most widely known type of multi-rotor UAV is probably the quad-rotor UAV. But in case of any failure in any one of the motor then the UAV cannot able to sustain in flight and it would led to a catastrophic damage. Moreover the mission will be unsuccessful. So in spite of the power requirement, it is necessary to go for a hexa-rotor configuration. In this paper a control law has been designed for the multi rotor UAV (more than four rotors) to sustain in the flight condition and to complete the mission in a successful manner even in the failure of any one of the motor. The paper is organized as follows. First the position and attitude control for multi rotor UAV is discussed. Then the next section deals with the adaptive control law that designed for the multi rotor UAV. The control law that we designed is applicable for both mini UAV and the MAV.

2. POSITION AND ATTITUDE CONTROL FOR MULTI ROTOR MAV

An attitude controller stabilizes the orientation of the aircraft in the 3D space. In other words the attitude controller tries to push the movement around each of the 3 axes (x, y, z) towards a set point (which can be given for example by the remote control).A position controller tries to make the aircraft fly towards a specific point in the world. In the case of a quad copter movement of the vehicle along the X, Y, Z are controlled by adjusting the force vector of the vehicle. In case of a quad copter, the thrust vector always points along the negative z direction of the body frame (upwards in a normal hovering position). By tilting the quad copter around the roll (x) or pitch (y) axis the orientation of the thrust vector can be changed. Therefore, to control position (via the force vector) we have to control the attitude of the quad copter. The fact that position is controlled by changing the attitude leads to a control setup where position is controlled in an outer loop and attitude in an inner loop.

If the system state is manual, attitude control is performed. The attitude controller reads the set points from the remote control values. If the system state is auto, attitude control is performed too. However, the attitude set points are generated in the outer position controller loop. Currently no altitude controller is implemented.

Rate controllers control the rate of a value (the time derivative of a value). The rate controller does not "see" the value itself. All it cares about is the rate which the controller tries to push towards a set point. It is common to use rate controllers for fast dynamics in an inner loop. The outer loop (sometimes referred to as position controller) controls the value itself and gives a set point for the rate controller (because the value is obviously changed by the rate). The inputs for the attitude controller are attitude set point (set e. g. by remote control): roll, pitch and yaw angles and current attitude. And the output will be motor control signals: roll, pitch and yaw and thrust. The attitude controller is realized using three separate SISO (single input, single output) control loops. We are allowed to simplify the control problem to separate SISO controllers as roll, pitch and yaw can be viewed decoupled, as long as we fly close to hover conditions and the angles are relatively small. Every SISO controller consists of a PID controller as shown in fig 2.1

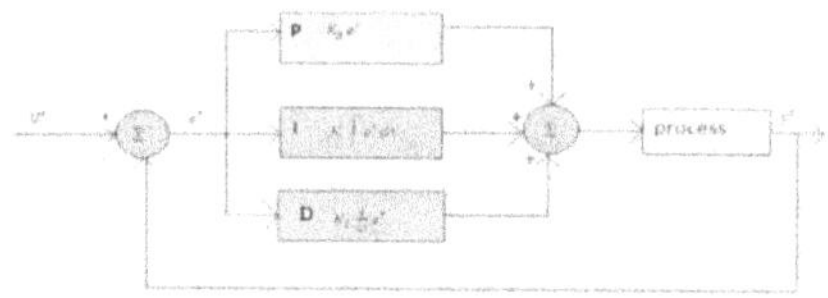

Fig 2.1 PID controllers

3. ADAPTIVE CONTROL LAW

The most widely known type of multi-rotor UAV is probably the quad-rotor UAV. But a failure in any one motor will lead to catastrophical damage and the mission will be failed. So here we have taken a micro hex-rotor platform as shown in the figure 3.1 and our

control algorithm is successfully implemented in it.

Fig 3.1 Micro Aerial Vehicle

The system architecture of the MAV and the controller overview is shown in figure 3.2 and figure 3.3 respectively. The control law implemented by the appropriate motor output.

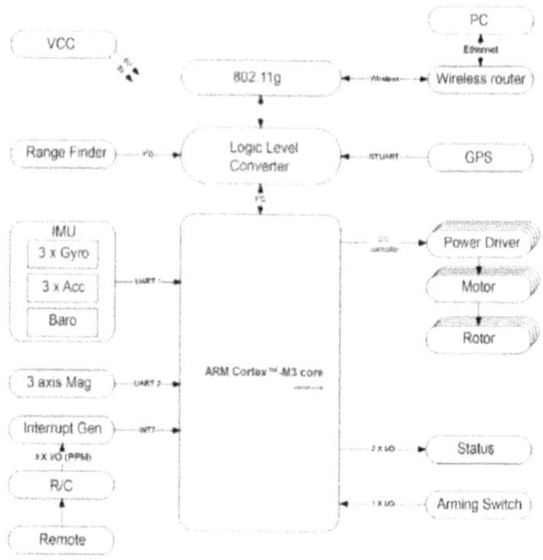

Fig 3.2 System architecture

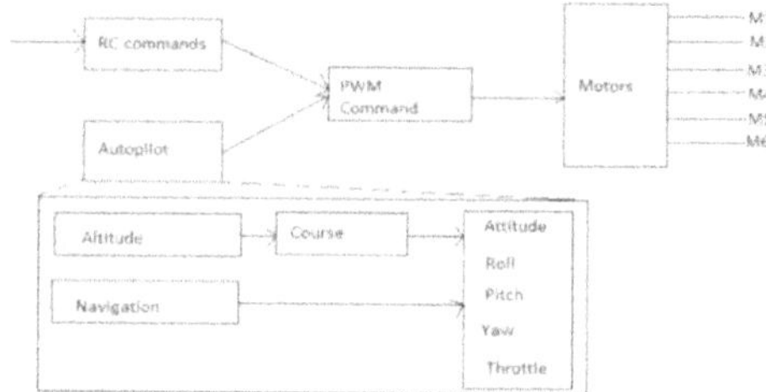

Fig 3.3 Controller over view for the 6 rotor MAV

3.1 CONTROL LAW DESIGN

The main aim is to prevent the multi rotor from falling down on the occasion of failure of one of its rotor. The algorithm provides an alternative of keeping the vehicle in air.

For a multi rotor, the roll, pitch, yaw and vertical axis are contributed by the combination of rotors outputs. This procedure is called Mixer control. The proportion of rotors output is decided by the mechanical structure.

- Motor output = Y × PWM COMMAND (B or E or T or R).
- Torque produced by motor = Motor output ×Force arm of motor (L) = Y × PWM COMMAND (B or E or T or R)×L
- The range of C is from -100% to 100%. Maximum C in the same column is 100%.The bigger absolute value of Y. The great effect of PWM COMMAND on motor output. PWM COMMAND will not affect motor output when C is 0, which means the motor output is fixed.
- Each motor has four different output coefficients: YT, YY, YP, YR. E.g.YY2representscoefficient of A2 in yaw control; YR5represents coefficient of A5 in roll control.
- Motor output is relative to its rotation speed. The bigger output, the faster rotation speed. Negative output does not represent counter rotation, but slower rotation speed. Motor is still spinning if its output is 0.
    a. Throttle PWM COMMAND (T): T<0,multi-rotor moves down; T>0multi-rotor moves up;
    b. Rudder PWM COMMAND (R): R<0,multi-rotor nose left; R>0,multi-rotor nose right;
    c. Elevator PWM COMMAND (E): E<0multi-rotor moves backward; E>0,multi-rotor moves forward
    d. Aileron PWM COMMAND (B): B<0, multi-rotor moves left; B>0, multi-rotor moves right.
- Multi-rotor should keep balance along all the other axis when moves along one axis:
- To keep throttle direction balance, sum of all motors output should be 0 when apply rudder or pitch or roll pwm command:

- To keep yaw direction balance, sum of counter clockwise motors 'output should be equal to sum of clockwise motors output when apply throttle or pitch or roll command;
- To keep pitch direction balance, total torques produced by motors at each side of pitch axis should be the same when apply throttle or rudder or roll command;
- To keep roll direction balance, total torques produced by motors at each side of roll axis
- To pitch or roll control, proportion of coefficients of the motors at the same side of pitch or roll axis should be equal to the proportion of force arms of those motors: $Ym/Yn= Lm/Ln$; Coefficient is 0% if the force arm of that motor is 0.

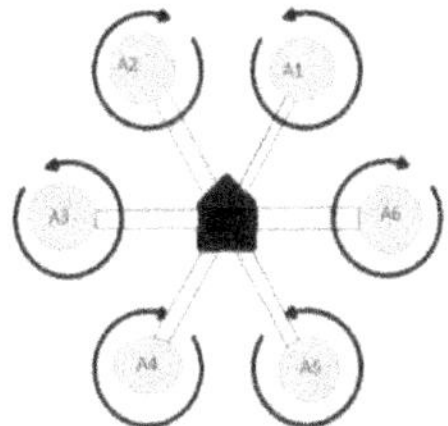

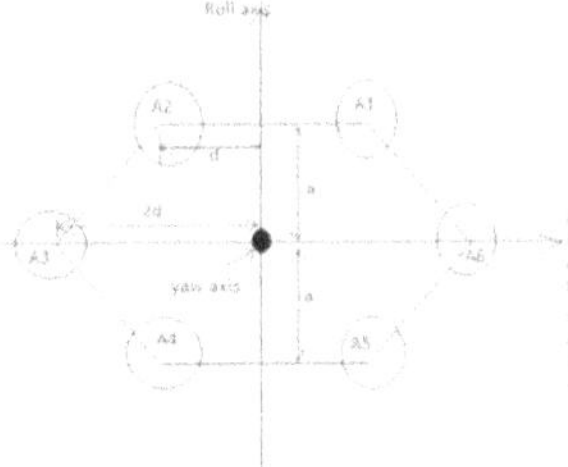

Fig 3.4 6 rotor UAV

Now we take the 6 rotor UAV as shown in figure 3.4 as an illustration to introduce customize motor out.

3.1.1 THROTTLE EQUATIONS:

When T<0, multi-rotor moves down; T>0 multi-rotor moves up, we can choose the following setup:

$$(Y_{T1} + Y_{T2} + Y_{T3}) \times T = (Y_{T2}+Y_{T4} +Y_{T5}) \times T \text{ (To keep yaw direction balance)} \quad (1)$$

$$(Y_{T1}+Y_{T2}) \times T \times a = (Y_{T4}+Y_{T5}) \times T \times a \text{ (To keep pitch direction balance)} \quad (2)$$

$(Y_{T2} + Y_{T4} + 2Y_{T3}) \times T \times d = (Y_{T1} + Y_{T5} + 2Y_{T6}) \times T \times d$ (To keep roll direction balance)   (3)

If while pushing the throttle command, the sum of all motors output $(Y_{T1}+Y_{T2} + Y_{T3}+Y_{T4}+Y_{T5}) \times T$ is positive, then multi-rotor moves up; the sum of all motors output $(Y_{T1}+Y_{T2} + Y_{T3}+Y_{T4}+Y_{T5}) \times T$ is negative, then multi-rotor moves down. And the balance along all the other axis can be derived by substituting the throttle command into equations set given above.

3.1.2 YAW EQUATIONS:

The movement about yaw axis is produced by the counter torque force from the rotation of propeller. In our example, A1 A3 A5 produce clockwise torque force; A2 A4 A6 produce counter clockwise torque force. When the hexa-rotor is hovering, all the rotors are spinning at the same angular velocity, which means the clockwise torque force equals to counter clockwise torque force, and this produces exactly 0 angular acceleration about yaw axis. Therefore, when the rotate speed of A1 A3 A5 is larger than A2 A4 A6, hexa-rotor spins clockwise; when the rotate speed of A1 A3 A5 is smaller than A2 A4 A6, hexa-rotor spins counter clockwise.

For the rotor to keep balance along all the other axes when apply the yaw command:

$(Y_{R1} + Y_{R2} + Y_{R3} + Y_{R4} + Y_{R4} + Y_{R5}+Y_{R6}) \times R - 0$
(To keep throttle direction balance)   (4)

$(Y_{R1} + Y_{R2}) \times R \times a = (Y_{R4} + Y_{R5}) \times R \times a$ (To keep pitch direction balance)   (5)

$(Y_{R2} + Y_{R4} + 2Y_{R3}) \times R \times d = (Y_{R1} + Y_{R5} + 2Y_{R6}) \times R \times d$ (To keep roll direction balance)   (6)

R<0, multi-rotor nose left; R>0.multi-rotor nose right

Now if move the yaw right, the sum of A1, A3, A5 output (YR1+ YR3+ YR5) ×R is positive, the sum of A2, A4, A6 output (YR2+ YR4+ YR6) ×R is negative, then the clockwise torque force is larger than counter clockwise torque force, multi-rotor nose right; if move the yaw left, the sum of A1, A3, A5 output (YR1+ YR3+ YR5) ×R is negative, the sum of A2, A4, A6 output (YR2+ YR4+ YR6) ×R is positive, then the clockwise torque force is smaller than counter clockwise torque force, multi-rotor nose left.

3.1.2 PITCH EQUATIONS:

The movement about the pitch axis is produced by the differential output of A1+A2 and A4+A5. Since A3 and A6 are on the pitch axis, they do not contribute and torque. You can just keep the rotation speed of A3 and A6 the same as hovering, so YP3and YP6are 0. Increase output of A4, A5 and decrease output of A1, A2, multi-rotor moves forward; decrease output of A4, A5 and increase output of A1, A2, multi-rotor moves backward. We also want multi-rotor to keep balance along all the other axis when apply the pitch command.

$(Y_{E1} + Y_{E2} + Y_{E3} + Y_{E4} + Y_{E5}+ Y_{E6}) \times E = 0$ (To keep throttle direction balance)   (7)

$(Y_{E1} + Y_{E3} + Y_{E5}) \times E = (Y_{E2} + Y_{E4}+ Y_{E6}) \times E$ (To keep yaw direction balance)   (8)

$(Y_{E2} + Y_{E4} + 2Y_{E3}) \times E \times d = (Y_{E1} + Y_{E5} + 2Y_{E6}) \times E \times d$ (To keep roll direction balance)   (9)

Also proportion of coefficients of the motors at the same side of pitch axis should be equal to the proportion of force arms of those motors: $Y_{E1} : Y_{E2} = Y_{E4} : Y_{E5} = a: a = 1: 1$.

As we defined before: E<0 multi-rotor moves backward; E>0, multi-rotor moves forward.

Now if push the pitch command, the sum of A1, A2 output (YE1+ YE2) ×E is negative, the sum of A4, A5 output (YE4+ YE5) ×E is positive, then multi-rotor moves forward; if pull the pitch command, the sum of A1, A2 output (YE1+ YE2) ×E is positive, the sum of A4, A5 output (YE4+ YE5) ×E is negative, then multi-rotor moves backward. And the balance along all the other axes can be derived by substituting the pitch command into the above equations set.

3.1.3 ROLL EQUATIONS:

The theory of movement about the roll axis is the same with pitch axis. However there is no motor on the axis in this case, no coefficient is 0%. We also want multi-rotor to keep balance along all the other axes when apply the roll command:

$(Y_{A1} + Y_{A2}+ Y_{A3} + Y_{A4} + Y_{A5} + Y_{A6}) \times A = 0$ (To keep throttle direction balance)   (10)

$(Y_{A1} + Y_{A3} + Y_{A5}) \times A = (Y_{A2} + Y_{A4} + Y_{A6}) \times A$ (To keep yaw direction balance)   (11)

$$(Y_{A1} + Y_{A2}) \times A \times a = (Y_{A4} + Y_{A5}) \times A \times a \quad (\text{To}$$

keep pitch direction balance) $\qquad$ (12)

Also proportion of coefficients of the motors at the same side of roll axis should be equal to the proportion of force arms of those motors:

$$Y_{E2} : Y_{E3} : Y_{E4} = Y_{E1} : Y_{E6} : Y_{E5} = d : 2d : d = 1 : 2 : 1.$$

As we defined before: A<0, multi-rotor moves left; A>0 multi-rotor moves right.

Now if move the roll, the sum of A2, A3, A4 output $(Y_{A2} + Y_{A4} + 2Y_{A3}) \times A$ is positive, the sum of A1, A5, A6 output $(Y_{A1} + Y_{A5} + 2Y_{A3}) \times A$ is negative, then multi-rotor moves right; if move the roll command , the sum of A2, A3, A4 output $(Y_{A2} + Y_{A4} + 2Y_{A3}) \times A$ is negative, the sum of A1, A5, A6 output $(Y_{A1} + Y_{A5} + 2Y_{A6}) \times A$ is positive, then multi-rotor moves left. And the balance along all the other axes can be derived by substituting the roll command into equations set.

4. CONCLUSION:

The above control law design was implemented and tested on a six rotor configuration unmanned aerial vehicle with all possible conditions including in flight motor failure. As expected the control law loop worked very clear and had a successful flight with all combination of in-flight motor failure. And thus the above control law concludes that a multirotor UAV with more than four rotors can sustain its flight in the air in case of engine failure, which ensures the system more reliable

REFERENCES

- "Modeling and altitude control of quad-rotor UAV", Keun Uk Lee, Dept. of Electr. & Electron. Eng., Yonsei Univ., Seoul, South Korea ,Young Hun Yun ; Wook Chang ; Jin Bae Park ; Yoon Ho Choi, Control, Automation and Systems (ICCAS), 2011 11th International Conference on, 26-29 Oct. 2011
- "An engineering development of novel hexrotor for 3D applications" D.Langkamp,G Roberts university of Manchester,UK. Proceeding on International Micro air vehicle 2011
- "Design and attitude control of Quad rotor,Tail-Sitter Vertical Takeoff and Landing Unmanned Aerial Vehicle", Atsushi Oosedo[a], Atsushi Konno[b], Takaaki Matsumoto[c],Kenta Go[d], Koji Masuko[e] & Masaru Uchiyama[f] Version of record first published: 13 Apr 2012
- Design and Control of an Indoor Micro Quadrotor, Samir Bouabdallah, Pierpaolo Murrieri, Roland Siegwart, Autonomous Systems Laboratory ,Swiss Federal Institute of Technology, Lausanne, Switzerland
- Multi-functional autopilot design and experiments for rotorcraft-based unmanned aerial vehicles David Hyunchul Shim', Hyoun Jin Kim', Hoam Chung', Shankur Sastry3 University of California, Berkeley, California
- AA Mian and W Daobo, "Modelling and Backstepping-based Nonlinear Control Strategy for a 6 DOF Quadrotor Helicopter", Chinese Journal of Aeronautics, volume 21, 2008
- P Castillo, R Lozano and AE Dzul, "Modelling and Control of Mini-Flying Machines", Springer-Verlag London Limited, 2005
- P Castillo, A Dzul and R Lozano, "Real-Time Stabilization and Tracking of a Four Rotor Mini-Rotorcraft", IEEE Transactions on Control Systems Technology, volume 12, 2004
- S Bouabdallah, "Advances in Unmanned Aerial Vehicles: Design and Control of Miniature Quadrotors", Springer Press, 2007

AN EFFICIENT AIR TRAFFIC MANAGEMENT FOR WEATHER PREDICTION USING HIDDEN MARKOV MODEL (HMM)

Ch Mahesh[1], K.Ravindra[2], V.Kamakshiprasad[3] and R.Rajeshwara Rao[4]

[1]Research Scholar, JNT University, Hyderabad.
[2]Principal,Malla Reddy Institute of Technology and Science ,Hyderabad.
[3]Prof. of Computer Science, School of Information Technology,JNT University, Hyderabad.
[4]Prof.of Computer Science, Mahatma Gandhi Institute of Technology, Hyderabad.

Abstract

Safe, efficient and regular air navigation is possible by providing necessary metrological information to the pilot, controller and other concerned with aviation[1]. Knowledge of weather information not only helps efficient takeoff and landing operations of aircraft which in turn aids the airline industry to conserve fuel and effective handling of air cargo. The root forecast (ROFOR) gives necessary weather information to the aircraft intended flow and influence in real time. Especially, the convective weather is very much stringent and plays a key role. Convective weather poses major challenges to the airport capability and demand. The major contributor to excessive air traffic delays and costs because of convective weather [2]. It is very much required for correct estimation of the airspace due to convective weather impact, which will minimize the delays and costs. To achieve this, a statistical model which accurately estimates the airspace by considering the convective weather is very much required. They should be capable to fit the actual and forecast the weather so it can predict airspace capability accurately. Although, many researchers contributed their work for better capability estimation. However, due to lack of probabilistic models in present methods for rooting decisions the performance is poor.This paper address the air traffic management problems using a Hidden Markov Model(HMM), where the evolution of the weather is modeled as a Markov chain model. In Our study, we propose an optimal state path sequence for aircraft in convective weather conditions using HMM.

Index Terms: Convective weather, hidden Markov model, viterbi search algorithm

INTRODUCTION

Properly predicting and modeling weather impact is very much critical in air traffic flow management (ATFM). Existing methods and forecasting are not sufficient for efficient AFTM. This is because of 1) lack of dynamic modeling of weather in spatial temporal characteristics 2) focus on describing weather rather than weather impact 3) computational complex algorithm. Knowledge of the location and the intensity of the hazardous of convective weather for three to four hours ahead are key to select air routes. The spatial temporal stochastic weather model can provide statistics of interest of various weather impacts.

Knowledge of the metrological parameters and advisors are very much important for the safety of aircraft as well as passengers. Metrological information in the form of forecasts, advices and warnings will help the efficient and cost effective operations of aircraft in all the phrases of flights i.e. Take off, En-Route and Landing phases. By ignoring the warnings, on contrary, leads to the heavy loss of lives and money [3].

The important parameters of metrological and weather systems with concerned to the aircraft are:

- Wind, Temperature, Pressure, Humidity
- Clouds
- Ceiling and Visibility
- Thunderstorms and our convective weather
- Turbulence
- Icing

The forces that effect the flight operations are Lift (L), Thrust (T), Drag (D), and the Weight (W) of the aircraft. This is shown in below Fig.1.

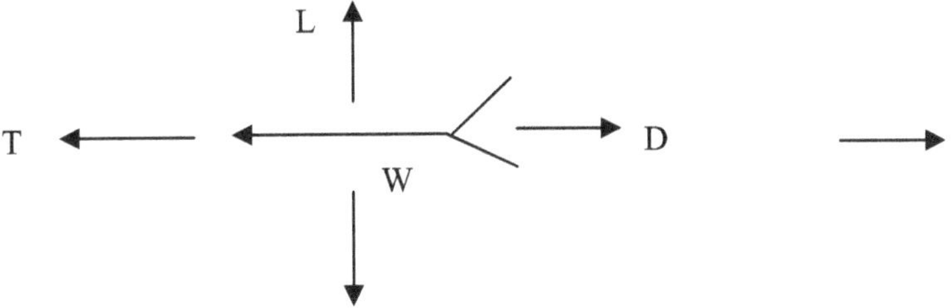

Fig. 1 Effects of weather on Aircraft

The Lift (L) depends on pressure, temperature and square of the velocity. Temperature is the prime factor for engine efficiency. The efficiency of the aircraft engine reduces when temperature increases and which in turn leads to the fuel consumption. If the temperature is high in the cruising level, more fuel is needed than normal. Therefore accurate measure of temperature is well in advance will aids the aircraft agencies to plan their cargo load.

Low ceiling and reduced visibility are safety hazards for all types of aviation. The low visibility, caused by fog or mist, Low clouds affects the takeoff and landing operations and it can be avoided by the use of category III Instrument Landing System along with proper rated pilot. The low visibility is not just a safety it can also severely degrade the capacity of an aircraft and cause delays, cancellations, diversions.

Take off and landing operations are usually against the directions of surface wind. Wind has two elements: direction and velocity, which are used to determine the head and cross wind. For takeoff and landing phases, the head wind is favorable while tail wind is for high altitude. Thunder storms and other convective weather pose a serious problem on aircraft and passenger. Thunder and Lighting interrupt the communication, navigational signals and affects the performance of onboard electronic equipment. In addition thunder storms clouds (Cumulonimbus (CB) clouds) cause bumpy movements of aircraft when an aircraft has no alternate to pass. The Clear Turbulence (CAT) is the other component in addition to the CB clouds, which causes bumping of aircraft and/or loosing of altitude. Accurate estimation and advance warning about *CB* clouds and CAT helps the pilot for safe operation of aircraft.

Wind shear is a change in wind speed and/or direction over a short distance. It is associated with the convective weather i.e. *CB* clouds. The Sophisticated aircraft sometimes reports loosing altitudes up to few 1000fts in a few seconds. The aircraft icing occurs when the aircraft flies in air at temperature just colder than freezing point. Due to this more lift is required as the weight of the aircraft increases. However with the introduction of an advance technology in an aircraft will overcome icing to a maximum extent. The wake turbulence is other parameter which imposes a hazardous to the aircraft during the landing and takeoff phases of flight.

To study the impact of convective weather on particular traffic flows, here we have considered Markov models. The Markov model, have different states/stages of weather or weather impacts. In this approach an optimal paths are explored when the aircrafts starts flowing from origin to destination.

MOTIVATION

Consider a two dimensional flight plan of an aircraft and the aircraft is in en-route phase. Here our aim is to find out the optimal path in en route phase. In our formulization the entire airspace is represented as two dimensional grid system as shown in Fig. 2,and each grid point represented as a way point.

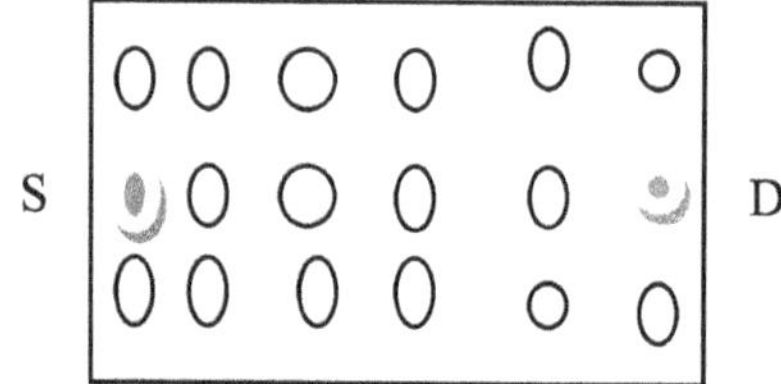

Fig. 2 2-D gridding system

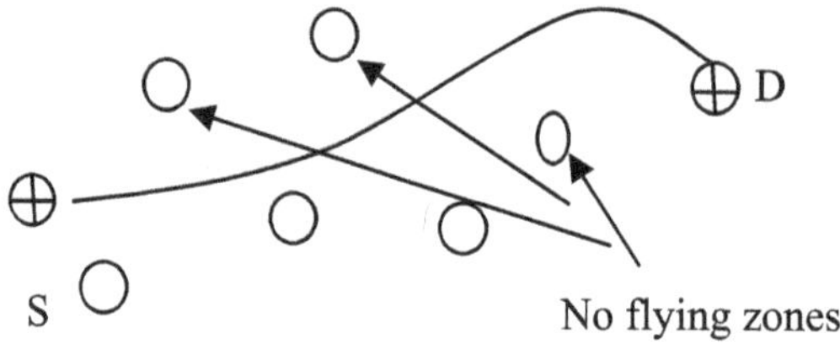

Fig. 3 2-D flight plan of an aircraft

As shown in Fig. 2, the aircraft at source 'S' want to reach a destination point 'D'. When the aircraft is progressing, it encounters many obstacles and it may deviate its original path as shown in Fig.3. The obstacle by nature two types: Deterministic and stochastic. The deterministic 'No fly zones' include military airspace and national security space. While the stochastic 'No fly zones' include storm zones, volcanic ash or strong wind zones.

The metrological department produces weather predictions of some zone in the airspace will be unusable for certain time and predictions are dynamically update with time. The unusable zones are predicted based on certain probability and often it turns out that the zones were perfectly usable. Since re-routing strategy does not uses these resources, the net effect will be the congestion in remaining air space.

WEATHER UNCERTAINITY

The metrological centre provides the probability of storm at particular place at particular time and updates dynamically for every 15 minutes. The later an event is from the prediction time, the more unreliable it becomes. Further, it is much reasonable to assume the aircraft has a perfect knowledge of weather in the regions that are during the period of 15 minutes i.e. each stage duration is 15 minutes. The entire time span of travel time of aircraft is discritized as 1, 2 ...n stages according to the weather update. The duration between each stage is 15 minutes and the aircraft get updating in this interval. Stage I corresponds to the time of 0-15 minutes from the current time, stage II corresponds to 15-30 minutes from the current time. Thus there are 'n' stages that accommodate the entire span of the time of the aircraft for worst case routing. The probability of storm (p) at a particular region is time varying and takes a value 0 or 1 when it reaches a 15 minutes away from the region. The dynamics of p will follow the path, is shown in Fig. 4, and p is the probability of storm in the next interval.

Let there are 1, 2... m storms that are predicted to happen in the region 'k'. Now we can define state '0' corresponds to the situation where there is no storm in the region and a state '1' corresponds to the storm in that region. The status of the storm at stage can be assigned 0 or 1 to every storm in that region. This problem can be formulated as simple two state Markov model as shown in Fig. 5.

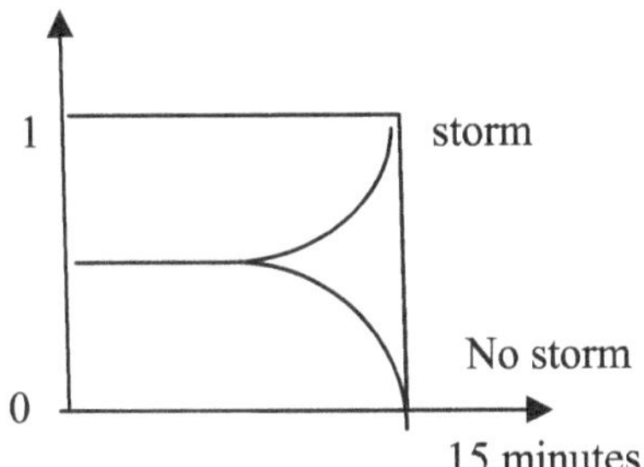

Fig. 4 Probability of storm with time

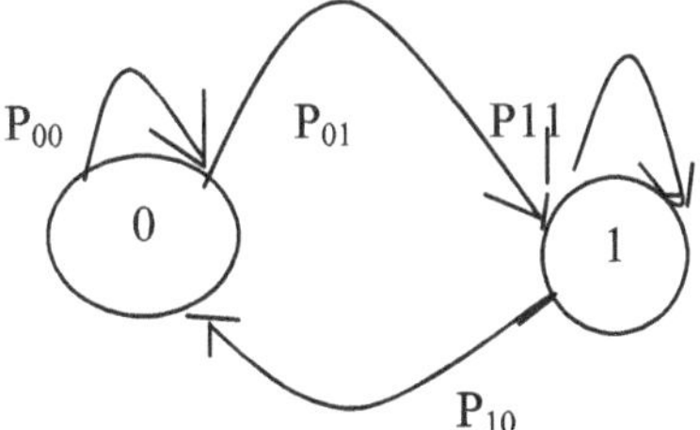

Fig. 5 Basic Two state markov model

We can define

$$P_{01} = 1 - p_{00} \quad\quad\quad\quad\quad .....(1)$$

$$P_{10} = 1 - p_{11} \quad\quad\quad\quad\quad .....(2)$$

Now we define,

p_{00} is the probability of no storm in the next stage if there is no storm in the current stage

p_{11} is the probability of having a storm in the next stage if there is a storm in the current stage.

The basic two state markov model is basis for our problem and it can be extended further to the 'm' storms. Because the actual outcome of predictions has many intermediate states rather than simple two binary states shown in Fig. 6. Therefore it is more appropriate to consider the M state markov chain.

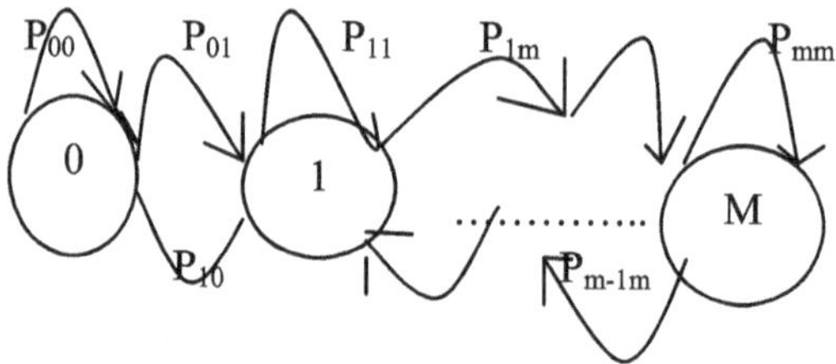

Fig. 6 hidden Markov model

If there are 'm' storms, the 2^m tupple vector can be describing the storm situation completely. i.e., [100….0] denotes that there is a storm in the zone K_1 and the rest of the predicted bad weather zones are storm free. We can define $q_1 = [0000...0]$, $q_2 = [100...0]$, $q_3 = [0100...0]....q_m = [m-1\ m-1....m-1]$. The transition matrix P for the markov chain is $2^m \times 2^m$ matrix and whose $(i,j)^{th}$ P denotes the

$$P = \begin{bmatrix} P_{11} & P_{12} & \cdots & P_{12^m} \\ \cdots & \cdots & \cdots & \cdots \\ \cdots & \cdots & \cdots & \cdots \\ P_{2^m 1} & P_{2^m 2} & \cdots & P_{2^m 2^m} \end{bmatrix} \quad\quad .....(3)$$

The conditional probability of markov model defined as

$$p\{q_{t+1} = j / q_t = i\}, 1 \le i, j \le M \quad\quad\quad .....(4)$$

It satisfy the following conditions

$$a_{ij} \ge 0, 1 \le i, j \le M \quad\quad\quad\quad .....(5)$$

$$\sum_{j=1}^{M} a_{ij} = 1, 1 \leq i \leq M \qquad \qquad .....(6)$$

In practice, there are only 3 to 4 predicted convective zones and therefore the number of states is not more than five.

PROBLEM FORMULATION: HIDDEN MARKOV MODEL

Here our problem is to find out the optimal path of an aircraft that receives the weather update at reasonable time duration of 15 minutes. In section II we described the time span and formulated 'n' number of stages. During the course of flight the aircraft will receive more information. Further the entire airspace is modeled as two dimensional grid system and each grid point is assumed as way point. This problem can accurately modeled using hidden markov model (HMM) and finding the optimal sequence is achieved by the use of viterbi algorithm.

The HMM is a finite set of states, each of which is associated with a probability distribution. Transitions among the states are governed by set of probability called transition probability. Since the only outcome is visible, not the state, the name hidden markov model[4].

A. Assumptions:

1. The 'm' storms are predicted in convective zone in the region 'K$_i$'
2. Each prediction have equal intensity and the coverage
3. No movement of storms in the predicted zone

B. Three problems of HMM

We can describe the complete set of HMM parameters for a given model.

There are three basic problems associated with HMM's

1. find $p(o/\lambda)$ for some $O = (o_1 o_2 .... o_T)$
2. Given some 'o' and 'λ', find the best state sequence $q = (q_1 q_2 .... q_T)$
3. Find $\lambda = \arg\max p(o/\lambda)$

Since our problem is very much relevant to finding the optimal state sequence here we consider viterbi algorithm and evaluated subsequently.

C. Viterbi searching algorithm

The aim of the viterbi algorithm is to find out the best sequence for, $q = (q_1 q_2 .... q_T)$, for given observation sequence , $O = (o_1 o_2 .... o_T)$

Now, we define,

$$\delta_t(i) = \max_{q_1 q_2 \cdots q_{t-1}} P[q_1 q_2 .... q_t = i, o_1 o_2 ... o_t / \lambda] \qquad .....(7)$$

At $t + 1$

$$\delta_{t+1}(j) = [\max_j \delta_t(i) a_{ij}] b_j(o_{t+1}) \qquad .....(8)$$

To find the best sequence, we can use the following recursion

1. Initialization: $\delta_t(j) = \pi_i b_j(o_1); 1 \le i \le N$

$$\psi_1(i) = 0$$

2. Recurssion: $\delta_t(j) = \max_{1 \le i \le N}[\delta_{t-1}(i)a_{ij}]b_j(o_t); 2 \le t \le T,\ 1 \le j \le N$

$$\psi_t(j) = \arg\max_{1 \le i \le N}[\delta_{t-1}(i)a_{ij}]; 2 \le t \le T$$

$$1 \le j \le N$$

3. Termination: $P^* = \max_{1 \le j \le N}[\delta_t(i)]$

$$q_T^* = \arg\max_{1 \le j \le N}[\delta_t(i)]$$

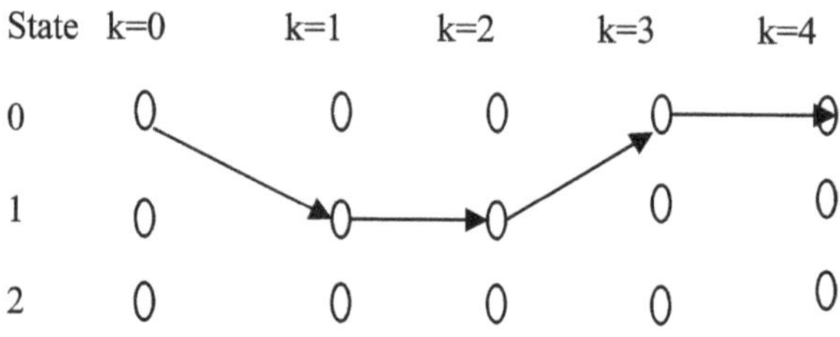

Fig. 7 optimal state path for aircraft for using viterbi algorithm

The evaluation process of viterbi algorithm graphically is shown in Fig.7. Viterbi search provides most likely state path for given states.

SUMMARY AND CONCLUSION

In this paper, we have explored to find the optimal best weather routing for an aircraft using viterbi algorithm. In our study, we have not made any experimental analysis but based upon the past experiments conducted by Hidden Markov models, it is said that by using HMM's we can provide optimal best path to the aircraft in convective weather.

ACKNOWLODGEMENTS

The author, Ch. Mahesh, expresses heartfelt acknowledgments to Mr. N.R Das., General Manager (CNS), Mr. G.S.Rao, JGM (CNS),AAI, Hyderabad , for their support and guidance.

REFERENCES

[1] ICAO. Annex. 3,"Metrological service for International Air Navigation" Montreal, July 2007.

[2] Arnab Nilim., Laurent El Ghaoui.,Vu Duong.,"Robust Dynamic Routing of Aircraft under uncertainty", in proceeding the 21st volume:1, Digital Avionics system conference,2002

[3] R. Suresh.," Significance of weather in Aviation", 14th All India conference of Aeronautical communication officers association, Chennai,India.

[4] Lawrence R. Rabiner.,"A tutorial on Hidden Markov Models and selected applications in speech recognition", Proceeding of the IEEE, 77(2), pp 275-286, Feb-1989, August 1980.

A LOGICAL APPROACH TO THE STUDY OF COLLISION AVOIDANCE PROBLEMS PERTAINING TO UNMANNED AERIAL VEHICLES MEANT FOR ELECTRONIC SUPPORT MEASURES (PART – 1)

Ch. Raja[1], M. Madhavi Latha[2], E. G. Rajan[3]

[1]Research Scholar,

E-mail: rajachaluvadi@yahoo.com

E-mail: mlmakkena@yahoo.com

[2]Professor, Electronics & Communication Engineering Dept.,

JNTUH,Kukatpally, Hyderabad-500062, India.

[3]Founder President, Pentagram Research Centre Pvt. Ltd.

#201, Venkat Homes, MIGH-59, Mehdipatnam, Hyderabad, India.

E-mail: rajaneg@yahoo.co.in

Abstract

Unmanned Aerial Vehicles (UAV) Autonomous Mobile Robots (UAVs) are used in different applications like aerial reconnaissance and space surveillance. The navigation problem is a constraint which has been treated in many ways, usually the difference lying in the knowledge base. Traditionally two basic approaches are being used for path generation of an UAV, (i) fixing of the total path in a known environment, and (ii) instantaneous fixing of path directions depending on situations in an unknown environment. The former approach is error free because the nature of the environment is a priori known, where as the latter approach is not. Most of the current research is being focused on to the second approach with an intention of developing a universal technique in order to meet the challenges of obstacle avoidance in any unknown environment. For the past few years, a number of path fixing techniques have been developed each having its own merits and demerits. Almost all the techniques involve the geometries, velocities, time segments and directions of motion of obstacles and UAV while fixing the path of the UAV instantaneously. The work carried out, as reported in this paper, centers around a novel path fixing technique in the framework of Cellular Logic Array Processing. Development of fast algorithms for the instantaneous fixing of collision free paths by UAVs in unknown environment had been the major output of this intended research. UAV control is modeled as a constructive system that operates on two regions (i) low risk region and (ii) high risk region as given in the text. The term *High Risk Region* (HRR) refers to, for example, the scenic coverage around the UAV whereas the *Low Risk Region* (LRR) refers to the front scenic coverage acquired by the vision system of the UAV. In a real life situation also, one could see that the risk due to obstacles approaching from the sides is more than the obstacles approaching from the front, hence head-on collisions are rare on roads when compared to collision from sides. It has been found that the collision avoidance problem amounts to say that there are 1,31,072 possibilities of collision that a UAV has to encounter if its dynamics is modeled in the square discrete space. Alternatively, if we consider the discretized space as a hexagonal model, then the number of collision possibilities to be involved while fixing the Escape Route (ER) would be reduced considerably. With the three dimensional Euclidean space being modeled as a discrete hexagonal grid, a UAV faces a total of 8,192 threat zones while undertaking a mission in this discrete space and so it is recommended that the Euclidean space E^3 is modeled as a hexagonal discrete space and the flight dynamics of a UAV is planned to be controlled by seven constructive formulas formulated in the logico-mathematical framework of Markov's languages of Я$_\alpha$.

Section 1

The problem at hand is to develop methodologies for instantaneous fixing of Mission Direction of UAVs (Unmanned Aerial Vehicles) in an unknown environment within the framework of Markov's Constructive Mathematical Logic [1] and Rajan's Cellular Logic Array Processing (CLAP)[2].

Objectives of the study

The major objectives of the study undertaken and reported in this paper are:
- To use various image processing and pattern recognition algorithms already developed in the framework of Cellular Logic Array Processing (CLAP).
- To develop techniques for instantaneous fixing of mission paths for UAVs in collision prone environment (Target Avoidance Problem, TAP).

The Target Avoidance Problem TAP is of concern to us, especially now when PSLVs are made and launched successfully in India. In the case of Autonomous Homing Weapons too, TAP plays and important role. The combat space in which such weapons and vehicles are placed could be viewed as the union of three classes of targets, viz. (i) class of wanted targets. (enemy), $\Re$, a class of rejected targets (friend) and (iii) U, a class of unknown targets.

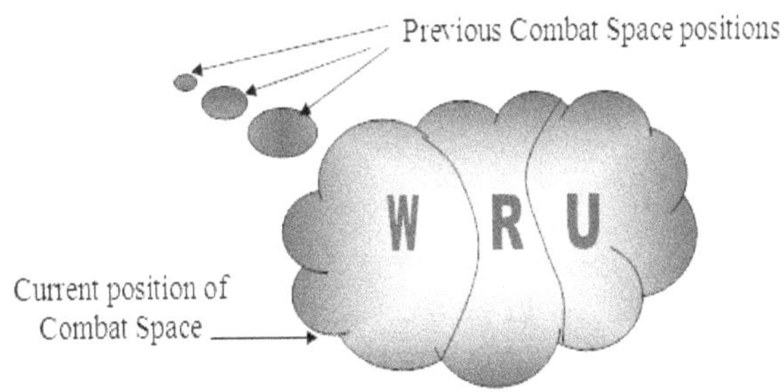

Fig. 1.1 Combat space model and its classification

First, the system (weapon) should recognize a target and classify it as a member of any one of these three classes. Next it should use TAP related strategy to avoid the target if it belongs to $\Re$ or use Target Seeking Problem (TSP) related strategy to home on to the target if it belongs to W.

Motivation

Traditional control systems theory advocates the basic technique of relating the reference input r(t) to a system with its output c(t). The formal representation of the input output relationship leads to what is called the system performance. Open loop systems are generally described by linear or non-linear relations between the inputs and the outputs, and the control of the outputs are usually done by reference input. On the other hand, closed loop systems work on feed back concepts, wherein, the desired output is achieved by feeding a part of it to a summing device, which finds the error by comparing it with the input. This error in turn drives the system to yield the desired output. In fact, the error is controlled to a minimum in order to obtain the desired result. Modern control systems do make use of digital devices and inevitably digital computers for their control action. ***Intelligent control systems are essentially logical systems, and this fact has been the motivating factor behind the work carried out and presented in this paper.***

Problem Formulation

The question of immediate concern to us now is how one can actually link *logic* with say ***Real Time Systems*** (RTS). ***The answer is simple.*** As we describe a system by its input output relations such as y = Fx, where y is the output, x is the input and F is some operator, we can describe RTS using logical model formulas like $\forall x(A(x) \Rightarrow \exists y B(x,y))$, where A and B are formulas of a first order language which is interpreted as a statement that for every value of the input x such that A(x) is valid, the output has the value y such that B(x,y) holds. Unlike the classical logical models, the accuracy of a logical model for RTS not only depends on the result of computations but also on time moment. So, analysis and synpaper of desired behavior of a system are always possible with the use of temporal logics of propositional calculus. This implies that we have to go in for some higher order logics, which use type quantifiers and are more expressible for applications. It is worth to mention here that PROLOG is one such logical programming language tool, which is more expressible. However, higher order logics are semi decidable and we know that many applications, especially, real time applications have restricted resources of time and memory. Moreover, it is a well known fact that automata networks are important components of a control system. So, it would be very useful to go in for a qualitative logical method instead of exhaustive numerical simulation. It is in this regard, an extensive research has been carried out and results compiled in the form of a report. A brief summary of the research carried out has been provided below.

Brief summary of the research

As outlined already Unmanned Aerial Vehicles (UAVs) are used in different applications like reconnaissance and space surveillance. In such cases, UAVs have to cope with static and moving obstacles. The navigation problem is a constraint which has been treated in many ways, usually the difference lying in the knowledge base. The robot motion is generally decided by taking into account obstacle configuration, robot kinematics and the result of the modification of the local perception. Subject to these basic constraints, several procedures and methodologies have been developed for fixing optimal paths of UAVs in known as well as unknown environment. Traditionally two basic approaches are being used for path generation of a UAV, (i) fixing of the total path in a known environment, and (ii) instantaneous fixing of path directions depending on situations in an unknown environment. The former approach is error free because the nature of the environment is a priori known, where as the latter approach is not. Most of the current research is being focused on to the second approach with an intention of developing a universal technique in order to meet the challenges of obstacle avoidance in any unknown environment. For the past few years, a number of path fixing techniques have been developed each having its own merits and demerits. Almost all the techniques involve the geometries, velocities, time segments and directions of motion of obstacles and UAV while fixing the path of the UAV instantaneously. The work carried out here centers around a novel path fixing technique in the framework of Cellular Logic Array Processing. Development of fast algorithms for the instantaneous fixing of collision free paths by UAVs in unknown environment had been the major output of this intended research. UAV control is modeled as a constructive system that operates on two regions (i) low risk region and (ii) high risk region as given below. The term **High Risk Region** (HRR) refers to, for example, the scenic coverage around the UAV whereas the **Low Risk Region** (LRR) refers to the front scenic coverage acquired by the vision system of the UAV. In a real life situation also, one could see that the risk due to obstacles approaching from the sides is more than the obstacles approaching from the front, hence head-on collisions are rare on roads when compared to collision from sides. Figure 1.4.1 shows the HRR and LRR with respect to the UAV [3].

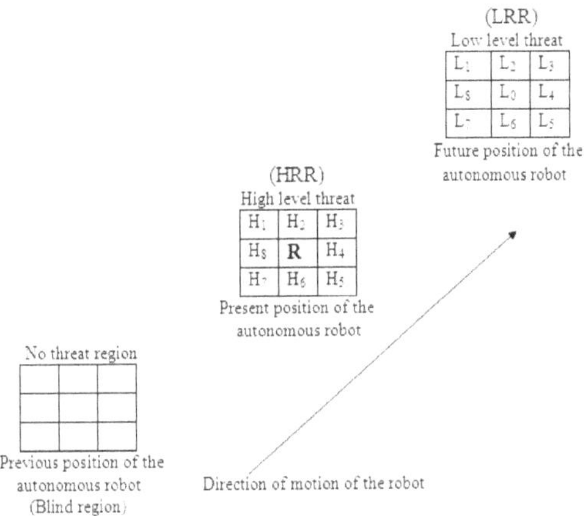

The state vector control is carried out by control formulas. For convenience, we denote the MD (Mission Direction) of a UAV by $R \rightarrow L_0$, where R denotes the current position of the robot and L_0 the next position of the robot in LRR. Now, we shall assume that there is no obstacle in the L_0 position. We shall also assume that there is no obstacle in the high risk region. Then certain control formulas are valid, which are described in section

Section 2

Modeling of three dimensional space

The three dimensional Euclidean space E^3 is discretized here as a 3-D grid of uniform cells, each cell taking the shape of a cube. The presence of an obstacle in a cell attributes a value 1 to that cell where as the absence of an obstacle attributes a 0 to it. It is important to note that a cell in the 3-D grid is an abstract model and so it is not limited by morphology and size. Usually the perception space of a UAV is classified as a disjoint union of two subspaces (i) one consisting of recognized entities, (ii) the other consisting of unrecognized entities. In our case, the perception space is treated as a disjoint union of three subspaces (i) one consisting of objectively recognized entities,

(ii) the other consisting of objectively unrecognized entities and (iii) the third consisting of subjectively recognized entities. It is clear from this classification that there is an in-built fuzziness in characterizing entities belonging to a perception space. It is to be noted here that a perception space of one UAV is not identical to that of another UAV. We are concerned here with the problem of instantaneous fixing of path of an autonomous mobile robot sent on a mission especially in an unknown space. The formulation of such a kind of technique has been tried theoretically in the framework of cellular logic array processing. The central idea behind this technique is that an UAV decides its direction of movement based on the neighborhood in which it is situated at a particular instant of time. The direction parameter of a UAV for next instant of time is decided by a formula that involves the position and intended direction of the UAV in a neighborhood consisting of obstacles, be them mobile or not, and their positions and estimated directions of motion. In order to do this, it is essential to model the three dimensional Euclidean space E^3 as a digital grid mathematically denoted as ZXZXZ, where Z denotes the system of integers. In fact there are many ways of discretizing the analog E^3 space: (i) using three dimensional *rectangular lattice* model, (ii) using three dimensional *hexagonal prism* model and (iii) using three dimensional *polyhedron* model, to name a few. The first two models are described in this paper in order to justify the need for discretizing the E^3 space.

Three dimensional rectangular lattice model of E^3 space

The three dimensional lattice is an array of nodes, each hypothetically connected to six more nodes in mutually orthogonal directions. Figure 2.1 shows such an arrangement where one node indexed as (i, j, k) is hypothetically connected to six more nodes (i-1, j, k), (i, j+1, k), (i, j-1, k). (i+1, j, k), (i, j, k-1) and (i, j, k+1).

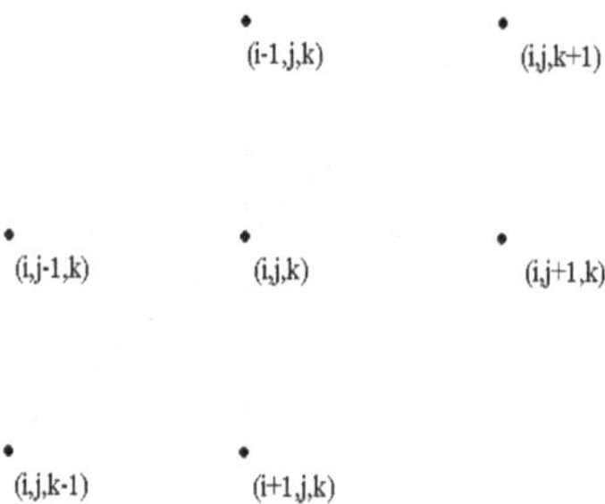

Fig. 1.2 Direct neighbors of 3-D rectangular lattice

These six nodes are called the direct neighbors of the node (i, j, k). This seven node arrangement is extended three dimensionally to form the lattice. Figure 1.2 shows the extended array of the one shown in figure 1.1. The nodes (i-1, j-1, k), (i-1, j+1, k), (i+1, j-1, k), (i-1, j+1, k), (i-1, j-1, k-1), (i-1, j+1, k-1), (i+1, j-1, k-1), (i+1, j+1, k-1), (i-1, j-1, k+1), (i-1, j+1, k+1), (i+1, j-1, k+1), (i+1, j+1, k+1) are called the indirect neighbors of (i, j, k).

Three dimensional hexagonal prism model of E^3 space

The three dimensional hexagonal prism is an array of nodes, each hypothetically connected to six more nodes in a zig-zag manner. In order to understand this basic model, first we shall see its two dimensional counterpart, a hexagonal lattice. The basic model of 2-D hexagonal lattice is obtained from rectangular 2-D array in the following manner. Figure 1.3 shows the hexagonal grid arrangement inside the rectangular array. The size of a hexagonal array is half of its corresponding rectangular array.

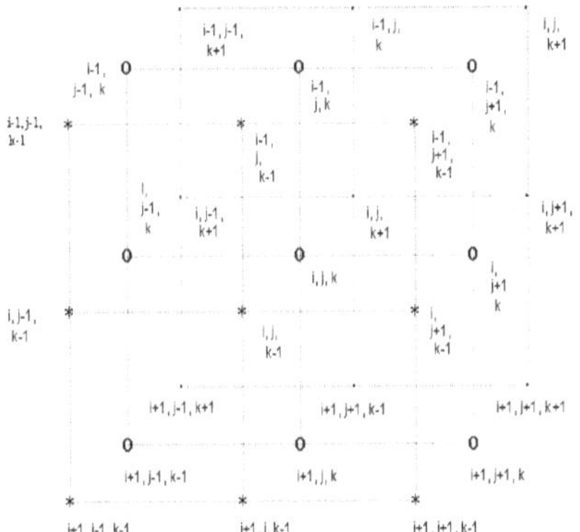

Fig. 1.3 neighborhood in a 3D rectangular lattice

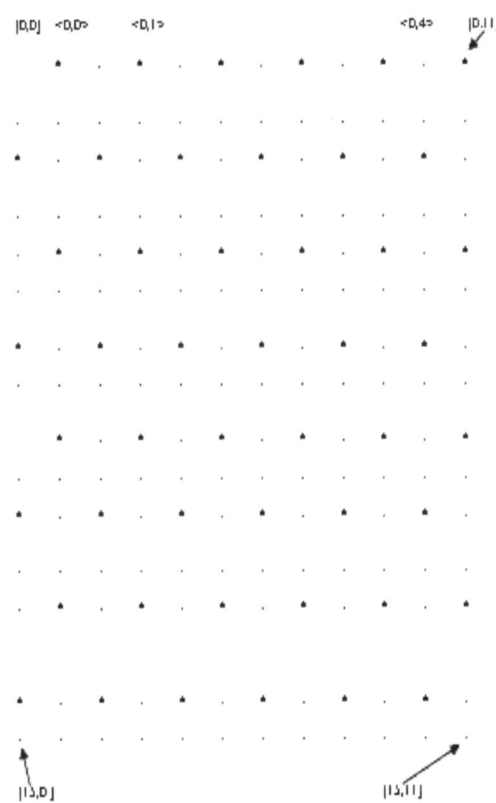

Fig. 1.4 Hexagonal array inside a rectangular array

[x, y] and <x, y> respectively denote rectangular and hexagonal coordinates. From the above figure, one can observe the following equalities:

	Rectangular Coordinates	Hexagonal Coordinates
[Row, Column]	<Row, zigzag>	
[0,1]	<0,0>	
[0,3]	<0,1>	
[0,5]	<0,2>	
...	...	
[2,0]	<1,0>	
[2,4]	<1,2>	
...	...	
[4,1]	<2,0>	
[4,3]	<2,1>	... ...

The rows of the hexagonal grid could be seen to be formed using the even numbered rows of the rectangular grid and the zigzags of the hexagonal grid using alternate odd and even numbered columns of the rectangular grid. The relationship between the two coordinate systems is as given below.

$$<2x, 2y+1> = [x, y] \text{ (for even numbered rows of hexagonal grid)}$$

$$<2x, 2y> = [x, y] \text{ (for odd numbered rows of hexagonal grid)}$$

Now one can construct a three dimensional hexagonal prism lattice as a grid consisting of a number of parallel lines of two dimensional hexagonal arrays. Due to the complexity involved in constructing such a 3-D lattice, it is not show here. In this paper, we restrict our study only to three dimensional rectangular discrete grid as the model of the Euclidean space E^3 and proceed with the formulation of certain methods and tools for developing collision avoiding techniques in the case of an UAV sent on a mission in space.

Section 3

Cellular logic techniques for collision avoidance - Basic philosophy

During the mission, the UAV acquires sequence of sampled image frames continually, each sampled image frame in the sequence is processed for detecting the presence of obstacles in the collision path and action taken. This procedure is called instantaneous path fixing. This is carried out in the following manner.

When no obstacle is detected in a sampled image frame, the UAV does not take any action. On the other hand, it continued in its state of uniform motion.

When an obstacle is detected in an image frame, the UAV applies the following *cellular logic fast algorithms* to the image: (i) *contouring*, (ii) *skeletonization* and (iii) *centroid detection*. Now the UAV has three processed images corresponding to a sampled image. From the contoured image S, the UAV estimates the size of the obstacles. From the skeletonized image O, it understands the orientation of the obstacle. From the centroid fixed image D, it compares the position of the centroid to its positions from the previous image frames and estimates the direction of motion of the obstacle. The data set consisting of these three images is represented as <S,O,D>.

After obtaining the data set <S,O,D>, the UAV decides whether the obstacle is approaching it or going away from it by checking the following. If the size grows from frame to frame, the UAV understands that the obstacle is approaching it. On the contrary, it the size goes on decreasing, it understands that the obstacle is moving away from it. In the event of the obstacle approaching in the collision path, the UAV changes its direction to avoid collision as per instantly decided strategy[4].

In any case, the UAV should have a powerful and reliable image acquisition system in order to make the best use of such high-throughput image understanding techniques. This is a problem related to *Computer Vision*. There are innumerable solutions for this. Yet this section proposes a novel technique for collision avoidance based on omni directional viewing and capturing of images by a conical projection image sensor. Subsequently, a heuristic algorithm called *Restorative Deviation* is also discussed, which is useful for tracking the *Mission Direction*.

Logical decision making for collision avoidance

The acquired composite image is processed in a computer for the purpose of pattern recognition and decision making. The composite 256 level gray image is of size 512x512. The digital image acquired by camera #1 is also of size 512x512 but it is reduced to the size of 128x128 in order to be superimposed exactly on the central blind region of the conical mirror. It is to be noted here that the COPIS has two blind regions (i) front blind region and (ii) rear blind region with solid angle ω radians. The central idea on which the decision making is done is that the moving objects in space are recognized either as harmful or as harmless. Harmful objects are those which approach towards the robot and harmless ones move away from the robot. Less harmful objects are those which are captured by camera #1, where as, those captured by the other camera are more harmful. The loci of harmful objects are evaluated using cellular logic principles. Then the logic of collision avoidance goes as follows:

The intended path of the robot (UAV) is given by R→L_0. Normal velocity of robot is denoted as V. A path is fixed by the robot depending on the positions and speeds of obstacles.

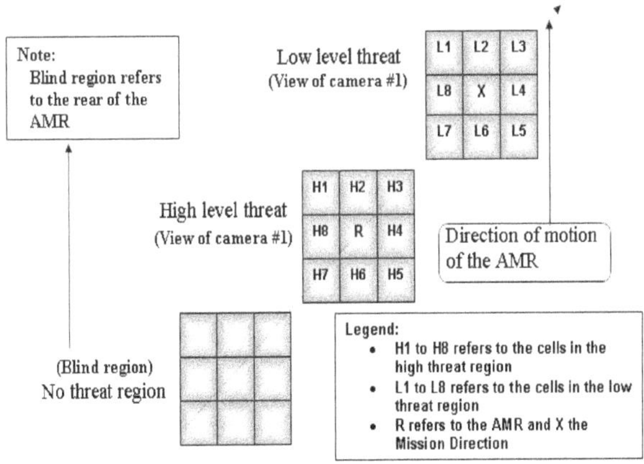

Fig. 1.5 Mission Direction in the rectangular lattice space model

At a particular time instant, the robot dynamics is determined by the ordered pair $< V, R{\rightarrow}L_i >$; $1 \leq i \leq 8$. There are 8 loci corresponding to less harmful objects approaching the robot from 8 directions: $L_j{\rightarrow}R$, $1{\leq}j{\leq}8$. Hence, $R{\rightarrow}L_i$ depends on $L_j{\rightarrow}R$, and the velocity V on $H_k{\rightarrow}R$; $1 \leq i \leq 8$. H_k is a cell in the high threat region and it is logical to see that UAV changes its velocity depending on the fact whether obstacles are there in the high threat region or not. Now the cellular logic scheme that fixes the instantaneous path is:

$$<V, R{\rightarrow}L_i> = \phi (L_j{\rightarrow}R ; 1{\leq}j{\leq}8, H_k{\rightarrow}R ; 1{\leq}i{\leq}8).$$

With this basic understanding, it was felt that a detailed case study on different space models would substantiate and justify the results obtained so far.

Section 4

Collision avoidance in rectangular lattice space model

This section discusses the results of a case study. Let us assume that $R{\rightarrow}L_0$ is the intended direction of an UAV's mission. The first *collision possibility* is due to a single obstacle approaching the UAV from the direction $L_0{\rightarrow}R$. This low threat position is pictorially represented in figure 1.5. ***The presence of a 1 in a cell indicates the presence of an obstacle in that region.***

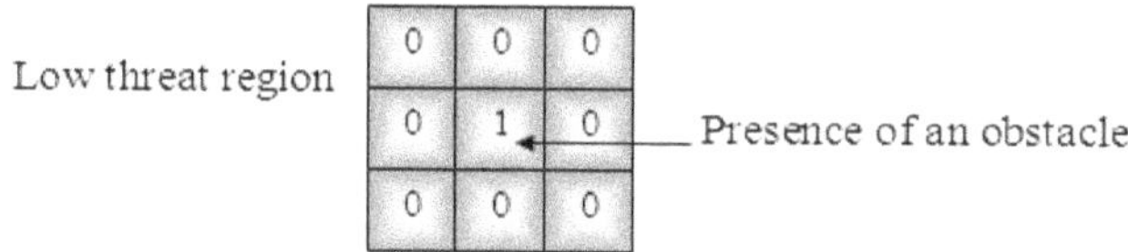

Fig. 1.5 Low threat position of an obstacle

Now the UAV can change its direction of motion to any one of the following ones: $R{\rightarrow}L_2$, $R{\rightarrow}L_3$, $R{\rightarrow}L_4$, $R{\rightarrow}L_5$, $R{\rightarrow}L_6$, $R{\rightarrow}L_7$, $R{\rightarrow}L_8$. Note that one can construct a total of eight 2-obstacles threat positions with an obstacle in the L_0 position as shown below.

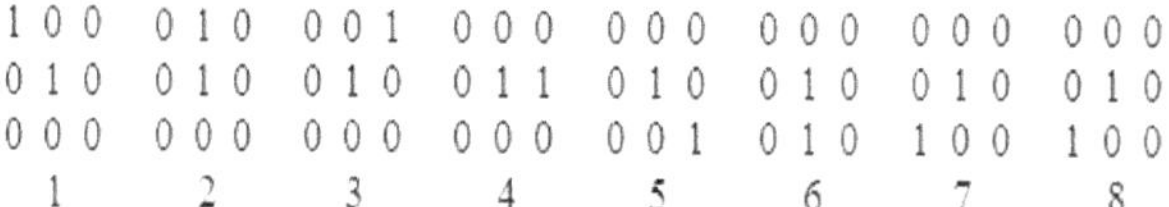

Threat positions with two obstacles

Logical control formulas for collision avoidance

Based on the threat positions of obstacles shown by images obtained by the acquisition system, COPIS, the UAV would make use of certain control formulas which would indicate the direction of least collision possibility for restorative deviation. The following table 1.1 gives a summary of ***high risk low threat positions*** for various groups of obstacles, ***with an obstacle in the cell*** $L_0[5]$. Table 1.1: High Risk Low Threat Positions (set #1)

Sl. No.	Number of obstacles (with an obstacle in L_0)	Threat positions (No. of obstacles)
1	(L_0) (No obstacle other than one in L_0)	$^8C_0 = 1$
2	(L_0, L_a/$1 \leq a \leq 8$)	$^8C_1 = 8$
3	(L_0, L_a, L_b/$1 \leq a,b \leq 8$)	$^8C_2 = 28$
4	(L_0, L_a, L_b, L_c /$1 \leq a,b,c \leq 8$)	$^8C_3 = 56$
5	(L_0, L_a, L_b, L_c, L_d /$1 \leq a,b,c,d \leq 8$)	$^8C_4 = 70$
6	(L_0, L_a, L_b, L_c, L_d, L_e /$1 \leq a,b,c,d,e \leq 8$)	$^8C_5 = 56$
7	(L_0, L_a, L_b, L_c, L_d, L_e, L_f /$1 \leq a,b,c,d,e,f \leq 8$)	$^8C_6 = 28$
8	(L_0, L_a, L_b, L_c, L_d, L_e, L_f, L_g /$1 \leq a,b,c,d,e,f,g \leq 8$)	$^8C_7 = 8$
9	(L_0,L_a,L_b,L_c,L_d,L_e,L_f,L_g,L_h /$1 \leq a,b,c,d,e,f,g,h \leq 8$)	$^8C_8 = 1$

The above set consists of a total of 256 high risk low threat positions. Similarly, one can construct table 4.1.2 summarizing 256 *low risk low threat positions without an obstacle in position* L_0. By *low risk threat*, we mean that the UAV can continue in its intended direction R→L_0 with low risk in spite of the fact there are obstacles present elsewhere in the neighborhood.

Table 1.2: Low Risk Low Threat Positions (set #2)

Sl. No.	Number of obstacles (without obstacle in L_0)	Threat positions (No. of obstacles)
1	( ) (No obstacle threat free)	$^8C_0 = 1$
2	(L_a/$1 \leq a \leq 8$)	$^8C_1 = 8$
3	(L_a, L_b /$1 \leq a,b \leq 8$)	$^8C_2 = 28$
4	(L_a, L_b, L_c /$1 \leq a,b,c \leq 8$)	$^8C_3 = 56$
5	(L_a, L_b, L_c, L_d /$1 \leq a, b,c,d \leq 8$)	$^8C_4 = 70$
6	(L_a, L_b, L_c, L_d, L_e /$1 \leq a,b,c,d,e \leq 8$)	$^8C_5 = 56$
7	(L_a, L_b, L_c, L_d, L_e, L_f/$1 \leq a,b,c,d,e,f \leq 8$)	$^8C_6 = 28$
8	(L_a, L_b, L_c, L_d, L_e, L_f, L_g /$1 \leq a,b,c,d,e,f,g \leq 8$)	$^8C_7 = 8$
9	(L_a, L_b, L_c, L_d, L_e, L_f, L_g, L_h /$1 \leq a,b,c,d,e,f,g,h \leq 8$)	$^8C_8 = 1$

From the above argument we see that an UAV faces a total of 512 low threats with 256 high risk zones (with an obstacle in L_0) and 256 low risk ones (without an obstacle in L_0). Note that the threat free zone is also considered here as a low risk low threat zone. In addition to the above low threat zones, an UAV is expected to face more 256 high threat zones from its sides.

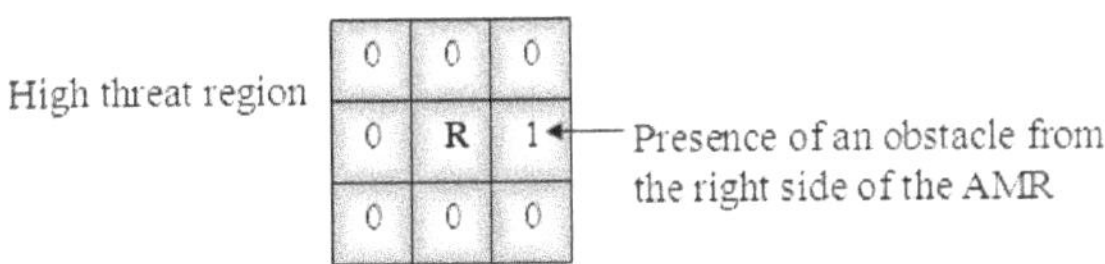

Fig. 1.6 High threat position of an obstacle on the right side of UAV

The UAV would change its speed depending on the high threat level. Table 4.1.3 provides nine formulas that define all the 256 high risk high threat positions. Note that the threat free zone of table 4.1.3 is also considered as a high risk high threat zone. To summarize, with the three dimensional space being modeled as a discrete grid, an UAV faces a total of 512 x 256, that is,

1,31,072 threat zones while undertaking a mission in the discrete space. The restorative deviation method is adopted by the UAV to avoid different types of collisions. Further precise modifications could be made if we resort to probabilistic methods in evaluating the relative velocities and assigning suitable priority values to different directions while fixing the path. With the basic model for a collision avoidance system, our efforts would now be focused on to the possibility of using logical control of the UAV in the framework of constructive mathematical framework. The constructive logical formulas govern the motion of UAV

Table 1.3 High Risk High Threat Positions

Sl. No.	No. of obstacles (UAV in H_0)	Threat positions (No. of obstacles)
1	(R) (UAV Position, threat free)	$^8C_0 = 1$
2	($H_a/1 \leq a \leq 8$)	$^8C_1 = 8$
3	($H_a, H_b /1 \leq a,b \leq 8$)	$^8C_2 = 28$
4	($H_a, H_b, H_c /1 \leq a,b,c \leq 8$)	$^8C_3 = 56$
5	($H_a, H_b, H_c, H_d /1 \leq a,b,c,d \leq 8$)	$^8C_4 = 70$
6	($H_a, H_b, H_c, H_d, H_e /1 \leq a,b,c,d,e \leq 8$)	$^8C_5 = 56$
7	($H_a, H_b, H_c, H_d, H_e, H_f /1 \leq a,b,c,d,e,f \leq 8$)	$^8C_6 = 28$
8	($H_a, H_b, H_c, H_d, H_e, H_f, H_g /1 \leq a,b,c,d,e,f,g \leq 8$)	$^8C_7 = 8$
9	$H_a, H_b, H_c, H_d, H_e, H_f, H_g, H_h/ 1 \leq a,b,c,d,e,f,g,h \leq 8$)	$^8C_8 = 1$

CASE #1

Assumptions:

The MD (Mission Direction) of an UAV is R→L_0. There is no obstacle in the high risk region. Then the following nine control formulas are valid:

1. $(^Lk_1=1) \Rightarrow \displaystyle\bigvee_{i=0}^{9,\ i \neq k_1} RL_i$

   (One obstacle in nine possible regions)

2. $(^Lk_1 \wedge {}^Lk_2 = 1) \Rightarrow \displaystyle\bigvee_{i=0}^{9,\ i \neq k_1,k_2} RL_i$

   (Two obstacles in nine possible regions)

3. $(^Lk_1 \wedge {}^Lk_2 \wedge {}^Lk_3 = 1) \Rightarrow \displaystyle\bigvee_{i=0}^{9,\ i \neq k_1,k_2,k_3} RL_i$

   (Three obstacles in nine possible regions)

4. $(^Lk_1 \wedge {}^Lk_2 \wedge {}^Lk_3 \wedge {}^Lk_4 = 1) \Rightarrow \displaystyle\bigvee_{i=0}^{9,\ i \neq k_1,k_2,k_3,k_4} RL_i$

   (Four obstacles in nine possible regions)

5. $(^Lk_1 \wedge {}^Lk_2 \wedge {}^Lk_3 \wedge {}^Lk_4 \wedge {}^Lk_5 = 1) \Rightarrow \displaystyle\bigvee_{i=0}^{9,\ i \neq k_1,k_2,k_3,k_4,k_5} RL_i$

   (Five obstacles in nine possible regions)

6. $(^{L}k_1 \wedge {}^{L}k_2 \wedge {}^{L}k_3 \wedge {}^{L}k_4 \wedge {}^{L}k_5 \wedge {}^{L}k_6 = 1) \Rightarrow$
$$\bigvee_{i=0}^{9,\ i \neq k_1, k_2, k_3, k_4, k_5, k_6} RL_i$$
(Six obstacles in nine possible regions)

7. $(^{L}k_1 \wedge {}^{L}k_2 \wedge {}^{L}k_3 \wedge {}^{L}k_4 \wedge {}^{L}k_5 \wedge {}^{L}k_6 \wedge {}^{L}k_7 = 1) \Rightarrow$
$$\bigvee_{i=0}^{9,\ i \neq k_1, k_2, k_3, k_4, k_5, k_6, k_7} RL_i$$
(Seven obstacles in nine possible regions)

8. $(^{L}k_1 \wedge {}^{L}k_2 \wedge {}^{L}k_3 \wedge {}^{L}k_4 \wedge {}^{L}k_5 \wedge {}^{L}k_6 \wedge {}^{L}k_7 \wedge {}^{L}k_8 = 1) \Rightarrow$
$$\bigvee_{i=0}^{9,\ i \neq k_1, k_2, k_3, k_4, k_5, k_6, k_7, k_8} RL_i$$
(Eight obstacles in nine possible regions)

9. $(^{L}k_1 \wedge {}^{L}k_2 \wedge {}^{L}k_3 \wedge {}^{L}k_4 \wedge {}^{L}k_5 \wedge {}^{L}k_6 \wedge {}^{L}k_7 \wedge {}^{L}k_8 \wedge {}^{L}k_0 = 1) \Rightarrow$
$$\bigvee_{i=0}^{9,\ i \neq k_1, k_2, k_3, k_4, k_5, k_6, k_7, k_8, k_0} RL_i$$
(Nine obstacles in nine possible regions)

CASE #2

Assumptions:

MD (Mission direction) of an UAV is assumed to be R→L$_0$. There are obstacles in the high threat region. Then the following eight control formulas are valid:

1. $(^{H}k_1 = 1) \Rightarrow$
$$\bigvee_{i=0}^{8,\ i \neq k_1} RH_i$$
(One obstacle in eight possible regions)

2. $(^{H}k_1 \wedge {}^{H}k_2 = 1) \Rightarrow$
$$\bigvee_{i=0}^{8,\ i \neq k_1, k_2} RH_i$$
(Two obstacles in eight possible regions)

3. $(^{H}k_1 \wedge {}^{H}k_2 \wedge {}^{H}k_3 = 1) \Rightarrow$
$$\bigvee_{i=0}^{8,\ i \neq k_1, k_2, k_3} RH_i$$
(Three obstacles in eight possible regions)

4. $(^{H}k_1 \wedge {}^{H}k_2 \wedge {}^{H}k_3 \wedge {}^{H}k_4 = 1) \Rightarrow$
$$\bigvee_{i=0}^{8,\ i \neq k_1, k_2, k_3, k_4} RH_i$$
(Four obstacles in eight possible regions)

5. $(^{H}k_1 \wedge {}^{H}k_2 \wedge {}^{H}k_3 \wedge {}^{H}k_4 \wedge {}^{H}k_5 = 1) \Rightarrow$
$$\bigvee_{i=0}^{8,\ i \neq k_1, k_2, k_3, k_4, k_5} RH_i$$
(Five obstacles in eight possible regions)

6. $(^Hk_1 \wedge ^Hk_2 \wedge ^Hk_3 \wedge ^Hk_4 \wedge ^Hk_5 \wedge ^Hk_6 = 1) \Rightarrow$

$$\bigvee_{i=0}^{8,\ i \neq k_1, k_2, k_3, k_4, k_5, k_6, k_7} RH_i$$

(Six obstacles in eight possible regions)

7. $(^Hk_1 \wedge ^Hk_2 \wedge ^Hk_3 \wedge ^Hk_4 \wedge ^Hk_5 \wedge ^Hk_6 \wedge ^Hk_7 = 1) \Rightarrow$

$$\bigvee_{i=0}^{8,\ i \neq k_1, k_2, k_3, k_4, k_5, k_6, k_7} RH_i$$

(Seven obstacles in eight possible regions)

8. $(^Hk_1 \wedge ^Hk_2 \wedge ^Hk_3 \wedge ^Hk_4 \wedge ^Hk_5 \wedge ^Hk_6 \wedge ^Hk_7 \wedge ^Hk_8 = 1) \Rightarrow$

$$\bigvee_{i=0} RH_i$$

(Eight obstacles in eight possible regions)

Refer to the diagram 4.1.2. Direct neighbors to the future position of the robot L_0 are L_2, L_4, L_6, L_8. The indirect neighbors to the robot are L_1, L_3, L_5 and L_7.

Low threat region

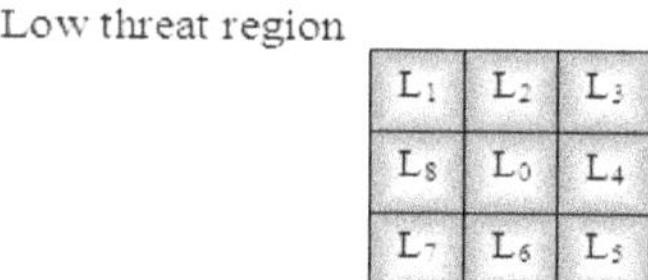

Fig. 1.7 Direct & indirect neighbors to L_0

We assume that the robot is moving towards L_0. Now, if there is an object in a direct neighborhood, then its distance metric from L_0 is assumed to be 1 unit. On the other hand, if an obstacle is found in an indirect neighborhood, then its distance metric from L_0 is assumed to be $2^{1/2}$ units. If one obstacle is found in a direct neighborhood and one obstacle in an indirect neighborhood, then the total distance from these two obstacles to L_0 would be calculated as $1+2^{1/2}$ units. There are 9 cells in the discrete region as defined in figure 4.1.2. The scalar metric (total distance) from all obstacles to any empty cell (a cell where there is no obstacle) L_i is denoted as $D(L_i)$ where $D(L_i) = \Sigma d(L_i, L_j);\ j \geq 1$ where L_i is the position of the empty cell and L_j is a cell containing an obstacle. For example consider an empty cell at L_4. Assume that obstacles are in L_1, L_3, L_5. Then $D(L_4) = d(L_4, L_1) + d(L_4, L_3) + d(L_4, L_5) = 2^{1/2} + 2^{1/2} + 2^{1/2} = 3(2^{1/2})$. Now the probability of collision due to an obstacle at L_1 is denoted as $P\{L_1\}$ or in short, as $P\{1\}$. With these basic assumptions, we now propose an algorithm to fix the change in the Mission Direction of the robot.

Low Threat Region:

Algorithm

Given number of obstacles and their respective positions, the scalar metric of each empty cell is calculated including L_0. The position with the largest distance D value is taken as the safest direction to avoid collision. There are nine control formulas and so, there are 512 possibilities the robot has to investigate before it proceeds further.

Equally likely probabilities

Assume that there is no obstacle in L_0. Assume that there is an obstacle in L_1. Now, the probability of collision due to this obstacle in L_1 with the UAV which is going to be at position L_0 is directly proportional to the D value of L_0, where, $D(L_0) = d(L_0, L_1) = 2^{1/2}$, which is the distance metric from L_1 to L_0. Let us denote this probability as $P\{1\}$. More precisely, $P\{1\}$ is the probability of collision due to an obstacle in position L_1. Note that $P\{3\} = P\{5\} = P\{7\}$, which are equally likely. In the same manner, if an obstacle is found in L_2, then D value of L_0 is 1. Then the probability of collision is $P\{2\}$ and it is equally likely with those of $P\{4\}$, $P\{6\}$ and $P\{8\}$. We shall extend this equally likely concept for multiple obstacle cases also. For instance, $P\{2,4\}$ is the probability of collision due to two obstacles in positions L_2 and L_4. After having equally likely collision spaces, the robot can choose any one of them depending on convenience. Assume that ***there is no obstacle in*** L_0. Then, one can easily verify that the probability of collision when no obstacle is found, is 0 and the probability of collision when all eight obstacles are found in all the positions L_1 to L_8 would lead to a value just less than 1. The remaining 254 possibilities form a part of the real time collision processes. ***All these 254 possibilities are categorized into 23 equally likely classes***. When the collision problem falls in a particular equally likely class, then any of the equally likely positions can be chosen for forward motion by the robot.

Assume that ***there is an obstacle in*** L_0. Then, the UAV has to necessarily change the Mission Direction because there is an obstacle in the cell L_0. In such a case the UAV has to face an additional 256 possibilities before taking a decision to change the Mission Direction. Note that the probability of collision would turn out to be exactly 1 when obstacles are found in all nine positions L_0 to L_8 of the low threat region.

The problem of collision is not fully solved by this strategy. In fact, the UAV has also to face serious threat from obstacles arriving from its sides, that is, from the high threat region. More about this is described below.

High Threat Region

The robot (UAV) does have risk from the high risk region also. In such a case, the eight control formulas corresponding to the high threat region are also considered before taking decision for onward motion. We shall review once again these eight formulas from the previous section.

1. $(^Hk_1 = 1) \Rightarrow \displaystyle\bigvee_{i=0}^{8,\ i \neq k_1} RH_i$

(One obstacle in eight possible regions)

2. $(^Hk_1 \wedge {}^Hk_2 = 1) \Rightarrow \displaystyle\bigvee_{i=0}^{8,\ i \neq k_1, k_2} RH_i$

(Two obstacles in eight possible regions)

3. $({}^{H}k_1 \wedge {}^{H}k_2 \wedge {}^{H}k_3 = 1) \Rightarrow \bigvee\limits_{i=0}^{8,\ i\neq k_1,k_2,k_3} RH_i$

(Three obstacles in eight possible regions)

4. $({}^{H}k_1 \wedge {}^{H}k_2 \wedge {}^{H}k_3 \wedge {}^{H}k_4 = 1) \Rightarrow \bigvee\limits_{i=0}^{8,\ i\neq k_1,k_2,k_3,k_4} RH_i$

(Four obstacles in eight possible regions)

5. $({}^{H}k_1 \wedge {}^{H}k_2 \wedge {}^{H}k_3 \wedge {}^{H}k_4 \wedge {}^{H}k_5 = 1) \Rightarrow \bigvee\limits_{i=0}^{8,\ i\neq k_1,k_2,k_3,k_4,k_5} RH_i$

(Five obstacles in eight possible regions)

6. $({}^{H}k_1 \wedge {}^{H}k_2 \wedge {}^{H}k_3 \wedge {}^{H}k_4 \wedge {}^{H}k_5 \wedge {}^{H}k_6 = 1) \Rightarrow \bigvee\limits_{i=0}^{8,\ i\neq k_1,k_2,k_3,k_4,k_5,k_6,k_7} RH_i$

(Six obstacles in eight possible regions)

7. $({}^{H}k_1 \wedge {}^{H}k_2 \wedge {}^{H}k_3 \wedge {}^{H}k_4 \wedge {}^{H}k_5 \wedge {}^{H}k_6 \wedge {}^{H}k_7 = 1) \Rightarrow \bigvee\limits_{i=0}^{8,\ i\neq k_1,k_2,k_3,k_4,k_5,k_6,k_7} RH_i$

(Seven obstacles in eight possible regions)

8. $({}^{H}k_1 \wedge {}^{H}k_2 \wedge {}^{H}k_3 \wedge {}^{H}k_4 \wedge {}^{H}k_5 \wedge {}^{H}k_6 \wedge {}^{H}k_7 \wedge {}^{H}k_8 = 1) \Rightarrow \bigvee\limits_{i=0}^{8,\ i\neq k_1,k_2,k_3,k_4,k_5,k_6,k_7,k_8} RH_i$

(Eight obstacles in eight possible regions)

Note that there are 256 such collision possibilities in high threat region. Based on the lines of the strategy adopted for the low threat region, one can visualize 23 collision avoidance strategies each with equally likely possibilities in the high threat region.

Till now we analyzed the collision possibilities for an UAV with the mission direction $R \rightarrow L_0$, from the low and high threat regions separately. In practice this would prove to be erroneous. An UAV in space would face collision threats from obstacles not only from low threat region but also from high threat region simultaneously. Consequently, we infer that the UAV with the mission direction $R \rightarrow L_0$, has to counter all these threats simultaneously and then only take decision for forward motion. *In short, the collision avoidance problem amounts to say that there are 512 x 256, that is, 1,31,072 possibilities of collision that an UAV has to encounter if its dynamics is modeled in the square discrete space.*

BIBLIOGRAPHY

1   Markov, A. A., Theory of Algorithms. The Israel Program for Scientific Translations Jerusalem, 1961.

2   Rajan, E.G., Cellular Logic Array Processing, Techniques for high throughput Image Processing Systems, SADHANA, Special Issue on Computer Vision, Volume 18, Part 2, June 1993, pp. 279-300, Indian Academy of Sciences.

3   Rajan, E. G., High-Throughput Cellular Logic Array Processing of Satellite Data for Geophysical Surveying, Paper No A.1-S.1.08. The World Space Congress, Washington, D.C., U.S.A., 28 August to 5 Sept. 1992.

4   Rajan, E. G., Fast Algorithm for collision avoidance by autonomous mobile systems in the framework of cellular logic array processing, 13th international conference on CAD/CAM Robotics & Factories of the Future, Universidad Technologica de Pereira, Columbia, South America, December 15-17, 1997.

5   Rajan, E. G., Symbolic Computing – Signal and Image Processing, Anshan Publications, Kent, United Kingdom 2003

A LOGICAL APPROACH TO THE STUDY OF COLLISION AVOIDANCE PROBLEMS PERTAINING TO UNMANNED AERIAL VEHICLES MEANT FOR ELECTRONIC SUPPORT MEASURES (PART – 2)

Ch. Raja
Research Scholar,
E-mail: rajachaluvadi@yahoo.com

M. Madhavi Latha
Professor,
Electronics & Communication Engineering Dept.,
JNTUH,Kukatpally, Hyderabad-500062, India
E-mail: mlmakkena@yahoo.com

E. G. Rajan
Founder President, Pentagram Research Centre Pvt. Ltd.
#201, Venkat Homes, MIGH-59, Mehdipatnam, Hyderabad, India
E-mail: rajaneg@yahoo.co

ABSTRACT

The work carried out, as reported in this paper, centers around a novel path fixing technique in the framework of Cellular Logic Array Processing. Development of fast algorithms for the instantaneous fixing of collision free paths by UAVs in unknown environment had been the major output of this intended research. UAV control is modeled as a constructive system that operates on two regions (i) low risk region and (ii) high risk region as given below. The term *High Risk Region* (HRR) refers to, for example, the scenic coverage around the UAV whereas the *Low Risk Region* (LRR) refers to the front scenic coverage acquired by the vision system of the UAV. In a real life situation also, one could see that the risk due to obstacles approaching from the sides is more than the obstacles approaching from the front, hence head-on collisions are rare on roads when compared to collision from sides. As outlined in part 1 of this paper, it has been found that the rectangular lattice model of Euclidean space has a rotational symmetry of 90°, where as that of a hexagonal lattice model has 60°. Hence, the latter would be a better choice for the study of collision avoidance of an UAV traveling in the Euclidean space. This paper deals with the state-of-the-art technique involved in developing control strategies for an autonomous mobile system in a hexagonal lattice model.

Hexagonal Lattice Model Of Space

This part gives the results of a study carried out in this regard. Figure 1.1 shows the hexagonal lattice model of space. The intended path of the UAV is given by $R \rightarrow L_0$. Normal speed of robot is denoted as S. A path is fixed by the robot depending on the positions and speeds of obstacles. At a given time instant, the robot dynamics is determined by the ordered pair $<S, R \rightarrow L_i>$, $0 \leq i \leq 6$. There are seven loci corresponding to less harmful objects approaching the robot from seven directions: $L_j \rightarrow R$, $0 \leq j \leq 6$. So $R \rightarrow L_i$ depends on $L_j \rightarrow R$, and S on $H_k \rightarrow R$, $1 \leq i \leq 6$.

Logical Control Formulas For Collision Avoidance

As it was done in the case of rectangular lattice model of Euclidean space, based on the threat positions of obstacles shown by images obtained by the acquisition system, COPIS, the UAV would make use of certain control formulas which would indicate the direction of least collision possibility for restorative deviation[1]-[3].

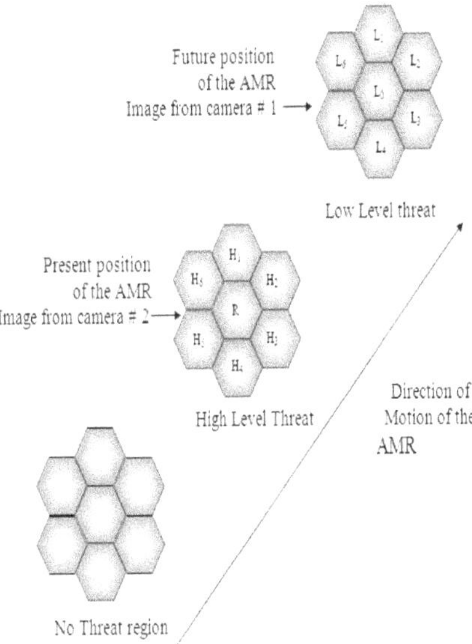

Fig. 1.1: Hexagonal lattice model of Euclidean space

Table 1.1.1 gives a summary of high risk low threat positions for various groups of obstacles, with an obstacle in the cell L_0. The set given in table 1.1.1 consists of a total of 64 high risk low threat positions.

Similarly, one can construct table 1.1.2 summarizing 64 low risk low threat positions without an obstacle in position L_0. By low risk threat, we mean that the UAV can continue in its intended direction $R \rightarrow L_0$ with low risk in spite of the fact there are obstacles present elsewhere in the neighborhood. Table 1.1.2 gives a summary of low risk low threat positions for various groups of obstacles, without an obstacle in the cell L_0.

Table 1.1.3 gives a summary of high risk high threat positions for various groups of obstacles, with UAV in the cell L_0.

Table 1.1.1: High Risk Low Threat Positions (set #1)

Sl.No.	Number of obstacles (with an obstacle in L_0)	Threat positions (No. of obstacles)
1	(L_0) (No obstacle other than one in L_0)	$^6C_0 = 1$
2	$(L_0, L_a/1 \le a \le 6)$	$^6C_1 = 6$
3	$(L_0, L_a, L_b/1 \le a,b \le 6)$	$^6C_2 = 15$
4	$(L_0, L_a, L_b, L_c /1 \le a,b,c \le 6)$	$^6C_3 = 20$
5	$(L_0, L_a, L_b, L_c, L_d /1 \le a,b,c,d \le 6)$	$^6C_4 = 15$
6	$(L_0, L_a, L_b, L_c, L_d, L_e /1 \le a,b,c,d,e \le 6)$	$^6C_5 = 6$
7	$(L_0, L_a, L_b, L_c, L_d, L_e, L_f /1 \le a,b,c,d,e,f \le 6)$	$^6C_6 = 1$

Table 1.1.2: Low Risk Low Threat Positions (set #2)

Sl. No.	No. of obstacles (without obstacle in L_0)	Threat positions (No. of obstacles)
1	( ) (No obstacle threat free)	$^6C_0 = 1$
2	$(L_a/1 \le a \le 6)$	$^6C_1 = 6$
3	$(L_a, L_b /1 \le a,b \le 6)$	$^6C_2 = 15$
4	$(L_a, L_b, L_c /1 \le a,b,c \le 6)$	$^6C_3 = 20$
5	$(L_a, L_b, L_c, L_d /1 \le a, b,c,d \le 6)$	$^6C_4 = 15$
6	$(L_a, L_b, L_c, L_d, L_e /1 \le a,b,c,d,e \le 6)$	$^6C_5 = 6$
7	$(L_a, L_b, L_c, L_d, L_e, L_f /1 \le a,b,c,d,e,f \le 6)$	$^6C_6 = 1$

From the above argument we see that an UAV faces a total of 128 low threats, that is, 64 high risk zones (with an obstacle in L_0) and 64 low risk ones (without an obstacle in L_0). Note that the threat free zone is also considered here as a low risk low threat zone.

Table 1.1.3: High Risk High Threat Positions

Sl.No.	No. of obstacles (UAV in H_0)	Threat positions (No. of obstacles)
1	(R) (UAV Position, threat free)	$^6C_0 = 1$
2	$(H_a/1 \le a \le 6)$	$^6C_1 = 6$
3	$(H_a, H_b /1 \le a,b \le 6)$	$^6C_2 = 15$
4	$(H_a, H_b, H_c /1 \le a,b,c \le 6)$	$^6C_3 = 20$
5	$(H_a, H_b, H_c, H_d /1 \le a,b,c,d \le 6)$	$^6C_4 = 15$
6	$(H_a, H_b, H_c, H_d, H_e /1 \le a,b,c,d,e \le 6)$	$^6C_5 = 6$
7	$(H_a, H_b, H_c, H_d, H_e, H_f /1 \le a,b,c,d,e,f \le 6)$	$^6C_6 = 1$

In addition to the above low threat zones, an UAV is expected to face more 64 high threat zones from its sides. The UAV would change its speed depending on the high threat level. Table 1.1.3 provides seven formulas that define all the 64 high risk high threat positions. Note that the threat free zone of table 1.1.3 is also considered as a high risk high threat zone.

To summarize, with the three dimensional Euclidean space being modeled as a discrete hexagonal grid, an UAV faces a total of 128 x 64, that is, **8,192** threat zones while undertaking a mission in this discrete space.

The restorative deviation method is adopted by the UAV to avoid different types of collisions. Further precise modifications could be made if we resort to probabilistic methods in evaluating the relative velocities and assigning suitable priority values to different directions while fixing the path. With the basic model for a collision avoidance system, we now proceed to examine the possibility of using logical control of the UAV in the framework of constructive mathematical framework discussed in detail in the previous chapters. The constructive logical formulas that govern the motion of an UAV are given below [4]-[6]:

CASE 1

Assumptions

The MD (Mission Direction) of an UAV is $R \rightarrow L_0$. There is no obstacle in the high risk region. Then the following seven control formulas are valid:

1. $(^L k_1 = 1) \Rightarrow \bigvee\limits_{\substack{i=0 \\ }}^{7,\ i \neq k_1} RL_i$

(One obstacle in seven possible regions)

2. $(^L k_1 \wedge {}^L k_2 = 1) \Rightarrow \bigvee\limits_{\substack{i=0 \\ }}^{7,\ i \neq k_1, k_2} RL_i$

(Two obstacles in seven possible regions)

3. $(^L k_1 \wedge {}^L k_2 \wedge {}^L k_3 = 1) \Rightarrow \bigvee\limits_{\substack{i=0 \\ }}^{7,\ i \neq k_1, k_2, k_3} RL_i$

(Three obstacles in seven possible regions)

4. $(^L k_1 \wedge {}^L k_2 \wedge {}^L k_3 \wedge {}^L k_4 = 1) \Rightarrow \bigvee\limits_{\substack{i=0 \\ }}^{7,\ i \neq k_1, k_2, k_3, k_4} RL_i$

(Four obstacles in seven possible regions)

5. $(^L k_1 \wedge {}^L k_2 \wedge {}^L k_3 \wedge {}^L k_4 \wedge {}^L k_5 = 1) \Rightarrow \bigvee\limits_{\substack{i=0 \\ }}^{7,\ i \neq k_1, k_2, k_3, k_4, k_5} RL_i$

(Five obstacles in seven possible regions)

6. $(^L k_1 \wedge {}^L k_2 \wedge {}^L k_3 \wedge {}^L k_4 \wedge {}^L k_5 \wedge {}^L k_6 = 1) \Rightarrow \bigvee\limits_{\substack{i=0 \\ }}^{7,\ i \neq k_1, k_2, k_3, k_4, k_5, k_6, k_7} RL_i$

(Six obstacles in seven possible regions)

7. $(^L k_1 \wedge {}^L k_2 \wedge {}^L k_3 \wedge {}^L k_4 \wedge {}^L k_5 \wedge {}^L k_6 \wedge {}^L k_7 = 1) \Rightarrow \bigvee\limits_{\substack{i=0 \\ }}^{7,\ i \neq k_1, k_2, k_3, k_4, k_5, k_6, k_7} RL_i$

(Seven obstacles in seven possible regions)

CASE 2

Assumptions

- MD (Mission direction) of an UAV is assumed to be $R \rightarrow L_0$.
- There are obstacles in the high threat region.

Then the following six control formulas are valid:

$$1.\ (^Hk_1=1) \Rightarrow \bigvee_{i=0}^{6,\ i\neq k_1} RH_i$$

(One obstacle in six possible regions)

$$2.\ (^Hk_1 \wedge {}^Hk_2 = 1) \Rightarrow \bigvee_{i=0}^{6,\ i\neq k_1,k_2} RH_i$$

(Two obstacles in six possible regions)

$$3.\ (^Hk_1 \wedge {}^Hk_2 \wedge {}^Hk_3 = 1) \Rightarrow \bigvee_{i=0}^{6,\ i\neq k_1,k_2,k_3} RH_i$$

(Three obstacles in six possible regions)

$$4.\ (^Hk_1 \wedge {}^Hk_2 \wedge {}^Hk_3 \wedge {}^Hk_4 = 1) \Rightarrow \bigvee_{i=0}^{6,\ i\neq k_1,k_2,k_3,k_4} RH_i$$

(Four obstacles in six possible regions)

$$5.\ (^Hk_1 \wedge {}^Hk_2 \wedge {}^Hk_3 \wedge {}^Hk_4 \wedge {}^Hk_5 = 1) \Rightarrow \bigvee_{i=0}^{6,\ i\neq k_1,k_2,k_3,k_4,k_5} RH_i$$

(Five obstacles in six possible regions)

$$6.\ (^Hk_1 \wedge {}^Hk_2 \wedge {}^Hk_3 \wedge {}^Hk_4 \wedge {}^Hk_5 \wedge {}^Hk_6 = 1) \Rightarrow \bigvee_{i=0}^{6,\ i\neq k_1,k_2,k_3,k_4,k_5,k_6,k_7} RH_i$$

(Six obstacles in six possible regions)

These $Я_{\omega|}$ - provable ($\exists$, $\forall$)-free constructive logical formulas form the control signals that change the state vectors of the UAV moving in space. The results and further discussions are given in the next section.

Analysis Of UAV Control Strategy In Hexagonal Lattice Model

In the previous sub section 1.1, the UAV control was modeled as a constructive system that operates on two regions (i) low threat region and (ii) high threat region. The term Low Threat Region (LTR) refers to, for example, the scenic coverage acquired by the camera #1 in COPIS, where as the High Threat Region (HTR) to the scenic coverage acquired by the camera #2 using the conical mirror of COPIS. The image acquired by COPIS would be a hexagonal array of pixels.

As it was done in the case of rectangular grid space model, the state vector control is treated here as a conjoint control taking into consideration both the low threat as well as high threat regions. However, with the idea of reducing the complexity, both the regions are treated independent of each other while calculating the respective possibilities of collision due to presence of obstacles in various sub-regions [7]. The results are discussed below in detail.

Collision Avoidance Control Strategy

CASE 1

For convenience, we review the seven control formulas.

1. $(^Lk_1=1) \Rightarrow \bigvee\limits_{i=0}^{7,\ i\neq k_1} RL_i$

    (One obstacle in seven possible regions)

2. $(^Lk_1 \wedge ^Lk_2 = 1) \Rightarrow \bigvee\limits_{i=0}^{7,\ i\neq k_1,k_2} RL_i$

    (Two obstacles in seven possible regions)

3. $(^Lk_1 \wedge ^Lk_2 \wedge ^Lk_3 = 1) \Rightarrow \bigvee\limits_{i=0}^{7,\ i\neq k_1,k_2,k_3} RL_i$

    (Three obstacles in seven possible regions)

4. $(^Lk_1 \wedge ^Lk_2 \wedge ^Lk_3 \wedge ^Lk_4 = 1) \Rightarrow \bigvee\limits_{i=0}^{7,\ i\neq k_1,k_2,k_3,k_4} RL_i$

    (Four obstacles in seven possible regions)

5. $(^Lk_1 \wedge ^Lk_2 \wedge ^Lk_3 \wedge ^Lk_4 \wedge ^Lk_5 = 1) \Rightarrow \bigvee\limits_{i=0}^{7,\ i\neq k_1,k_2,k_3,k_4,k_5} RL_i$

    (Five obstacles in seven possible regions)

6. $(^Lk_1 \wedge ^Lk_2 \wedge ^Lk_3 \wedge ^Lk_4 \wedge ^Lk_5 \wedge ^Lk_6 = 1) \Rightarrow \bigvee\limits_{i=0}^{7,\ i\neq k_1,k_2,k_3,k_4,k_5,k_6,k_7} RL_i$

    (Six obstacles in seven possible regions)

7. $(^Lk_1 \wedge ^Lk_2 \wedge ^Lk_3 \wedge ^Lk_4 \wedge ^Lk_5 \wedge ^Lk_6 \wedge ^Lk_7 = 1) \Rightarrow \bigvee\limits_{i=0}^{7,\ i\neq k_1,k_2,k_3,k_4,k_5,k_6,k_7} RL_i$

    (Seven obstacles in seven possible regions)

Refer to the diagram 1.2.1.1. Direct neighbors to the future position of the robot L_0 are L_1, L_2, L_3, L_4, L_5, L_6.

Low Threat Region

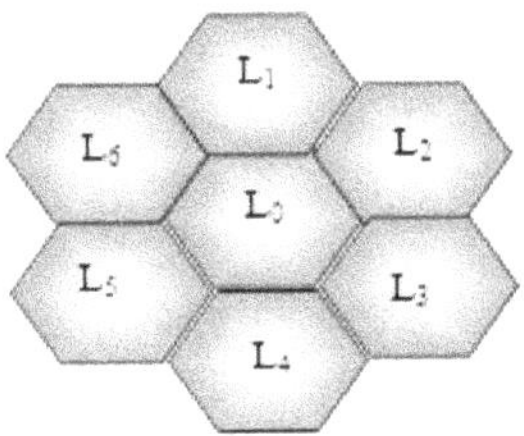

Fig. 1.2.1.1: Neighbors to L_0

We assume that the robot is moving towards L_0. Now, if there is an object in a neighborhood, then its distance metric from L_0 to L_1, L_2, L_3, L_4, L_5 and L_6 are assumed to be 1 unit each. There are 7 cells in the discrete region as defined in figure 1.2.1.1.

The scalar metric (total distance) from all obstacles to any empty cell (a cell where there is no obstacle) L_i is denoted as $D(L_i)$ where $D(L_i) = \sum d(L_i, L_j)$; $j \geq 1$ where L_i is the position of the empty cell and L_j is a cell containing an obstacle. For example consider an empty cell at L_4. Assume that obstacles are in L_1, L_3, L_5. Then $D(L_4) = d(L_4,L_1) + d(L_4,L_3) + d(L_4,L_5) = 2 + 1 + 1 = 4$. Now the probability of collision due to an obstacle at L_1 is denoted as $P\{L_1\}$ or in short, as $P\{1\}$.

With these basic assumptions, we now propose an algorithm to fix the change in the Mission Direction of the robot.

Low Threat Region

Algorithm

Given number of obstacles and their respective positions, the scalar metric of each empty cell is calculated including L_0. The position with the largest distance D value is taken as the safest direction to avoid collision [8]. There are seven control formulas and so, there are 128 possibilities the robot has to investigate before it proceeds further.

Equally likely probabilities

Assume that there is no obstacle in L_0. Assume that there is an obstacle in L_1. Now, the probability of collision due to this obstacle in L_1 with the UAV which is going to be at position L_0 is directly proportional to the D value of L_0, where, $D(L_0) = d(L_0,L_1) = 1$, which is the distance metric from L_1 to L_0. Let us denote this probability as $P\{1\}$. More precisely, $P\{1\}$ is the probability of collision due to an obstacle in position L_1.

Note that $P\{1\} = P\{2\} = P\{3\} = P\{4\} = P\{5\} = P\{6\}$, which are equally likely. We shall extend this equally likely concept for multiple obstacle cases also. For instance, $P\{2,4\}$ is the probability of collision due to two obstacles in positions L_2 and L_4[9].

After having equally likely collision spaces, the robot can choose any one of them depending on convenience. Assuming that *there is no obstacle in* L_0, the following list shows such equally likely probabilities:

No. of obstacles	Equally likely probabilities	Total formulas	Scalar metric (D) value
0	0	1	0
1	P{1}=P{2}=P{3}=P{4}=P{5}=P{6}	6	1
2	P{1,2}=P{2,3}=P{3,4}=P{4,5}=P{5,6}=P{1,6}=P{1,3}=P{2,4}=P{3,5}=P{4,6}=P{1,5}=P{2,6}=P{1,4}=P{2,5}=P{3,6}	15	2
3	P{1,2,3}=P{2,3,4}=P{3,4,5}=P{4,5,6}=P{1,5,6}=P{1,2,6}=P{1,3,4}=P{2,4,5}=P{3,5,6}=P{1,4,6}=P{1,2,5}=P{2,3,6}=P{1,3,5}=P{2,4,6}=P{1,2,4}=P{2,3,5}=P{3,4,6}=P{1,4,5}=P{2,5,6}=P{1,3,6}	20	3
4	P{1,2,3,4}=P{2,3,4,5}=P{3,4,5,6}=P{1,4,5,6}=P{1,2,5,6}=P{1,2,3,6}=P{1,2,4,5}=P{2,3,5,6}=P{1,3,4,6}=P{1,3,4,5}=P{2,4,5,6}=P{1,3,5,6}=P{1,2,4,6}=P{1,2,3,5}=P{2,3,4,6}	15	4
5	P{1,2,3,4,5}=P{2,3,4,5,6}=P{1,3,4,5,6}=P{1,2,4,5,6}=P{1,2,3,5,6}=P{1,2,3,4,6}	6	5
6	P{1,2,3,4,5,6}	1	6

From the above list, one can easily verify that the probability of collision when no obstacle is found, is 0 and the probability of collision when all six obstacles are found in all the positions L_1 to L_6 would lead to a value just less than 1. The remaining 62 possibilities given in the above list form a part of the real time collision processes. All these 62 possibilities are categorized into **5** equally likely classes. When the collision problem falls in a particular equally likely class, then any of the equally likely positions can be chosen for forward motion by the robot.

Assume that there is an obstacle in L_0. Then, the UAV has to necessarily change the Mission Direction because there is an obstacle in the cell L_0. In such a case the UAV has to face an additional 64 possibilities before taking a decision to change the Mission Direction. Note that the probability of collision would turn out to be exactly 1 when obstacles are found in all seven positions L_0 to L_6 of the low threat region.

Note that the problem of collision is not fully solved by this strategy. In fact, the UAV has also to face serious threat from obstacles arriving from its sides, that is, from the high threat region. More about this is described below.

High Threat Region

The robot (UAV) does have risk from the high risk region also. In such a case, the eight control formulas corresponding to the high threat region are also considered before taking decision for onward motion [10]. We shall review once again these eight formulas from the previous section.

1. $({}^H k_1 = 1) \Rightarrow \displaystyle\bigvee_{\substack{i=0}}^{6,\ i \neq k_1} RH_i$

(One obstacle in eight possible regions)

2. $({}^H k_1 \wedge {}^H k_2 = 1) \Rightarrow \displaystyle\bigvee_{\substack{i=0}}^{6,\ i \neq k_1, k_2} RH_i$

(Two obstacles in eight possible regions)

3. $({}^H k_1 \wedge {}^H k_2 \wedge {}^H k_3 = 1) \Rightarrow \displaystyle\bigvee_{\substack{i=0}}^{6,\ i \neq k_1, k_2, k_3} RH_i$

(Three obstacles in eight possible regions)

4. $({}^H k_1 \wedge {}^H k_2 \wedge {}^H k_3 \wedge {}^H k_4 = 1) \Rightarrow \displaystyle\bigvee_{\substack{i=0}}^{6,\ i \neq k_1, k_2, k_3, k_4} RH_i$

(Four obstacles in eight possible regions)

5. $({}^H k_1 \wedge {}^H k_2 \wedge {}^H k_3 \wedge {}^H k_4 \wedge {}^H k_5 = 1) \Rightarrow \displaystyle\bigvee_{\substack{i=0}}^{6,\ i \neq k_1, k_2, k_3, k_4, k_5} RH_i$

(Five obstacles in eight possible regions)

6. $({}^H k_1 \wedge {}^H k_2 \wedge {}^H k_3 \wedge {}^H k_4 \wedge {}^H k_5 \wedge {}^H k_6 = 1) \Rightarrow \displaystyle\bigvee_{\substack{i=0}}^{6,\ i \neq k_1, k_2, k_3, k_4, k_5, k_6, k_7} RH_i$

(Six obstacles in eight possible regions)

Note that there are 62 such collision possibilities in high threat region. Based on the lines of the strategy adopted for the low threat region, one can visualize 5 collision avoidance strategies each with equally likely possibilities in the high threat region[11]-[12].

Till now we analyzed the collision possibilities for an UAV with the mission direction R→L_0, from the low and high threat regions separately. In practice this would prove to be erroneous.

A UAV in space would face collision threats from obstacles not only from low threat region but also from high threat region simultaneously.

Consequently, we infer that the UAV with the mission direction R→L_0, has to counter all these threats simultaneously and then only take decision for forward motion[14].

In short, the collision avoidance problem amounts to say that there are 128 x 64, that is, 8,192 possibilities of collision that an UAV has to encounter if its dynamics is modeled in the hexagonal discrete space.

Collision Avoidance Strategy In A Hexagonal Lattice Model

This section proposes a logical method of avoiding collision when the space is viewed as a hexagonal lattice model. Figure 1.2.2.1 shows the low threat region.

Low Threat Region

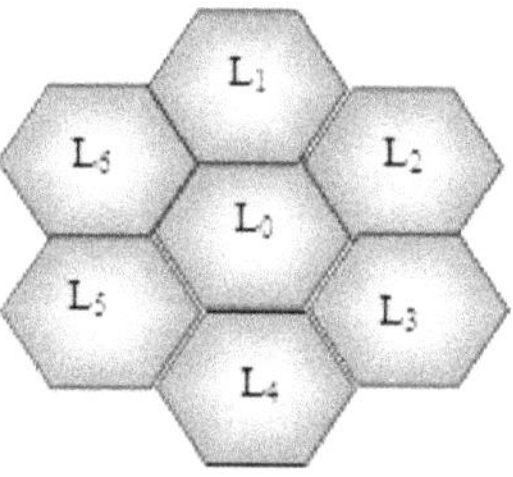

Fig. 1.2.2.1: Neighbors to L_0
(Same as figure 1.2.1.1)

Let us assume that there is an obstacle in the cell L_1. Now we calculate the distance metric of each of the remaining 6 cells and tabulate the values as shown below.

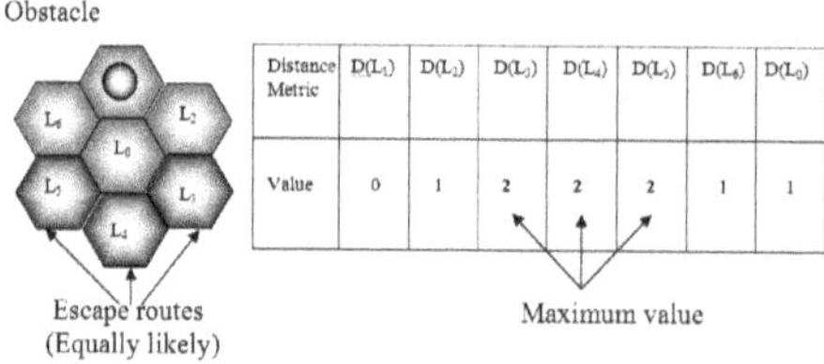

Distance Metric	$D(L_1)$	$D(L_2)$	$D(L_3)$	$D(L_4)$	$D(L_5)$	$D(L_6)$	$D(L_0)$
Value	0	1	2	2	2	1	1

The value corresponding to the cells L_3 L_4 and L_5 is seen to be the same and maximum. This amounts to saying that these cells are appropriate *Escape Routes* (ER) for the UAV when it confronts an obstacle in L_1. The UAV could change the mission direction to R→L_5 for example, in order to avoid collision with the obstacle in L_1 and regain its mission direction R→L_0 using the restorative deviation technique discussed earlier in this paper.

As another example let us consider that there are two obstacles in cells L_1 and L_3. Then the distance metrics of the remaining 7 cells are calculated and tabulated as given below.

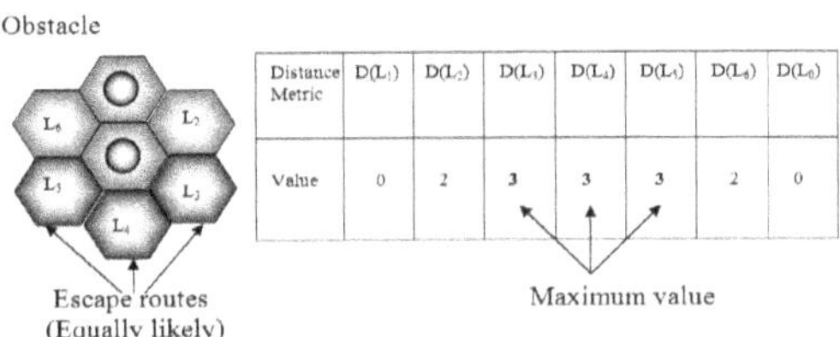

Distance Metric	D(L₁)	D(L₂)	D(L₃)	D(L₄)	D(L₅)	D(L₆)	D(L₀)
Value	0	2	3	3	3	2	0

Escape routes
(Equally likely)

Maximum value

From the above table, one can observe that there are three escape routes L_5, L_6, and L_7 and an UAV can choose any of these three for changing its mission direction when it confronts two obstacles from the cells L_1 and L_3.

In what follows, we provide sample combinations of obstacles (red spheres) (we shall call each combination as a *scene*) and the corresponding distance metrics for each *escape route* (green cell). One can determine the best escape route for a scene and change mission direction (MD) of an UAV based on distance metrics[16].

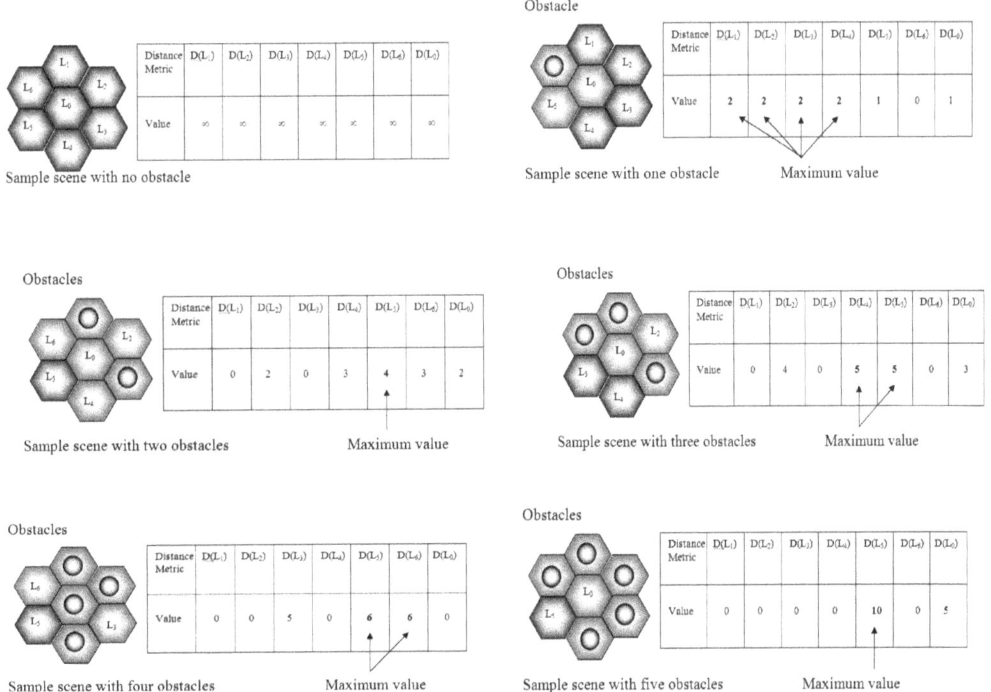

Distance Metric	D(L₁)	D(L₂)	D(L₃)	D(L₄)	D(L₅)	D(L₆)	D(L₀)
Value	∞	∞	∞	∞	∞	∞	∞

Sample scene with no obstacle

Distance Metric	D(L₁)	D(L₂)	D(L₃)	D(L₄)	D(L₅)	D(L₆)	D(L₀)
Value	2	2	2	2	1	0	1

Sample scene with one obstacle — Maximum value

Distance Metric	D(L₁)	D(L₂)	D(L₃)	D(L₄)	D(L₅)	D(L₆)	D(L₀)
Value	0	2	0	3	4	3	2

Sample scene with two obstacles — Maximum value

Distance Metric	D(L₁)	D(L₂)	D(L₃)	D(L₄)	D(L₅)	D(L₆)	D(L₀)
Value	0	4	0	5	5	0	3

Sample scene with three obstacles — Maximum value

Distance Metric	D(L₁)	D(L₂)	D(L₃)	D(L₄)	D(L₅)	D(L₆)	D(L₀)
Value	0	0	5	0	6	6	0

Sample scene with four obstacles — Maximum value

Distance Metric	D(L₁)	D(L₂)	D(L₃)	D(L₄)	D(L₅)	D(L₆)	D(L₀)
Value	0	0	0	0	10	0	5

Sample scene with five obstacles — Maximum value

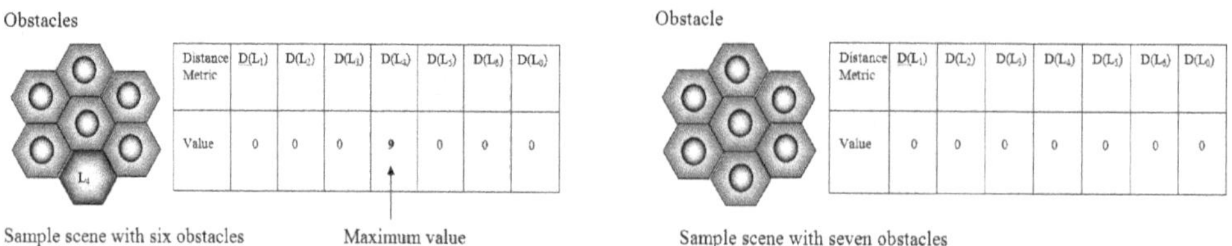

Distance Metric	D(L₁)	D(L₂)	D(L₃)	D(L₄)	D(L₅)	D(L₆)	D(L₀)
Value	0	0	0	9	0	0	0

Distance Metric	D(L₁)	D(L₂)	D(L₃)	D(L₄)	D(L₅)	D(L₆)	D(L₀)
Value	0	0	0	0	0	0	0

RESULTS AND DISCUSSIONS

As described in part 1 of this paper, a thorough study was carried out by simulating in a computer, the control strategy of UAV to avoid collisions with obstacles approaching from different directions in the Euclidean space E^3, which is modeled as a hexagonal lattice of discrete space cells. The following observations were made.

1. The Euclidean space E^3 was modeled as a hexagonal lattice of cells and 128 scenes with and with out obstacles were contemplated.
2. Seven control formulas were formulated in the logico-mathematical framework of Markov's languages of $Я_\alpha$.
3. Metric distance of an escape route has been viewed as the maximum value of the total distance calculated from each obstacle in a particular scene. This has already led to the possibility of framing concepts like the ***probability of collision avoidance*** and the ***probability of collision***.

Now, let us consider the scene shown below in figure 1.2.3.1.

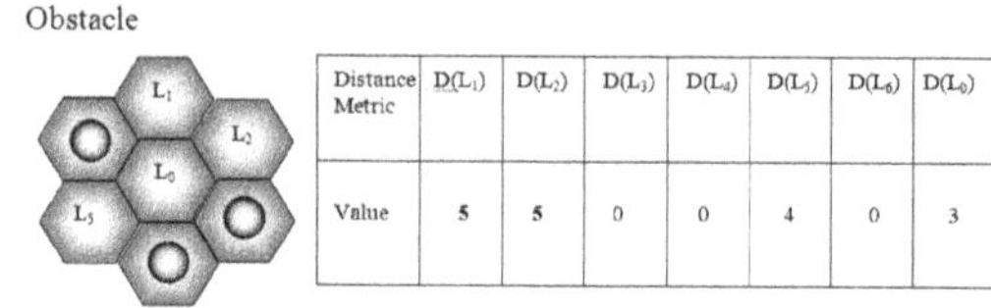

Distance Metric	D(L₁)	D(L₂)	D(L₃)	D(L₄)	D(L₅)	D(L₆)	D(L₀)
Value	5	5	0	0	4	0	3

Figure 1.2.3.1: Sample scene with three obstacles

The probability of collision avoidance is calculated in the following manner. The cell L_1 or L_2 which has the largest distance metric value of 5, provides escape route for the UAV. Let us sum up the distance metrics of all the seven cells, which amounts to be equal to 17. Now the ratio of the largest distance metric to the sum of all distance metrics would yield a measure of ***probability of collision avoidance*** (CAP)[20]. So, in the example scene shown above, the probability of collision avoidance is calculated as 5/17. The sum of all collision avoidance probabilities corresponding to all the nine cells would be seen to be equal to 1. Alternatively one can construct a measure for the ***collision probability*** (CP) [21] also in the following manner. It should be note here that the collision probability and the collision avoidance probability are not directly related to each other like the sum of these probabilities is equal to 1. On the other hand, they could be verified to be related to each other by the formula:

$$\text{Collision probability in a cell (CP)} = \frac{1 - \text{collision avoidance probability (CAP)}}{\text{Number of escape routes} - 1}$$

For the above example, the collision probability of L_1 is calculated as:

$$\text{Collision probability of cell } L_1 = \frac{1 - (5/17)}{4 - 1} = \frac{12}{51}$$

The following list provides sample scenes and values of the collision avoidance and collision probabilities of various cells.

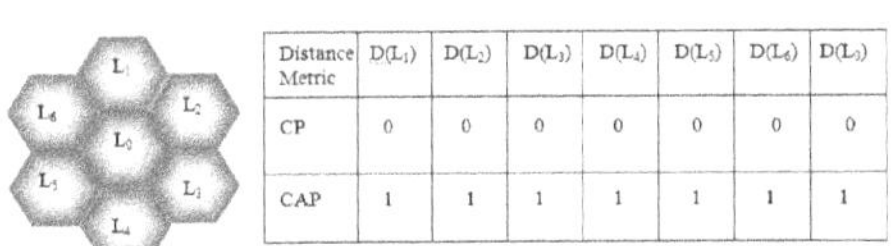

Distance Metric	D(L_1)	D(L_2)	D(L_3)	D(L_4)	D(L_5)	D(L_6)	D(L_0)
CP	0	0	0	0	0	0	0
CAP	1	1	1	1	1	1	1

Sample scene with no obstacle

Obstacle

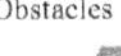

Distance Metric	D(L_1)	D(L_2)	D(L_3)	D(L_4)	D(L_5)	D(L_6)	D(L_0)
CP	2/10	2/10	2/10	2/10	1/10	0	1/10
CAP	8/50	8/50	8/50	8/50	9/50	1	9/50

Sample scene with one obstacle

Obstacles

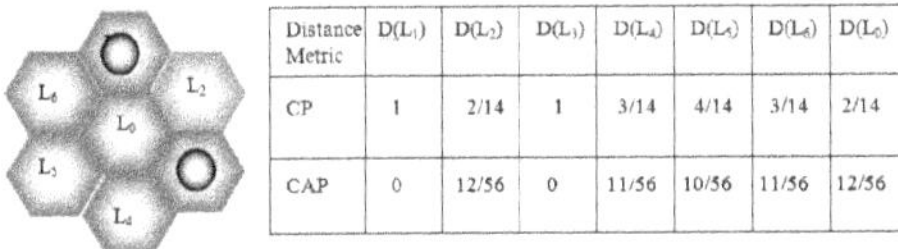

Distance Metric	D(L_1)	D(L_2)	D(L_3)	D(L_4)	D(L_5)	D(L_6)	D(L_0)
CP	1	2/14	1	3/14	4/14	3/14	2/14
CAP	0	12/56	0	11/56	10/56	11/56	12/56

Sample scene with two obstacles

Obstacles

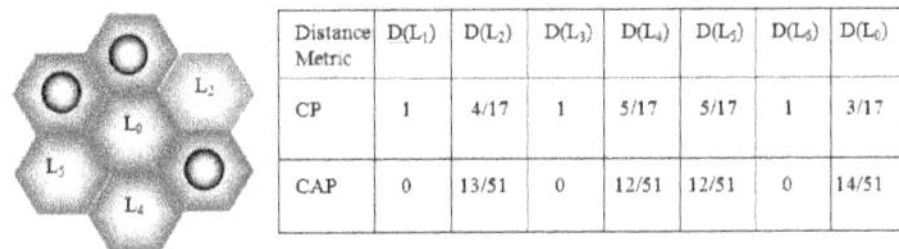

Distance Metric	D(L_1)	D(L_2)	D(L_3)	D(L_4)	D(L_5)	D(L_6)	D(L_0)
CP	1	4/17	1	5/17	5/17	1	3/17
CAP	0	13/51	0	12/51	12/51	0	14/51

Sample scene with three obstacles

Obstacles

Distance Metric	D(L_1)	D(L_2)	D(L_3)	D(L_4)	D(L_5)	D(L_6)	D(L_0)
CP	1	1	5/17	1	6/17	6/17	1
CAP	0	0	12/34	0	11/34	11/34	0

Sample scene with four obstacles

Obstacles

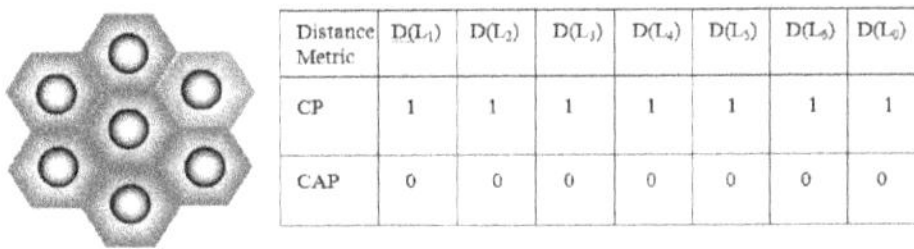

Distance Metric	D(L_1)	D(L_2)	D(L_3)	D(L_4)	D(L_5)	D(L_6)	D(L_0)
CP	1	1	1	1	10/15	1	5/15
CAP	0	0	0	0	5/15	0	10/15

Sample scene with five obstacles

Obstacles

Distance Metric	D(L_1)	D(L_2)	D(L_3)	D(L_4)	D(L_5)	D(L_6)	D(L_0)
CP	1	1	1	~1($\neq$1)	1	1	1
CAP	0	0	0	~0($\neq$)	0	0	0

Sample scene with six obstacles

Obstacles

Distance Metric	D(L_1)	D(L_2)	D(L_3)	D(L_4)	D(L_5)	D(L_6)	D(L_0)
CP	1	1	1	1	1	1	1
CAP	0	0	0	0	0	0	0

Sample scene with seven obstacles

Thus, the UAV has to encounter a total of 8,192 possibilities of collision if its dynamics is modeled in the hexagonal discrete space. This is very much less when compared to 1,31,072 collision possibilities in a rectangular discrete space.

To conclude, it is recommended that the Euclidean space E^3 is modeled as a hexagonal discrete space and the flight dynamics of an UAV is planned to be controlled by seven constructive formulas formulated in the logico-mathematical framework of Markov's languages of $Я_\alpha$.

Bibliography

1. Giardina, C. R., and Dougherty, E.R., Morphological Methods in Image

2. Grzegorczyk, A., An Outline of Mathematical Logic, D. Reidel Pub. Company, 1974.

3. Kushner, B. K., Lectures on Constructive Mathematical Analysis, AMS, providence, Phode Island, Vol 60, 1984.

4. Markov, A. A., Theory of Algorithms. The Israel Program for Scientific Translations Jerusalem, 1961.

5. Markov, A. A., The Language of $Я_\alpha$, Soviet Math. Dokl., 5, 1964.

6. Markov, A. A., Normal Algorithms which Compute Boolean Functions, Soviet Math. Dokl., 5, 1964.

7. Pin, J.E., Varieties of Formal Languages, Plenum Press, New York, 1986.

8. Rajan, E. G., Modelling of Collision Processes, RESPOND program proposal submitted to VSSC, Department of Space, Trivandrum, by the Advanced Centre for Automation Research, Vasavi College of Engineering, Ibrahimbagh, Hyderabad, 1998.

9. Rajan, E.G., Symbolic Computing - Signal and Image Processing, B. S. Publications, Hyderabad, 2003.

10. Rajan, E.G., and Ramprasad, V.V., Pattern-Directed Array Processing, Technical Report, TR/IP-5-91, Department of Electrical Engineering, Indian Institute of Technology, Kanpur, January, 1991.

11. Rajan, E.G., Cellular Logic Array Processing, Techniques for high throughput Image Processing Systems, SADHANA, Special Issue on Computer Vision, Volume 18, Part 2, June 1993, pp. 279-300, Indian Academy of Sciences.

12. Rajan E. G., High-Throughput Cellular Logic Array Processing of Remote Sensed Imageries, International Conference on Remote Sensing and GIS, ICORG-92, J.N.T. University, Hyderabad, 1992, Proceedings, Tata McGraw Hill Publishing Company Pvt. Ltd., pp-347-440.

13. Rajan, E. G., High-Throughput Cellular Logic Array Processing of Satellite Data for Geophysical Surveying, Paper No A.1-S.1.08. The World Space Congress, Washington, D.C., U.S.A., 28 August to 5 Sept. 1992.

14. Rajan E. G., Fast Algorithm for Detecting Volumetric and Superficial Features in 3D Images, Proceedings of the Intl. Conference on Recent Advances in Biomedical Engineering Osmania University, January 6-8, 1994.

15. Rajan, E. G., Object Recognition and Tracking in a Multiple Target Environment, 1994 IEEE International Conference on Systems, Man and Cybernetics, San Antonio, Texas, U.S.A., October 2-5, 1994, pp-2390-2395.

16. Rajan, E. G., A Genetic Algorithmic Fixing of Attack Points in a Moving Target, 1995 IEEE International Conference on Systems, Man and Cybernetics, Vancouver, B.C., Canada, October 22-25, 1995, pp. 217-222.

17. Rajan, E. G., Cellular Logic Array Processing, Invited paper, World Congress for Nonlinear Analysts, organized by the international Federation of Nonlinear Analysts, Florida,\ Institute of Technology, July 10-17, 1996, Athens, Greece

18. Rajan, E. G., Nonnumerical signal processing using Markov's algorithms, World Multiconference on Systemics, Cybernetics and, Caracas, Venezuela, 1997.

19. Rajan, E. G., An Elementary Formal System for representing constructive systems, World Multiconference on Systemics, Cybernetics and Informatics, Caracas, Venezuela, 1997.

20. Rajan, E. G., Fast Algorithm for collision avoidance by autonomous mobile systems in the framework of cellular logic array processing, 13th international conference on CAD/CAM Robotics & Factories of the Future, Universidad Technologica de Pereira, Columbia, South America, December 15-17, 1997.

21. Rajan, E. G., Symbolic Computing – Signal and Image Processing, Anshan

22. Publications, Kent, United Kingdom 2003.

TECHNICAL SESSION – III

MAB APPLICATION

APPLICATION OF MICRO AIR VEHICLE TECHNOLOGY FOR CROP MANAGEMENT STUDIES

Giridhar M.V.S.S[1]

[1]Asst Prof in Water Resources and Addl Controller of Examinations, J.N.T.U. H, Hyderabad,

Chandra Bose A.S[2]

[2]Lecturer, Dept. of Civil Engineering, Govt. Polytechnic, Warangal – 506 001

Viswanadh G.K[3]

[3]Professor of Civil engineering and Director Academic and Planning, J.N.T.U. H, Hyderabad
email:mvssgiridhar@gmail.com

Pradeep Gorthi[4]

[4]B.Tech Student, Indian Institute of Information Technology, Hyderabad

ABSTRACT

Satellite imageries are extensively used for water resources development and management including the areas of crop management. Considerable research has already been conducted using satellite and aerial imagery to observe cropping areas. However, these imagery plat forms have limitations like repeatability, cloud cover, cost and poor spatial resolution. Remote sensing satellites are appropriate for static land imageries. For acquisition of instant and dynamic information such as canopy density, other land cover types, unplanted areas, micro-relief, matured and unmatured crop growth, weed identification and growth, crop stress, the technology of low altitude unmanned micro sized air vehicles is gaining significance. The information acquired from low altitude platform sensor system is potentially useful for many precision agriculture applications, farm planning, crop yield prediction, drought monitoring, spraying of pesticides, environmental, climate change studies, monitoring of air quality, meteorological studies and predictions, forest monitoring, fire detection, flood monitoring, surface water quality studies in reservoirs. The aim of the paper is to illustrate the potential of detecting and mapping crop attributes using digital imagery acquired from a Micro sized unmanned Air Vehicles.

Keywords: Remotely controlled aircraft imaging, Micro Aerial Vehicle, crop quality evaluation, low altitude remote sensing, crop yield prediction.

INTRODUCTION

The conventional method of compiling statistics on crop management is generally carried out by the staff from the respective departments. These traditional surveys are both time-consuming and expensive. In addition, the information collected is often imprecise and unreliable, leading to inaccurate crop acreage and yield forecasts and subsequent difficulties for agricultural planners and managers on both regional and national scale. The reliable method of real time assessment of crop condition is the analysis of remotely sensed imagery which can be sourced from either an aerial or a satellite platform. Considerable research has been conducted using satellite imagery to observe cropping areas which include matching multi-temporal yield and image data (Layrol et al., 2000). Despite the advantages, satellite remote sensing has limitations as well, such as timeliness, cloud cover, cost and poor spatial resolution (Zhang et al., 2002), that reduces its usefulness for evaluations of small areas and objects. Other aerial imagery applications include crop stress investigations in cotton (Roth, 1993) and peanut (Wright et al., 2002), the prediction of grain yield (Staggenborg and Taylor, 2000) and to map within-field crop variability in wheat (Wood et al., 2003). Airborne sensors offer much greater flexibility than satellite platforms by being able to operate under clouds and having a much finer spatial resolution (Lamb and Brown, 2001). However, aerial imagery is still costly when dedicated mobilization of the aircraft is required to acquire data for small areas. When the area imaged per flight is large, the cost per hectare is relatively inexpensive.

Significance of the study:

It is commonly admitted that precision agriculture and crop management encompasses all the techniques and methods of crop and field information gathering that help taking into account the local and site-specific heterogeneity [Nemenyi et al., (2007), Murakami et al.,, (2007)]. The main drawback of the current remote sensing technology is that the quantity and quality information can only be used retrospectively, and thus cannot be used to rectify deficiencies encountered during the growing season of the crop or plan harvesting strategies for consistent quality segregation. In many instances, as it has also been proved that satellite sensors did not meet the requirements in increasing image temporal frequency and spatial resolution for such application like crop monitoring, many airborne photographic or video systems have been developed to compensate spatial lack of opportunities (Oppelt (2007) et al.,). Micro aerial technology (MAV) or Unmanned aerial vehicle (UAV) is now a good compromise between the high performances that such sensors can provide and cost-effectiveness of data acquisition. The increasing capabilities of such remotely controlled vehicles, of on-board altitude-control or positioning systems, and of digital cameras technology, combined to improve remote sensing techniques, have recently pushed the MAV/UAV into the precision farming field. This would give the operator more flexibility to allow for unfavourable weather conditions, intensive and frequent acquisitions, varying spatial and spectral resolutions, and the ability to perform a quick preliminary inspections. The present study has got wide significance as it has got ample scope for application in various scientific, engineering and technological areas in general such as agriculture planning and management, water resources development and management, environmental management, planners and finally for decision makers for future studies and development of the nation. Keeping in view the above facts, application of Micro Air Vehicle Technology for Crop Management Studies such as Canopy density, Other land cover types, Unplanted areas, Micro-relief, Matured and unmatured crop growth, Weed identification and growth, Crop stress, Crop acreage and production estimation, Crop yield models, Crop condition assessment and Agricultural drought assessment has been presented.

Materials and Methodology:

The following are the steps involved in the crop management studies using MAV for any selected study area as given in the following sub sections:

Study area extraction:

For the purpose of study area extraction, Survey of India topographic maps on a scale of 1:25,000 will be collected. The collected topographic sheets will be scanned and registered with tic points and rectified in Arc map of ArcGIS 9.3. Further, the rectified maps will be projected and merged together as a single layer. The study area boundary will be delineated along with village boundaries in the study area.

Crop discrimination/identification from MAV data:

Once the boundary of the study area has been defined and mapped, major crops will be identified in image acquired by the micro air vehicle technology. Spectrometer reflectance data will be collected from the field for major crops such as Paddy, Cotton, Redgram and Sorghum at each of four important crop growing stages i.e., initial, growing, maturity and harvesting stages. In particular, image acquisition system incorporating the digital camera sensor of micro air vehicle technology can be beneficially utilized to study individual crop growth, Individual rows and plants, weed growth, ploughed field, influence of trees, soil type change etc.,. The data acquired through micro air vehicle technology shall be validated against the collected reflectance data for major crops obtained using spectrometer in the field. The description and specifications of the digital still camera which is generally utilized as the image sensor are given in Table 1.

Estimation of area under each of the crop in the study area:

After the crop identification is validated with the field data, then the areas under each crop will be estimated for Kharif and Rabi seasons using respective MAV imageries for a period of three years.

Assessment of accuracy of crop identification and area estimates:

Finally the crop growth at each stage will be estimated using MAV data and the same will be validated for accuracy assessment. Further, the validity of the derived relationships and the precision of the produced maps, to conclude about the relevance of MAV technology platform will be carried out for crop monitoring and management studies. Finally, the yield will be estimated row wise, plant/crop wise and for the study area.

Table 1 Specifications of the digital still camera to be used as the image sensor

CCD resolution: 1344 × 971 pixel	**Picture resolution:** 1152 × 864 pixels
Colours: 24-bit	**File format:** JPEG
Glass lens: F/3.6 (Max)	**Focal length:** 39 mm
Fixed focus: 0.6 m to infinity	**Dimensions:** 113 mm × 54 mm × 81 mm
Weight (without batteries): 215 g	**Size of batteries:** 4 AA

Conclusions:

A cost-effective multispectral sensor with digital cameras with relevant filters and fitted on light MAVs are being used to perform aerial acquisitions of micro plots at each stage of the individual crop growth during the growing season, at very low altitude. Cost-effective MAV multispectral devices are relevant for quantitative and qualitative crop monitoring with a good precision. MAV technology has got ample scope for application in various scientific, engineering and technological areas in general such as agriculture planning and management, water resources development and management, environmental management, planners and finally for decision makers for future studies and development of the nation. In particular, image acquisition system incorporating the digital camera sensor of micro air vehicle technology can be beneficially utilized to study individual crop growth, Individual rows and plants, weed growth, ploughed field, influence of trees, soil type change etc.,. The data acquired through micro air vehicle technology are generally validated against the collected reflectance data for major crops obtained using spectrometer in the field. In the thematic point of view, other relationships shall also be explored to evaluate the possibility of estimation of different agronomic parameters through indices to increase the crops characterization.

REFERENCES

1. Lamb, D. W. and Brown, R. B. (2001) 'Remote-Sensing and Mapping of Weeds in Crops', Journal of Agricultural Engineering Research, 78, 117–125.
2. Layrol, L., Hedoin, E., Lepoutre, D. and Francois, O. (2000) 'Matching multitemporal yield and images data', In 5th International Conference on Precision Agriculture,,(Eds, Robert, P. C., Rust, R. H. and Larson, W. E.) American Society of Agronomy, Madison, USA, Bloomington, Minnesota, USA,, pp. 1–16
3. Murakami, E.; Saraiava, A.M.; Ribeiro, L.C.M.; Cugnasca, C.E.; Hirakawa, A.R.; Correa, P.L.P. An infrastructure for the development of distributed service-oriented information system for precision agriculture. *Computers and Electronics in Agriculture* 2007, *58*, 37-48.
4. Nemenyi, M.; Milics, G. Precision agriculture technology and diversity. *Cereal Research Communications* 2007, *35*, 829-832.
5. Oppelt, N.; Mauser, W. Airborne visible/near infrared imaging spectrometer AVIS: design, characterization and calibration. *Sensors* 2007, *7*, 1934-1953.
6. Roth, G. W. (1993) 'Agronomic Measurements to Validate Airborne Video Imagery for Irrigated Cotton Management', University of Canberra, Canberra, pp. 140
7. Staggenborg, S. A. and Taylor, R. K. (2000) 'Predicting grain yield variability with infrared images', In 5th International Conference on Precision Agriculture,(Eds, Robert, P. C., Rust, R. H. and Larson, W. E.) American Society of Agronomy, Madison, USA, Bloomington, Minnesota, USA
8. Wood, G. A., Taylor, J. C. and Godwin, R. J. (2003) 'Calibration Methodology for Mapping Within-field Crop Variability using Remote Sensing', Biosystems Engineering, 84, 409–423.
9. Wright, G. C., Lamb, D. W. and Medway, J. (2002) 'Application of aerial remote sensing technology for the indentification of aflatoxin affected areas in peanut fields', In 11th Australasian Remote Sensing and Photogrammetry Conference, Brisbane, Queensland, Australia, pp. 3693–371
10. Zhang, N., Wang, M. and Wang, N. (2002) 'Precision agriculture—a worldwide overview', Computers and Electronics in Agriculture, 36, 113–132.

DESIGN, FABRICATION AND TESTING OF MAV QUADROCOPTOR FOR FREQUENCY JAMMING AND COMMUNICATION APPLICATIONS

Junaidh. S
Noorul Islam University,India
junnugamer@gmail.com

Abstract

This paper describes the design, fabrication and testing aspects of a strong, sturdy and versatile Quadracoptor. It is named as Kurai Chasseur (Kurai Chasseur means **Dark Hunter).** It is a MAV Quadrocopter that is capable of helpingat war zone operations and assistssoldiers to be successful by effectively jamming enemy communication systems and allowing selective communication channel. Itis designed, built and tested and the details are described in the paper. They use components like mobile frequency jammer, Wi-Fi hotspot, mini camera. It is a 50*50 cm sized Quadrocopter powered by 1400 KV motor with a 5A, 11.1V battery giving a fly time of 30-40mins and has an ability to lift a payload of 1.5Kg.The KC with all the above mentioned payloads on board will function as a hunter, first it will search the enemies then cut their radio and mobile communication using jammer which is of less than 2.14GHz for 2G, 3G and serve as a hub of communication using the Wi-Fi router which works in 2.4GHz frequency with a closed voice chat support in it only for its soldiers who activated it. KC will give no room for the enemies to get away when its soldiers surround them depriving of any communication medium or escape plans. KC will act as an eye in the vast sky giving feed only to those who send him, blinding others and locates the enemy's positions while they are trying to run.

INTRODUCTION

The Quadrotor is a 4 motor powered flying machine that has much more stabilized flight and low power consumption compared to other Air Vehicles. Quadrotors are classified as rotorcraft, as opposed to fixed-wing aircraft, because their lift is generated by a set of revolving narrow-chord airfoils. Unlike most helicopters, quadrotors generally use symmetrically pitched blades; these can be adjusted as a group, a property known as 'collective', but not individually based upon the blade's position in the rotor disc, which is called 'cyclic'. Control of vehicle motion is achieved by altering the pitch and/or rotation rate of one or more rotor discs, thereby changing its torqueload and thrust/lift characteristics.

Control of Quadrotors flight

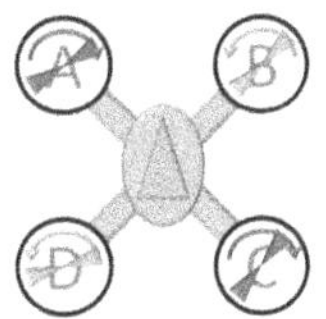

Fig: 1

Schematic of reaction torques on each motor of a quadrotor aircraft, due to spinning rotors. Rotors A and C spin in one direction, while rotors B and D spin in the opposite direction, yielding opposing torques for control.Each rotor produces both a thrust and torque about its center of rotation, as well as a drag force opposite to the vehicle's direction of flight. If all rotors are spinning at the same angular velocity, with rotors one and three rotating clockwise and rotors two and four counterclockwise, the net aerodynamic torque, and hence the angular acceleration about the yaw axis is exactly zero, which implies that the yaw stabilizing rotor of conventional helicopters is not needed. Yaw is induced by mismatching the balance in aerodynamic torques (i.e., by offsetting the cumulative thrust commands between the counter-rotating blade pairs).

Angular accelerations about the pitch and roll axes can be caused separately without affecting the yaw axis. Each pair of blades rotating in the same direction controls one axis, either roll or pitch, and increasing thrust for one rotor while decreasing thrust for the other will maintain the torque balance needed for yaw stability and induce a net torque about the roll or pitch axes. This way, fixed rotor blades can be made to maneuver the quad rotor vehicle in all dimensions. Translational acceleration is achieved by maintaining a non-zero pitch or roll angle.

Kurai chasseur (KC) a MAV

The KC is a Quadrocopter powered with 4 1400KV motors that are capable of lifting a load of 1.2KG in air vertically. It is powered by Li-Po(Lithium Polymer) battery of 4800mAh which will give a flight time of 28-42 minutes. It also has a Wi-Fi router and a mobile frequency jammer that work with 5V input is connected to the main board and can be switched on when necessary. It has a mini camera mounted in the top to give a good view for controlling and also for the surveillance of an area.

Table 1

S.No	Components	Quantity	Description
1	Brushless motor	4	1400 KV
2	ESC (Electronic Speed controller)	4	30A
3	Battery 11.1V	1	4800 mAh
4	Carbon Fiber Frame	4	25 Cm
5	Mini Wi-Fi router	1	2.4 GHz
6	Mini Mobile Jammer	1	<2.4 GHz
7	Propellers	4	10X4"
8	Stabilization Board	1	Gyroscope operating

The KC prototype has been designed, fabricated and tested for the main purpose of its built.

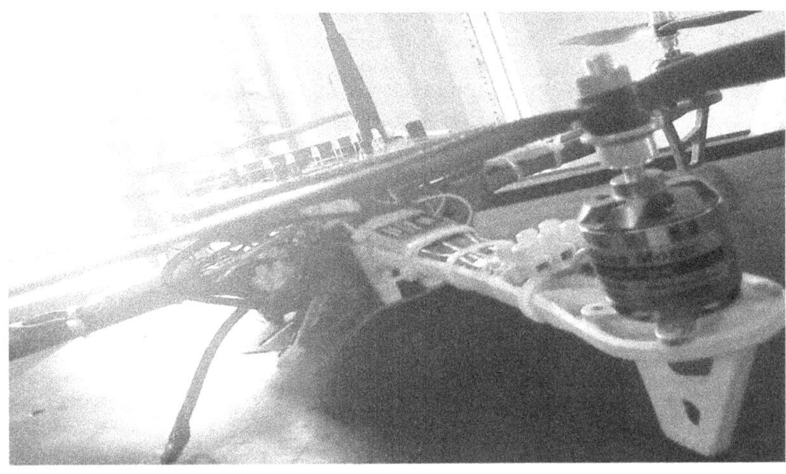

Fig: 2

The KC weighs about 1.35kg and tested, this resulted after a testing flight time of 36 minutes. When the on board payloads and camera is powered on expected flight time is about 25-30 minutes.

The Purpose of Built

The KC is designed for utilizing its payloads in a combat situation or for a secret mission. Its flight over the area will provide surveillance on the ground and it is designed to be used during night missions to keep away from visibility from ground. When the payloads are switched on the mobile jammer will jam the frequency that is less than 2.4 GHz this includes all 2G, 3G and CDMA signals.   The Wi-Fi router working above the frequency 2.4 GHz and above is not jammed by the mobile frequency jammer giving an active communication. Using the active Wi-Fi communication two or more mobiles connected to the KC network can communicate with each other.

Examples of Combat Mission Operation and KC's Use Supported Missions

When the intelligence department gives any report on enemy's spot, KC could be used as the first step of plan. Let it fly over the active zone and place itself on a tall tower, tree or building to suppress its sound then search for enemies hide out. It then activates the jammer and jams the enemy's communication with the outside world and the landline connections could be taken out by the soldiers. Then it switches on the Wi-Fi and establishes a closed voice communication link between its soldiers. With it as a guide from the sky for visual and audio our soldiers can have the whole area surrounded thereby trapping the enemies even before they could realize what happened. Then they can be further blinded by all the smokes and fumes it have and can be caught in a less violent attack. This will help in seizing the enemies without causing heavy damage to our soldiers. The trapped people will not be able to contact their fellow mates and loss of life could be avoided.

Self-supported mission

With no men around as back up KC powered with flame thrower and faint gas could be sent to houses or closed buildings where we can trap mobs. When the mob is in a single room KC could be used to jam the signals not letting them to contact others and trapping them by burning the doors thus not letting them to escape. Smoking them up by shooting faint gas is done first then KC's camera will monitor the activity in the area till the policemen gets the upper hand by locking up the bad guys with no tension of crime situation.

Fig: 3

Conclusion

The KC is fully compact for secret situation and with its less noise it will be hard to be spotted in dark. The support of KC for the soldiers will be very much helpful for their missions to reduce the life loses in both the sides by avoiding a critical situation. The MAV's are the future generation combat weapons which will revolutionize the combats by giving advantage to the soldiers and becoming threat to the enemies.

Reference:

1) http://en.wikipedia.org/wiki/Quadrotor
2) http://www.rchelicopterfun.com/quadrocopter.html
3) http://en.wikipedia.org/wiki/Wireless_router
4) http://en.wikipedia.org/wiki/Mobile_phone_jammer

APPLICATION OF NANOTECHNOLOGY AND MICRO ELECTRO MECHANICAL SYSTEMS (MEMS) IN MAV'S

Aparna.A.R.[1]
Malla Reddy College Of Engineering and Technology, Hyderabad
Contact number: +919900197254
Email-id: aparna_agnihothri@yahoo.co.in

T.V.Karthikeyan[2]
DRDO, Hyderabad

A.G.Sarwade[3]
DRDL, Aero. Engg. Hyderabad

ABSTRACT

The two futuristic technologies for our technology advancement in respect of miniaturization of aerospace components are:

MEMS - Microelectromechanical systems

NEMS- Nanoelectromechanical systems

The miniaturization of aeronautical components is a great advantage for the MAV flight because both, the size and weight will come down. The above systems need developing special manufacturing processes such as surface micro machining. The polymers and composites are the favorite materials for MAV's. Polymers are the economical choice, but their strength properties are low. Carbon composites have strength but they are too costly.

The mechanical properties needed to withstand the flight environment like vibrations, accelerations can be enhanced by incorporation of nanoparticles into polymers. Properties like improved modules, fracture toughness can be enhanced this way. Suitable nanoparticles of TiO_2, SiO_2 etc. are candidate additives. The major applications of miniaturization in MAV's are towards the sensors. MAV's have to transmit signals and images to ground continuously. This includes their position (x,y,z co-ordinates with reference to their launch point) information. Some of these applications and their details are described in the text of this paper. The main requirement in MAV's is that the components should be small and light weight. These requirements are met by the modern technologies indicated above. The telemetry system in MAV for two way communications between ground and MAV has to be reliable which ensures the successful mission of MAV.

INTRODUCTION

Micro- and nano air vehicles are defined as "extremely small and ultra-lightweight air vehicle systems" with a maximum wingspan length of 15 cm and a weight less than 20 grams. MAVs are defined as small flying systems which are designed for performing useful operations. MAVS applications span a wide range, and the majority of them are military. MAVS are capable to perform both indoor missions and outdoor missions in very challenging environments. The main applications are intelligence, surveillance, and reconnaissance (ISR) missions. These systems can provide a rapid overview in the area around the personnel, without exposing them to danger. The two futuristic technologies for our technology advancement in respect of miniaturization of aerospace components are: MEMS and NEMS.

Emerging from the "science of the small" are a wide array of theoretical and experimental approaches to engineering tiny machines, ranging from smaller microchips to molecular robots that we cannot even see. Prominent among the new miniaturization technologies are MEMS(Micro electromechanical system) devices and nanotechnology.

As their name implies, MEMS devices are a mechanical extension of microelectronics. Often described as the "second silicon revolution," these devices are distinguished from conventional microchips by their built-in mechanical functions, which allow them to be used in building integrated sensors and actuators.

MEMS devices
Electromechanical devices on the micron to millimeter scale, known as Micro Electro Mechanical System (MEMS).

Key features
Miniature mechanical systems with micron feature size

Batch fabricated with no assembly required

Exploits micro electronics infrastructure for common technology base for sensors, actuators and electronics.

About MEMS
MEMS is a blend of semi-conductor processing and mechanical engineering at a very small scale. MEMS refers to integration of sensors, actuators and electronics using techniques originating semi-conductor microelectronics to realize miniature, high performance and low cost mechanical systems. Manufactured using micro-machining processes.

Fabrication techniques of MEMS
The three more widely used fabrication techniques are:
Surface micro-machining (photolithography), LIGA process (lithography, Galvanoforming, Abforming) and Bulk micro-machining

Application and future techniques of MEMS
Micro fabricated sensors, actuators and electronics are the most critical components in implementing a complete system for a specific function.

Flight control system
Accelerometer
Micro size pressure gauges and accelerometers are currently available [C. Ajluni and B.E. Boser et al., 1996, and R. Allan, 1997] and miniature magnetic compasses may also be feasible soon. Most useful for this application, however, are rate sensors. Microchip angular-rate sensors are now being produced [J. Bernstein et al., 1993 andM. Weinberg, 1996], and will be useful for MAVs as

soon as they are mated to miniaturized readout electronics. Drift rates from these sensors will be adequate for vehicle stabilization applications.

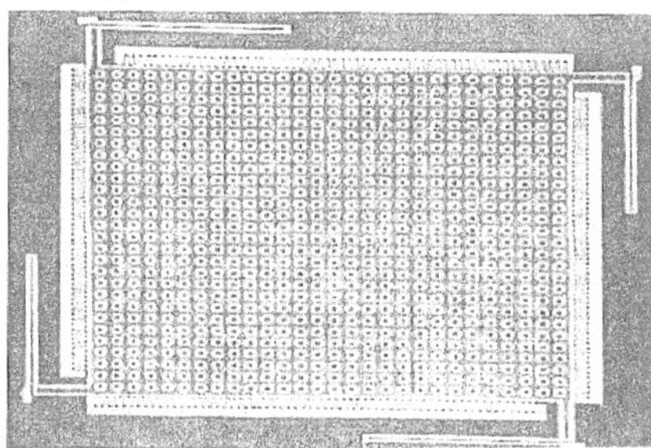

Fig 1 SEM micrograph of a polysilicon Surface - micromachined z-axis Accelerometer

Actuators

The ability to generate aerodynamic forces and moments is also required to stabilize and maneuver the MAV. These controls could be achieved with conventional discrete hinged surfaces such as ailerons and elevators; distributed micro-actuated control surfaces; or wings that change shape or warp. All methods require micromechanical actuators. Because of recent advances in MEMS, a number of different actuator candidates should be available in the next one to two years [P. Dario et al.,1992, C. Liu et al.,1995, C.H. Ahn et al.,1992, K. Matsuzaki et al.,1994, H. Guckel et al,1993, W.P. Robbinset al 1991, D.L. Polla et al.,1994, T. Niinoet al.,1994, S.M. Bobbio et al.,1993 and J.D. Jacobson et al.,]. Examples include integrated force arrays, which generate electrostatic attraction force, and several approaches using piezoelectric crystals. These actuators can generate linear forces or be used in the construction of rotary machines that produce torque. They have the advantage of employing fabrication approaches that lend themselves to high production rates. Tiny conventional electromagnetic actuators, such as those used in watches, may also be tapped for some first-generation MAVs [M.A. Gottschalk 1997]. The flight-control sensors and actuators must be integrated into the flight-control system by using a digital processor with the necessary signal interfaces.

A custom microcontroller chip that also serves as the central processor for the communications and optical-sensor subsystem will accomplish this function.

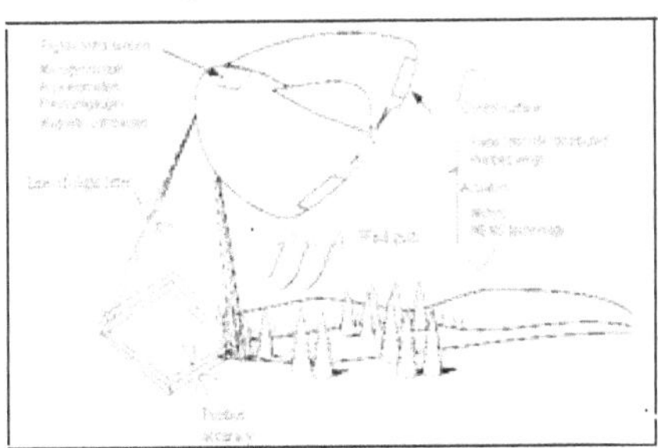

Fig. 2 MAV flight-control system. Flight control requires sensors that measure motion (roll, pitch, and yaw) of the MAV, and aerodynamic control inputs that stabilize and maneuver the MAV in wind gusts and turbulence.

Sensors

Many of the system functions depicted will be provided by microelectronics or MEMS-based components. Even so, separate modules for each function would consume more volume than may be available. From an electronics perspective, the on-board processor and communications electronics form the core of the vehicle. They provide critical links between the sensor systems and the ground station, and they are vital to the flight and propulsion control systems. Power generation and propulsion subsystems support critical electronics and flight control functions in addition to flight propulsion power. The multifunctionality required by the MAV weight and power budgets may be achieved only by a highly integrated design, with physical components serving multiple purposes, or accomplishing multiple and often diverse functions. For example, the wings may also serve as antennae or as sensor apertures. The power source may be integrated with the fuselage structure, and so on. The degree of design 'synergy' required has never been achieved in a flight vehicle design.

MEMS pressure sensor has more advantage than conventional pressure sensor because of its low weight, low cost, reliable, smart function and occupies less space [L. Lin et al.,1998].Capacitive pressure sensors provide high sensitivity to pressure, low power consumption, low noise, large dynamic range and low thermal sensitivity than piezoresistive pressure sensors [Y. Lee et al.,1982]. Capacitive differential pressure sensor measures changes in pressure by the deflection of a conductive diaphragm due to the applied pressure. Parallel plate differential pressure sensors typically have spacer (dielectric separator) between the two electrodes and the deflection in the diaphragm due to change in pressure produces a change in capacitance [J. Han and et.al 2007]. The differential pressure sensor uses the ambient pressure as reference pressure and the external pressure as other source. This proposed technique reduces the package cost, eliminating the need of vacuum sealing of the diaphragm; moreover it has high sensitivity for static and dynamic pressure measurements.

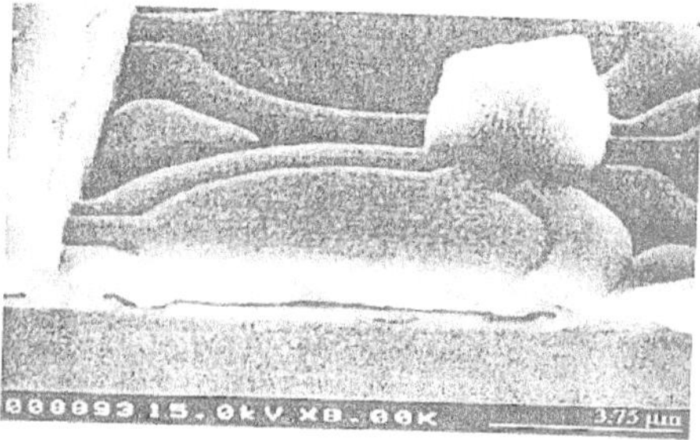

Fig 3 SEM micrograph of a close-up view of a polysilicon surface-micromachined capacitive pressure sensor.

Global positioning system (GPS)/Inertial navigation system(INS)
The interest in micro aerial vehicles (MAV) for surveillance and reconnaissance purposes is growing constantly. For many mission scenarios, vertical take-off and landing, as well as the ability to hover are desired. This is offered by MAVs of helicopter-type. However, while fixed-wing MAVs often show an inherent stability, the helicopter attitude requires permanent control. The key to such an attitude control is reliable, accurate and especially continuously available attitude information.

In principal, attitude information can be obtained from a multi-antenna GPS system. Due to the availability of low-cost, light-weight GPS receivers, this is an interesting option for a variety of applications. Concerning a MAV, the possible baselines are very short, which has a negative influence on the attitude accuracy that can be achieved. Additionally, this technique works only if the carrier phase integer ambiguities can be estimated which is not a trivial task, especially with single-frequency receivers. As attitude information is required in situations without any GPS signals available, a multi-antenna GPS system is not an option for the VTOL-MAV considered here. Currently, the usage of computer vision for navigation purposes is receiving considerable interest. Different approaches are investigated [C. Schlaile, et al.,2004, B. Sinopoli et al.,2001,, S. Winkler et al.,2004 and S. Winkler et al.,2004], including the usage of image processing as the main aiding source for an inertial navigator. However, there are several problems connected to computer vision that are not easily overcome. Image processing requires a significant processing power, which may not be available onboard of a MAV. If the image processing is performed on a ground station, the continuous availability of an appropriate radio link has to be assured. Additionally, besides the possible scaling problem associated with image processing, there is often a sensitivity to lighting conditions and contrast. A well-known and widely used technique to obtain attitude information is the combination of GPS with inertial navigation. The continuous availability of the navigation solution, together with a high data rate, is assured by the inertial navigation system(INS), while the INS drift is compensated using the GPS measurements.

Furthermore, the GPS measurements can be used to calibrate the inertial measurement unit (IMU), which improves the inertial navigation performance during GPS outages. Some systems use several GPS receivers in order to increase the availability of GPS measurements during dynamic maneuvering. Examples of GPS/INS systems applied to unmanned aerial vehicles (UAV) can be found in [J.-H. Kim et al.,2002, J. Kim et al., 2003 and C.-S. Yoo et al.,2003]. However, while the attitude accuracy offered by a GPS/INS system is impressive when GPS aiding is available, the usage of MEMS inertial sensors with significant scalefactor nonlinearities, misalignment, noise, and temperature-varying biases, is a problem during GPS outages: Without aiding information available, a sufficient attitude accuracy can be maintained for a short time only. Therefore, for MAV and UAV applications, an integration strategy has become popular that is significant different from the usual GPS/INS integration known from missile and aircraft applications. Hereby, the GPS measurements are used to calculate the acceleration of the vehicle. Using the estimated attitude, this trajectory-caused acceleration can be removed from the accelerometer measurements, leading to an approximate measurement of the gravitational acceleration in body frame coordinates. Using this measurement of the local gravity vector, the roll and pitch angle estimates can be corrected.

Finally, for fixed-wing MAVs, the yaw-angle estimate is corrected by exploiting assumptions on the vehicles aerodynamics, or a magnetometer can be used. Examples for this approach can be found in [D.B. Kingston et al.,2004 and M. Musial et al., 2004]. The advantage of this approach is the insensitivity to GPS loss. Without GPS measurements available, the acceleration caused by the trajectory dynamics cannot be removed, which leads to a reduced attitude accuracy, but a growth of attitude errors without bounds is still prevented. Unfortunately, when GPS measurements are available, they are used in a sub-optimal way, as the accelerometer biases cannot be estimated. Uncompensated accelerometer biases directly lead to systematic errors in the attitude estimates.

NEMS

Nano electro mechanical systems(NEMS) are made of electromechanical devices that have critical structural dimensions on the submicron scale.

These devices are attractive for applications where structures of very small mass provide essential functionality, such as force sensors, chemical sensors, biological sensors and ultra high frequency resonators

NEMS Fabrication techniques
E- beam and Nano imprint fabrication
Epitaxy and strain engineering
Scanned probe techniques
Self assembly and template manufacturing

NEMS Applications
NEMS Bi-stable device
NEMS for Bio sensing application
Nano-mechanical electro-thermal probe array
Nanotube based NEMS devices

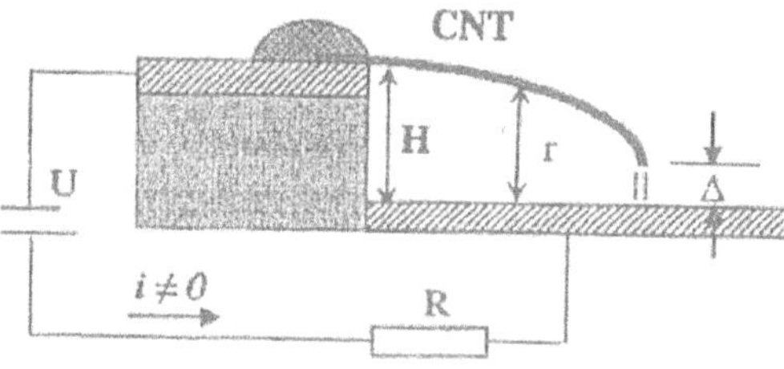

Fig 4 CNT based NEMS Bi-stable device with feedback control

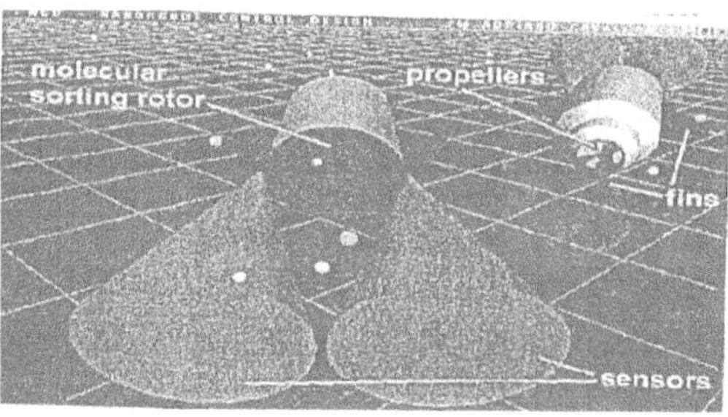

Fig 5 The depicted blue cones show the sensors "touching"areas that triggers the nanorobots behavior

MEMS/NEMS Conclusion
The advancement of MEMS/NEMS will open many new potential application areas to the technology for the improvement and development of efficient MAVS.

In most cases, MEMS/NEMS will be one of several alternatives available for implementation.

For cost sensitive applications, the tradeoff between technical capabilities and cost will challenge those who desire to commercialize the technology.

The biggest challenge to the field will be to identify application areas that are well suited for MEMS/NEMS technology and have no serious challengers.

Now you may not find physically the role of MEMS in our life but in coming days it is going to play a major role in our lives to make our life more fasyer, safer and easier.

The second being NEMS and micro-manufacturing revolution, the first being MEMS.

References

1. C. Ajluni, "Low-Pressure Sensor Opens Wide Applications Frontier," *Electron. Des.,* 16 Dec. 1996, pp. 59–64.

2. B.E. Boser and R.T. Howe, "Surface Micromachined Accelerometers,"*IEEE J. Solid-State Circuits.* 31 (3), 1996, pp. 366–375.

3. R. Allan, "Silicon MEMS Technology Is Coming of Age Commercially,"*Electron. Des.,* 20 Jan. 1997, pp. 75–88.

4. J. Bernstein, S. Cho, A.T. King, A. Kourepenis, P. Maciel, and M. Weinberg, "A Micromachined Comb-Drive Tuning Fork Rate Gyroscope," *Proc. IEEE Conf. on Micro Electro Mechanical Systems, Fort Lauderdale, Fla., 7–10 Feb. 1993,* pp.143–148.

5. M. Weinberg, J. Bernstein, J. Borenstein, J. Campbell, J.Cousens, B. Cunningham, R. Fields, P. Greiff, B. Hugh, L.Niles, and J. Sohn, "Micromachining Inertial Instruments," *SPIE* 2879, 1996, pp. 26–36.

6. P. Dario, R. Valleggi, M.C. Carrozza, M.C. Montesi, and M. Cocco, "Microactuators for Microrobots: A Critical Survey,"*J. Micromech. Microeng.* 2 (3), 1992, pp. 141–157.

7. C. Liu, T. Tsao, Y.-C. Tai, T.-S. Leu, C.-M. Ho, W.-L. Tang, and D. Miu, "Out-of-Plane Permalloy Magnetic Actuators for Delta-Wing Control," *Proc. IEEE Conf. on Micro Electro Mechanical Systems, Amsterdam, 29 Jan.–2. Feb. 1995,* pp. 7–12.

8. C.H. Ahn and M.G. Allen, "A Fully Integrated Micromagnetic Actuator with a Multilevel Meander Magnetic Core,"*IEEE Solid-State Sensor and Actuator Workshop, Hilton Head, S.C., 22–25 June 1992,* pp. 14–18.

9. K. Matsuzaki, T. Matsuo, and Y. Mikuriya, "Comparison of Electrostatic and Electromagnetic Motors Based on Fabrication and Performance Criteria," *1994 5th Int. Symp. on Micro Machines and Human Science Proc., Nagoya, Japan, 2–4 Oct.1994,* pp. 77–81.

10. H. Guckel, T.R. Christenson, K.J. Skrobis, T.S. Jung, J. Klein, K.V. Hartojo, and I. Widjaja, "A First Functional Current Excited Planar Rotational Magnetic Micromotor," *Proc. IEEE Conf. on Micro Electro Mechanical Systems, Fort Lauderdale, Fla., 7–10 Feb. 1993,* pp. 7–11.

11. W.P. Robbins, D.L. Polla, T. Tamagawa, D.E. Glumac, and W. Tjhen, "Design of Linear-Motion Microactuators Using Piezoelectric Thin Films," *J. Micromech. Microeng.* 1 (4), 1991, pp. 247–252.

12. D.L. Polla, P.J. Schiller, and L.F. Francis, "Microelectromechanical Systems Using Piezoelectric Thin Films," *SPIE* 2291, 1994, pp. 108–124.

13. T. Niino, S. Egawa, H. Kimura, and T. Higuchi, "Electrostatic Artificial Muscle: Compact, High-Power Linear Actuators with Multiple-Layer Structures," *Proc. IEEE Conf. on Micro Electro Mechanical Systems, Oiso, Japan, 25–28 Jan. 1994*, pp. 130–135.

14. S.M. Bobbio, M.D. Kellam, B.W. Dudley, S. Goodwin- Johansson, S.K. Jones, J.D. Jacobson, F.M. Tranjan, and T.D. DuBois, "Integrated Force Arrays," *Proc. IEEE Conf. on Micro Electro Mechanical Systems, Fort Lauderdale, Fla., 7–10 Feb. 1993*, pp. 149–154.

15. J.D. Jacobson, S.H. Goodwin-Johansson, S.M. Bobbio, C.A. Bartlett, and L.N. Yadon, "Integrated Force Arrays: Theory and Modeling of Static Operation," *J. Microelectromech. Syst.* 4 (3), pp. 139–150.

16. M.A. Gottschalk, "Miniature Motors Deliver Big Performance," *Design News* 52 (9), 5 May 1997, pp. 67–68.

17. L. Lin and W. Yun, "MEMS Pressure Sensors for Aerospace Applications," *Proceeding IEEE Aerospace Conference*, vol. 1, Colorado, 21-28, pp. 429- 436, March 1998.

18. Y. Lee and K. D. Wise, "A Batch-Fabricated Silicon Capacitive Pressure Transducer with Low Temperature Sensitivity," IEEE *Transactions on Electron Devices*, vol. Ed-29, no.1, pp. 42-48, Jan. 1982.

19. J. Han and et.al, "Smooth Contact Mode Capacitive Pressure Sensor with Polyimide Diaphragm," *Proceeding of IEEE conference on Sensors*, Atlanta, 28-31, pp. 1468-1471, Oct. 2007.

20. C. Schlaile, J. Wendel, G.F. Trommer, Stabilizing a four-rotor helicopter using computer vision, in: First European Micro Air Vehicle Conference and Flight Competition EMAV 2004, 13–14 July, Braunschweig, Germany,2004.

21. B. Sinopoli, M. Micheli, G. Donate, T.J. Koo, Vision based navigation for an unmanned aerial vehicle, in: IEEE International Conference on Robotics and Automation, vol. 2, pp. 1757–1764, 2001.

22. S. Winkler, H.W. Schulz, M. Buschmann, T. Kordes, P. Vörsmann, Improving low-cost GPS/MEMS-based INS integration for autonomous MAV navigation by visual aiding, in: ION GNSS 2004, 21–24 September, Long Beach, CA, USA, 2004, pp. 1069-1075.

23. S. Winkler, H.W. Schulz, M. Buschmann, T. Kordes, P. Vörsmann, Horizon aided low-cost GPS/INS integration for autonomous micro air vehicle navigation, in: First European Micro Air Vehicle Conference and Flight Competition EMAV 2004, 13–14 July, Braunschweig, Germany, 2004.

24. J.-H. Kim, S. Sukkarieh, Flight test results of GPS/INS navigation loop for an autonomous unmanned aerial vehicle (UAV), in: ION GPS 2002, 24–27 September, Portland, OR, USA, 2002, pp. 510–517.

25. J. Kim, S. Sukkarieh, S.Wishart, Real-time navigation, guidance and control of a UAV using low-cost sensors, in: International Conference of Field and Service Robotics (FSR'03), Yamanashi, Japan, 2003, pp. 95–100.

26. C.-S. Yoo, I.-K. Ahn, Low cost GPS/INS sensor fusion system for UAV navigation, in: The 22nd Digital Avionics Systems Conference 2003, vol. 2, 2003, pp. 8.A.1–8.1-9.

27. D.B. Kingston, R.W. Beard, Real-time attitude and position estimation for small UAVs using low-cost sensors, in: AIAA 3rd Unmanned Unlimited Systems Conference and Workshop, 2004.

28. M. Musial, C. Deeg, V. Remuß, G. Hommel, Orientation sensing for helicopter UAVs under strict resource constraints, in: First European Micro Air Vehicle Conference and Flight Competition EMAV 2004, 13–14 July, Braunschweig, Germany, 2004.

DESIGN AND DEVELOPMENT OF FLAPPING WING MICRO AERIAL VEHICLE

Kaviyarasu. A*

*Research Scholar, Dept.of Aerospace Engg. MIT, Anna university, Chennai

E-Mail: isrokavi@gmail.com

Dr.K.Senthilkumar**

** Associate professor, Dept.of Aerospace Engg. MIT, Anna university, Chennai

Email: ksk_mit@rediffmail.com

ABSTRACT

From the past few years, there is a rapid increase in mimic of bird like Micro Air Vehicle known as Ornithopter. Ornithopter is nothing but a flying vehicle that it flies like a bird. This paper presents the design and development of flapping wing micro aerial vehicle (MAV). The heart of the Ornithopter is its flapping mechanism, it converts the rotating motion from the brushless motor into a flapping mechanism. The flapping rate of the wing determines the lift and the forward propulsion of the MAV. The pitch and the yaw of the flapping wing MAV can be controlled by Elevator as well as Rudder. The Aerodynamic properties of Ornithopter are very complex because steady state Aerodynamics do not accurately account for the required forces. The mathematical dynamics of the flapping wing Micro Air Vehicle using the wind tunnel study is impossible because of the unsteady aerodynamic forces which act on the platform. The IMU and GPS embedded into the flapping wing micro air vehicle gives the attitude and the position of vehicle through micro wireless modem which should be monitored in the ground control station. The flapping wing Micro Aerial vehicle can control through the RC wireless Transmitter as well as Ground controlled station.

KEYWORD: MAV, IMU, GPS

INTRODUCTION

An ornithopter is an aircraft that uses flapping wing motion to fly. This type of flight offers potential advantages over fixed-wing flight, such as maneuverability, at slow speeds (1-40m/s). Natural Ornithopter range in size from small flying insects to large birds, and flap their wings from about 5 to 200Hz. The Defense Advanced Research Projects Agency (DARPA) largely motivates Micro Aerial/Air Vehicle development in military application for reconnaissance missions in confined spaces or under dangerous circumstances. The discreetness of a flapping wing MAV adds appeal for these types of covert operations. Employment in civilian search and Created by Im Backrescue missions under dangerous or questionable circumstances such as fire or earthquake also fuels MAV development. Basic open-loop flight control is possible through a radio frequency remote controller with joystick. Closed-loop control is desired to make the ornithopter easier to maneuver and control for the average user. Because MAVs are restricted both in size and weight, an embedded system used to control the ornithopter must be low-power and lightweight. This limits the processing power available for control calculations and generates the need for simple control loops. The Fig shows how lift and the forward motion is created during flapping of wing.

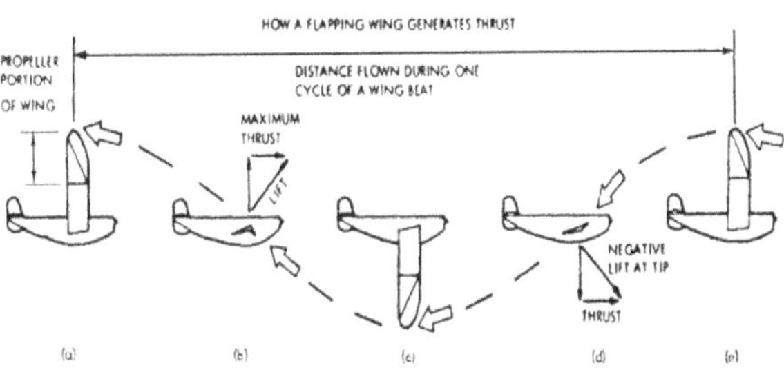

Fig1 Various Phases of wing flapping

These five sketches show the wing tip, at the top, middle and bottom of its beat during one cycle. The dotted line traces the path of the wing tip during this cycle. The wide arrows show the angle at which the air stream apparently meets the wing tip at each instant. The leading edge of the propeller portion of the wing is caused to move up and down, synchronized with the wing position so as to allow the air stream to meet the leading edge cleanly.

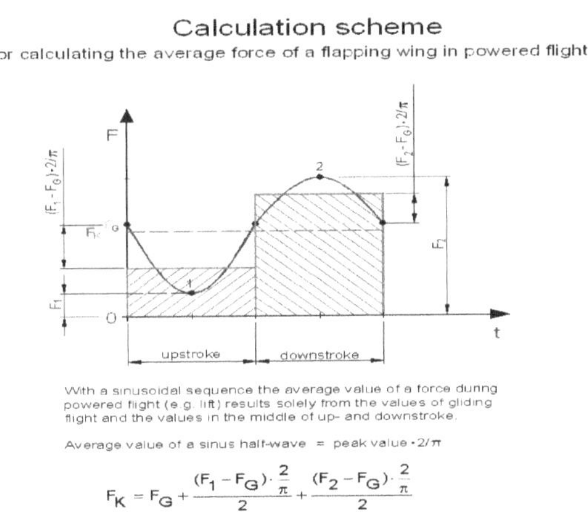

$$F_K = F_G + \frac{(F_1 - F_G) \cdot \frac{2}{\pi}}{2} + \frac{(F_2 - F_G) \cdot \frac{2}{\pi}}{2}$$

Fig 2 Calculating Average force of flapping wing

Strange as it may seem, the more action of oscillating a wing up and down whilst moving forward, actually produces thrust by virtue of the fact that lift is always at right angles of the air stream. One cycle of wing beat is shown in fig 2

Gear Mechanism

The most critical part of the ornithopter is the drive mechanism the converts the electric power from the battery to the apping motion of the wings. This system is the most complex to design and fabricate because it must withstand very large forces which reverse direction several times a second while at the same time being extremely light and durable.

The drive system can be further broken down into four sections, the electric motor, a gear reduction stage, a linkage to convert the high torque rotation into a reciprocating motion, and the connection to the wing spars.

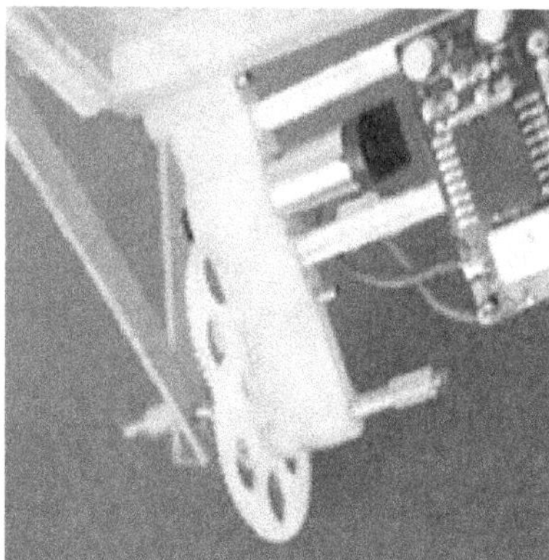

strut type gearbox

plate type gearbox

Fig 3 Gearbox

Mono and Bi wings

The choice of mono or bi wings primarily depends on how powerful the motor is and required stability. A single set of wings (mono) will cause the body to oscillate because of the shifting forces caused by the down stroke and upstroke.

If the wings are larger because frequency is limited, then forces on each stroke would be larger and last for a longer period of time.

This could easily cause problems in this design because the effective lift and thrust component need to be effectively stable. A bi wing configuration (similar to a dragonfly) would cancel out this oscillation effect. These configurations are modeled in Fig. 4

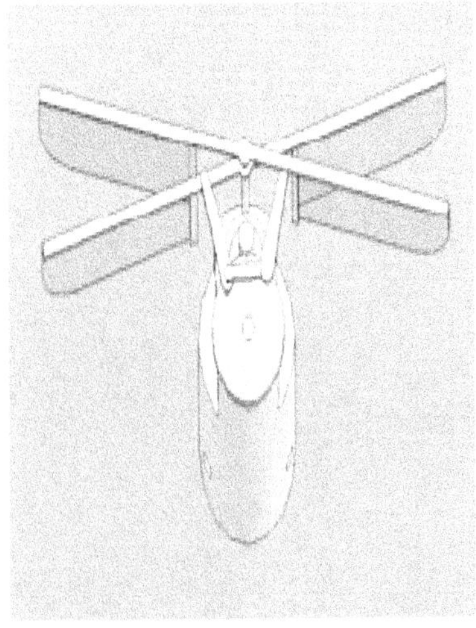 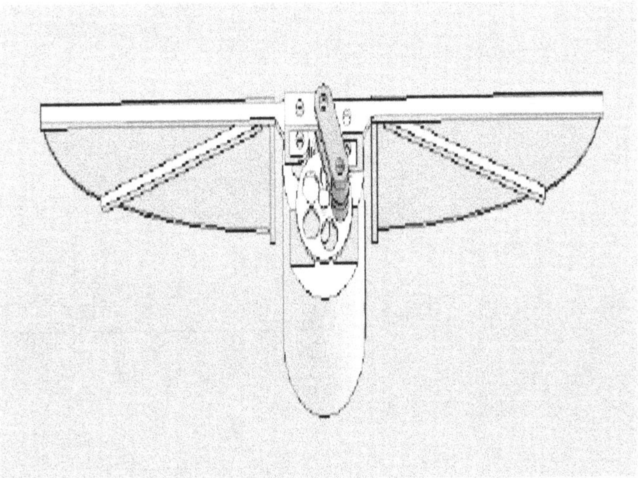

Fig 4 a) Bi-Wing   b) Mono Wing

Aerodynamic Coefficients

The lift and thrust coefficients can be expressed as

$$\begin{cases} C_L = \dfrac{2L}{\rho A U^2} \\ C_T = \dfrac{2T}{\rho A U^2} \end{cases}$$

where L, T, U, A, ρ, are lift, thrust, flight speed, wing area, and air density, respectively. The advance ratio J is the ratio of the flight speed to the speed of the wingtip and is given by

$$J = \frac{U}{2\Phi f b}.$$

where Φ, f, b are stroke angle, flapping frequency, and wing semi-span, respectively. In order to estimate the flying speed, necessary to calculate the advance ratio J.

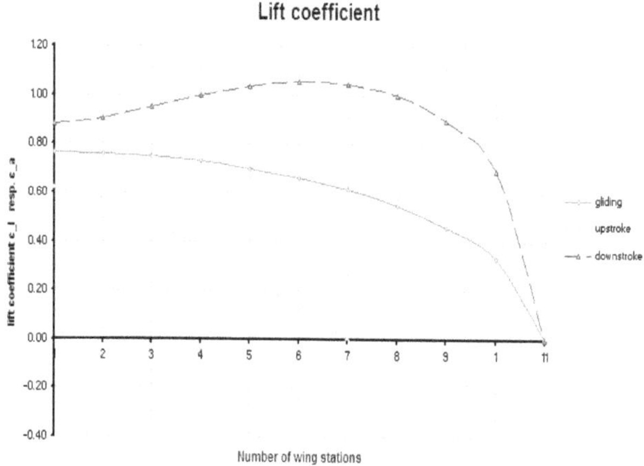

Fig 5 C_L vs C_D Value

Data Acquisition

Data was collected from the ornithopter while traveling down a linear bearing rail by Eric A. Johnson[1] and Charles Fisher. The ornithopter was strapped to a slider on the rail, restricting the ornithopter to movement in the forward horizontal direction.

Acceleration and position sensor information were collected at different throttles. Here, throttle refers to the position of the joystick on the remote control device. This is approximately proportional to speed of the brushless DC motor sourcing the wings.[2] Acceleration data was measured by an accelerometer strapped to the body of the Ornithopter angular rotation data were collected with 3 axis gyroscopes.

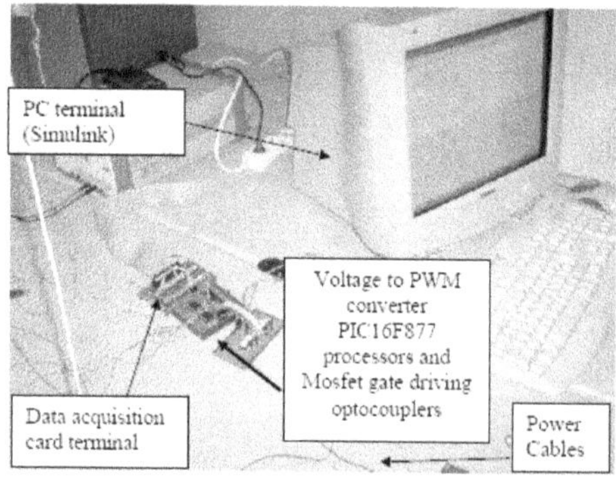

Fig 6 View of the Data Acquisition system

The bias level of the accelerometer data was adjusted until it agreed with the second derivative of the position data. In this way, the sensor drift of the accelerometer is accounted for. Approximate velocity data were found from the integral of the tuned acceleration data. Figures 2.1, 2.2, and 2.3 show the acceleration data filtered with a low pass filter with cutoff frequency of 40Hz, velocity, and position data found from the accelerometer.

Fig 7 6DOF IMU Sensor

TESTS AND RESULTS

ELIMINATING SENSOR NOISE

Vibration Noise is intrinsically present in the accelerometers. In order to filter that noise 8th order Butterworth filters are placed at the outputs of the accelerometer in the SIMULINK code. Before and after filtering of the accelerometer outputs are given in Fig 8 & 9. Note that the disturbance is given by pushing the system on x-y plane

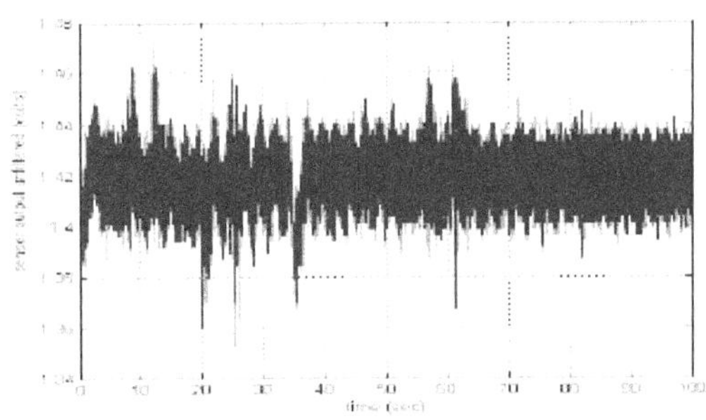

Fig 8 Unfiltered sensor output for Pitch angle

In Figure 8 it is clearly seen that the signal is almost disappeared under the noise. After the signal filtered with an 8th order Butterworth low pass filter the signal becomes distinguishable and the pattern of motion can be seen (Figure 9) with a 8th order Butterworth low pass filter with 10 Hz bandwidth.

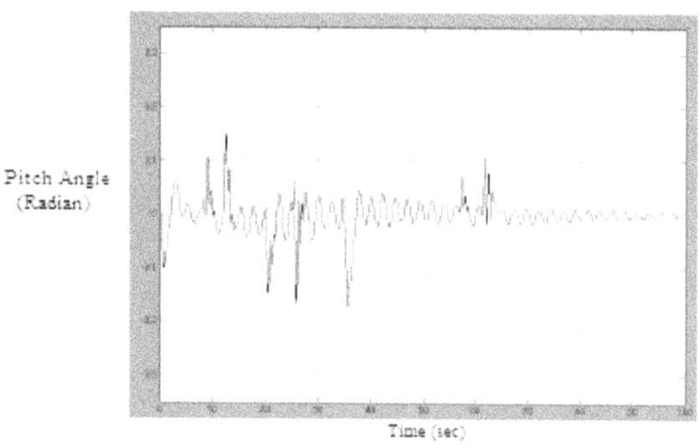

Fig 9 Filtered Pitch angle measurement

Ground Control Station

The last essential part of the overall unmanned system is the ground control station. Its main responsibility is to realize effective communications between the avionic system and the ground users and pilots. To fulfill this aim, the ground station is generally required to have the following fundamental capabilities: (i) displaying and monitoring the in-flight status, (ii) generating and updating flight trajectories, (iii) sending control commands to the avionic system, (iv) facilitating the ground piloted control or automatic control, especially in unexpected situations such as emergency landing and cruise, and (v) logging in-flight data.

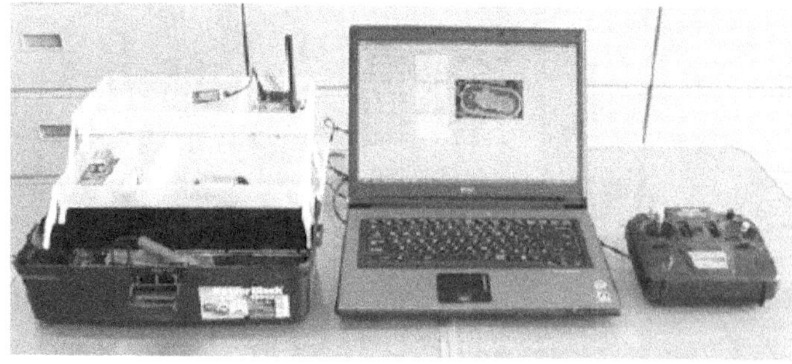

Fig 10 Ground control station set up

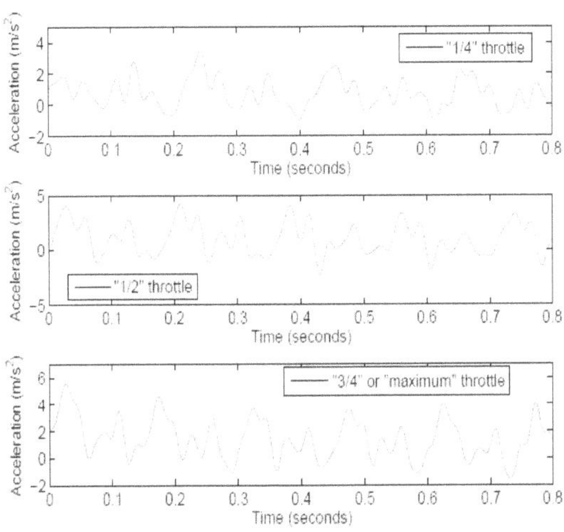

Fig 11 Pitch, roll and yaw acceleration

Data

Throttle	Frequency of Flapping
1/4	4.6HZ
1/2	5.6HZ
3/4	6.6HZ

Fig 12 Throttle vs Estimated frequency of flapping

ACKNOWLEDGEMENT

The authors are grateful to **The Vice Chancellor, Anna University** and The Dean, MIT for providing organizational facilities for carrying out this research work at Madras Institute of Technology. We would like to express our sincere thanks & gratitude to all my professors and avionics students to give their valuable suggestions to make the project as successful one.

Reference

[1] T. N. Pornsin-Sirirak, S. W. Lee, H. Nassef, J. Grasmeyer, Y. C. Tai, C. M. Ho, and M. Keennon, "Mems wing technology for a battery-powered ornithopter," 2000, pp. 799–804.

[2] R. Zbikowski, "Sensor-rich feedback control: a new paradigm for flight control inspired by insect agility," Instrumentation & Measurement Magazine, IEEE, vol. 7,no. 3, pp. 19–26, 2004.

[3]   Z. A. Khan and S. K. Agrawal, "Force and moment characterization of flapping wings for micro air vehicle application," 2005, pp. 1515–1520 vol. 3.

[4]   R. Madangopal, Z. A. Khan, and S. K. Agrawal, "Energetics based design of small flapping wing air vehicles," vol. 3, 2004, pp. 2367–2372 Vol.3.

[5]   J. Tang and M. Sun, "Force and flow structures of a wing performing flapping motion at low reynolds number," Acta Mechanica, vol. 152, no. 1-4, pp. 35–48, 2001.

[6]   R. S. Fearing, K. H. Chiang, M. H. Dickinson, D. L. Pick, M. Sitti, and J. Yan, "Wing transmission for a micromechanical flying insect," in Robotics and Automation, 2000. Proceedings. ICRA '01. IEEE International Conference on, vol. 2, 2000, pp. 1509– 1516.

[7]   R. J. Wood and R. S. Fearing, "Flight force measurements for a micromechanical flying insect," vol. 1, 2001, pp. 355–362 vol.1.

[8]   J. Yan, R. J. Wood, S. Avadhanula, M. Sitti, and R. S. Fearing, "Towards flapping wing control for a micromechanical flying insect," vol. 4, 2001, pp. 3901–3908 vol.4.

[9]   J. Yan and R. S. Fearing, "Wing force map characterization and simulation for the micromechanical flying insect," vol. 2, 2003, pp. 1343–1349 vol.2.

[10]   X. Deng, L. Schenato, and S. S. Sastry, "Model identification and attitude control for a micromechanical flying insect including thorax and sensor models," vol. 1, 2003, pp. 1152– 1157 vol.1.

DESIGN DEVELOPMENT AND EVALUATION OF LONGITUDINAL AND LATERAL AUTOPILOT FOR AN MICRO AIR VEHICLE USING X-PLANE/SIMULINK

Kaviyarasu   A K.Senthilkumar   P.Sivaprakash
Research Scholar, Associate. Professor, PG Student
Dept of Aerospace Engg. MIT, Anna university. Chennai

Abstract

Presently there is vast interest in MAV (Micro air vehicle) development   given   its civilian and military applications. One of the main MAV components is the autopilot system. Its development invariably demands several lab simulations and field tests. Generally after an MAV crash few parts remain usable. Thus, before embedding an autopilot system, it has to be exhaustively lab tested. With educational and research purposes in autopilot control systems development area, a test platform is herein proposed. It employs Matlab/Simulink to run the autopilot controller under test, the flight simulator X-Plane with the aircraft to be commanded, a microcontroller to command model aircraft flight control surfaces and a servo to drive these control surfaces. These resources are interconnected through data communication buses. So that, the autopilot controller designed on Matlab/Simulink is tested by controlling an aircraft on X-Plane. The inputs given to the aircraft flight control surfaces in the X-Plane are simultaneously sent to the microcontroller which translates these commands into effective servo movement control. Through this platform, designed autopilot systems can be applied into models similar to real aircraft minimizing risks and increasing flexibility for design changes. As study case, tests results from a roll attitude autopilot system are presented.

Index Terms- Autopilot, X-Plane.

INTRODUCTION

The fast advancement of modern aircraft design has required the development of many technologies such as aerodynamics, structures, materials, propulsion and flight controls. Currently, the aircraft design strongly relies on automatic control systems to monitor and control many of the aircraft subsystems. Those control systems can also provide artificial stability to improve the flying qualities of an aircraft.

The autopilot systems were firstly conceived to primarily stabilize the aircraft after being disturbed from its wing-level equilibrium flight attitude [1].The modern autopilot systems are much more complex and are essential to flight aiding crew in navigation, flight management, stability augmentation and landing operations, for example.

With the development of modern aircraft new problems arose due to their dynamic stability. For instance, some oscillations inherent to the aircraft can be adequately damped or controlled by the pilot if their period is around 10 seconds or more. On the other hand, if the period of oscillation is 4 seconds or less, the pilot's reaction time is not short enough. Thus, such oscillations have to be artificially well damped [3]

The oscillations named "Short period" in pitch for longitudinal motion, and "Dutch roll" for lateral-directional motion fall into the category of a 4-second oscillation. In more conventional aircraft that kind of oscillation may be damped by handling their constructive characteristics. However, for modern jets artificial damping is required. It is provided by an automatic system [3].

The above concepts are also applicable to MAVs (Micro air vehicle)   whose developments are presently a vast interest given the increasing possibilities of its civilian and military applications. Some of the applications are reconnaissance, surveillance, search and rescue, remote sensing, traffic monitoring missions, etc [4].

One of the main components of MAVs is the autopilot system. An autopilot is a mechanical, electrical, or hydraulic system used to guide a vehicle without assistance from a human being. The autopilot system development invariably demands several lab simulations and field tests. The later ones are of high risk particularly when conducted on MAVs. Generally after an MAV crash few parts remain usable. Thus, before embedding an autopilot system, it has to be exhaustively lab tested.

Based on this necessity, with educational and research purposes in autopilot control systems development area, a test platform is herein proposed and various control laws are verified before make it to the final integration process.

Objective

This present article intends to describe a test platform developed to aid in the autopilot design process.

The test platform herein proposed provides an environment where the designed autopilot system can be applied into models very similar to real aircraft. Several flight situations can be simulated. Parameters of flight as well as aircraft responses can be monitored and easily analyzed. These features increase the flexibility to implement changes and immediately check out the results

Basically, this test platform employs the MATLAB/SIMULINK to run the autopilot controller to be tested, the flight simulator X-plane with the aircraft to be commanded, a microcontroller to command model aircraft flight control surfaces and a servo to drive these control surfaces. All this resources are interconnected through data buses in order to exchange information.

As study case, this article presents the results obtained from tests conducted over an autopilot control system designed for lateral movement, specifically for roll mode control. This system was conceived for Cessna aircraft.

AIRCRAFT TRANSLATIONAL MOTIONS

An airplane in free flight has three translational motions (vertical, horizontal and transverse), three rotational motions (pitch, yaw and roll) and numerous elastic degrees of freedom.

In order to reduce the complexity of this mathematical modeling problem, some simplifying assumptions may be applied. First, it is assumed that the aircraft motion consists of small deviations from its equilibrium flight condition. Second, it is assumed that the motion of the airplane can be analyzed by separating the equations into two groups [2]:

- Longitudinal Equations: composed by X-force, Z-force and pitching moment equations.

- Lateral Equations: composed by Y-force and yawing and rolling moments equations.

Figure 1 shows the above mentioned aircraft reference axes, forces and moments.

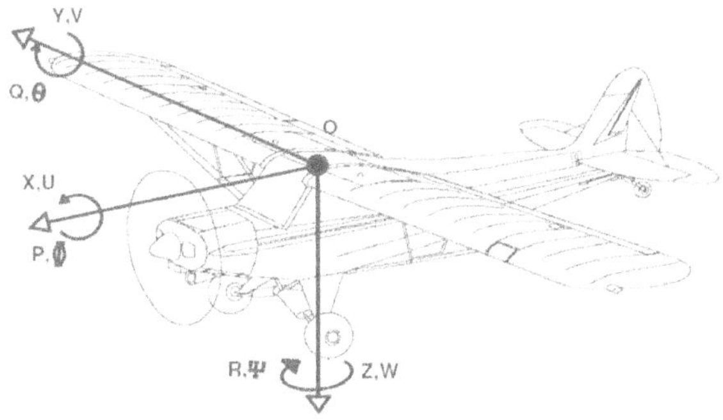

Fig 1 Aircraft Axes, Forces and Moments

Lateral –directional motion

The lateral-directional motion of an aircraft disturbed from its equilibrium state is a combination of rolling, yawing and side slipping motions. This interaction between roll and yaw produces a coupled motion [1].

The lateral-directional motion, specifically the roll motion, is the movement intended to be simulated on the proposed test platform as demonstration example. To simplify the studied case, it was chosen to address specifically the roll motion in this work. Even though this simplifies the dynamic model equations, the platform functionality and characteristics are very well demonstrated.

Pitch motion closed-loop control system

The aircraft pitch motion can be controlled by an autopilot system block diagram as shown in the figure 2.

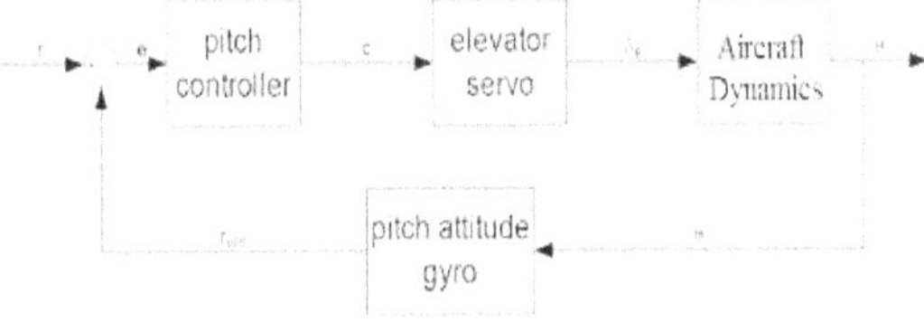

Fig 2 A Pitch Attitude Autopilot System

Through this simplified system, the pitch angle (Θ) can be controlled by a reference angle applied as input. Primarily, the pitch attitude autopilot is designed to maintain the aircraft with wings leveled, or with $\Theta = 0$. Usually, this simplified system does not fulfill the control design performance requirements in relation to damping ratio, overshoot and undamped natural frequency.

So that, in the study case herein presented, a more efficient control system is employed . It is shown in Figure 3.

In this case, an accurate aircraft pitch to elevator transfer function for Cessna aircraft is employed in the pitch autopilot system design. This transfer function was calculated based on aircraft physical characteristics and stability derivatives as detailed in [6]. Moreover, a typical elevator actuator transfer function is employed.

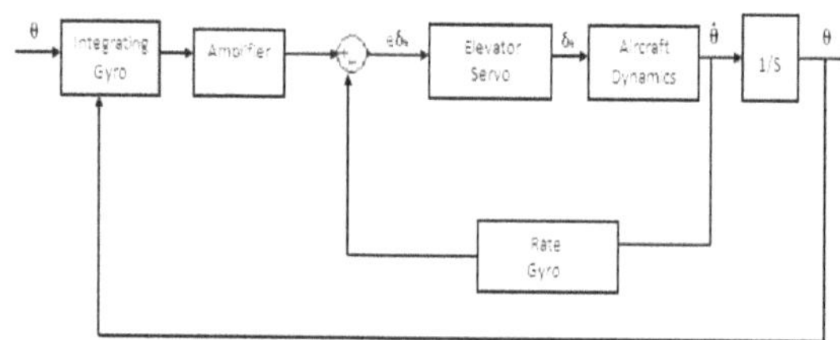

Fig 3 Pitch Attitude Autopilot System

The pitch attitude autopilot system gains *KG, Ka* and *KRG* are calculated through the root locus method considering the above rationale.

THE TEST PLATFORM

The main components of the test platforms are as follows.

- Matlab/Simulink containing the autopilot control system.

- Flight Simulator X-Plane containing the aircraft model to be controlled.

- Microcontroller to command the aircraft   flight control surfaces.

The test platform concept is based on the block diagram presented in Figure 4. In the test platform, the block "Controller" is replaced by the designed autopilot system model. This model runs into the Matlab/Simulink environment.

Similarly, the block "Aircraft Dynamics" is replaced by the flight simulator X-Plane with the aircraft to be controlled.

Thus, the basic principle of the test platform consists in establishing the communication between Matlab/Simulink, X-Plane, microcontroller and servo in the following mode:

The parameters calculated by the autopilot control system are sent to X-Plane in order to command the aircraft flight control surfaces. The X-Plane calculates the new aircraft attitude according to the inputs received from Matlab/Simulink. The X-Plane sends those new aircraft attitude parameters back to Matlab/Simulink closing the loop. Matlab/Simulink restarts the process by providing updated commands to X-Plane aircraft. The inputs given to the aircraft flight control surfaces in the X-Plane are simultaneously sent to a microcontroller which translates these commands received from Matlab/Simulink into effective servo movement control. The microcontroller calculates the deflection angle to be imposed to the flight control surfaces and generates a proportional PWM signal to command the servo. The model aircraft flight control surfaces reproduce the same deflection observed on the X-Plane aircraft.

The communication between Matlab/Simulink and X-Plane is made through UDP (User Datagram Protocol). Between Matlab/Simulink and microcontroller it used RS-232 serial communication. Block diagram shown by Figure 4 summarizes the platform concept.

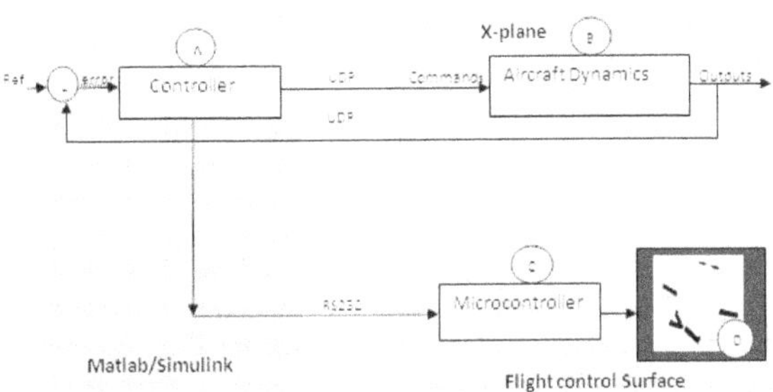

Fig 4 Test platform block diagram

The flight simulator X-Pane provides very accurate aircraft models and has a very important feature: the possibility to exchange data with external systems [7]. Giving its realistic simulations capability, X-Plane is also FAA (Federal Aviation Administration) approved for pilot training. The aircraft models simulated in X-Plane are built based on their real physical dimensions, power and weight among other characteristics. X-Plane is not considered a game, but can be categorized as an engineering tool that can be used to predict the flying qualities of fixed and rotary wing aircraft [7].

Most other flight simulators use the stability derivatives method to compute how an airplane flies. This technique involves simply forcing the aircraft nose to return to a centered position along the flight path with certain acceleration for each degree of offset from straight-ahead flight of the airplane. This is too simplistic to be used across the entire flight envelope of the airplane. Stability derivatives will not normally take into proper account the asymmetric effects of engine failures, the chaotic effects of turbulence, stalls, spins and the myriad of dynamic effects that airplane flight generates. In other words, the commonly used stability derivatives are gross over simplifications of how an airplane flies. In summary, those simulators can not predict how the airplane will fly. Basically, the airplane designer teaches the simulator how the airplane should fly, and the simulator reproduces that information right back to the user [7].

X-Plane instead, assimilates the geometric shape of any aircraft and then figures out how that aircraft will fly. It does this by an engineering process called "blade element theory", which involves breaking the aircraft down into many small elements and then finding the forces on each little element many times per second. These forces are then converted into accelerations which are then integrated to velocities and positions. This method of computing the forces on the airplane is much more detailed, flexible, and advanced than the flight model that is used by most other flight simulators. By doing this process, X-Plane accurately predicts what will be the performance and handling qualities of an airplane of a given geometry [7]. The microcontroller represented by block C on Figure 4 symbolizes the experimenter board Arduino-Ethernet shield shown in Figure 5.

Fig 5 Microcontroller unit

This is a development kit based Arduino Ethernet Shield with a clock of 16 MHz, low power consumption, featured with a variety of I/O (digital, PWM), besides RAM and flash memory space. The choice for this device is justified by its in build Ethernet microcontroller chip, serial port (communicate between i/p and o/p devices in the external world) as well as easy integration with digital servos through PWM I/O ports. On block D of Figure 4 it is represented the digital servo that commands model aircraft flight control surfaces. A typical servo is shown in Figure 6.

Fig 6 Model aircraft digital servo

It receives PWM signals from the microcontroller unit and converts them into proportional axis movement.

X-PLANE DATA INTERFACE

Flight Simulator X-Plane has an important feature that is essential to this test platform development. It is its capacity of sending and receiving data to and from other devices. One way to implement this communication is by employing protocol UDP. Data packets are sent and received through the microcomputer Ethernet port. Figure 4 shows UDP protocol being used to establish the communication between Matlab/Simulink (Controller) and X-Plane (Aircraft Dynamics).

UDP uses a simple transmission model without implicit hand-shaking dialogues for guaranteeing reliability, ordering, or data integrity. Thus, UDP provides an unreliable service and data packets may arrive out of order, appear duplicated or go missing without notice. Error checking and correction are considered either not necessary or performed in the application. This way, UDP avoids the overhead of such processing at the network interface level being extremely fast [5]. Time-sensitive applications such as this test platform often use UDP once dropping packets is preferable to waiting for delayed packets, which it is not an option in real-time systems.

Hence UDP speed constitutes a key point in the test platform once the communication between Matlab/Simulink and X-Plane must be fast enough to synchronize commands, data processing and system responses.

X-Plane is able to send or receive up to 99.9 data packets per second via UDP. Each data package may be configured to carry aircraft parameters data that are selected on X-Plane Data

Input and Output interface.

For example, in the case of lateral motion, parameters such as roll, yaw, pitch, altitude and speed shall be selected for transmission. Each parameter receives a numeric label for proper identification.

Figure 7 shows the interface that serves to select data for input and output from X-Plane.

Besides parameters selection for UDP data transmission or receiving, it also serves for selecting parameters to be shown during aircraft flight simulation, file recording and graphs construction.

Fig 7 X-Plane parameters user interface

Once the parameters are selected on X-Plane, through UDP it is possible to receive them at Matlab/Simulink environment and after processing, to send commands back to X-Plane aircraft. In the case of roll attitude autopilot proposed example, Matlab/Simulink generates the roll angle reference and control signal sending them to X-Plane. In its turn, X-Plane commands the simulated aircraft according to the inputs received. The new aircraft roll angle position is sent from X- Plane back to Matlab/Simulink, restarting the process

The X-Plane data package follows a pattern shown on Figure 8.

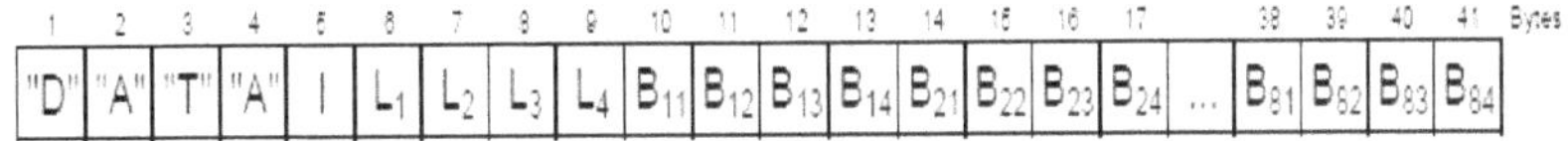

Fig 8 X-Plane data package pattern

Basically it is composed by a sequence of bytes that need to be properly interpreted. X-Plane uses what are known as single precision floating point variables for just about everything sent over the network. This means that the numbers can be stored using four bytes. The first four bytes of the packet shown on Figure 7represents the characters "DATA" used to indicate that this is a data packet. The fifth byte is an internal code (I). The next four bytes represent the parameter label (L1, L2, L3, and L4). Following there are 8 sets of 4 bytes (B11, B12, B13, B14 to B81, B82, B83, B84) representing the data itself in single precision floating point. Taking each set of 4 bytes, the first bit is the sign bit. It tells whether the number is positive or negative. The next 8 bits are the biased exponent. The remaining 23 bits represents the mantissa. The next seven sets of 4 bytes complete the data. So, the Matlab/Simulink model has to decode this data packet accordingly. It also has to properly encode the data to be sent to X-Plane. Most of the blocks of the test platform implemented on Matlab/Simulink environment are dedicated to decode and encode data packets from and to X-Plane.

IMPLEMENTATION

Figure 9 illustrates the test platform implemented.

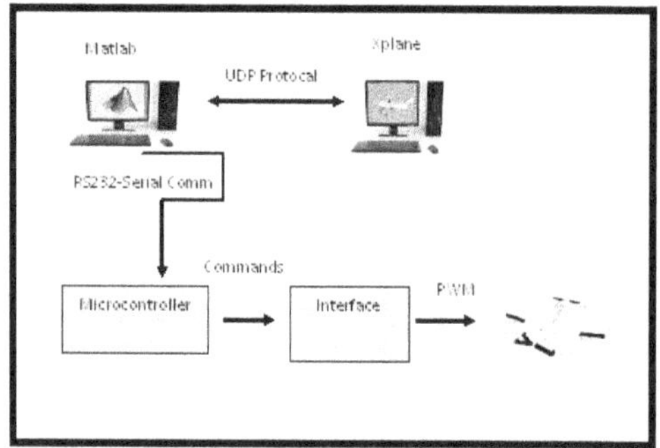

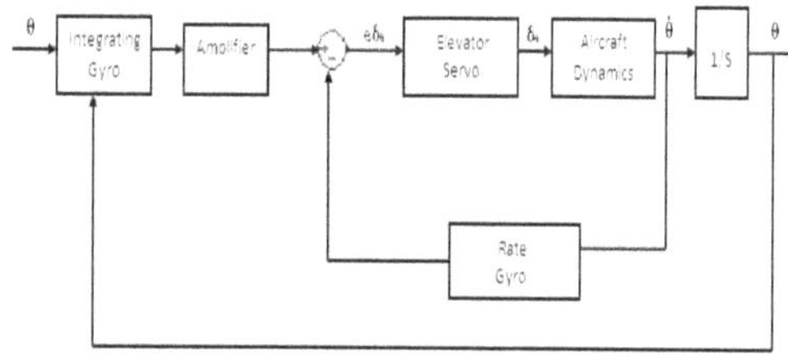

Fig 9 Autopilot Test Platform Implementation

Fig 10 Pitch Attitude Autopilot System for Cessna

In one computer Matlab/Simulink runs a model containing the autopilot control system and other blocks responsible for UDP data packets decode and encode process as well as serial communication with microcontroller. In the other computer X-Plane simulates the aircraft to be controlled. Both computers communicate to each other through their Ethernet port using UDP protocol. Through serial bus the microcontroller receives the same commands Matlab/Simulink sent to X-Plane. It decodes this data converting them into PWM signal. An interface amplifies this signal to apply it to the servo that commands the model aircraft flight control surfaces.

APPLICATION EXAMPLE – TEST RESULTS

In order to demonstrate how the test platform could be applied in laboratory classes and also help autopilot systems development, the roll attitude autopilot system was designed and integrated into the platform. The dynamic system used as the case to be studied and tested using the proposed platform can be represented by Figure 3. The gains KG, Ka and KRG of that system were calculated through the root locus method using the systems dynamics as in (2) and (3), which are the Cessna pitch to elevator and typical Pitch servo transfer functions.

Pitch Dynamics:
Pitch Servo:

$$S(s) = \frac{10}{s+10}$$

The designed system result is shown on Figure 10. It is also shown which blocks will be simulated by X-Plane and which ones will be integrated into the test platform model at Matlab/Simulink.

The Pitch to elevator and servo transfer functions are simulated by the Cessna at X-Plane software. The other autopilot blocks are distributed into the test platform model. Figure 11 shows the entire test platform model implemented at Matlab/Simulink environment.

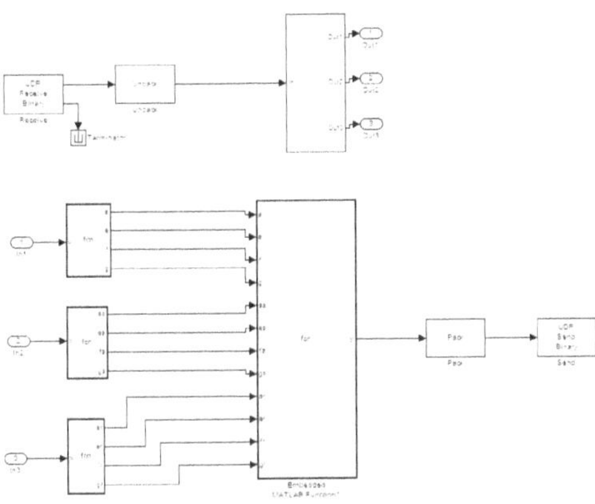

Fig 11 Autopilot Test Platform Model At Matlab/Simulink

In a typical lab class, the student should find the test platform ready to be used with the communication interfaces appropriately configured. The gains KG, Ka and KRG shall then be fed into the controller implemented at Matlab/Simulink. The gains calculation may be performed at the lab or be previously done, being the lab classes preferably used for test at the platform.

To initialize the simulation, X-Plane is loaded with aircraft Cessna in a cruise flight at 40,000 ft altitude. As soon as the test platform model shown on Figure11 runs, the designed autopilot system takes the aircraft control. In principle, for any pitch angle applied as reference into the autopilot system, the aircraft at X-Plane shall respond following that reference.

The reference signals applied as system input, the system response and also the commands sent to the aircraft flight control surfaces actuators can be monitored through real time graphs at Matlab/Simulink. In the proposed study case, the input is the reference pitch angle, the system response is the new aircraft Pitch angle attitude and the commands are the deflections to be imposed to the ailerons. Figure 12 shows the results for the proposed pitch attitude autopilot system implemented at the test platform.

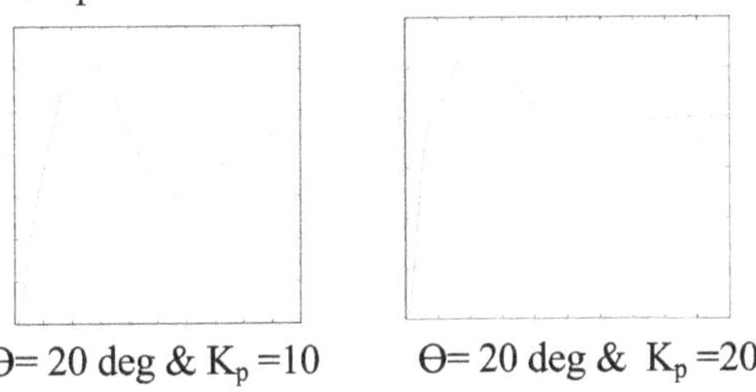

$\Theta = 20$ deg & $K_p = 10$ $\Theta = 20$ deg & $K_p = 20$

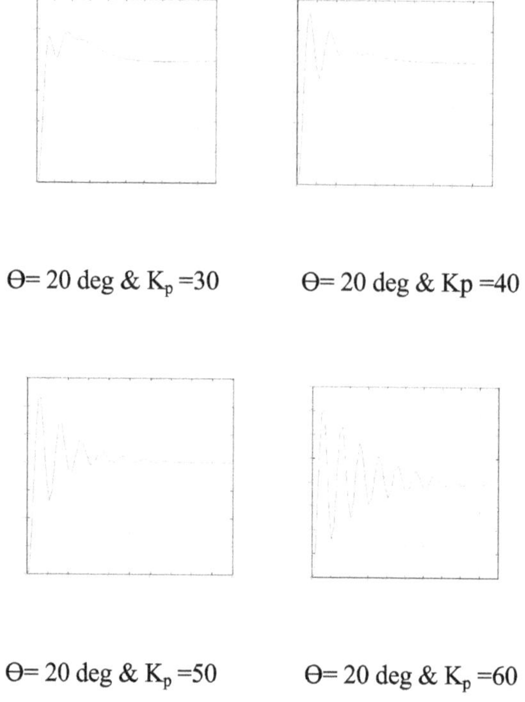

Fig 12 Pitch Reference and Response

The graph on Figure 12 presents the autopilot Pitch angle reference 20deg applied into the system. During simulation it is possible to see the Cessna make an 20 degrees pitch angle attitude and performing a slight pitch up. In parallel it is possible to verify the servo movements in response to these commands in the aircraft model. Interesting to notice that even with the aircraft leveled and keeping a certain pitch attitude there are small commands to the flight control surfaces. This is something that only was possible to observe due to the realism provided by X-Plane simulation which introduces small perturbations to aircraft flight reproducing a real atmosphere with wind, turbulences, etc. One exercise that can also be performed on the designed autopilot system is to vary the loop gains and check it out the aircraft responses. This can be used to properly tune the loop gains to optimize system performance.

Figure 14 below shows the output .The output below shows that the value such as pitch, roll and yaw set in Matlab/Simulink will produce the same change in their attitude of the aircraft in the X-Plane.

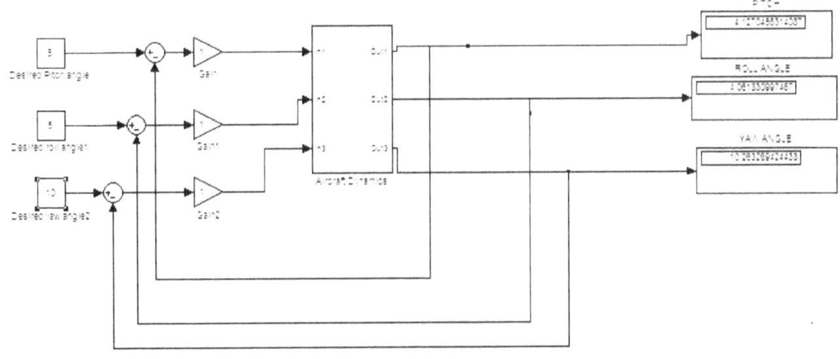

Fig 13 3 Axis Closed Loop System

Fig 14 Simulated Output

CONCLUSION

The development of this test platform resulted in a valuable tool for the students and professionals to aid autopilot systems study and design. It allows monitoring the aircraft responses for a designed autopilot system with high degree of realism. It is possible to change control system parameters very easily and check the results out in a friendly environment. This possibility easies design task as well minimizes risks of embedding the system for field tests.The test platform also permits the study of longitudinal and lateral-directional movement of different fixed and rotary wing unmanned aerial vehicle platform for which the autopilot system is being designed. A possible extension of this study is to test the platform and the design with the Hardware in loop system (HILS) to simulate MAV in the Real-time.

REFERENCES

[1]   Lucio R. Ribeiro and Neusa Maria F. Oliveira "MAV Autopilot Controllers Test Platform Using Matlab/Simulink and X-Plane"40th ASEE/IEEE Frontiers in Education Conference, 2010.

[2] NELSON, R. C. "Flight Stability and Automatic Control" McGraw-Hill Second Edition, 1998

[3] BLACKLOCK J. H. "Automatic Control of Aircraft and Missiles" Wiley Inter science Publication, 1991

[4] ADANG SUWANDI AHMAD1, JAKA SEMBIRING, "Hardware in the Loop Simulation for Simple Low Cost Autonomous MAV (Unmanned Aerial Vehicle) Autopilot System Research and Development" Institute Technology Bandung, Indonesia, 2007

[5] DANIEL ERNST, "Development of Research Platform for Unmanned Vehicle Controller Design, Evaluation, and Implementation System: From MATLAB to Hardware Based Embedded System", Department of Computer Science and Engineering College of Engineering University of South Florida, Florida, 2007

[6] ROSKAM, J. "Airplane Flight Dynamics and Automatic Flight Controls" PART I. Roskam Aviation and Engineering Corporation, 1982

[7] LAMINAR RESEARCH, "X-Plane Description", X-Plane Manual, 2009

MAV'S FOR ATMOSPHERIC PROPERTIES MONITORING FOR AEROSPACE APPLICATIONS

K.Durga Rao   A.G.Sarwade
[1]Associate Professor, [2]Professor,
Dept of Aeronautical engineering,
Malla Reddy college of Engineering &Technology, Hyderabad
durgaraok9@gmail.com
a.g.sarwade@gmail.com

S.Srinivas Prasad
[3]Professor, Dept of Aeronautical engineering
Marri Laxman Reddy Institute of Technology, Hyderabad
sanakaprasad@yahoo.com

T.V.Kartikeyan
[4]Sr. Scientist, ASL, DRDO, Hyderabad
dr.tv.karthikeyan@gmail.com

ABSTRACT

Atmospheric properties vary from time to time. Pressure, Temperature and Density are the dominant variables. During launch of short range rockets or missiles for experimental purposes, their trajectory gets affected by the atmosphere properties near the ground. The properties of air at an altitude of 0.5 to 1 km are the useful data to predict the trajectories of the launch of short range missiles during tests. The atmosphere properties near ground can be well assessed by the outputs of sensor signals received from MAV's launched for this purpose. MAV's provide a low cost atmosphere monitory. This paper describes the sensors layout, various sensors that are to be used in the MAV's. The miniature pilot static probes mounted at some distance ahead of wing tip measure static and dynamic pressures accurately. The static temperature can be measured by miniature thermocouples mounted on the top and bottom side of fuselage of MAV. If the wings are made flapping, then it is not advisable to mount the pressure probes on wing tips, but the probe can be mounted on a small pilot static tube protruding from the nose tip. Strain gauges to monitor stress levels, these can be mounted on the top and bottom side of fuselage and on wings top and bottom surfaces. Vibration gauge (measures 'g' levels) as optimal ones its extra information from the data of pilot. Telemetry provision to transmit sensor data to ground is required. This paper also covers the image capture of events on the ground for surveillance.

KEYWORDS: MAV, Measurement of atmospheric air properties, sensors, MEMS

INTRODUCTION TO MICRO AIR VEHICLE

It is a class of unmanned aerial vehicles (UAV) that has a size restriction and may be autonomous. Modern craft can be as small as 15cm. Development is driven by commercial, research, government, and military purposes with insect-sized aircraft reportedly expected in the future. The small craft allows remote observation of hazardous environments inaccessible to ground vehicles. MAVs have been built for hobby purposes, such as aerial robotics contests and aerial photography. In contrast to higher-level reconnaissance assets like satellites and high altitude UAVs, MAVs will be operated by and for the individual soldier in the field as a platoon-level asset, providing local reconnaissance or other sensor information on demand, where and when it is needed. Police and Intelligence departments can use them for the maintenance of law and order and for the national security. MAVs may also be used for tagging, targeting, and communications, and may eventually find application as weapons as well.

The reconnaissance application is a primary driver behind the first generation of MAVs. Micro sensors like those mentioned earlier suggest the possibility of reduced latency and greatly enhanced situational awareness for the small unit or individual soldier. This is partly attributed to the direct connectivity envisioned between these systems and the "user" in emerging operational concepts. Direct connectivity means the user has to carry it. So the MAV must trade favorably with other soldier assets - like water and ammunition. The system must also be affordable. It must have vanishingly small logistics geometry and for many missions it must be intrinsically covert.

Additionally, the MAV's ability to operate in constrained environments like urban canyons and, eventually, even the interior of buildings, gives these systems a level of uniqueness unmatched by other concepts. MAVs are not replacements for previously manned air vehicle missions. Because of their size, MAV's will be capable of completely new missions not possible with any existing systems. In this paper the main emphasis of MAV is for atmospheric monitoring like air pressure, cross winds etc. which is normally required just before the launch of large rockets, to ensure a trouble face launch.

MEASUREMENT OF ATMOSPHERIC PROPERTIES

The significant atmospheric parameters measurement methodology has been elucidated in this section.

PRESSURE MEASUREMENT

The pressure measurement on MAVs allows to estimate the current altitude and, if two sensors are used, the current airspeed.

AIRSPEED

The airspeed can be measured by comparing the pressure in a pitot tube and the ambient pressure, measured. The indicated airspeed is the true airspeed at Density can be constant on all altitudes, the true air speed (travelled distance, neglecting wind effects) will be higher at high altitudes. The indicated airspeed thus directly relates to the pressure in front of aircraft, which in turn relates to the speed at which stall occurs. So to prevent stall, always keep the indicated airspeed above the stall speed, no matter at which altitude the MAV is flying. Consequently this will be at a lower true air speed near sea level and at a high true air speed at very high altitudes in thin air.

Calculation of the indicated airspeed

The speed in a medium is

$V_{in} = [2(P_t-P_s)/\rho_o]^{1/2}$

Where all variables are in SI units:

V_{in} = The indicated airspeed (m/s)

$\rho_o = 1.225$ kg/m³, the air density at sea level and 15 degrees Celsius

P_t = Stagnation or Total Pressure inside the pitot tube(Pascal's)

P_s = Static pressure (N/m²)

Calculating True Airspeed for low speeds

The true airspeed is the vehicle's forward ground speed combined with wind speed. The wind has a direction, so the true airspeed contains only the part of the wind speed along the pitot tube axis. Please also note that the overall ground speed (including movements to the side through wind drift) is not equal to the true airspeed. Wind from the side can induce a speed with respect to the ground, which is not measured by the true airspeed.

The formula below applies for low speeds (lower subsonic range).

$V_{true} = V_{in} * (\rho_o/\rho)^{1/2}$, Where

V_{in} = The indicated airspeed (m/s, which depends on altitude)

V_{true} = The true airspeed (m/s)

$\rho_o = 1.225$ kg/m³, the air density at sea level and 15 degrees Celsius

ρ = The air density at the current altitude (kg/m³)

calculated from GPS altitude and/or barometric absolute pressure

Air Density for a specific temperature

This measure is used to compensate the speed calculation at different altitudes

$\rho = P/(R.T)$, Where

ρ = Density at this temperature in kg/m³

P= Static pressure in Pascal

micro fabrication technology enables fabrication of large arrays of devices, which individually perform simple tasks, but in combination can accomplish complicated functions.

R= Gas constant for air, 8.31432 N·m /(mol·K)

T= Absolute Temperature in Kelvin Altitude

V_{in} = The indicated airspeed (m/s, which depends on altitude)

V_{true} =The true airspeed (m/s)

V_{in} = The indicated airspeed (m/s, which depends on altitude)

V_{true} = The true airspeed (m/s)

$\rho_o = 1.225$ kg/m³, the air density at sea level and 15 degrees Celsius

Calculating the pressure altitude

To calculate the pressure altitude, the barometric formula has to be used. If the MAV is only active between 0 and 11000 meters, it can be simplified. For almost all MAVs the simplified version is recommended to use.

Sensors for MAV's

The MAV's are small objects. To monitor atmospheric pressure and winds (by means of dynamic pressure gauging) miniature gauges are to be mounted on the surfaces of the MAV's.

Role of MEMS in miniature sensors and actuators

Microelectro mechanical systems (MEMS) are small integrated devices or systems that combine electrical and mechanical components. They range in size from the sub micrometer (or sub micron) level to the millimeter level, and there can be any number, from a few to millions, in a particular system. MEMS extend the fabrication techniques developed for the integrated circuit industry to add mechanical elements such as beams, gears, diaphragms, and springs to devices.

These systems can sense, control, and activate mechanical processes on the micro scale, and function individually or in arrays to generate effects on the macro scale. The micro fabrication technology enables fabrication of large arrays of devices, which individually perform simple tasks, but in combination can accomplish complicated functions.

More recently, the MEMS research and development community has demonstrated a number of microactuators including: microvalves for control of gas and liquid flows; optical switches and mirrors to redirect or modulate light beams; independently controlled micromirror arrays for displays, microresonators for a number of different applications, micropumps to develop positive fluid pressures, microflaps to modulate airstreams on airfoils, as well as many others.

Surprisingly, even though these microactuators are extremely small, they frequently can cause effects at the macroscale level; that is, these tiny actuators can perform mechanical feats far larger than their size would imply. For example, researchers have placed small microactuators on the leading edge of airfoils of an aircraft and have been able to steer the aircraft using only these microminiaturized devices.

MEMS Fabrication Technologies

The three characteristic features of MEMS fabrication technologies are miniaturization, multiplicity, and microelectronics. Miniaturization enables the production of compact, quick-response devices. Multiplicity refers to the batch fabrication inherent in semiconductor processing, which allows thousands or millions of components to be easily and concurrently fabricated. Microelectronics provides the intelligence to MEMS and allows the monolithic merger of sensors, actuators, and logic to build closed-loop feedback components and systems. The successful miniaturization and multiplicity of traditional electronics systems would not have been possible without IC fabrication technology. Therefore, IC fabrication technology, or micro fabrication, has so far been the primary enabling technology for the development of MEMS. Micro fabrication provides a powerful tool for batch processing and miniaturization of mechanical systems into a dimensional domain not accessible by conventional metal removal techniques. Furthermore, micro fabrication provides an opportunity for integration of mechanical systems with electronics to develop high-performance closed-loop-controlled MEMS. Advances in IC technology in the last decade have brought about corresponding progress in MEMS fabrication processes. Manufacturing processes allow for the monolithic integration of micro electro-mechanical structures with driving, controlling, and signal-processing electronics. This integration promises to improve the performance of micromechanical devices as well as reduce the cost of manufacturing, packaging, and incrementing these devices. IC Fabrication, Bulk Micromachining and Wafer Bonding, Surface Micromachining, Micromolding, LIGA Process are the manufacturing Processes of MEMS.

Applications of MEMS

Pressure Sensors, Accelerometers, Inertial Sensors, Microengines, Automotive, Bio-MEMS, Computer Hardware, Optical MEMS, Defence/Environment, Telicommunication are the applications of MEMS. The MEMS are used as sensors for the guidance and navigation of MAV. In the designing of algorithm, the outputs of accelerometer were compensated by airspeed meter, then the gravitational and geomagnetic field vectors were used to correct the attitude solved from gyroscopes through a fifteen-state Extended Kalman Filter. The measurement values of Kalman filter were calculated from the attitude errors obtained through introducing the magnetic yaw and horizontal attitude. Furthermore, the stochastic errors of the gyroscope and accelerometer were set into state vector, which could correct the outputs of the inertial sensors and improve the measurement accuracy. The advantage with present approach is that the state equations and measurement equations are linear which making it easy to implement. The simulation of dynamic flight tests demonstrated that the estimated error of yaw, pitch and roll less than $1.0°$, $1.2°$ and $0.5°$ respectively. It also proved that Kalman filter could improve the accuracy of attitude estimation effectively.

Applications of MEMS

Pressure Sensors, Accelerometers, Inertial Sensors, Microengines, Automotive, Bio-MEMS, Computer Hardware, Optical.

MEMS, Defence/Environment, Telicommunication are the applications of MEMS. The MEMS are used as sensors for the guidance and navigation of MAV. In the designing of algorithm, the outputs of accelerometer were compensated by airspeed meter, then the gravitational and geomagnetic field vectors were used to correct the attitude solved from gyroscopes through a fifteen-state Extended Kalman Filter. The measurement values of Kalman filter were calculated from the attitude errors obtained through introducing the magnetic yaw and horizontal attitude. Furthermore, the stochastic errors of the gyroscope and accelerometer were set into state vector, which could correct the outputs of the inertial sensors and improve the measurement accuracy. The advantage with present approach is that the state equations and measurement equations are linear which making it easy to implement. The simulation of dynamic flight tests demonstrated that the estimated error of yaw, pitch and roll less than $1.0°$, $1.2°$ and $0.5°$ respectively. It also proved that Kalman filter could improve the accuracy of attitude estimation effectively.

Pressure Sensors

Two types of sensors are used in MAVs: Absolute pressure sensors for altitude and differential pressure sensors for airspeed. Differential sensors have two inputs.

Pressure sensors for physiological applications

Sketches are enclosed of the sensors developed abroad, for medical applications. If these have to be updated for applications on hardware like MAV, they have to be suitably ruggedized.

Strain Gauges

An integrated Structural Health Monitoring (SHM) diagnostics/prognostics system will improve the affordability, survivability, and service life of Unmanned Air Vehicles (UAVs) by ensuring the performance and reliability of high value, critical components on-board the UAV. The detection, monitoring and prediction of structural and material degradation are critical diagnoses for maintaining efficiency and battle readiness. Using its existing SHM system platform presently being field tested by Boeing and Delta Air Lines, Analatom proposes to develop an integrated central health management system for UAVs using a combination of MEMS strain gauges, Linear Polarization Resistance (LPR) corrosion sensors and a Texas Instruments (TI) MSP430 microprocessor for lower weight, power and cost requirements and higher sensitivity than conventional sensors. Data transmission and downloading will be accomplished through either a hardwired approach or through wireless data links when assessment for final system implementation is completed. Use of Motorola neuRFonTM chip for a wireless, self-organizing network that has low power requirements would be investigated and pursued. The envisioned system will meet the cost and weight requirements for UAV applications by multiplexing low cost, microfabricated MEMS strain gauges, corrosion sensors; and data acquisition, processing and subsequent data transmission via wired or wireless interface for data downloading.

Temperature gauging

Static temperature can be measured by miniature foil type thermocouples. Vibration level is estimated by the data of acceleration spectral density. The main use of MAV's is for surveillance by capturing object images and transmitting them to ground. SWIR (Short wave Infra red) sensors can reliably capture images. They cannot be jammed as in the case of radio images. Laser imaging is an alternative technique. SWIR technique is planned to be perfected in phases as Phase1: Identify materials for SWIR detector arrays considering less cost, weight, power drawn etc.Phase2: Test the prototype for the efficiency of the materials. Integrate the imagery detection in a compact camera for further performance testing.

Conclusions

1. Lot of useful data on the atmosphere can be obtained by tiny sensors mounted on the MAV's. The variations in the atmospheric properties can be easily monitored. So far balloons are being used for this purpose but the usage of balloons is tedious and needs preparation time, besides being costly.

2. MEMS technology helps in miniaturizing the sensors. The technology is in the initial stages in India. In the near future it is envisaged that the indigenous sensors can be used in the MAV's.

3. Since the MAV size is small, it is advised that the sensors mounted on the MAV surface do not disturb the air flow.

4. The modern   Infra red imagery capture and its transmission to ground avoid the pitfalls faced in radio imaging, which is susceptible to jamming.

References

1. A.H. Epstain and S.D.Senturia 2006"Macro Power from Micro Machinary" Science 276,23.P.1211.

2. W.J. Pisano and D.A Lawrence, 2007."Autonomous UAV Control using a 3-sensor autopilot", In AIAA Infotech@Aerospace conf, Sunomia,CA.

3. J.Zheng and M.J.Lee, 2006 A Comprehensive Performance study of IEEE 802, 15.4, IEEE Press Wiley Interscience.

4. A.Mokrini, M.A.Huneault,2006 "Proton exchange Membrances based on PVDF/SEBS Blends", Journal of Power sources 154:51-58.

5. K.Chebrolu, B.Raman, and S.Sen, 2006 Long distance802.11b links: "performance measurements and experience.In ACM mobicom",

6. S.Lupashin, A.Scholling, M.sherback and R.D.Andrea,2010. "A simple learning strategy for high-speed quadrocopter multi-flips" in IEEE International conference on Robotics and Automation May

7. S.ahrents D.Levine, G.andrews and J.How, 2009"Vision-based guidance and control of a hovering vehicle in unknown gps-denied environment," in International conference on robotics and Automation, May.

8. M. Ettenberg, M. Lange, M. Cohen, G. Olsen1998, IEEE Lasers and Electro-Optics Society Annual Meeting, Vol. 2, pp. 71-2 4.

9. Onat, B., Huang, W., Masaun, N., Lange, M., Ettenberg, M. , and Dries, C. 2007"Ultra Low Dark Current InGaAs Technology for Focal Plane Arrays for Low-Light Level Visible-Shortwave Infrared Imaging," Infrared Technology and Applications XXXIII, edited by Bjørn F. Andresen, Gabor F. Fulop, Paul R. Norton, Proc. of SPIE Vol. 6542, 65420L

10. Dr. K.Ramji , 2008 "AU Autonomous college of Engg, Lecture on Over View and its emerging applications of Nano technology" during Seminor on Nano Technology at SNIST,HYDERABAD.

11. R.O. Hundley and E.C.gritton,1992 "Future Technology-Driven Revolutions in military applications".

APPLICATIONS AND THE USABILITY OF MICRO AIR VEHICLES IN REAL-LIFE SITUATIONS IN INDIAN SCENARIO

Pradeep Gorthi

Student of Electronics and Communication Engineering, International Institute of Information Technology, Hyderabad, India

Anupam Gorthi

Student of Information and Communication Technology, Dhirubhai Ambani Institute of Information and Communication Technology, India.
Presently working as consultant with Deloitte US

ABSTRACT

The idea of Micro Air Vehicle existed since the day flight system had been invented. The equivalent support from the technical side was absent. But, in recent decades, advancement in technology led to developments in the design of Micro Air Vehicle. This advancement was recognized by the Indian Defence agencies and consequently design and developments are seen over the years. MAV has many applications in various fields like disaster management, commercial usage and defence applications. Each task in every field has its own specific set of needs and require different materials for the specific set of purposes. The main aim of the paper is to demonstrate the applications of MAV in these fields and their future development. The ideology stated in the paper is restricted to Indian scenario. A main issue with the mobility of the air vehicle is the power source. A new way of source is proposed in the paper to prolong the time of flight.

INTRODUCTION

Bots have claimed a significant part in human life for past few decades. Advancement in technology has resulted in design and development of different kinds of bots for varied purposes. One of these is Micro Air Vehicle. Objects designed for flight have been present for decades, but as we decrease the size of the objects, it is more and more difficult to design the air vehicles. Varied research is being done all over the world in the development of unmanned micro air vehicles as the MAVs come with their own set of advantages. Even though remotely operated MAVs are present and are in use, popularity for these small sized devices comes from their capability of autonomy. This technology finds way into the military and research as loss of life is minimal if UAV is ever used in a combat. But at the same time, the importance of UAV is seen in non-military fields such as educational research, situations of disaster management and government usage. These UAVs are extensively used by the military and paramilitary forces to look for explosives and even defuse bombs if possible.

MICRO AIR VEHICLES PRESENTLY IN THE INDIAN SCENARIO

Aeronautical Development Establishment collaborated with National Aerospace Laboratories and built three models of Micro Air Vehicles. These three are named Golden Hawk, Black Kite and Pushpak. These three Unmanned Micro Air Vehicles have sizes ranging from 300-450 mm and weighed between 0.3 to 0.5 kilograms. These prototypes have endurance of a maximum of thirty minutes while carrying a small sized video camera. These micro air vehicles were built to be carried as a soldier's backpack which can be hand launched and recovered through soft landing. The fully autonomous vehicles have been programmed with waypoint navigation and the way points can also be changed through tele-command during the flight. With a ground tracker system, it is capable of providing continuous imagery of the on-board camera irrespective of the attitude of the aircraft. This have many applications ranging from defense to disaster management.

The work to date has produced prototypes of fixed, rotor and flapping-wing MAVs with further work needed on a range of enabling technologies including navigation, power supply and launch and recovery.

Related government-funded research that could be applied to MAVs includes India's National Aerospace Laboratories' work on DC micromotors, lithium ion batteries, fuel cells and microwave energy sources. It is also working on microturbines that could be used for MAVs as small as 150 millimetres across. Research on fluid flow through microchannels at the Indian Space Research Organization was also identified as having MAV applications.

Indian scientists from Defence Research and Development Organization (DRDO) are presently developing Micro Air Vehicles (MAV) for varied defence applications such as surveillance and disaster management. These MAVs are made of a unique mix of material and are not easily detectable by radars. The research and development on the MAVs are promoted by the National Design and Research Forum (NDRF), with support of Aeronautical Research and Development Board (AR&DB), Defence Research and Development Organization (DRDO), Council for Scientific and Industrial Research (CSIR) and various private groups.

Since the MAVs are not entirely made of metal and carbon fibres, they are hard to intercept by radars and sensors. These MAVs have a range of two to five kilometers and they fly some 100 to 200 metres above the ground for 30-40 minutes and capture images. MAV may be of the order of $16,000 and the sensors alone will take up 30 to 40 per cent of the cost. As of now, the MAV programme has become a national research initiative and a proposal of $19.6 million has been submitted to the Government to approach the project in an integrated manner. Demonstration of MAVs designed by ADE has been done to many potential user agencies, like National Disaster Management Agency, the Indo-Tibetan Border Police, police forces of a few State governments and the Central Reserve Police Force. A package of four MAVs with ground control costs around Rs. 8 lakh.

MICRO AIR VEHICLE AS A COMMUNICATION DEVICE

MAV as small as the objects which can be handled by an average sized human hand are created and used for varied real life situations. These objects in the size of birds, can discreetly peep into windows standing on a branch of a tree. These unmanned micro air vehicles are already used in the corporate campuses as surveillance to prevent the involvement of third party. MAVs have can find their way to the media in a few years. Instead of using helicopters to get the footage of large crowds at election time, these air vehicles can be used to transfer real-time footage of the news.

In the coming years, MAVs have greater probability of being the eyes and ears of the nation's civil, paramilitary as well as military affairs, and are needed in large numbers. The Services have squadrons of the bigger Nishant UAVs or drones, as also the imported Israeli Herons and Searchers.

There is a chance that the history would have been significantly different only if the MAVs were in use at the time of attack on Mumbai in the year 2008. The exact locations of the terrorists in the hotel would be easier to find and also accurate and appropriate action could have been taken to avoid the situation hours before it was worsen. Remote sensing satellites were just right for static land imageries. But for instant and dynamic information during conflicts and counter-terrorism events, the micro air vehicles would be game changing. A five-year National Programme on Micro Air Vehicles which is sponsored by Aeronautical Development Establishment(ADE), Defence Research and Development Organisation and the Department of Science and Technology is in progression at present. The National Design and Research Forum is also involved in packing stealth, silence and versatility into these micro air vehicles.

MICRO AIR VEHICLE AND DISASTER MANAGEMENT

Disasters are of various types consisting of both manmade and natural causes. Managing the disasters is a very important aspect to control the wastage of materials and to reduce the loss of human lives. National Disaster Management Authority is working together with ADE to design MAV that are unmanned to use in disaster issues. Disasters are of various types and each of these can be minimized with the help of micro air vehicle designed separately for each type of disaster. When fires occur, it is better to send in remotely piloted or autonomous micro air vehicles to identify the source of fire and eradicate it to minimize the loss. For these purposes, different types of materials are to be used so as to resist the heat caused due to fire and also the material has to be lightweight as flying objects need flexibility and mobility. Compolite-LF is a lightweight material composed by compression moulding of a web of glass fibre, hollow micro-spheres and thermosetting resin, showing outstanding properties of mechanical strength and fire resistance. This can be used in MAVs specifically designed for fire eradication. MAV can also be used in flood control by using heavy material to fix down in the flood areas to control the flow. Also, MAV can be used to identify the locations of victims at different locations and transmitting this data to a control centre. This will be more helpful if the MAV is autonomous. MAV can also be designed to sense chemical leaks. For instance, if there is a leak of any chemical in a designated industrial area, patrolling MAV can immediately send an indication to nearby emergency centre and necessary precautions can be taken. Designs are presently being explored in this specific area. MAV can be adopted in search and rescue missions and in earthquake situations to locate victims in areas difficult to search for a human. Autonomous MAV are designed for these purposes by University of Pennsylvania and are successful in achieving accuracy and effectiveness.

In the past few years, tsunami has occurred at various different geographical regions. It is very important to know its occurrence beforehand to minimize the loss. Recently, it had affected Japan and crippled a nuclear plant which spewed radiation. Micro Air vehicles can be used in this very situations. These can help the rescue team to design contingency plans at the right time. India has very few nuclear plants at present, but plan to expand in the future. Safety features are very important in case of occurrence of a disaster. Humans are affected by the radiation, but it is not the same with the MAVs. The MAVs can be used at that time to pass through inaccessible areas and transmit images and videos to central control system to manage the disaster and minimize human life loss.

MAV AND ITS COMMERCIAL USAGE

Commercially, Micro Air Vehicle has its own set of advantages. Camera can be integrated into the MAV to get video, audio and imagery data from the device. For real-time access, wireless communication can be set up in the MAV to transmit the data in real-time. This can be used by television channels for instant news update. This integrated device can be used to get imagery data which will be difficult to get otherwise. Its main advantage over the other options is its cost-effectiveness. Due to its reduction in size, cost of production is reduced and so is more effective. MAV also found its way into agriculture. India being an agricultural country, can find many applications for the current MAV designs. Harvesting crops is a long process and MAVs can be used to maximize the utility. MAV can be used for hyperspectral imaging so that appropriate process for management of crops can be used. Generally, satellite imagery can be used, but the height from which the images are taken are disadvantageous as they can not be used optimally. MAV flying at a low height are advantageous as they give a more clear imagery. They have low image acquisition costs per image and lower aircraft maintenance due to their small sizes and less fuel usage. Also, man power is reduced as they are unmanned and so costs are even minimized. But, there are limitations too as the initial setup costs are more and also due to the training that has to be given to non-sufficient knowledgeable farmers.

MAV IN DEFENCE

Micro Air Vehicles can be used at various levels in India for Defence purposes. Use of MAV in surveillance is an optimal option. The size of MAV is very much suitable for surveillance. The prolonged usage of devices in surveillance needs optimal power utilization. MAVs suitable for this sole purpose are designed and used by many corporate companies. Communication relay can also be done by MAV designed for communication purposes. Reconnaissance missions are carried out by MAVs for military purposes. Unmanned Aerial Vehicles, used by US forces to track down Taliban militants, successfully flew over over the dense forests of Bastar in the first trial run for anti-naxal operations. The trials, which assumed urgency after the Dantewada massacre in which seventy six security personnel were killed by Maoists, were aimed at generating real-time intelligence information to help ground forces in any offensive. The first trial involved an American UAV. The decision to have UAV flights was taken by the Union home ministry after the April 6, 2010 attack and their field trials were ordered immediately. An UAV of US' Honeywell, whose pilotless planes are used successfully by allied forces in the hunt for targets in war-hit Afghanistan and Iraq, flew during the night trial. The compact UAV, weighing nearly 10 kgs, was put through the rugged terrains of the hills overlooking Kanker after its take off from Counter Terrorism and Jungle Warfare College in Kanker. The UAV was checked for providing thermal images of any movement on the ground, detection of Improvised Explosive Devices (IED) and ammunition dumps. The UAV known as T-MAV (Micro Air Vehicle) is a compact machine manufactured by Honeywell. It has been useful to the US forces in tracking down Taliban militants in high mountain passes and dense Waziristan area of Pakistan. The UAVs are urgently required as the forces engaged in anti-naxal operations need real-time information to achieve greater success. It can go up to a height of 10,000 feet, fly at a speed of 70 kms per hour and can provide 240 minutes of sensor imagery to the ground station. With intelligence gathering still a problem in Naxal areas, the UAVs are expected to help in gathering advanced reconnaissance and situational awareness functions would be critical in protection of security personnel.

The police force in India, particularly Chandigarh Police Department has taken a bold and much needed step by adopting an Unmanned Micro Air Vehicle called Golden Hawk developed by ADE. It can be useful in maintaining law and order during important visits and rallies. Also, this is very important in handling disasters and reduce human life loss. The Golden Hawk can fly up to the height of 100 metres and can do continuous recording for 30 minutes. The Golden Hawk runs on battery and can fly at a stretch for 30 minutes. Night vision cameras are also installed on it. Special training is provided to Chandigarh police officials to operate the device. As the MAV is used for long periods of time, it will be sensible to design devices that can operate on solar power and can switch the source to battery whenever needed. Solar power usage is more sensible option in India in accordance to the climatic conditions.

DESIGN AT THE NEXT LEVEL

Shaping up in this realm are the flapping wing MAVs, tiny helicopters, bio-inspired or bio-mimicking intelligent birds, butterflies and insects. Work is being done on models that can sense chemicals, gas leaks, explosives or nuclear, biological or chemical threats. While applications are being worked on, the time of usage is equally important. Any application needs more time of flight for better usage and applicability. Various designs are being proposed to prolong the time of flight. One design can be the usage of solar power and other is to just use the aerodynamics. Another way is to switch between multiple power sources to increase the time of flight. A default way of source can be battery and the power source can be switched between solar power source and just use the aerodynamics, if possible. This design can increase the time of flight for a considerable amount.

REFERENCES

[1]   http://www.ndrfindia.org

[2]   Honeywell Wins FAA Approval for MAV, Flying Magazine, Vol. 135., No. 5, May 2008

[3]   Benchergui, Dyna, "The Year in Review: Aircraft Design," Aerospace America, December 2009, Volume 47, Number 11, American Institute of Aeronautics and Astronautics

[4]   http://timesofindia.indiatimes.com

[5]   Encyclopedia of Astrobiology, Astronomy, and Space Flight: Kettering Bug

[6]   http://www.thehindu.com

[7]   http://www.defence.pk

WIRELESS AUDIO AND VIDEO NOISE TRANSMISSION FOR MICRO AIR VEHICLES

Mr. V. Ravi Kumar, Dr. K. Ravindra, D. Tagore

ABSTRACT

Onboard processing of audio and video is currently outside the capabilities of power limited micro air vehicles (MAVs), which forces researchers to transmit video and telemetry to a ground control station for capture and offline processing. Unfortunately, wireless AV transmission can introduce structured noise, which can corrupt image processing algorithms if the noisy frames are not identified and rejected from further processing. In this paper, we describe the design and evaluation of a supervised learning based classifier for labeling frames of AV "noisy" or "clean". We pose the classification problem as one of texture classification in AV, where texture is represented using a feature set including informative statistics of steerable pyramid coefficients and the principal components of the chromatic histogram. The main contribution of this paper is the definition of this feature set as determined from analysis of noisy AV. We evaluate nine different binary classifiers using cross validation on a MAV AV training set, with additional evaluation on two validation sets collected from different times, days and locations.

INTRODUCTION

Micro Air Vehicles (MAVs) are small, lightweight, and autonomous aerial systems that can fit in a backpack, and promise to enable on-demand intelligence, surveillance and reconnaissance tasks in a near-earth environment. Such tasks for military operations may include: "over the hill" recon naissance, "perch and stare" surveillance, covert imaging, biological and chemical agent detection, tagging and targeting, precision strike missions and bomb impact indication. Civil and commercial applications for MAVs are not as well developed, although potential applications are extremely broad in scope.

A common problem in wireless video and audio transmission on the MAV, the synchronization between audio and video data has to be preserved in order to offer the best perceptual quality. However, the timestamp-based synchronization methodology in MPEG/System layer suffers data packet loss during transmission, and the resulted absence of synchronization will be unacceptable for users. In order to address this issue, an information hiding based synchronization methodology has been proposed. In this methodology, audio data is embedded within the corresponding video frames by means of high bit rate information hiding techniques. At the receiver, the embedded audio data is extracted and played with the host video frames to achieve the synchronization. With this approach, significant advantages have been obtained: (1) the communication channel for audio data transmission is avoided; (2) the synchronization between audio and video data is robust to packet loss.

Processing and Transmitting Section

The audio and video synchronization, DCT based high bit rate hiding information algorithm is proposed. At the receiver, the embedded audio data is extracted and played with the host video frames to achieve the synchronization. Some significant advantages in this approach is (a) the communication channel for audio data transmission is avoided; (b) The synchronization between audio and video data is robust to packet loss (c) Complex task of multiplexing, de-multiplexing and synchronization in MPEG system have been avoided.

The video processing with acceleration sensitive function was designed. The system can record and store the analog video to H.264 bit stream files when acceleration exceeds the limitation value. The acceleration sensitive algorithm has been derived.

Proposed System

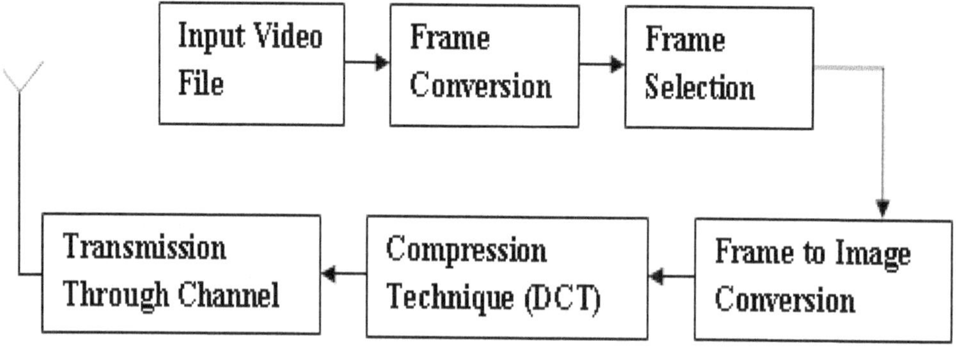

In this paper, the above proposed block diagram explains that based on the surveillance system this recorder continuously stores the video streams. The stored video streams are converted into frames. From that sampled frames are converted into images by DCT compression technique. Then transmission takes place through the channel.

Methodology

The proposed work involves the simulation of the images to frame conversion. From that sampled frames are took and analyzed signal to noise ratio (SNR) Vs bit error rate (BER) of first three frames as well as Bits/Sec Vs PSNR for Frames.

Simulation Results

The following figures shows that comparison of signal to noise ratio Vs bit error rate and bits/sec Vs peak signal noise ratio(PSNR).

Bit-Error-Rate:

Each time run a bit-error-rate simulation, then transmit and receive a fixed number of bits. We determine how many of the received bits are in error, then compute the bit-error-rate as the number of bit errors divided by the total number of bits in the transmitted signal.

Using Matlab, we compute the bit-error-rate, 'BER', as:

BER = Total no of Bit Error/length(tx),

   where 'tx' is the transmitted bit vector.

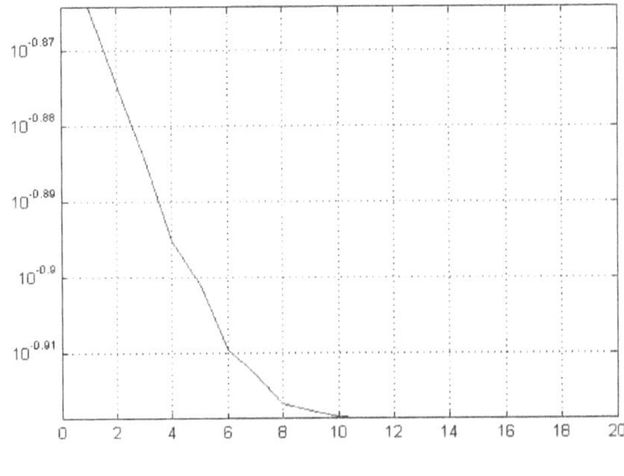

SNR Vs BER for Frame -1

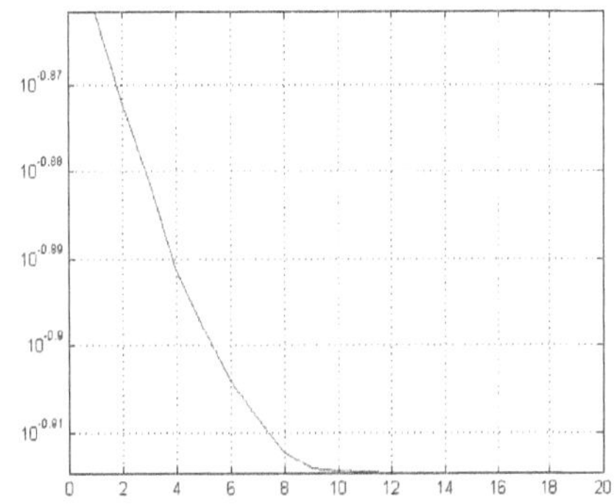

SNR Vs BER for Frame -2

Function of PSNR: To find PSNR between two images

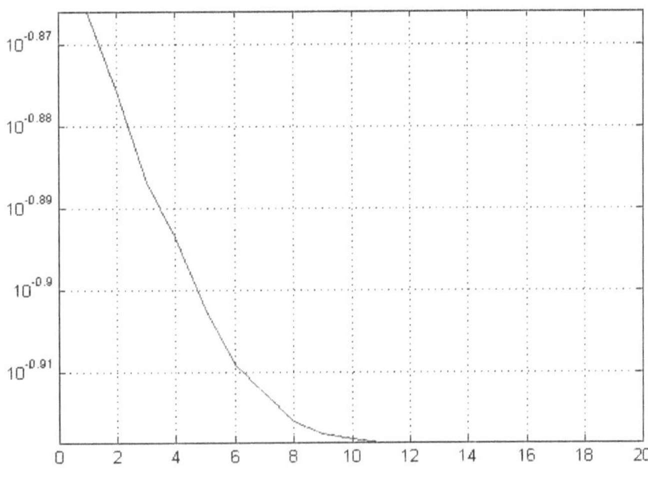

SNR Vs BER for Frame -3

Function of PSNR: To find PSNR between two images
PSNR = 20 * log10 (b/rms)

Where b is the largest possible value of the signal (typically 255 or 1), and rms is the root mean square difference between two images. The PSNR is given in decibel units (dB), which measure the ratio of the peak signal and the difference between two images

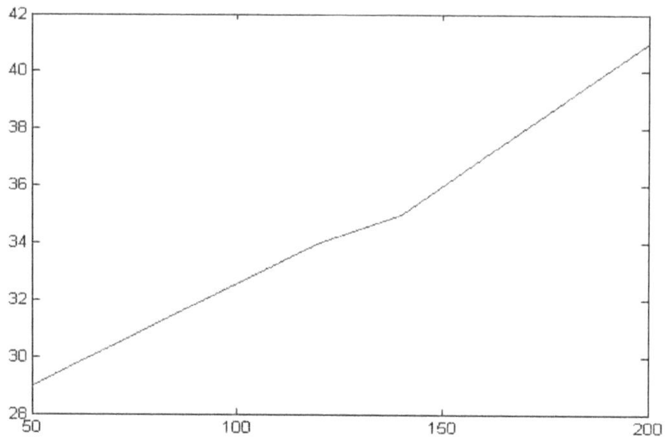

Bits/Sec Vs PSNR for Frames

Conclusion

This paper summarizes how video and audio synchronization file is converted into frames and what is the bit error and signal error rate is analyzed. Finally a prototype has been designed and developed which enables continuous video and audio transmission process from source to target.

References

[1] R.J. Wood et. al, "An autonomous palm-sized gliding micro air vehicle: Design, fabrication, and results of a fully integrated centimeter-scale mav," IEEE Robotics and Automation Magazine, vol. 4, no. 2, pp. 82–91, June 2007.

[2] Randal Beard et. al, "Autonomous vehicle technologies for small fixed wing uavs," AIAA Journal of Aerospace Computing, Information, and Communication, vol. 2, no. 1, pp. 92–108, January 2005.

[3] Paul Y. Oh William E. Green and Geoffrey Barrows, "Flying insect inspired vision for autonomous aerial robot maneuvers in near-earth environments," in IEEE International Conference on Robotics and Automation (ICRA), New Orleans, LA, May 2004, vol. 3, pp. 2347–2352.

MICRO AIR VEHICLE NAVIGATIONAL SYSTEM

C. R. Prakash, Vineet Chandan, J. Venkatesh
Centre for Spatial Information Technology
IST, JNTUH, Hyderabad-500085

ABSTRACT

MAV is an airborne vehicle with control from within or outside its system. This vehicle has to be of very minimal weight and at the same time able to carry certain essential electronics in it as a payload, along with self weight. The MAV Navigational system has to be Vision based or GPS controlled. The Flight path has to be well planned on a predefined path taking into account of the Turbulence, crab due to the wind effects and other atmospheric interactions. For telecontrol, an onboard miniaturized GPS is added to the existing computer system with manual Ground/Base Control depending on the end use/applications. GPS signal is to tell where they are, where to go, what to look for and where to land. MAV that uses a camera pointed at the ground to navigate and pick landing spots, with uncanny navigation and real-time mapping capabilities being used, zipping through indoor and outdoor spaces, running reconnaissance missions. The current scenario revealed that a size of a MAV is of about 100 to 150 mm in length and weighs about 8 to 15gms. With such low values for dimension & weight, it is very difficult to design a sensor system and data recording/ transmission for multi use, therefore nanotechnology sensor is to be designed and developed, which can be fitted in the cockpit /belly of the vehicle. The MAV using the camera and onboard Hardware/Software will enable to build 3D/Stereo maps of its surroundings. It can be planned to fit the sensor system to an object of size of an insect/ cockroach. It can also avoid obstacles and detect surfaces above a predetermined height as possible landing zones. Once it selects a place to put down, it maps the site's dimensions, moves overhead and lands.

Keywords: GPS, MAV Navigational System, Real-time Mapping

MATERIALS, DEVICES, MECHANISMS & CONTROLS OF NANO AIR VEHICLES –CHALLENGES AHEAD

B.R.R. Narasimham
Research fellow, Innovative Technologies Centre, JNTUH, Hyderabad -500072.

V. Brahmajirao
Senior Professor, Department of Nanoscience & Technology, School of Bio-Technology,
MGNIRSA, A unit of D. Swaminathan Research Foundation, Hyderabad -500072, A.P.

V. Kamakshi Prasad
Director, Innovative Technologies Centre, JNTUH, Hyderabad -500072.

ABSTRACT

Micro- and Nano air vehicles are defined as "extremely small and ultra-lightweight air vehicle systems" with a maximum wingspan length of a few cm and a weight less than 20 grams. Obviously such systems demand the need of Nano materials and devices in the development of the modules, mechanisms, and controls to achieve the objectives. In the first part of the Review we provide Rudiments of the current state of the art and attempt to identify the challenges of Nano Air Vehicles design [NAVD] and fabrication.The problem on hand about the challenges ahead of the technologist can be broadly classified into three parts, namely (1) Materials (2) Mechanisms and Robotics and Assembly (3) Complexities in assembly, monitoring and application. In the second part tasks that the Materials & Electronics Engineers should be addressed in the NAVD and Fabrication of the Latest Nano Electronic Gadgets, like Nano composites, Nano materials to develop Nano gates& Nano processing systems relevant to the NAVD are discussed.In the Third part Robotics and related aspects about the software needs, like Artificial Intelligence, and neural networks, essential to overcome several hurdles thatwe come across in the mechanisms about mobility and task performance of the NAVD are presented.An aviating device essentially should be assembled in strict obedience to fundamental Physical principles enunciated in Motions of Translation, Rotation, and Oscillation. For example: An NAVD embarks upon wings, which consumes a lion's share of our attention. Several Differentconfigurations such as (1) fixed wings,(2) rotary wings, and (3)flapping wings, each of which has its own story, are in a process of development, around them. In defence it is essential to provide mobility Special attention is given to rotary-wing vehicles (helicopter concept). The main advantages and drawbacks for each type are identified and several features are discussed.Including a review of their main structures, such as the airframe, energy storage, controls, and communications systems. In addition, a review of relevant sensors is also included. Examples of existing and future systems are also included. Micro- and nano-vehicles with rotary wings and rechargeable batteries are dominating in the latest development of Avionics. Some aspects in this line of approach are touched upon.

INTRODUCTION

At the turn of 20th century, following the suggestion of Feynman, 'Nano' word came to practical existence .Nano structures were identified, Quntumdots turned another leaf in to the story of electronic gadgets. Several approaches for the manufacture of Nano-materials areavailable namely,

(1) physical methods (2)chemical methods (3)Biological methods and(4) Mechanical methods

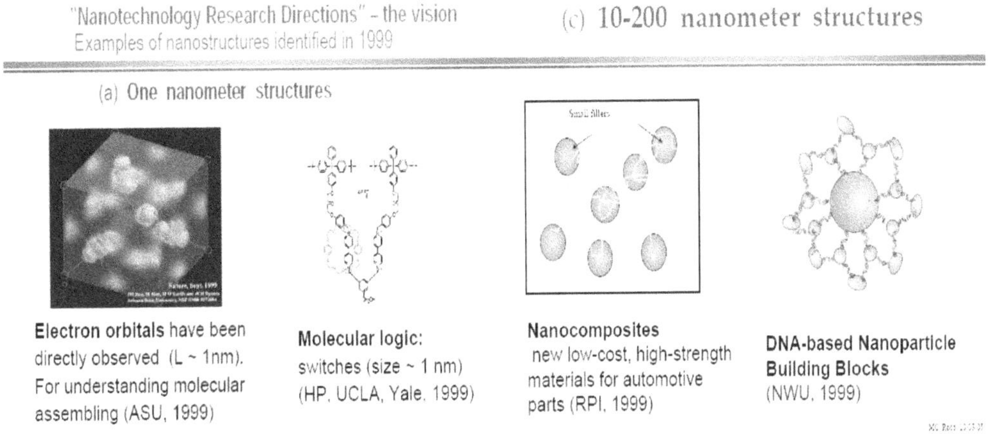

Fig: 1. Nanostructures , Molecular orbitals ,Nano composites,
DNA based Nano particle building blocks

Precision tools like Atomic force microscope make it possible to conceive electronic gadgets of very high precision. Present technology facilitates the production of nanomaterial with conspicuous characteristics' which can be achieved at the finest requirement that suits the need. With the advent of Quantum dots the story became still more interesting.

A **quantum dot** is a portion of matter (e.g., semiconductor) whose excitons are confined in all three spatial dimensions. Consequently, such materials have electronic properties intermediate between those of bulk semiconductors and those of discrete molecules.[1.2.3] They were discovered at the beginning of the 1980s by Alexei Ekimov[4] in a glass matrix and by Louis E. Brus in colloidal solutions. The term "quantum dot" was coined by Mark Reed.[5] . Researchers have established vital technologies for the usage of quantum dots in transistors, solar cells, LEDs, and diode lasers. They have also investigated quantum dots as agents for medical imaging and hope to use them as qubits in quantum computing.

There are several ways to confine excitons in semiconductors, resulting in different methods to produce quantum dots. In general, quantum wires, wells and dots are grown by advanced epitaxial techniques in nanocrystals produced by chemical methods or by ion implantation, or in nanodevices made by state-of-the-art lithographic techniques.[6].Quantum dot manufacturing in bulk relies on a process called "high temperature dual injection" which has been scaled by multiple companies for commercial applications that require large quantities (100's of kgs to tonnes) of quantum dots. This is a reproducible production method that can be applied to a wide range of quantum dot sizes and compositions.

In robotics where the processing devices size is to be reduced by a very large scale ,quantum dots become vital, especially for the development of quantum computers . In electronic applications they have been proven to operate like a single electron transistor and show the Coulomb blockade effect. Quantum dots have also been suggested as implementations of qubits for quantum information processing. In the developed countries rapid strides are ahead and Nano robotics is a fast developing field contributing for the growth of gadgets like Micro air vehicles , Nano air vehicles   and Pico air vehicles.

MICRO AIR VEHICLES

Micro- and Nano air vehicles are defined as "extremely small and ultra-lightweight air vehicle systems" with a maximum wingspan length of 15 cm and a weight less than 20 grams. Both of them put together are referred as *Air Vehicle systems (AVS).*Development of insect-sized aircrafts called Pico Air Vehicles (PAV) is reportedly expected in the very near future. . Potential military use is one of the driving factors, although MAVs are also being used commercially and in scientific, police and mapping applications. The current state of the art is fast developing and poses several challenges of design and fabrication.The design of the wing, the robotics, the ability to Trans communicate with the MAV during the flights, are some of the main tasks at hand. If a rotary wing system is incorporated, it would provide a facility for rechargeable batteries can be accommodated. This in its turn enhances the flight time from the present one hour. Fuel cells and ultra-capacitors are promising alternative energy supply technologies for the future. Technology improvements, mainly based on micro- and nanotechnologies, are expected to continue in an evolutionary way to improve the capabilities of future micro- and Nano air vehicles, giving improved flight times and payload capabilities. AVS are capable to perform both indoor missions and outdoor missions in very challenging environments. The main applications are Intelligence, Surveillance, and Reconnaissance (ISR) missions.

These systems can provide a rapid overview [9] in the area around the personnel, without exposing them to danger.Infrared (IR) cameras can give detailed images even in the darkness. Furthermore, NAVs, thanks to their reduced dimensions, are perfect for reconnaissance inside buildings, providing a very useful tactical advantage. They can carry specific sensors [10] such as gas, radiation or other sensors used to locate biological, nuclear, chemical, or other threats. SUCH SYSTEMS can, fly inside toxic clouds and transmit data or even bring samples back to the base station, and, thus, provide vital information on the composition and extent of gaseous clouds and improve the assessment of danger.

The police and the fire brigade find good use ofthe capability of indoor flights for inspecting unsafe or collapsed buildings [11] in order to search for survivors or simply do a safety check of the building structure.

Figure2 : The Hornet 2-b (Prox Dynamics) MAV, complete with camera and video transmitter [Prox Dynamics, http://www.proxdynamics.com.].

TRANS -COMMUNICATION WITH THE MAV & ROBOTICS

Control signals are mainly transmitted from the ground station to the vehicle while the data is sent from the vehicle to the user. An example of control communication systems on board is given by H.Wu et.al, [8], where they developed a home-made RF transmitter for use onboard an MAV. With a weight of8 g, it could transmit at 56mW. The transmitter operated at a frequency range between 1.18 and 1.45GHz. Furthermore, a microdemodulator operating at 50MHz and weighing 5.4 g was used at the receiver end [8].

Substantial progress has been made recently towards designing, building and test-flying remotely piloted Micro Air Vehicles (MAVs) and small UAVs. We seek to complement this progress in overcoming the aerodynamic obstacles to flight at very small scales with a vision-guided flight stability and autonomy system, based on a robust horizon detection algorithm. the use of computer vision for MAV autonomy, arguing that given current sensor technology, vision may be the only practical approach to the problem, is the approach of Scott M. Ettinger et.al,,for which *'statistical vision-based horizon detection algorithm'*, is used by them .Important obstacles for such a computer vision are (1) no visibility of the horizon,(2) Horizon estimation errors, due to external factors such as video transmission noise.

MAV flight stability and control presents some difficult challenges. The low moments of inertia of MAVs make them vulnerable to rapid angular accelerations; a problem further complicated by the fact that aerodynamic damping of angular rates decreases with a reduction in wingspan. Another potential source of instability for MAVs is the relative magnitudes of wind gusts, which are much higher at the MAV scale than for larger aircraft. In fact, wind gusts can typically be equal to or greater than the forward airspeed of the MAV itself. Thus, an average wind gust can immediately affect a dramatic change in the vehicle's flight path.Suitable feedback control device is mandatory to achieve   self-stabilized flight.

ORNITHOPTERS ARE MAV S DEVELOPED FROM BIRD AND INSECTS

Birds, the biological counterpart of mechanical MAVs, can offer some important insights into how one may best be able to overcome these problems. In studying the nervous system of birds, one basic observation holds true for virtually all of the thousands of different bird species: Birds rely heavily on sharp eyes and vision to guide almost every aspect of their behaviour [12, 13, 14, 15, and 16]. Biological systems, while forceful evidence of the importance of vision in flight, do not, however, in and of themselves warrant a computer-vision based approach to MAV autonomy.

REMOTE SURVEILLANCE APPLICATION FOR GPS GUIDED, AUTONOMOUS MAV AND SMALL UAV

Onboard of a typical 24 inch UAV With the antenna and related electronics and transmitter, the GPS subsystem weighs approximately 30 grams, which will allow the incorporation of this GPS on platforms with wingspans of approximately 12 inches or less. Once basic guidance, control and navigation are achieved one of the most important missions for these systems, namely, remote surveillance is fulfilled (see Figure 10).

On the computer vision front, horizon tracking, while very useful, does not make full use of this important sensor. Currently, multi-resolution methods for detecting man-made and other artificial structures on the ground [17]are in the stage of development. Structure-from-motion (SFM) algorithms for recovering the 3d scene structure in the vicinity of the UAV during flight are also in the process of development. [12] This will eventually enable more advanced mission profiles that include flying in complex 3d environments.

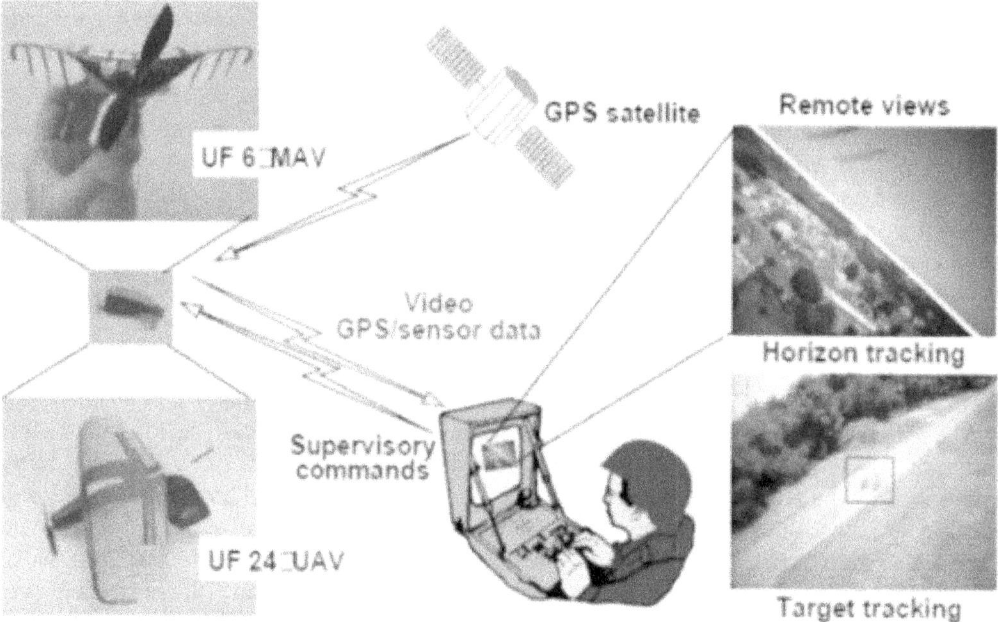

Fig 3 MAV based remote surveillance and sensor data collection

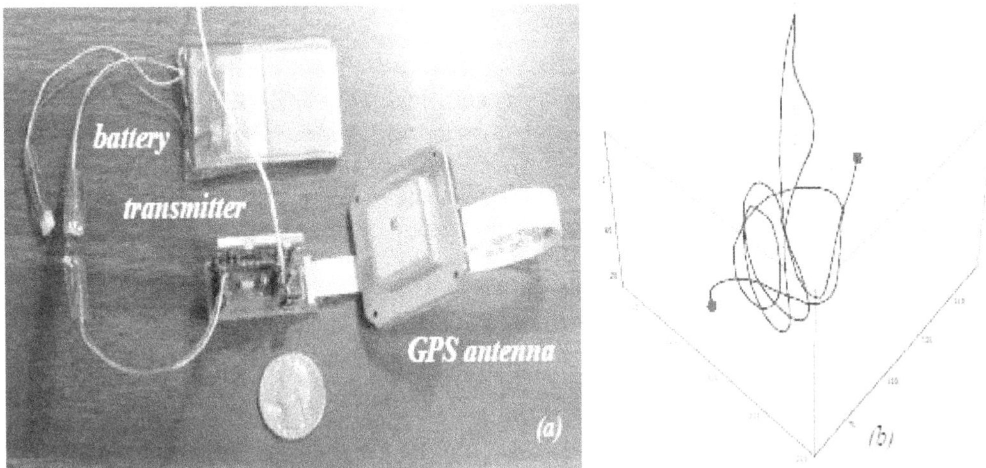

Fig 4 (a) On board GPS System   Fig 4(b) Data from remotely test flight

REYNOLDS NUMBER AND SIMULATION STUDIES IN AIR-FOIL DESIGN

The design of MAVs essentially is dependent on a thorough understanding of the flow physics of very small aircraft flying at low speeds. Complete analytical methods are yet unavailable [18, 19] .So the design optimization approach and the expensive computational numerical methods are the practical alternatives to the problem. The emergence of remotely piloted vehicles for military surveillance missions during the late seventies led to an increase in research of lower Reynolds numbers aerodynamics (in the range below 500,000). Comprehensive literature surveys of this area of research can be found in publications of Mueller [18] and Lissaman [19].Miguel R. Visbal et.al, [20] highlighted results derived from the application of a high-fidelity simulation technique to the analysis of low-Reynolds-number transitional flows over moving and flexible canonical configurations motivated by small natural and man-made flyers. Their work addressed three separate fluid dynamic phenomena relevant to small fliers, including: laminar separation and transition over a stationary air foil, transition effects on the dynamic stall vortex generated by a plunging air foil, and the effect of flexibility on the flow structure above a membrane air foil. The specific cases were also selected to permit comparison with available experimental measurements.

P.C.Babu Salapakkam [21] in his doctoral thesis discussed work towards three fundamental turbulent flow problems: (a) passive scalar transport in turbulent flows, (b) development of the simulation capability for bubble transport and, (c) aerodynamics of flow around cactus shaped cylinder. Jesse Wells [22] in his thesis quantified the effects of RANS turbulence modelling on the resolution of free shear vertical flows. The simulation of aerodynamic wing-tip vortices is used as a test bed. Alejandra Uranga [23] ,in his doctoral work predicted the formation of laminar separation bubbles at low Reynolds numbers and related transition to turbulence . A preliminary Transition model suitable for such flows is introduced and its underlying concept was proven valid.

DETAILS OF MAVS SO FAR DEVELOPED

Expertise spans all differing types of MAVs: namely (1)fixed wing models,(2) insect-like (flapping wing) models, and (3) rotary wing models.Honeywell T-HAWK MAV has particular design and deployment success in rotary (or first ducted fan) models. The Honeywell RQ-16A T-Hawk (for "Tarantula hawk", a wasp species) is a ducted fanVTOLmicroUAV. Developed by Honeywell, it is suitable for backpack deployment and single-person operation.In 2007, the T-HAWK MAV was deployed to Iraq where it assisted US troops by identifying improvised explosive devices (IEDs) from the air. The program has been an overwhelming success. The T-HAWK MAV is in evaluation by the Miami-Dade police department for civilian applications. Highly specialized and integrated design approach required by the unique aerodynamics of ducted fan vehicles. Critical high fidelity models of the vehicle aerodynamics for flight control developmentFeaturing vertical take-off and landing, the lightweight and portable T-Hawk is a combat-proven unmanned micro air vehicle that can be quickly deployed. T-Hawk is easy to assemble and can be airborne within 10 minutes. It is simple to fly with minimal training. With unique hover and stare capability, T-Hawk supports advanced intelligence, surveillance, and reconnaissance (ISR) with real time video documentation.The Honeywell Honeywell RQ-16T-HAWKMAV andKestrel are developed as a part of DARPA Advanced Concept Technology Demonstrator (ACTD) in OAV/MAV programs

Fig 8 (a) Honeywell RQ-16A T-HawkFig8(b) The Honeywell Kestrel

This aircraft, shown in Figure , is capable of carrying a multiple camera payloads at density altitudes up to and over 10,000 feet for 45 minute flights [25]. Since 2007, different versions of this basic aircraft have gone through thousands of test flights, demonstrations, operational exercises, as well as use in theater with different US military customers. More recently,[26] the T-HawkTM was used in the spring of (2011) to provide detailed, close up video imaging and radiation sensing of damaged nuclear power facilities in Japan (the Fukushima Daiichi facility).

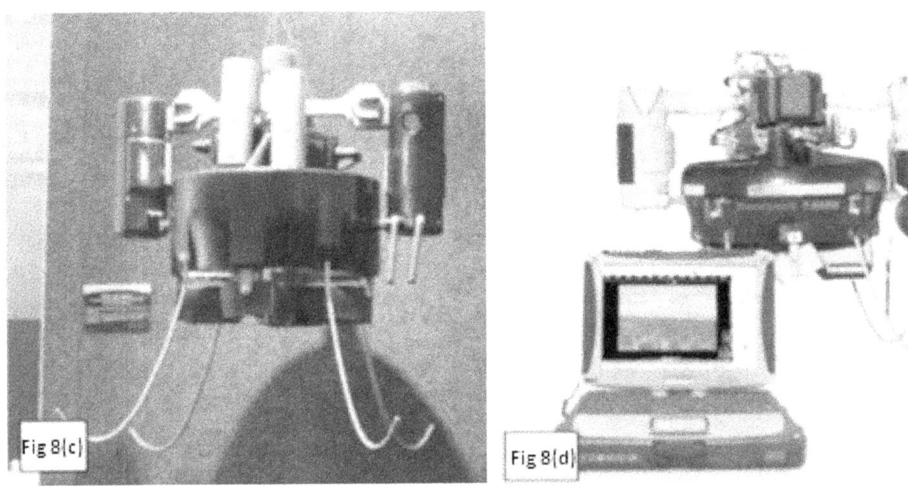

Fig 8 (a) The forerunner of the Honeywell T-Hawk™ is the Honeywell Micro Air Vehicle (MAV) ducted fan UAV (from AUVSI North America Exhibition- 2005), and Fig 8(b) current HoneywellT-Hawk™ vehicle with ground station computer. A typical system includes two vehicles, one ground station, and related support equipment for vehicle operations (Photograph courtesy of Honeywell International Inc., 2010.

Fig8(e):Tiny insect size drones known as a "tactical advantage in war " by U.S military

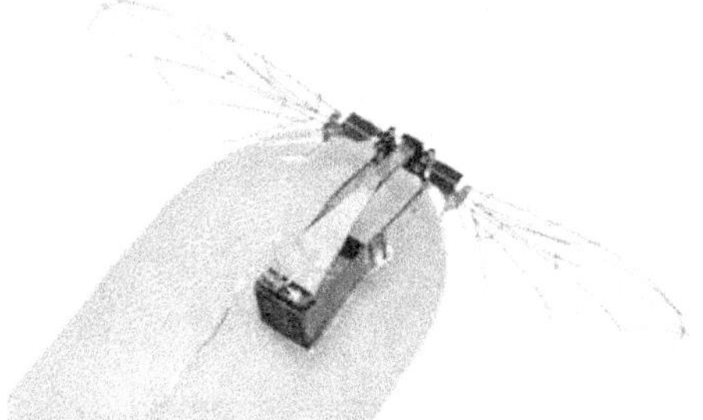

Fig8(f): Airborne Remote Controlled MAV

CONCLUSIONS

Starting from Simulation &Modelling [27], several aspects of the development of the MAVs and NAVs (and more so for PAVs) are still in a stage of development. The aerodynamics of the MAVs continue to be unsteady [28].Several facets (a few are cited) like Information Systems Technology, System Analysis and Studies, Concepts and Integration,Sensors and Electronics,Relevant Nanomaterial and technology of Nano compositesdevelopment ,Robotics and Relevant Hardware, Analysis for Unsteady Air foil and Wing Aerodynamics, studies of Flapping ,rotary and other mechanisms of wings for MAVs, NAVs and PAVs, are the essential fields open for development. No doubt sophisticated tools like 'Wind Tunnels', Simulation Hard and Software, are the prime requirements to do sensible research.

REFERENCES

1. L.E. Brus (2007). "Chemistry and Physics of Semiconductor Nanocrystals". http://www.columbia.edu/cu/chemistry/fac-bios/brus/group/pdf-files/semi_nano_website_2007.

2. D.J. Norris (1995). "Measurement and Assignment of the Size-Dependent Optical Spectrum in Cadmium Selenide (CdSe) Quantum Dots, PhD thesis, MIT". http://hdl.handle.net/1721.1/11129.

3. C.B. Murray, C.R. Kagan, M. G. Bawendi (2000). "Synthesis and Characterization of Monodisperse Nanocrystals and Close-Packed Nanocrystal Assemblies". Annual Review of Materials Research30 (1): 545–610.

4. Ekimov, A. I. &Onushchenko, A. A. (1981). "Quantum size effect in three-dimensional microscopic semiconductor crystals". JETP Lett.34: 345–349.

5. Reed MA, Randall JN, Aggarwal RJ, Matyi RJ, Moore TM, Wetsel AE (1988). "Observation of discrete electronic states in a zero-dimensional semiconductor nanostructure". Phys Rev Lett60 (6): 535–537.

6. C. Delerue, M. Lannoo (2004). Nanostructures: Theory and Modelling. Springer. p. 47. ISBN3-540-20694-9.

7. U. Yearbook, UAS: The Global Perspective, vol. 164, UAS, Yearbook, 7th edition, 2009/2010.

8. H.Wu, D. Sun, and Z. Zhou, "Micro air vehicle: configuration, analysis, fabrication, and test," IEEE/ASME Transactions on Mechatronics, vol. 9, no. 1, pp. 108–117, 2004.

9. R. J. Bachmann, "Biologically inspired mechanisms facilitating multimodal locomotion for areal micro-robot," in Proceedings of the 24th International Unmanned Air Vehicles Conference, Bristol, UK, 2009.

10. S. V. Serokhvostov, "Ways and technologies required for MAV, miniaturization," in Proceedings of the European Micro Air Vehicle Conference (EMAV '08), Braunschweig, Germany, July2008

11. D. H. Paulsen, "Nano UAS- an upcoming reality," in Proceedings of the 24th International Unmanned Air Vehicles Conference, Bristol, UK, 2009.

12. S. M. Ettinger, M. C. Nechyba, P. G. Ifju, and M. Waszak. Vision-guided flight stability and control for Micro Air Vehicles. In Proc. IEEE Int. Conf. on Intelligent Robots and Systems, volume 3, pages 2134–40, Laussane, Switzerland, 2002.

13. P. R. Ehrlich, D. S. Dobkin, and D. Wheye. Adaptions for flight. World Wide Web, http://www.stanfordalumni.org/birdsite/text/essays/Adaptions.html, June 2001.

14. P. R. Ehrlich, S. Dobkin, and D. Wheye. Flying in vee formation. World Wide Web, http://www.stanfordalumni.org/birdsite/text/essays/Flying_in_Vee.html, June 2001.

15. R. Fox, S. W. Lehmkule, and D. H. Westendorf. Falcon visual acuity. Science, 192:263–5, 1976.

16. Northern PrairieWildlife Research Center. Migration of birds: Orientation and avigation. World Wide Web, http://www.npwrc.usgs.gov/resource/othrdata/migration/ori.htm, June 2001.

17. S. Todorovic and M. C. Nechyba. Multiresolution linear discriminant analysis: efficient extraction of geometrical structures in images. In Proc. IEEE Int. Conference on Image Processing, volume 1, pages 1029–32, September 2003.

18. Mueller, T. J., "Low Reynolds Number Vehicles", AGARDograph No. 288, 1985.

19. Lissaman, P. B. S., "Low-Reynolds-Number Airfoils", Annual Review of FluidMechanics, Vol. 15, 1983, pp. 223-239.

20. Miguel R. Visbal et.al,(2009), AFRL-RB-WP-TP-2010-3026, 'High Fidelity Simulations of ......Flexible Airfoils at low Reynolds numbers', AFRL-RB-WP-TP-2010-3026 Exp Fluids (2009) 46:903–922

21. P.C.Babu Salapakkam(2007),Ph.D.,Thesis : 'Simulation &Modeling of Three Turbulent Flow Problems', submitted to THE UNIVERSITY OF MINNESOTA,U.S.A.

22. Jesse Wells(2009),Thesis :'Effects of Turbulence Modeling on RANS Simulations of TipVortices' , submitted to Virginia Polytechnic Institute and State University, Blacksburg, Virginia,U.S.A.

23. Alejandra Uranga (2011), Ph.D.,Thesis: "Investigation of transition to turbulence......." submitted to submitted to Massachusets Institute of technology(MIT),

24. NATO- TECHNICAL REPORT (2010), on 'Unsteady Aerodynamics for Micro Air Vehicles', RESEARCH AND TECHNOLOGY ORGANISATION, TR-AVT-149.

25. T-HawkTM Data Sheet,(june2010)http://www.thawkmav.com/ downloads/T-hawk_Data_Sheet.pdf, June 2010

26. Pasztor, Andy,(2011) "A New Role for Honeywell's T-Hawk", Wall Street Journal, April 20, 2011.

27. *Pradeep C Babu Salapakkam(2007),Ph.D., Thesis entitled: 'Simulation &Modelling of ...' Submitted to the GRADUATE SCHOOL ofTHE UNIVERSITY OF MINNESOTA,U.S.A.,*

28. *Unsteady Aerodynamicsfor Micro Air Vehicles (2010) [Aérodynamique instable pourmicro-véhicules Aériens],Final Report of Task Group AVT-149, RESEARCH AND TECHNOLOGYORGANISATION, NATO.*

An ROI Based Image Compression System for SAR Imagery in Micro Unmanned Air Vehicles for Border Security

N. Udaya Kumar[1]
Professor, Department of ECE
SRKR Engineering College,
Bhimavaram
India
n_uk2007@yahoo.com

K. Padma Vasavi[2]
Associate Professor, Department of ECE
SVECW
Bhimavaram
India
Padmavasavi1973@yahoo.com

Dr. M. Madhavi Latha[3]
Professor, Department of ECE
JNTUCE,
Hyderabad,
India
mlmakkena@yahoo.com

Dr. E. V. Krishna Rao[4]
Professor and Dean, Department of ECE
Andhra Layola Institute of Engg and
Technology
Vijayawada, India
krishnaraoede@yahoo.co.in

Abstract

To reduce the intensive human involvement in the border security systems, high tech devices such as Unmanned Aerial Vehicles (UAV) are being installed by the defense and military services. Synthetic Aperture Radars (SAR) deployed on UAV play important role in warfare because, these systems can operate in any weather conditions and at any time of day or night and can generate images of ground in real time. SAR imaging payloads most often operate on a store and forward mechanism, wherein the captured images are stored and transmitted to the ground at a later time. With the increase in spatial resolution, space missions handle huge image data. So, a smart image compression system is required to transmit images over band limited channels for SAR payloads deployed on UAV. An ROI based image compression system is presented in this paper, which uses a multi-scale and multi directional analysis for automatic target identification. For achieving high compression ratios without losing the image quality, the target areas in the SAR image are encoded with more number of bits and the background area is coded with less number of bits. The improved performance of the proposed system is demonstrated by comparing with the existing ROI based compression systems.

Key words: Unmanned Aerial Vehicle. Synthetic Aperture Radar, Automatic Target Detection, ROI Based Compression

1. Introduction

Border Patrol system is a very important issue of concern about national security. The major challenge in protecting long stretches of borders is the need for intensive human involvement in guarding the area [1]. More often, security check points are setup on the international roads where all vehicle traffic is stopped to detect and stop illegal foreigners, drugs and other illicit activity. To do so, each border area is assigned with a border troop. Each border troop watches and controls a specific division of the border. The troops monitor the border according to a pre-defined route and interval. This process requires an exhaustive human resources if manual patrolling alone is considered. To monitor the border in real time with high accuracy and minimize the need for human support, micro unmanned air vehicles are being deployed by the defense community [2].

Micro unmanned vehicles are fully functional small flight vehicles with a size less than 15cm in length, width or height [3]. Synthetic Aperture Radar (SAR) payloads deployed on UAV are very useful because, these systems can operate in all weather conditions at any time of day or night and generate images of ground in real time.UAV directs imagery data streams to the ground station and image analyst through the use of satellite communication.

As sensor resolution increases, more than one million pixels per second are produced by the SAR system. However, satellite communication system channels to transmit image data from UAV are band limited. Image data is compressed on- board and then transmitted to the ground stations analysis by automatic target recognition (ATR) system. Conventional image compression algorithms introduce distortion by coding the entire image uniformly which severely degrades the ATR system's performance. Furthermore, due to the physical limitations of on-board SAR, an efficient AT detection algorithm should also be incorporated in the encoder [4].

The first on-board image compression system is SPOT-1. This algorithm used DPCM with a fixed compression ratio [5]. This algorithm was followed by SPOT-2 and SPOT-4. The Eastman Kodak company developed a proprietary adaptive DPCM image compression algorithm for commercial remote sensing applications [6]. Golomb coding, proposed by Solomon.W. Golomb uses a tunable parameter for image compression [7]. The Rice algorithm is the most popular on- board image compression system. The Rice algorithm together with a lossy image waveket image compression algorithm is called H- Compress algorithm [8]. A Discrete Cosine Transform based algorithm with a scalar quantizer and fixed length coding was implemented in one of the camera payloads of Phobas [9]. Between 1990 and 1991, the image compression module (ICM) was developed as a DCT based image compression Application Specific Integrated Circuit (ASIC). JPEG200 is the most recent DWT based image compression standard, that can operate at higher compression ratios without generating the block and blurry artifacts of the DCT based JPEG- baseline. Recent works have been devoted to vector quantization (VQ) or variable length entropy coding. However, both VQ and entropy coding are susceptible to channel errors [10].

To obtain a high compression ratio, while maintaining high quality of target area, a smart ROI based image compression system is proposed in this paper, where the image is compressed heavily in the background and lightly in the target area, since targets occupy small parts of SAR's image.

The rest of the paper is organized as follows: Section 2 provides the details of the proposed method. Section 3 discusses the experimental setup, results obtained and also the comparison with the existing technologies. Finally, section 4 concludes the paper.

2. Methodology

The overall ROI based image compression for SAR imagery is shown in Fig. 1.

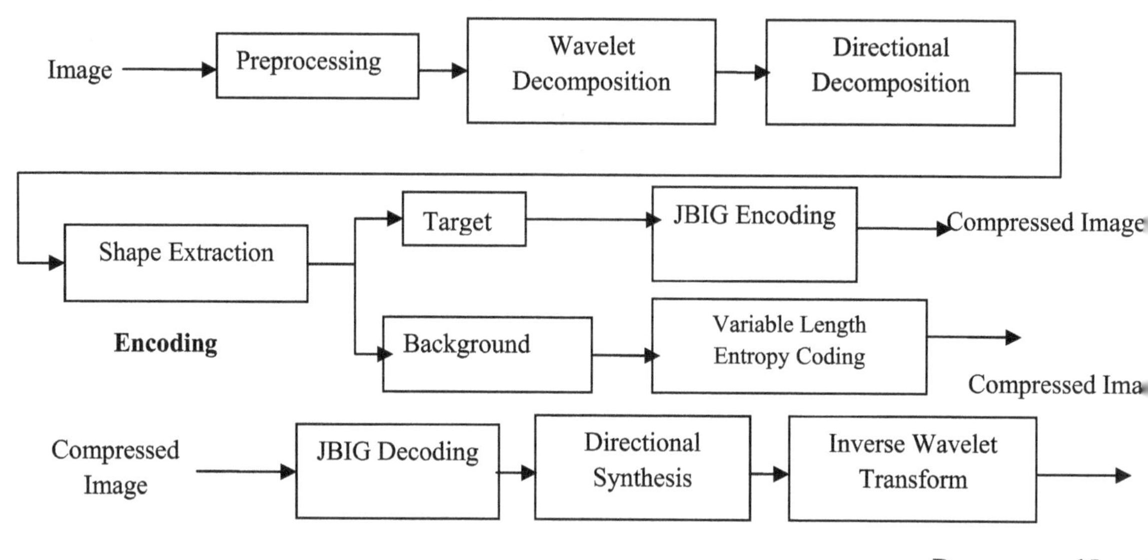

Fig.1. Block Diagram of the Proposed Method

A. Encoding

When a SAR image is formed by coherently processing the backscatter returns from successive radar pulses, this effect causes a pixel-to-pixel variation in intensity, which manifests itself as a salt-and-pepper granular pattern called speckle. Speckle noise can be reduced by spatial filtering. The spatial filters are categorized in to adaptive and non-adaptive filters. Non-adaptive filters do global processing of the image and leave out the local properties of the sensor. The non-adaptive filters are not suitable for stationery scene signals. Conversely, adaptive filters adjust themselves for changes in the local properties of the image [11]. These filters reduce speckles while preserving the edge information of the images. These filters modify the images based on the statistics extracted from the local neighborhood of each pixel. They vary the contrast for each pixel depending on the Digital Number (DN) values in the surrounding. As a result, adaptive filtering causes a better enhancement in the quality of an image. A number of speckle reduction filters are available in the literature. The mean filter is supposed to be the most classical one amongst them. The mean filter does not remove the speckle but it averages in to the data. The application of mean filter results in loss of details and resolution. The Median filter is also a simple filter that can remove pulse and speckle noise. Pulse functions which have less than fifty percent width than that of the kernel width are efficiently removed by using this filter. However, step and ramp functions will be retained. Therefore, it is essential to select a filter that is simple to realize but is efficient in avoiding the speckle noise completely. To select the appropriate filter, experiments are conducted on various SAR images using different filters and the results are shown in Fig.2.

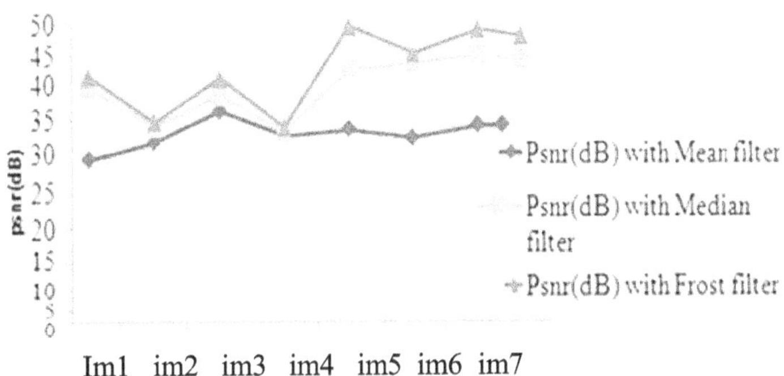

Fig. 2. Performance Evaluation of Filters

As the performance of the Frost filter is the best among the three filters, this filter is chosen in the proposed algorithm to remove the speckle noise

Then, the image is subjected to decomposition by applying wavelets. As the wavelets can give the information along only four directions like horizontal, vertical and diagonal directions, a directional decomposition of each sub-band of wavelet decomposed image is performed to obtain the directional information along 8 directions. Now, for each direction in the sub band a statistical thresholding is applied to extract the edge information. All the directional edge maps are integrated to get the final edge map.

The edge map thus obtained gives very useful information about the boundaries of the image. The details of the edge detection used in the proposed method are explained elaborately by the authors in the reference[12].

Further, it is understood that blocking artifacts are predominant in block based compression methods. In this paper, it is proposed that instead of using a fixed size and shaped blocks, arbitrarily shaped blocks obtained by means of edge detection will be coded. The edge detection gives the information about only the boundaries of the image. So, a watershed segmentation algorithm is applied on the edge map to get the arbitrarily shaped blocks in the image.

The watershed transform is the method of choice for image segmentation in the field of mathematical morphology. The morphological image segmentation often produces more stable segmentation and provides a simple frame work for incorporating knowledge constraints.

An image can visualized in three dimensions; viz, the two spatial dimensions and third, the gray level of an image. Any gray tone image can be considered as a topological surface. The topographical surface consists of points belonging to regional minimum, catchment basin or watershed and divide lines or watershed lines. The main aim of the watershed segmentation is to find watershed lines. The procedure is visualized as punching the regional minimum and floods the entire topography at uniform rate below. Then, a dam is built to prevent the rising water from distinct catchment basins from merging. Eventually only the tops of the dams are visible above the water line. These dam boundaries correspond to the divide lines of the watersheds.

The regions of the image characterized by small variations in gray levels have small gradient values, so watershed segmentation is applied on the gradient of the image rather than the actual image. In this way, the regional minima of catchment basins correlate nicely with the small value of the gradients corresponding to the objects of interest.

However, direct application of the watershed segmentation algorithm generally lead to over segmentation of an image due to noise and other local irregularities of the gradient. This problem can be overcome by limiting the number of regional minima by making use of markers to specify the allowed regional minima. Internal markers specify the region of interest while external markers specify the pixels that belong to the background.

In the watershed based image segmentation, internal markers are used to obtain watershed lines of the gradients. The watershed lines are then used as external markers. Each region defined by the external markers contains a single internal marker and a part of the background. Then partition the image into two parts: object which contains internal markers and a single background, which contains external markers [13].

The watershed segmentation algorithm is used in this paper to discriminate between object and the background. Furthermore, it is also used for the target identification which is essential for further processing.

As the target and the boundary has been identified using the watershed segmentation algorithm, the JBIG coding is applied to the target and a variable length entropy coding is applied to the background.

The current international coding standard JBIG [14] provides efficient lossless coding of bi-level images. It also provides progressive coding by resolution reduction. The technique applies sequential arithmetic coding of the image pixels conditioned on the value of nearby, previously encoded pixels. JBIG achieves impressive compression ratios by adapting to the information content of the image data being encoded. An adaptive arithmetic coder is used to predict and code future data symbols based on the characteristics of the data currently being encoded. Sequentially encoded images are stored in a single layer at full resolution and without other lower resolution images being stored in the same data stream. Progressively encoded images start with the highest resolution image and end with the lowest. The high-resolution image is stored in a separate layer and is then used to produce a lower resolution image, also stored in its own layer. Each layer after the first layer is called a *resolution doubling*. An image with three layers is said to have two doublings. The JBIG coding that uses progressive scanning is used in this paper.

B. Decoding

The inverse process of obtaining a decoded image is rather simple when compared to the encoded process. The encoded image of the target is subjected to JBIG decoder and the background image is decoded using variable length entropy decoder. JBIG decoder is almost lossless. So, approximately the total information in the edge map is decoded. Then, an inverse of directional filter bank is applied on the image. Then an inverse discrete wavelet transform is applied on to the image to get the final decoded image. As the Directional Filter Banks and wavelet transforms are over complete and JBIG compression is a lossless compression technique, the error in the reconstructed images is almost zero and the peak signal to noise ratio is very high. Thus, a very high compression ratio without losing much information can be achieved by using the proposed method.

The algorithm for the proposed method is presented below:

Algorithm ROI Edge Based Coding
Input: A Gray level image of size N x N

Outputs: A coded Gray level image of size N x N

Step 1: Pre processing

   i.   Apply a frost filter to remove speckle noise
Step2: Wavelet Decomposition

      Decompose the given image using wavelet transforms

Step 3: For each Sub band do

   i.   Identify the direction with maximum frequency
   ii.   Determine the threshold to identify an edge based on the covariance between the neighboring pixels
   iii. Obtain the edge map
Step 4: Integrate all the edge maps

Step 5: Obtain the arbitrarily shaped boundaries from the edge map using watershed segmentation

Step 6: Encode the target and the background using JBIG coding

Step7: Apply JBIG decoding to the target and variable length entropy coding to the background

Step8: Apply Inverse Discrete Wavelet Transform to get the decoded image

3. Experiments and Results

The proposed method is tested for its performance by conducting experiments with several SAR images illustrating border security. The size of the images is 512x512. The images acquired are at different resolutions ranging from few feet to several meters. At first, the tanks at a border image taken by the SAR image from a distance of one foot are taken into consideration and are shown in Fig. 3.

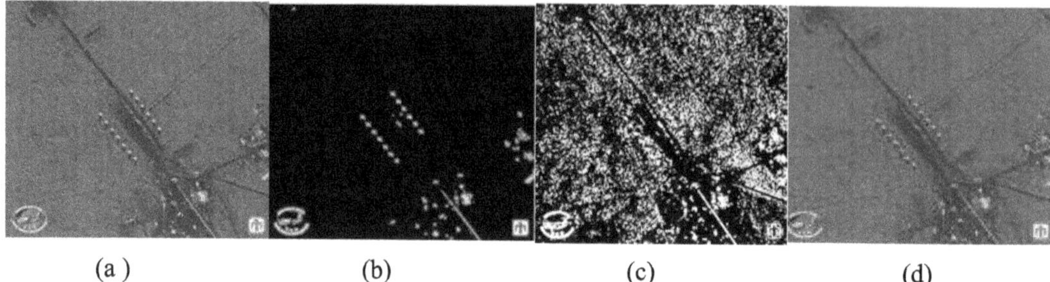

(a)   (b)   (c)   (d)

Fig. 3. (a) Original Tank Image (b) Target Image (c) Background Image (d) Decoded Image

The first task of the algorithm is to segment the target from the background and the target image of the tankers is shown in Fig. 3.(b) and the background of the original image is shown in Fig. 3. (c). The target image is separately encoded with a JBIG encoding scheme which is a lossless compression technique. The texture image is encoded with variable length entropy coding which achieves very high compression ratio. Both the encoded images are transmitted in the channel separately. Finally, they are combined together after decoding them separately at the receiver end. The decoded image at the receiver end is shown in Fig. 3. (d). The compression ratio obtained for the target image is 1: 20 while the compression ratio is 1: 32 for the background image. Finally, a good quality decoded image is transmitted to the receiving end through the band limited channel.

The results obtained for the same image captured by SAR at a resolution of four inches is shown in Fig. 4.

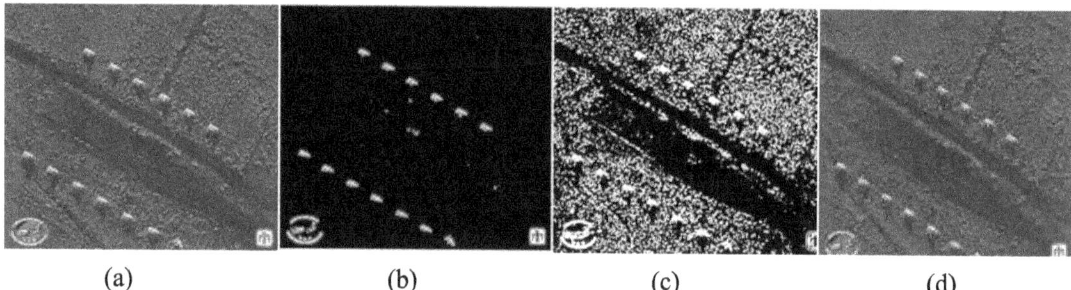

(a)   (b)   (c)   (d)

Fig.4. (a) Original Tank Image at four inch distance (b) Target Image (c) Background Image (d) Decoded Image

The results of a man crossing the border image are shown in Fig. 5. The original image is shown in Fig. 5.(a), the target image is shown in Fig. 5.(b), the background image is shown in Fig. 5. (c) and the decoded image is shown in Fig. 5. (d) respectively.

340

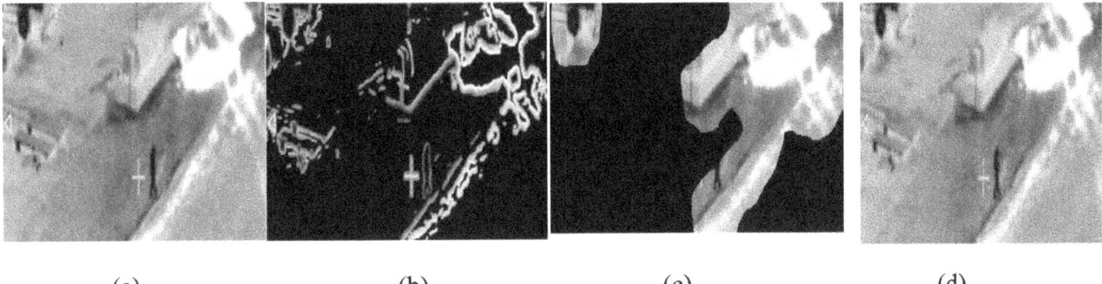

(a)   (b)   (c)   (d)

Fig. 5. (a) Original Man Crossing Border (b) Target Image (c) Background Image (d) Decoded Image

A quantitative evaluation is also made to test the performance of the proposed method. Various performance metrics like Peak Signal to Noise Ratio (PSNR), Mean Square Error (MSE) are used to analyze the performance of the proposed method in quantitative fashion. The PSNR (dB) and the MSE are given by the equations 1 and 2 respectively.

$$e_{rms} = \sqrt{\frac{1}{MN} \sum_{x=0}^{M-1} \sum_{y=0}^{N-1} (E(x,y)-O(X,y))^2}$$

$$\tag{1}$$

$$SNR_{peak} = 10 \log_{10} \frac{(L-1)^2}{\frac{1}{MN} \sum_{x=0}^{M-1} \sum_{y=0}^{N-1} \left[E(x,y)-O(x,y)\right]^2}$$

$$\tag{2}$$

4. Conclusion

An ROI based image compression system is presented in this paper, which uses a multi-scale and multi directional analysis for automatic target identification. For achieving high compression ratios without losing the image quality, the target areas in the SAR image are encoded with more number of bits and the background area is coded with less number of bits. The improved performance of the proposed system is demonstrated by comparing with the existing ROI based compression systems.

Table 1 Performance Evaluation of Proposed Method

Sl.No.	Image	MSE			PSNR(dB)		
		EZW	SPIHT	Proposed Method	EZW	SPIHT	Proposed Method
1	Tanks at 1ft	0.3651	0.2771	0.2452	26.21	35.229	36.41
2	Tanks at 4 inches	0.4521	0.3121	0.0024	21.793	30.25	43.61
3	Border crossing	0.4325	0.4569	0.3906	24.43	26.1066	35.56
5	Farm House	0.3661	0.0045	0.3164	26.5642	39.62	47.46
6	Bridge	0.5469	0.3012	0.2494	20.7284	28.68	39.59

References

[1] . J. Laycock, V. Van der Zel, " High Resolution Optical Imaging with Low Cost Topsat Small Satellite", International Astronautical Conference 2004, Vancouver, Canada, October 2004

[2]. P. Brooks,"Topsat- High Resolution Imaging from a Small Satellite", IAA symposium on Small Satellites for Earth Observation", April 2-6, 2001

[3]. Srijaya Mohan, G. Sridharan, " Emerging Technologies for Micro Unmanned Air Vehicles", Defense Science Journal July 2004, pp 223-228

[4]. Guioxia Yu, "Image Compression Systems on Board Satellites", Acta Astronautica, Elsevier Publications 2009 pp 988-1005

[5]. P. S. Yeh, " Implementation of CCDS Lossless Compression- recommendation: Development and Status", Proceedings of SPIE vol 4790,2002,pp-302-313

[6]. J. Serra Sagrista, "Lossy Coding technique for High Resolution Image ,Proceedings of SPIE,vol 5238, 2003, pp-250-261

[7]. R. F. Rice, "Adaptive Variable Length Coding for Efficient Compression of Space Craft television Data", IEEE transactions on Communications 1983 889-897

[8]. Zhi Sun, Pu Wang, " Border Sense: Border Patrol through Advanced Wireless Sensor Networks", Journal Of Adhoc Networks, Elsevier Publications, September 2010

[9]. X. H. Yuan, Z. D. Zhu, "Target Aided SAR Image Intelligent Compression", Progress in Electromagnetic Research B, Vol.20, 285-320, 2010

[10]. Kim. A, Hierarchial Stochastic Modelling of SAR Imagery fro Segmentation/Compression", IEEE Transactions on Signal Processing, vol.47, No.2, 458-468,1999

[11]. V. S. Frost, et al., "A Model for Radar Images and Its Application to Adaptive Digital Filtering of Multiplicative Noise," IEEE Trans. Pattern Anal., Machine Intell., vol. 4, no. 2, pp. 157-166, Mar. 1982.

[12]. K. Padma Vasavi, N. Udaya Kumar, E. V. Krishna Rao, M. Madhavilatha "A Novel Statistical Thresholding in Edge Detection Using Laplacian Pyramid and Directional Filter Bank",Vol1, pp 589-593, WCECS 2010

[13]. Phan T.H. Truc, Md. A.U. Khan, Young-Koo Lee, "Vessel Enhancement Filter Using Directional Filter Bank" Computer Vision and Image Understanding, Vol 113, issue 1, pp 101-112, 2008

[14]. Anton Brezina, Jaraslav Polec, "Region Based Texture Coding at Very Low Bit Rates", Journal of Electrical Engineering, Vol. 56, No1-2, 2005 pp 36-40

APPLICATION OF PARTIAL RECONFIGURATION OF FPGA TO SHARE RESOURCES IN A THRONG OF MICRO AIR VEHICLES

Pradeep AVVP Gorthi

Student of Electronics and Communication Engineering at International Institute of Information Technology, Hyderabad

ABSTRACT

A throng of Micro air Vehicles is defined as a group of vehicles connected as a network to complete a given task. As the vehicles are small in size, it is difficult to have long lasting power sources. To overcome this, a method of sharing the resources between different vehicles in a throng is described. For this to be accomplished, partially reconfigurable FPGA can be used to activate a device whenever needed, deactivate and direct the resources to another device in the throng at another time. Applications can be redirected to devices, which are not heavily loaded in computation from devices that are overloaded and so reduce power consumption and heat loss. The device need not be reset when reconfiguring the device and so partial dynamic reconfiguration is useful. The possibility of this design is described in the paper.

INTRODUCTION

Field Programmable Gate Arrays are generally abbreviated as FPGA. The FPGA is a semiconductor device that can be programmed after manufacturing. An FPGA allows you to program product features and functions, adapt to new standards, and reconfigure hardware for specific applications even after the product has been installed in the field instead of being restricted to any predetermined hardware function. You can use an FPGA to implement any logical function that an application-specific integrated circuit (ASIC) could perform, but the ability to update the functionality after shipping offers advantages for many applications.For example, updates can be done over the air without any manual update that is very useful. One of the uses of FPGA is partial reconfiguration. Partial reconfiguration is the process of reconfiguring only a part of the device partially. There is also the concept of partial dynamic reconfiguration where a part of the device is reconfigured dynamically, i.e. a part of the device is configured while the rest of the device is running. It is very useful particularly, in the case of Air vehicles where partial dynamic reconfiguration helps in updating the device of the air vehicle without the need to stop the vehicle, which results in continuous fly of the air vehicle.

Unmanned Micro Air Vehicle belong to the section of the air vehicles where size of the vehicle is significantly reduced and also the vehicle is autonomous or is controlled from a base station. Objects designed for flight have been present for decades, but as we decrease the size of the objects, it is more and more difficult to design the air vehicles. Varied research is being done all over the world in the development of unmanned micro air vehicles as the MAVs come with their own set of advantages. Even though remotely operated MAVs are present and are in use, popularity for these small sized devices comes from their capability of autonomy. This technology finds way into the military and research, as loss of life is minimal when UAV is used in a combat. But at the same time, the importance of UAV is seen in non-military fields such as educational research, situations of disaster management and government usage. The military and paramilitary forces to look for explosives and even defuse bombs if possible extensively use these UAVs.

A throng of micro air vehicles is defined as a group of micro air vehicles connected as a network to work in various fields, whether it is in military or non-military. As the vehicles are small in size, this group of vehicles is helpful in maintaining sufficient power resources and computing power whenever needed.

DESIGN

For the tasks mentioned above for an unmanned micro air vehicle, the source of power must be long lasting. If it is not, then these applications are of no practical use. To increase the life of a MAV, either the computing power can be reduced or source of power can be given a boost to last long in a single life of flight. One solution is to form a group of micro air vehicles, which are interconnected in a way to share the resources to increase the computing power. The same solution can also be applied to increase the battery life and so the life of flight of the MAV. Application of the concept of partial reconfiguration to the design of micro air vehicles is the design solution presented in this paper.

As technology advances, more processor cores on chip are built and so the problems will become more complex and so need better tools. Communication issues between the micro air vehicles must be considered, as we want to limit congestion on system bus between cores. In accordance to communication latencies, unpredictability in the performance must be minimized which is important for real time systems.

The idea of partially reconfiguring the device is proposed where the group of micro air vehicles shares a very less number of resources i.e. number of FPGAs used is minimized using partial reconfiguration. If less FPGAs are used, then a protocol must be described for sharing the resources between different MAVs to implement the optimized throng of micro air vehicles. In a given throng of micro air vehicles each MAV or a group of MAVs may have a different goals to complete. The goals have to be organized by a centralized processor. The design proposed is centralized sharing of resources, which is controlled from a central processor, which is designed to be included in a throng of micro air vehicles to deploy functions of each MAV to its designated device. Effectively, the central processor is sharing the resources for computation. This processor also shares the goals of each vehicle, so as to reduce the computation power and in turn reduces the power consumed by a single device.

The whole system must address allocation, partitioning, placement and routing for the partial reconfiguration of FPGA. The central processor can be placed in one of the micro air vehicles and the processor has to be protected from any external disturbance or attack. If the processor is attacked, then the whole throng of micro air vehicles is destroyed. The central processor must have abilities to act autonomously. No external agent must be involved in the computation process and no directions must be given to the processor. There is no need for inter device communication as the communication is taken care by the central processor. There are deterministic and realistic changes in the course of flight of the micro air vehicles. The processor must be able to respond to the deterministic as well as realistic changes. The concept of partial dynamic reconfiguration is important and useful as every task need not be loaded and executed at the same time in the throng of micro air vehicles. Job scheduling algorithms can be used and implemented to decide on optimized scheduling of jobs between different micro air vehicles. Some of the job scheduling algorithms are shortest job first, early start time, early finish time, etc. Early finish time is the optimum algorithm, which can be used to schedule the jobs between different micro air vehicles. This reduces cost of implementation when compared to other scheduling algorithms. The operating system that is used to co-ordinate the tasks is constraint-based dynamic FPGA configuration writing, a two-dimensional geometric packing problem. Fast packing heuristics must be used some of which are Best Fit, Bottom Left, Minkowski Sum. Out of these, Minkowski sum is determined to be the best of the above considered Sums.

MINKOWSKI SUM

Given two sets $A,B \in \mathbb{R}^d$, their *Minkowski sum*, denoted by $A \oplus B$, is the set $\{\, a + b \mid a \in A, b \in B \,\}$. Minkowski sums are used in many applications, such as motion planning and computer-aided design and manufacturing. Computing the Minkowski sum of two convex polygons P and Q with m and n vertices respectively is very easy, as $P \oplus Q$ is a convex polygon bounded by copies of the $m + n$ edges, and these edges are sorted by the angle they form with the x-axis. As the two input polygons are convex, their edges are already sorted by the angle they form with the x-axis. The Minkowski sum can therefore be computed in $O(m + n)$ time, by starting from two bottommost vertices in P and in Q and performing ``merge sort" on the edges.

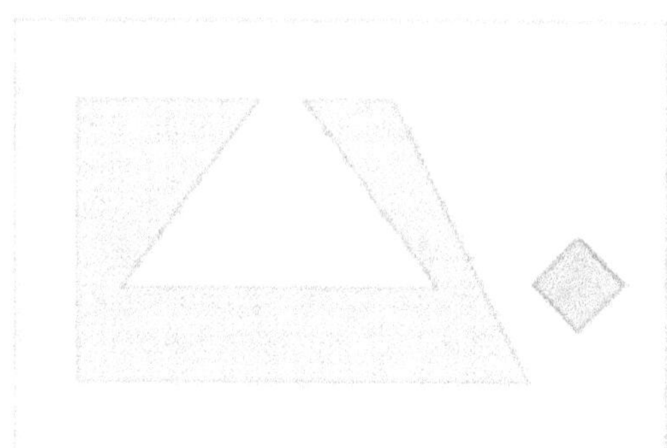
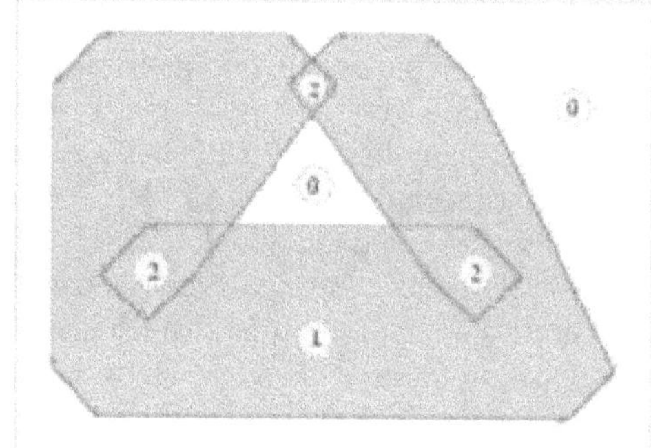

DECOMPOSITION

We decompose P and Q into convex sub-polygons, namely we obtain two sets of convex polygons $P_1, \cdots, P_k$ and $Q_1, \cdots, Q_\ell$ such that $\cup_{i=1}^{k} P_i = P$ and $\cup_{i=j}^{\ell} Q_j = Q$. We then calculate the pairwise sums $S_{ij} = P_i \oplus Q_j$ using the simple procedure described above, and compute the union $P \oplus Q = \cup_{ij} S_{ij}$. This approach relies on a decomposition strategy that computes the convex decomposition of the input polygons and its performance depends on the quality of the decomposition.

CONVOLUTION

Let us denote the vertices of the input polygons by $P = (p_0, \cdots, p_{m-1})$ and $Q=(q_0, \cdots, q_{n-1})$. We assume that both P and Q have positive orientations (i.e. their boundaries wind in a counterclockwise order around their interiors) and compute the convolution of the two polygon boundaries. The *convolution* of these two polygons denoted $P * Q$, is a collection of line segments of the form $[p_i + q_j, \; p_{i+1} + q_j]$,[1] where the vector $p_i\,p_{i+1}$ lies between $q_{j-1}\,q_j$ and $q_j\,q_{j+1}$, and symmetrically of segments of the form $[p_i + q_j, p_i + q_{j+1}]$, where the vector $q_j\,q_{j+1}$ lies between $p_{i-1}\,p_i$ and $p_i\,p_{i+1}$.

The following code constructs the Minkowski sum of two triangles. As the input polygons may not be convex, their Minkowski sum may not be simply connected and contain polygonal holes; see for example Figure 24.1. S is therefore an instance of the Polygon_with_holes_2<Kernel,Container> class-template, defined in the Boolean Set-Operations package: The outer boundary of S is a polygon that can be accessed using S.outer_boundary(), and its polygonal holes are given by the range [S.holes_begin(), S.holes_end()) (where S contains S.number_of_holes() holes in its interior).

//Program

```
#include "ms_rational_nt.h"

#include<CGAL/Cartesian.h>

#include<CGAL/minkowski_sum_2.h>

#include<iostream>

#include "print_utils.h"

typedef CGAL::Cartesian<Number_type>        Kernel;

typedef Kernel::Point_2                 Point_2;

typedef CGAL::Polygon_2<Kernel>             Polygon_2;

typedef CGAL::Polygon_with_holes_2<Kernel>      Polygon_with_holes_2;

int main ()

{

// Construct the first polygon (a triangle).

  Polygon_2  P;


P.push_back (Point_2 (0, 0));

P.push_back (Point_2 (6, 0));

P.push_back (Point_2 (3, 5));
```

```cpp
// Construct the second polygon (a triangle).
Polygon_2  Q;

Q.push_back (Point_2 (0, 0));
Q.push_back (Point_2 (2, -2));
Q.push_back (Point_2 (2, 2));

// Compute the Minkowski sum.
Polygon_with_holes_2  sum = minkowski_sum_2 (P, Q);

CGAL_assertion (sum.number_of_holes() == 0);

std::cout<< "P = "; print_polygon (P);
std::cout<< "Q = "; print_polygon (Q);
std::cout<< "P (+) Q = "; print_polygon (sum.outer_boundary());

return (0);
}
```

The function minkowski_sum_2 (P, Q) accepts two simple polygons P and Q, represented using the Polygon_2<Kernel,Container> class-template and uses the convolution method in order to compute and return their Minkowski sum S = P Q.

Minowski sum can be used to allocate non-rectangular cores on FPGA. It has good runtime performance characteristics.

Memory network topologies can be Bus, star, ring, tree, etc. Memory network topology must have good ease of implementation, less wire routing cost, support concurrency. Each network topology has some level of favorability in accordance to ease of implementation, wire routing cost, etc. Considering all these, tree topology is considered to be the optimum choice as it favors wire routing cost, latency and scalability. It does not affect ease of implementation and concurrency.

Sensors and cameras are used at different nodes to acquire information of different nodes and this information is sent to the central processor. Also, a token can be used to transmit information from one node to the other and also to the central processor. This token carries and is responsible for the information of a particular node. A token should give the position of other agents, which is sent to the central processor. The central processor provides best possible routing and forwarding mechanisms and sends it back through the corresponding token.

ADVANTAGES OF MICRO AIR VEHICLES WITH PARTIAL RECONFIGURATION CAPABILITY

In the areas of military, cheaperand more expendable groups of Micro Air Vehicles can approach a target and collect data from closer proximity with less risk as there are more number of micro air vehicles and so if any one of the vehicle is disturbed, immediately a new vehicle can be inserted from the token from central processor. Multiple micro air vehicles can be used to precisely locate a target by combining each of their sensor readings and employing geolocation techniques. Due to its increase in life of flight, continuous surveillance can be implemented. Some of the low on battery vehicles can be replaced periodically with newly powered vehicles and so the throng can perform indefinitely. Also, computation can be distributed among the devices, which will not only speed up the process but also reduces the power consumption if properly designed.

CONCLUSION

Applying the concept of partial reconfiguration is certainly advantageous as seen by the design presented in the paper. It can be further improvised by changing the protocols to be used to attain optimum results for specific tasks.

REFERENCES

[1] http://en.wikipedia.org/wiki/Field-programmable_gate_array

[2] http://en.wikipedia.org/wiki/Least_squares

[3] http://en.wikipedia.org/wiki/Riemann_sum

[4] http://en.wikipedia.org/wiki/Minkowski_addition

[5] http://www.cgal.org/Manual/latest/doc_html/cgal_manual/Minkowski_sum_2/Chapter_main.html

[6] International Symposium on Flying Insects and Robots, Monte Verità, Switzerland

[7] Nembrini J, Winfield A, Melhuish C (2002) Minimalist coherent swarming of wireless net-worked autonomous mobile robots. In: From Animals to Animats 7, Proceedings of the 7thInternational Conference on Simulation of Adaptive Behavior, MIT Press, Cambridge

[8] Using Simulated Partial Dynamic Run-Time Reconfiguration to Share Embedded FPGA Compute and Power Resources across a Swarm of Unpiloted Airborne Vehicles John David Eriksen Jamie Unger-Fink

NANO TECHNOLOGICAL & COMPUTATIONAL AERODYNAMICS OF MICRO AIR VEHICLES

B.R.R.Narasimham

Research fellow, Innovative TechnologiesCentre, JNTUH, Hyderabad -500072.

V. Brahmajirao

Senior Professor, Department of Nanoscience&Technology, School of Bio-Technology, MGNIRSA, A unit of D. Swaminathan Research Foundation, Hyderabad-500072, A.P.

V.Kamakshi Prasad

Director, Innovative Technologies Centre, JNTUH, Hyderabad -500072.

ABSTRACT

Micro- and Nano air vehicles are defined as "extremely small and ultra-lightweight air vehicle systems" with a maximum wingspan length of a few cm and a weight less than 20 grams. Obviously such systems demand the need of Nano materials and devices in the development of the modules, mechanisms, and Artificial Intelligence for controls to achieve the objectives. Nano Air Vehicles (NAVs) have the very great potential to revolutionize our sensing and information gathering capabilities in environmental monitoring and homeland security areas. Due to the □ very small size, limited flight regime, and versatile modes of operation of the NAVs" a very significant scientific research & advancement is needed to create these revolutionary objects of unlimited capability. To mention few of the challenges in the design of the NAV's, Aerodynamics, structural dynamics, and flight dynamics of natural flyers intersects with some of the richest problems in NAVs, including massively unsteady three-dimensional separation, transition in boundary layers and shear layers, vertical flows and bluff body flows, unsteady flight environment, aero elasticity, and nonlinear and adaptive control are just a few examples that require attention of research.

Computational projects: Insufficient knowledge, predictive capabilities, and experimental data exist regarding the fundamental unsteady aerodynamics. This leads to a need for the development of artificial intelligence to simulate models that can fulfil this requirement. A challenge is that the scaling of both fluid dynamics and structural dynamics between smaller natural flyer and practical flying hardware/lab experiment (larger dimension) is fundamentally difficult. In this paper, we offer an overview of the challenges and issues, along with sample results illustrating some of the efforts made from a computational modelling angle. Artificial intelligence will be very helpful in the developmental efforts of modelling of these mechanisms, for trials before going to the proto type development In the later part: tasks that the Materials & Software and Hardware, Electronics Engineers should be addressed in the NAVD and Fabrication of the Latest Nano Electronic Gadgets, like Nano materials & devices required to Nano processing systems relevant to the NAV Design are discussed.

INTRODUCTION

…..Danger please

When one of the bad guys opens his apartment door, a tiny robo-bug, looking like a garage door opener with wings, sneaks in to spy. In another scene, a bug—the Air Force calls them Micro Air Vehicles, or MAVs—creeps into a sniper's roost and delivers a deadly shot to the back of his head.

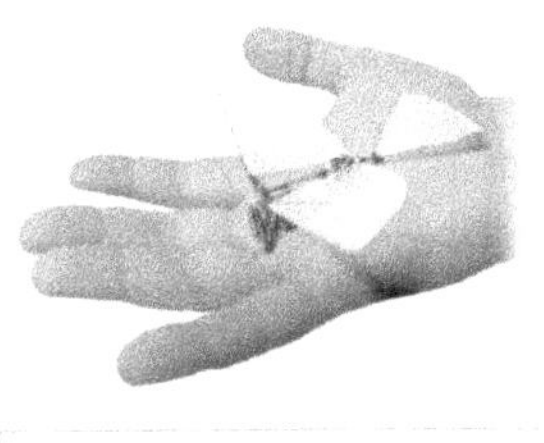

Fig 1: MAV Ornithopters based on Leonardo Da Vinci's famous desians.

These are the targets of research in U.S.Defence, most of which are ready for the mass production it seems [1]. Micro- and Nano air vehicles are defined as "extremely small and ultra-lightweight air vehicle systems" with a maximum wingspan length of a few cm and a weight less than 20 grams. Obviously such systems demand the need of Artificial Intelligence along with nano materials and Nano -devices in the development of the modules, mechanisms, and controls to achieve the objectives.

SYNCHRONIZATION AND QUORUM SENSING IN.."A SWARM OF HUMANOID ROBOTS", A COMPLEX SYSTEM NECESSARY FOR ACHIEVING GROUP TASKS [2].

One of these is collective behavior of groups of humanoid robots, and in particular robot synchronization and swarming. The goal of this work is to robustly synchronize a group of humanoid robots, and to demonstrate the approach experimentally on choreography of 8 robots. We aim to be robust to network latencies and to allow robots to join or leave the group at any time (for example a fallen robot should be able to stand up to rejoin the choreography). Contraction theory is used to allow each robot in the group to synchronize to a common virtual oscillator, and quorum sensing strategies are exploited to fit within the available bandwidth. The humanoids used are NAO's, developed by Aldebaran Robotics.The roachbots and the swarm of MIT humanoid robotsdancing in sync, as well as "disposable" quarter-sized kilobots are "cheap enough to swarm in the thousands".

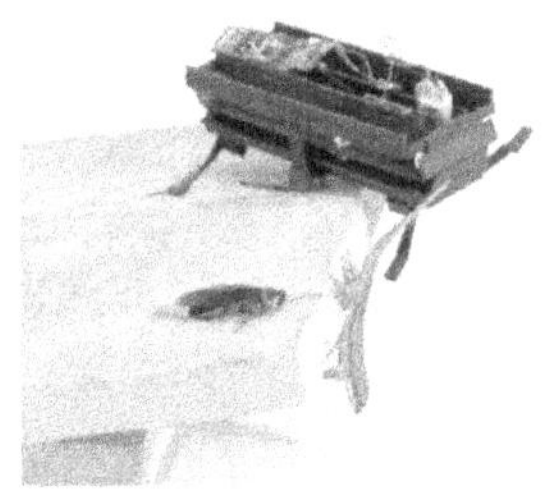

Fig (2) DASHHexapedal Cockroach-Inspired Robot

DASH, Hexapedal Cockroach-Inspired Robot developed by University of California , Berkeley's (from PolyPEDAL Lab) is a 10-centimeter long, 16-gram Dynamic Autonomous Sprawled Hexapod, [3] has learned a new trick: the robot can now perform "rapid inversion" manoeuvres, dashing up to a ledge and then swinging itself around to end up underneath the ledge and upside-down. This replicates behaviours in cockroaches and geckos, and may lead to a new generation of acrobatically-inclined insectobots.DASH Hexapedal Cockroach-Inspired Robot Survives Large Falls, (as reported by Anne-Marie Corley).Wing-flapping micro robots, unmanned helicopters, formation flight algorithms -- there were lots of cool UAV projects [4] ,at the International Conference on Intelligent Robots and Systems (IROS)[EricoGuizzo / Mon, November 05, 2007] in San Diego, Calif. Too many, in fact, to describe them all here.

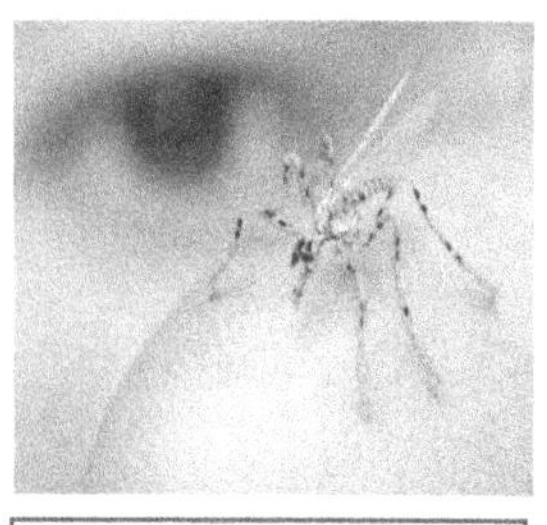

Fig(3) DARPA-like tiny insect cyborg

The "Defense Advanced Research Projects Agency (DARPA)",tiny insect cyborg drones [5] are "designed to go places" that soldiers cannot to work as spies or as swarm weapons[6]. Is this a mosquito micro air vehicle (MAV)? "Such a device could be controlled from a great distance and is equipped with a camera, microphone. It could land on you and then use its needle to take a DNA sample with the pain of a mosquito bite. Or it could inject "a micro RFID tracking device under your skin.[7] " While DNA-sucking, RFID-chip-injecting mosquito drones are currently a bunch of bunk, a Bing image search shows a multitude of MAVs that aren't simply CGI mockups.

IBM PRODUCES FIRST WORKING CHIPS MODELED ON THE HUMAN BRAIN[8]

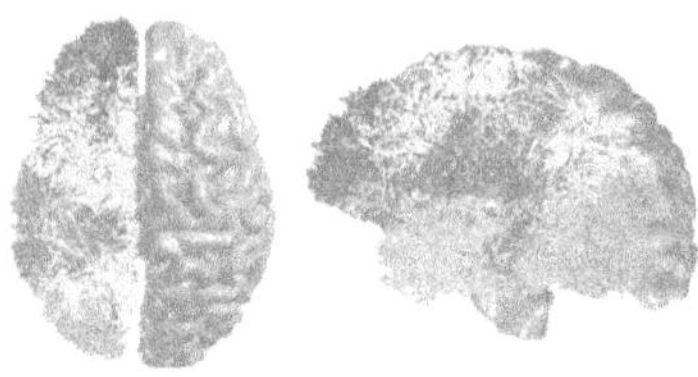

Fig4: Correlative imagination of a model of brain by IBM TEAM

An IC chip is in the market soon, to facilitate the Artificial intelligence, rescue the tasks of pico- miniaturizing the hardware and it is beyond the imagination of any one as to what is ready tomorrow, for the Air vehicles. Big Blue [8], along with four universities and the Defense Advanced Research Projects Agency (DARPA), have created the basic design of an *"experimental computer chip"* that emulates the way the brain processes information. IBM's so-called cognitive computing chips could one day simulate and emulate the brain's ability to sense, perceive, interact and recognize — all tasks that humans can currently do much better than computers can. The DARPA project, called Synapse (Systems of Neuromorphic Adaptive Plastic Scalable Electronics, or SyNAPSE) is actively working at thisat the IBM Almaden Research Center in San Jose, Calif. "This is the seed for a new generation of computers, using a combination of supercomputing, neuroscience, and nanotechnology"-says Modha [8] in an interview with VentureBeat [9]

But nano-biomimicry MAV design has long been studied by DARPA. DARPA's 2008 symposium discussed "bugs, bots, borgs and bio-weapons." The Pentagon's "cyborg moth" is now defunct tech and bat drone bots are also old surveillance news. Researchers have developed bio-inspired drones with bug eyes, bat ears, bird wings, and even honeybee-like hairs to sense biological, chemical and nuclear weapons.

IROS CONFERENCE, AT SANDIEGO

Robert Wood (Harvard) and Xinyan Deng (University of Delaware) [10] presented The Coolest Flying Robot Projects at IROS Conference. Wing-flapping micro robots, unmanned helicopters, formation flight algorithms, and so on.... At San Diego, California.This little MAV had a 3 centimeter wingspan and that was back in 2007. When the U.S. government was accused of making insect spy drones in 2007.

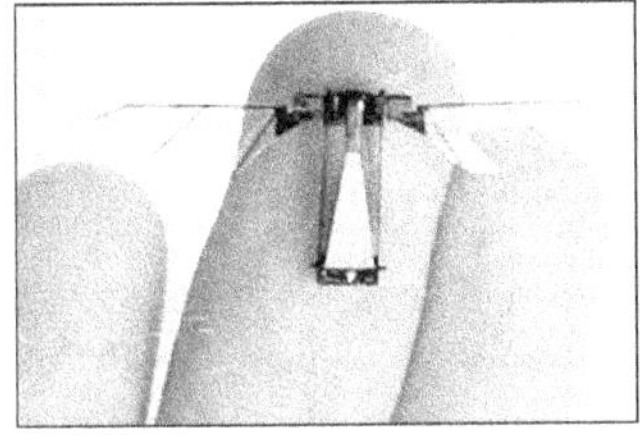

Fig5:Mechanical fly Developed by Harvard College.al fly

Tom Ehrhard,[11], a retired Air Force colonel and expert on unmanned aerial craft, **told the Telegraph,** "America can be pretty sneaky." The article also mentioned a dragonfly drone the CIA had developed in the 1970s While reading people's comments concerning spy drones flying overhead, there have been many comments about "skeet shooting" drones down from the sky. *That* would most likely be destroying government property and make a person a "terrorist." Besides, would you really see a tiny part bot, part bug "cyborg insect" drone from a distance if it was spying on you? The answer is **an obvious NO.**

MICRO AERIAL VEHICLE BY MICROPROPULSION THAT RESEMBLES A LARGE INSECT

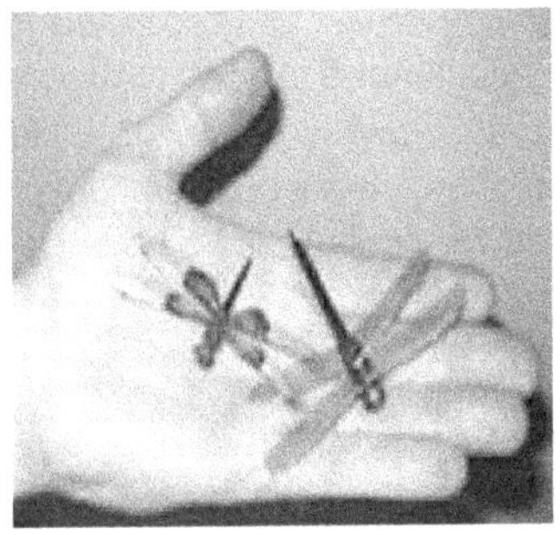

Fig :6(a) Micro Propulsion MAV

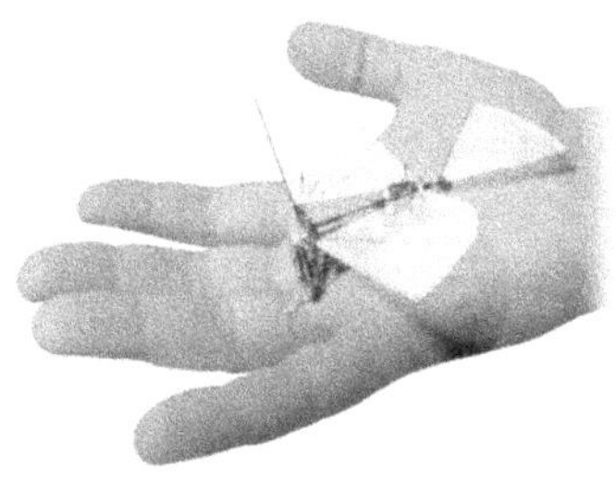

Fig: 6(b)The MAV Ornithopters Lethal Mini Drone

This Micro Aerial Vehicle (MAV) Ornithopters [12] is based on Leonardo Da Vinci's famous designs. It has a wingspan of less than 5" and has an on-board rechargeable lithium polymer battery, which allows for 7 minutes of flying time from a 5 minute charge.However their objective with a proper Nano charging system is to be able to operate them for several days without charging. The MAV Ornithopters , or the so-called *"lethal mini drones,"* were being developed outside of Dayton, Ohio, and were set to roll-out by 2015.Wright –Patterson Air Force Base at Ohio is of the opinion: that Bad guys of indiscernible origin being shadowed, from a careful distance, by small robotic drones designed to resemble birds and insects. It might sound far-fetched. But top Air Force officials believe that MAVs could be a significant part of the Defence Department's arsenal in the not-so-distant future

SAMARAI"MAPLE-SEED-LIKE" DRONES FROM LOCKHEED MARTIN'S SCIENTISTS

Lockheed Martin's Intelligent Robotics Laboratories unveiled "maple-seed-like" drones called **Samarai** that also mimic nature.[13]. U.S. troops could throw them like a boomrang to see real-time images of what's around the next corner, the **Navy Times reported.** It could also be "useful for the military and police" to look inside buildings.

CYBORG MOTH MAV OF PENTAGON

Fig 7 U.S. Defence developing hairy wings and bug eyes& Insects with Tiny Hairs for their MAVs

Nano-biomimicry MAV design has long been studied by DARPA. DARPA's 2008 symposium discussed "bugs, bots, borgs and bio-weapons." The Pentagon's "cyborg moth" is now defunct tech and bat drone bots are also old surveillance news. Researchers have developed bio-inspired drones with bug eyes, bat ears, bird wings, and even honeybee-like hairs to sense biological, chemical and nuclear weapons.

For years, the military has turned to the birds and the bees for inspiration, churning out mechanical hummingbirds and remote-controlled insect cyborgs. Now the Pentagon wants its mini-drones to have hairy wings and bug eyes, too [14]. It'll help the tiny machines spy on — and creep out — any enemies, military researchers promise.The realization of cyborgs with most of the machine component inside the insect body will provide stealthy robots that use muscle actuators which have been developed over millions of years of evolution. For years, now, Pentagon-backed researchers have been trying to create cyborg insects that could serve as living, remote-controlled spies. The problem is, those modified bugs never survived long enough to be useful. Now, Georgia Tech professor Robert Michelson says he's managed to get the bug 'borgs to live into adulthood.[15]

INSERTION OF MICRO-MECHANICAL SYSTEMS [MEMS] INTO ABDOMINAL SEGMENTS OF MANDUCA MOTH

For years, now, Pentagon-backed researchers have been trying to create cyborg insects that could serve as living, remote-controlled spies. The problem is, those modified bugs never survived long enough to be useful.*Flight International*reports that, in his latest work, Michelson truncated a Manduca moth's thorax "to reduce its mass." Then he put in "a MEMS component... where abdominal segments would have been, during the larval stage.That way, as the bugs get older, tissues grow around — and fuse together with — the tiny machines.Robert Michelson says he's managed to get the bug 'borgs to live into adulthood

Fig 8: MANDUCA Moth carrying MEMS component In the abdominal segments

Images taken by x-ray of insects with these changes and others found that tissue growth around the inserted probes was good. One DARPA goal is to show that during locomotion the heat and mechanical power generated by the thorax could be harnessed to power the MEMS. For years, now, Pentagon-backed researchers have been trying to create cyborg insects that could serve as living, remote-controlled spies. The problem is, those modified bugs never survived long enough to be useful.

BRAIN IMPLANT INTO A PIEGEON IN CHINA

The *People's Daily Online*reported that scientists in China have successfully used brain implants in pigeons to control the birds' movement: Scientists with the Robot Engineering Technology Research Center of east China's Shandong University of Science and Technology [16] implanted micro electrodes in the brain of a pigeon so they can command it to fly right or left or up or down. The implants stimulated different areas of the pigeon's brain according to signals sent by the scientists via computer, and forced the bird to comply with their commands. "It's the first such successful experiment on a pigeon in the world," said the lead researcher, who hopes the work will have "practical use" in the future

Fig9 : The pigeon that underwent the brain Transplant in CHINA

CONCLUSIONS

Artificial Intelligence is the prime content for the coming generation of PICO Air Vehicles. Nano Technology can revolutionize the hard ware content using quantum gates. QBITS have special characteristics over bits and bytes. Consequently suitable Nano robotic software to replace present digital software is to be developed simultaneously. Most of the algorithms regarding the tasks are to be developed without hardware on hand. This requires simulation of models to generate shapes, sizes, plans and distributions of components. Also Nano robotics and hardware is to be programmed and manipulated to incorporate with a required size and shape with least possible mass into the systems. All these tasks require enormous amount ofskill and knowledge to be developed in collaboration between teaching and research.

REFERENCES

[1] http://warnewsupdates.blogspot.in/2008/09/smaller-is-better-future-of-warfare.html

[2] http://arxiv.org/abs/1205.2952

[3] http://spectrum.ieee.org/automaton/robotics/diy/uc-berkeley-dash-roachbot-acrobatic-flips

[4]   http://spectrum.ieee.org/automaton/robotics/robotics-software/the_coolest_flying_robot_projects

[5]   [http://www.photosfan.com/insects/]

[6]   https://plus.google.com/u/0/103255130479497734964/posts/TT1sSkff7Hm#103255130479497734964/posts/TT1sSkff7Hm

[7]   http://biology-forums.com/index.php?action=gallery;sa=view;id=6268

[8]   <http://venturebeat.com/2011/08/17/ibm-cognitive-computing-chips/#CIZQ7ZDVCf412SaF.99>

[9]   http://venturebeat.com/2011/08/17/ibm-cognitive-computing-chips/

[10]   EricoGuizzo / Mon, November 05, 2007 at :http://spectrum.ieee.org/automaton/robotics/robotics-software/the_coolest_flying_robot_proje

[11]   Tom Ehrhard in http://www.telegraph.co.uk/news/worldnews/1565879/US-accused-of-making-insect-spy-robots.html

[12]   <Thttp://defense.about.com/od/weaponry/ig/Micro-Aerial-Vehicles-Image-Gallery/MicroPropulsion-MAV.htm>

[13]   http://www.youtube.com/watch?v=5LqSWiatV0Q&feature=youtu.be

[14]   http://www.wired.com/dangerroom/2011/10/drones-tiny-hairs/

[15]   http://www.flightglobal.com/news/articles/cyborg-insects-born-in-darpa-project-222271/

[16]   http://www.wired.com/dangerroom/2007/02/cyborg_flying_r/

TECHNICAL SESSION – IV

MODELLING AND SIMULATION & FUSELAGE AND PROPULSION SYSTEMS

GRAY IMAGE SEGMENTATION: EXPECTATION MAXIMIZATION APPROACH

Chillarige Adithya and Prof V Kamashi Prasad

ABSTRACT

Image Segmentation is one of the important tasks in image understanding, in various applications. There are various approaches deterministic, fuzzy and non-deterministic in nature. One of the probabilistic approach is Expectation Maximization has been studied in this paper to understand and demonstrate in-centric issues in image segmentation as well as the interpretations. In this paper a novel initialization method is proposed and integrated with EM Algorithm in MATLAB Environment. The same has been experimented on various image datasets like benchmark images, natural, outdoor, medical, synthetic images etc., A layered representation of the given image based on these segments proposed in this paper show promising outcome to asses domain experts.

KEYWORDS: Expectation, Posterior, Gaussian, Mixture models

INTRODUCTION

IMAGE

An image (from Latin imago) or picture is an artifact, usually two-dimensional, that has a similar appearance to some subject-usually a physical object or a person. Images may be two-dimensional, such as a photograph or screen display, or three-dimensional such as a statue. They may be produced by optical devices-such as cameras, mirrors, lenses, telescopes, microscopes, etc. and natural objects and phenomena, such as the human eye or water surfaces. In our daily life we see world in the world map, what all we can see are images, these images are Asia, India, Antarctica and America etc., even oceans, sea, rivers, etc. can also be represented as an image.

ROLE OF SEGMENTATION

If we want to see India in the world maps how we can see it? Just by observing the region that is covered, which is given by its boundaries. This way of finding the boundaries and finding out the particular required regions in images is called image segmentation. Image boundaries can be identified by the contrast. The background and foreground will not be the same. This way of finding the boundaries by observing the contrast that will help us to partition the images into segments. The images will change their properties on basing on the frequency. So it is difficult to prepare a priori information for the contents and image segmentation methods on the basis of the observed data information are required for discrimination of objects. Image segmentation is majorly used in many fields in our daily life like geographical observations, Genetics, and many other fields, majorly used in medical for diagnosis, detection of the affected region of the body like cancers etc. Even this can be used for the detection of the particular region of picture, just like finding the eye of the face. Image segmentation methods can be categorized into edge detection base and pixel classification base. A disadvantage of the former is that unclosed regions may appear, whereas a disadvantage of the latter is that isolated classification may appear. The unclosed regions sometimes become crucial for object detection. On the other hand the isolated classification can be removed by some technique of filtering.

Several general-purpose algorithms and techniques have been developed for image segmentation2. Since there is no general solution to the image segmentation problem, these techniques often have to be combined with domain knowledge in order to effectively solve an image segmentation problem for a problem domain.

ABOUT IMAGE SEGMENTATION

If we want to see India in the world map how we can see it? Just by observing the region that is covered, which is given by its boundaries. This way of finding the boundaries and finding out the particular required regions in images is called image segmentation. Image boundaries can be identified by the contrast. The background and foreground will not be the same. This way of finding the boundaries by observing the contrast that will help us to partition the images into segments. The images will change their properties on basing on the frequency. So it is difficult to prepare a priori information for the contents and image segmentation methods on the basis of the observed data information are required for discrimination of objects. Image segmentation is majorly used in many fields in our daily life like geographical observations, Genetics, and many other fields, majorly used in medical for diagnosis, detection of the affected region of the body like cancers etc. Even this can be used for the detection of the particular region of picture, just like finding the eye of the face. Image segmentation methods can be categorized into edge detection base and pixel classification base. A disadvantage of the former is that unclosed regions may appear, whereas a disadvantage of the latter is that isolated classification may appear. The unclosed regions sometimes become crucial for object detection. On the other hand the isolated classification can be removed by some technique of filtering.

EM ALGORITHM

It is a generic iterative algorithm for calculating the maximum-likelihood or maximum-a-posteriori estimates when the observations can be viewed as incomplete data. Algorithm consists of an expectation step followed by a maximization step for every iteration.

$$P(x|\Theta) = \sum_{i=1}^{M} \alpha_i p_i(x|\Theta_i)$$

Where

$$\Theta = (\alpha_1, \dots, \alpha_M, \theta_1, \dots, \theta_M)$$

Such that

$\alpha_i > 0 \text{ And} \sum_{i=1}^{M} \alpha_i = 1$

POSTERIOR PROBABILITIES

$$P^{new}(j) = \frac{\sum_{i=1}^{M} p_i(j|x_i, \theta^{old})}{\sum_{j=1}^{M} \sum_{i=1}^{N} p_i(j|x_i, \theta^{old})}$$

The present study is confining to Gaussian distribution i.e., the probability density function is normal with mean μ and variance σ^2 (here $\theta = (\mu, \sigma^2)$) the corresponding expressions for evaluating these parameters is as follows

$$\mu_j^{new} = \frac{\sum_{i=1}^{N} p(j|x_i, \theta^{old})x_i}{\sum_{i=1}^{N} p(j|x_i, \theta^{old})}$$

$$(\sigma_j^{new})^2 = \frac{1}{d}\frac{\sum_{i=1}^{N} p(j|x_i, \theta^{old})||x_i - \mu_j^{new}||^2}{\sum_{i=1}^{N} p(j|x_i, \theta^{old})}$$

$$\alpha_j = P^{new}(j)$$

PROPOSED ARCHITECTURE

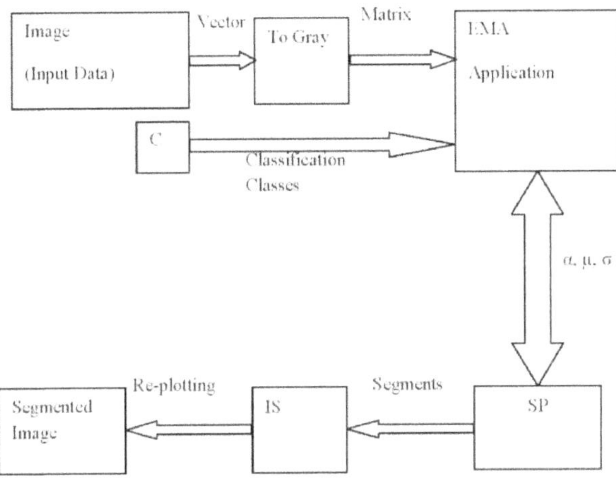

This is single tire architecture; here we will give only two inputs to the system. The inputs are image (which is to be segmented) and the number of segments (number of classes that we expect). For the ease of the application we have provided with the GUI developed using GUIDE. In this GUI we are going to provide an input file option, number of segments, and number of iterations.

ASSOCIATED GUI & DEMONSTRATIONS

It takes Image, Number of Segments and Iterations as input.

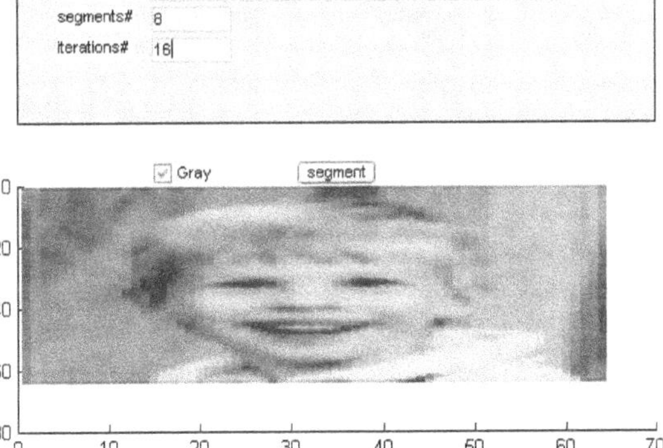

LEENA

INPUT SELECTION

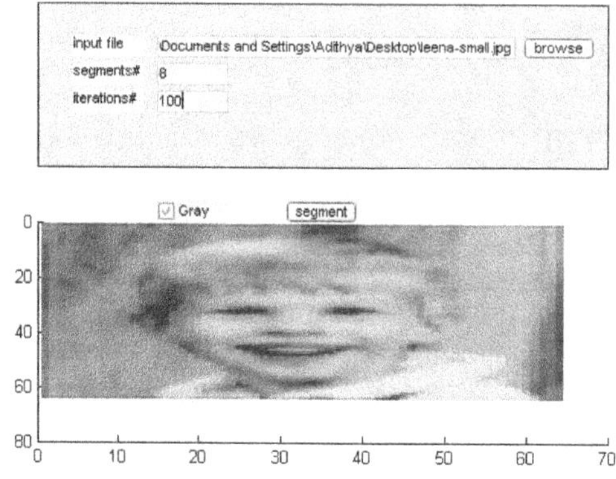

OUTPUT OVERVIEW

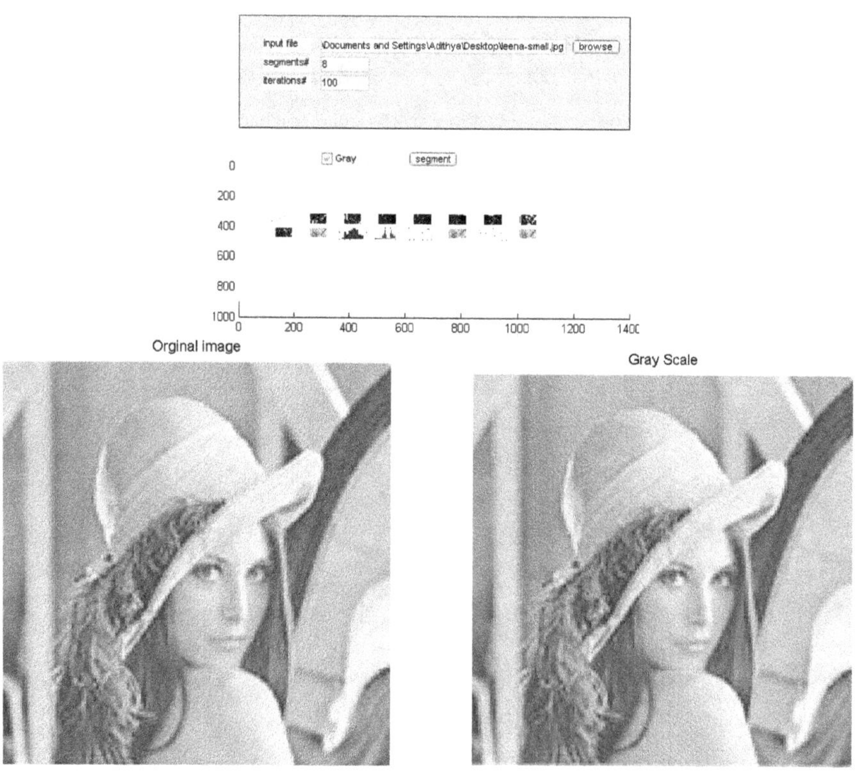

HISTOGRAM

The following picture shows the gray image of Leena

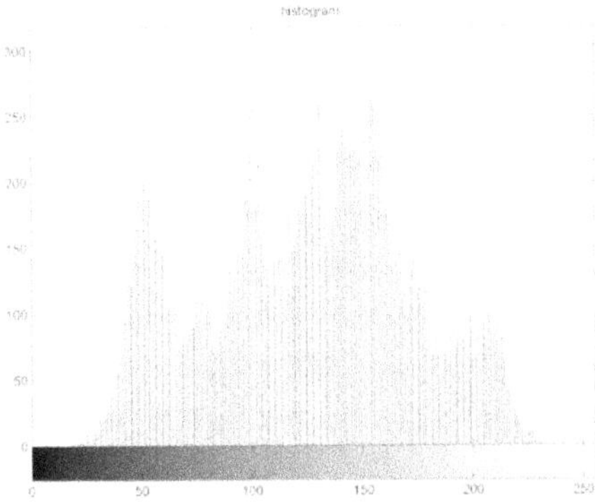

ALPHA

Modeling aspect of Leena is expected with 8 Gaussian components
The following figure depicts, convergence pattern of the mixing coefficients over the iterations.

MEANS

The following figure depicts, convergence pattern of the means over the iterations.

STANDARD DEVIATIONS

The following figure depicts, convergence pattern of the standard deviations over the iterations.

LAYERS OF LEENA

The following 8 pictures shows the corresponding asymptotic segments of Gray Scale Leena as an output of the proposed EM approach with 8 EM components which are referred as layers.

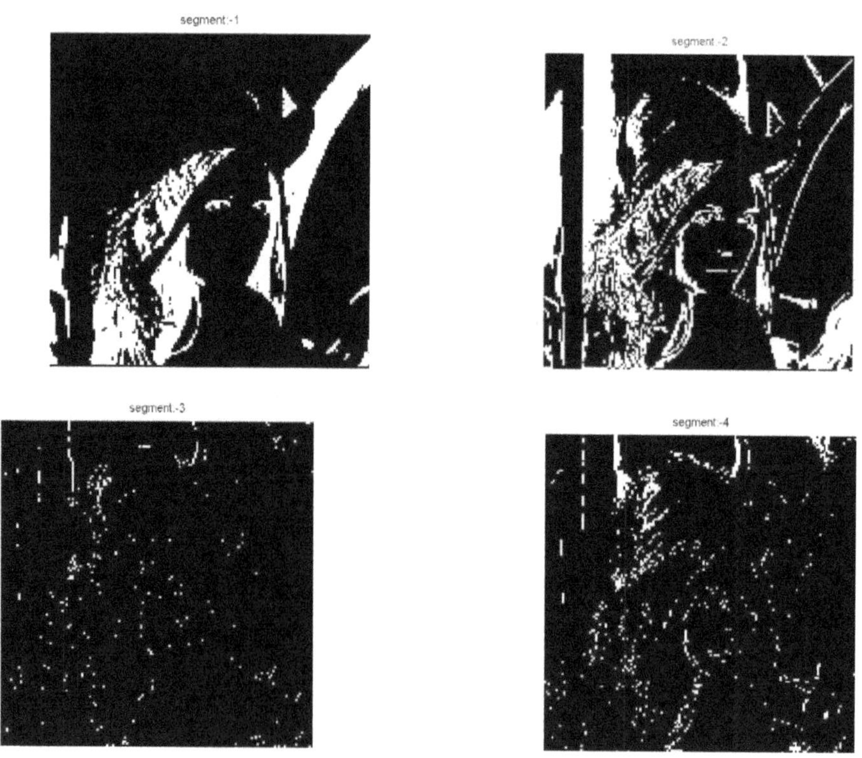

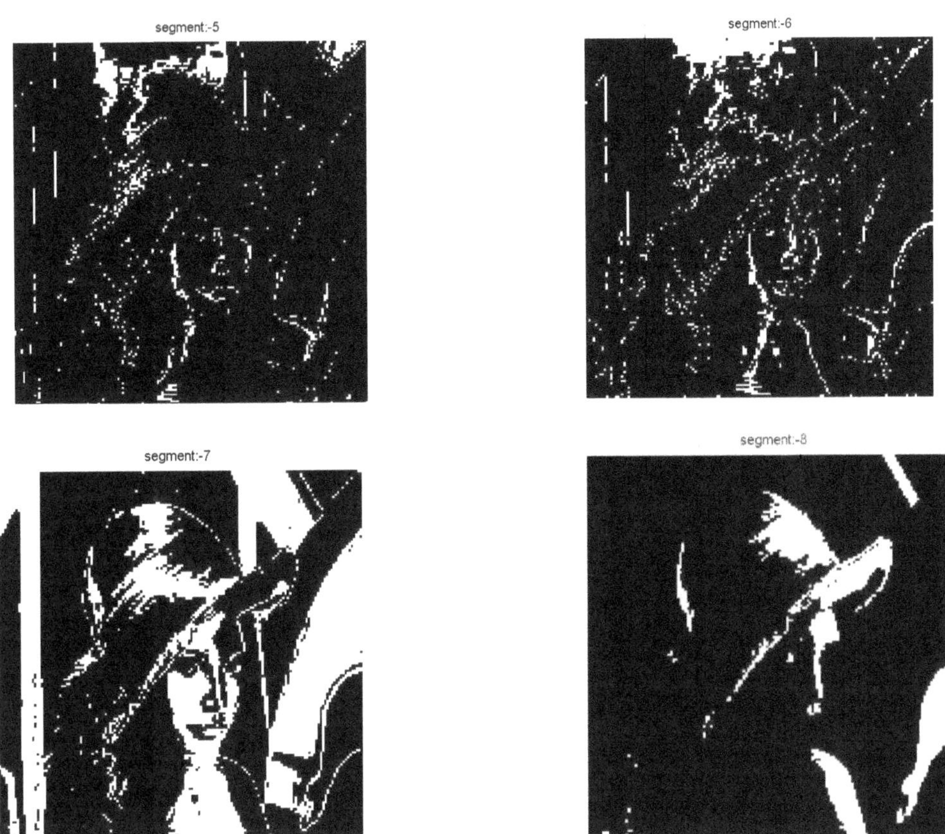

SEGMENTED IMAGE

Below picture depicts the modelled synthetic Leena image and its EM diagnostics.

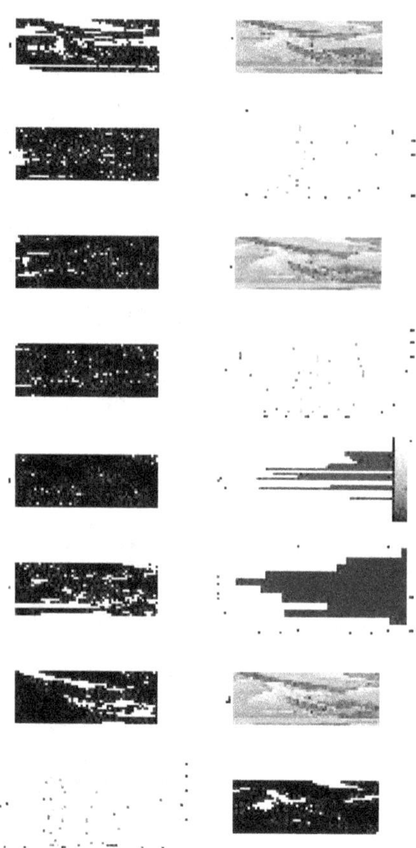

BENCHMARK ANALYSIS
WORLD MAP

With components = 4.

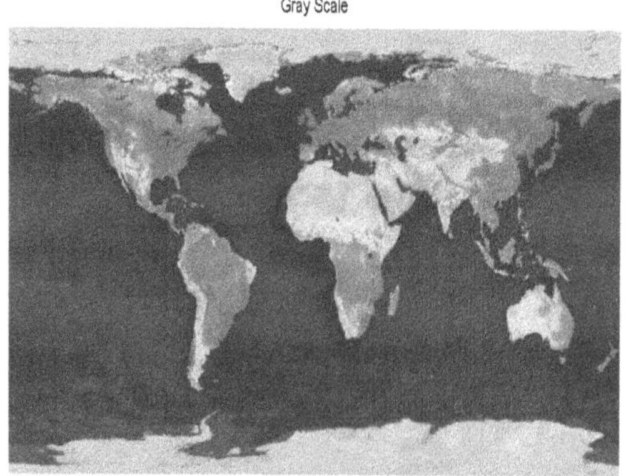

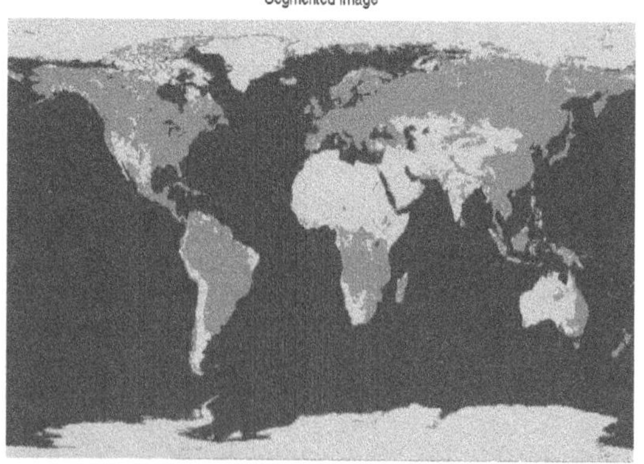

Below are the 4 layers which depict the salient features of the image.

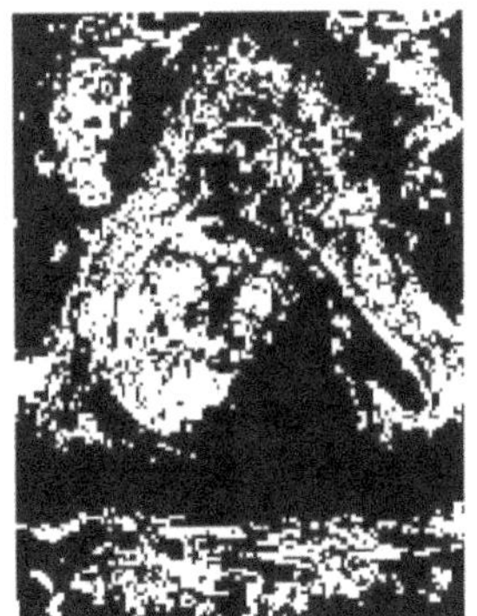

GRAPHICAL IMAGES

MEDICAL IMAGES

Below image depicts how the EM segmentation assists the medical professionals in diagnosing problems with the skin in particular with cancer patch.

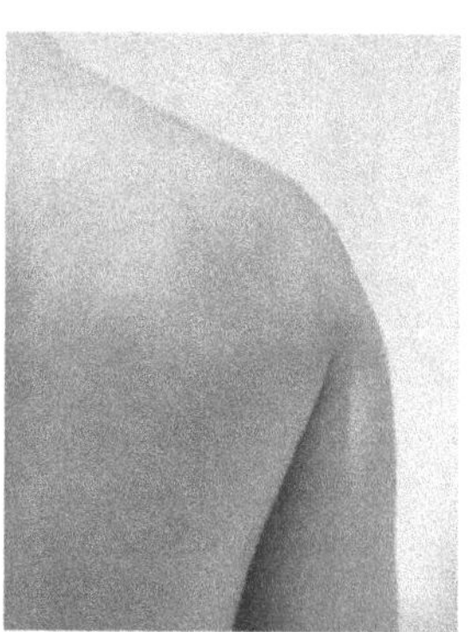 

CONCLUSIONS

By apt component configuration the regions obtained through this EM segmentation found useful for highlighteing contasting features with its neighbours in particular medical diagnostics. Further segmentations with varying components(large) is expected to provide knowledge for edge detection, regions detection and hence fore building highlevel knowledge.

REFERENCES

1. Alexis Roche "EM algorithm and variants: an informal tutorial" CEA { Service Hospitalier Fr_ed_eric Joliot 4, place du G_en_eral Leclerc, 91401 Orsay, France roche@shfj.cea.fr

2. Dana Elena Ilea and Paul F. Whelan "COLOR IMAGE SEGMENTATION USING A SELF-INITIALIZING EM ALGORITHM" Proceedings of the Sixth IASTED International Conference Visualization, Imaging , and Image Processing August 28-30, 2006.

3. Dimitris Karlis "An EM algorithm for multivariate Poisson distribution and related models"Journal of Applied Statistics, Volume 30, Issue 1 January 2003 , pages 63 – 77 http://www.informaworld.com/smpp/content~content=a713675079

4. F. Dellaert"Monte Carlo EM for Data-Association and its Applications in Computer Vision" doctoral dissertation, tech. report CMU-CS-01-153, Computer Science Department, Carnegie Mellon University, September, 2001 http://www.ri.cmu.edu/pubs/pub_3851.html

5. Fatih Gelgi —Expectation-Maximization□ ASU, 2005.

6. Frank Dellaert "The Expectation Maximization Algorithm" Technical Report number GIT-GVU-02-20 College of Computing, Georgia Institute of Technology, February 2002

7. Jeff A. Bilmes —A Gentle Tutorial of the EM Algorithm and its Application to Parameter Estimation for Gaussian Mixture and Hidden Markov Models□ INTERNATIONAL COMPUTER SCIENCE INSTITUTE TR-97-021 April 1998.

8. Jimbo H. C and Suzuki T "A Comparative View on the EM Algorithms" Technical Report No.2004-3, Advanced Research Institute for Science and Engineering, Waseda University May, 2004

9. M.G. Mostafa,T.F. Gharid, M.F. Tolba, and M.A. Megeed "Medical Image Segmentation using Wavelet-Based Multiresolution EM Algorithm" IEEE International Conference on

Industrial Electronics, Techonology & Automation, IETA'2001, Cario, 19th – 21st Dec., 2001.

10. Mário A. T. Figueiredo, and Robert D. Nowak "An EM Algorithm for Wavelet-Based Image Restoration" IEEE TRANSACTIONS ON IMAGE PROCESSING, VOL. 12, NO. 8, AUGUST 2003.

11. Michael I. Jordan and Robert A. Jacobs "Hierarchical Mixtures of Experts and the EM Algorithm" A. I. Memo No. 1440, C.B.C.L. Memo No. 83 MASSACHUSETTS INSTITUTE OF TECHNOLOGY ARTIFICIAL INTELLIGENCE LABORATORY And CENTER FOR BIOLOGICAL AND COMPUTATIONAL LEARNING DEPARTMENT OF BRAIN AND COGNITIVE SCIENCES, August 6, 1993.

12. Sankar K. Pal and Pabitra Mitra, "Multispectral Image Segmentation Using the Rough-Set-Initialized EM Algorithm" IEEE TRANSACTIONS ON GEOSCIENCE AND REMOTE SENSING.

13. Sean Borman —The Expectation Maximization Algorithm A short tutorial☐ Sean Borman July 18 2004.

14. Tatsuya Yamazaki "Introduction of EM Algorithm into Color Image Segmentation" ATR Adaptive Communications Research Laboratories, 2-2 Hikaridai, Seika-cho, Soraku-gun, Kyoto 619-0288 Japan yamazaki@acr.atr.co.jp

15. http://en.wikipedia.org/wiki/Segmentation_(image_processing)

16. http://en.wikipedia.org/wiki/Graphics_file_format

17. http://en.wikipedia.org/wiki/EM_algorithm

18. http://www.gps.caltech.edu/~tapio/imputation/

19. Rafael C.Gonzalez, Richard E. Woods and Steven L.Eddins "Digital Image Processing Using MATLAB" Low Price Edition, Pearson Education, 2007

20. Anil K.Jain " Fundamentals Of Digital Image Processing" Eastern Economy Edition, Prentice – Hall of India 1995

21. Earl Gose, Richard Johnsonbaugh,, and Steve Jost "Pattern Reorganization and Image analysis" Eastern Economy Edition, Prentice – Hall of India 1999

22. Milan Sonka, Vaclav Hlavac, and Roger Boyle —Image Processing, Analysis, and Machine Vision" Thomson Brooks/Cole, Vikas Publication House.

23. Rafael C. Gonzalze, and Richard E. Woods "Digital Image Processing" Addison Wesley, Pearson Education 2000.

HARDWARE IN LOOP SIMULATIONS FOR CO-OPERATIVE MISSIONS USING ARDUPILOT

Bharat Tak
Research Assistant, email: bharat.tak8@iitb.ac.in

Swaroop Hangal
Graduate Student, email: swaroop.hangal@iitb.ac.in

Hemendra Arya
Associate Professor, email: arya@aero.iitb.ac.in
Department of Aerospace Engineering, Indian Institute of Technology Bombay, Mumbai-400076

ABSTRACT

A framework for Hardware in Loop Sim- ulation (HILS) of cooperative missions in autonomous miniature aerial vehicles (MAV) using open source Ardupilot-Mega [1] based hardware and software platform has been implemented. Ardupilot Mega (APM) is an Arduino [2] based commercial autopilot board for stabilization and navigation of individual MAVs. Our framework extends its features to support HILS of cooperative missions for multiple MAVs by facilitating inter-aircraft and aircraft-ground station communication. The framework has a loosely coupled design which isolates the high level planning algorithms from the rapidly evolving low level autopilot routines. It is designed as a natural extension to the various flight modes (Manual/Return-to-Launch/Circle/Waypoint navigation) already present in Ardupilot-Mega platform and can be switched between them seamlessly. Inter-aircraft communication is done using Xbee API protocol which facilitates unicasting custom message packets to-and-from any aircraft of the cooperative system. The Xbees are also used by all the aircrafts in the HILS to communicate with the QGroundControl Station (qGCS) [3] using the widely compatible MAVlink protocol [4]. This makes the framework compatible with any other GCS supporting MAVlink protocol. Since all the communication between various nodes of simulation is through Xbees in API mode, it makes the framework readily extendable for flight testing of the cooperative algorithms in real flights. The hardware in this HIL simulation comprises of the APM autopilot controller board and Xbee modules in API mode. The APM board runs both the low level autopilot software and the high-level cooperative algorithms. Xbees perform the inter-aircraft and aircraft-GCS communication. The sensors and actuators are being simulated in this HILS framework. Open source flight simulation software Flightgear [5] is used to simulate the flight dy namics. The sensor/actuator information is exchanged between the autopilot board and Flightgear over a serial port. The paper demonstrates the framework implementing Leader-Follower and Orbit follower co- operative algorithms.

KEYWORDS: Hardware-in-Loop Simulation, Coop-erative Flight, MAV, Ardupilot Mega, Qgroundcon-trol, Flightgear.

INTRODUCTION

A lot of research and development has occurred in the field of design and control of MAVs in the recent past owing to the plethora of hobby, civil and defense applications they are suitable for. The growth in the number of widely available cheap, lightweight and increasingly powerful micro controllers and sensors has fueled the rise of numerous community owned commercially available Autopilot boards that provide features like stabilization and way-point navigation of MAVs. ArduPilotMega, pxIMU, SLUGS, Paparazzi, UAVDevBoard are a few of this rapidly expanding group of open source Autopilots. Since the firmware of these Autopilot boards are open source, they can be modified or extended as required. Various Do It Yourself (DIY) communities have spawned consisting of members from the academia, serious hobbyists and end users alike collaborating to develop better controllers.

Traditionally, these Autopilot systems provide support for communication and control of a single MAV with the ground station. The autopilot system consist of an onboard controller with an IMU shield, a wireless modem for communication and a Ground control station (GCS) software. The GCS software comes bundled up with the autopilot firmware or in some cases can be separately downloaded. Common features of most GCS include realtime GPS trajectory plotting, data logging, MAV attitude display and inflight parameter updates to the MAV. With the rise in the number of such Autopilot boards and their corresponding groundcontrol station (GCS) software, there is a trend towards using common communication protocols for cross compatibility of Autopilots and GCS. MAVlink protocol is one such cross compatible Autopilot-GCS communication protocol that is becoming increasingly popular with most open source Autopilot systems.

Given the size and resource constraints on MAVs that limit their range and endurance, operating them in teams hugely multiplies their usability for missions such as search and rescue, surveillance, data gathering etc. Autonomous control of multiple MAVs in cooperative missions has thus been a hot topic of research in the scientific community. Min- imizing the communication overhead while sharing information between MAVs and the use of decentral-ized control schemes rather than centralized control schemes has been preferred to increase robustness in case of failure of individual entities. Implementing such algorithms on MAVs requires several stages of testing to verify the autopilot routines, communica-tion network and planning algorithms. Hardware. In Loop Simulations (HILS) play an integral part in this process of testing the system in near-realistic environment thus increasing the confidence in the system for real world implementation. In this paper, a Hardware In Loop Simulation framework based on open source autopilot hardware and software platform capable of simulating cooperative missions of multiple MAVs has been presented.

LITERATURE SURVEY

Use of HILS simulation for reliable testing and verification of hardware and software modules of complex autonomous systems is common in lit- erature. An engineering approach to designing a HILS framework for UAV control system has been discussed in [7]. The author also discusses actuator emulator system construction based on NI PXI 1042 platform for testing of the HILS system. Dspace, a MATLAB based real time testing platform, is used for implementing the

HILS system. The setup consists of an embedded control system, attitude sensors and actuator emulation system using Dspace for virtual sensor simulation. [8] discusses system architecture for design of realtime HIL simulation for multiple UAVs. The author also presents Comm- LibX, a communication framework for simulation modules to communicate over multiple virtual channels. [9] discusses the design and development of a Hardware in Loop simulator in LabVIEW. [10] discusses the design and development of HILS for cooperative missions of upto 8 MAVs. It uses XPC target real time operating system provided by MAT- LAB for simulating the flight dynamics of MAVs in real time. The design of communication network for inter aircraft communication in the HILS for various cooperative missions has also been discussed.

PROBLEM STATEMENT

It is clear from the literature survey that most HILS setups for cooperative flight have been implemented on custom designed hardware and software platforms involving their own specific communica-tion protocols. Although the design and implemen-tation of these HILS setups might be well suited to their applications, they have many drawbacks. It requires time and effort to design, implement and test the Autopilot controller, the low level autopi- lot stabilization and navigation routines, the design and development of ground control station (GCS) software, the design of interfacing communication protocol, creating the flight dynamics model etc before the eventual design and implementation of the cooperative mission control. It also restricts the design and development of the HILS setup to those who have access to the specific hardware and software used. In a scenario where a large number of rapidly developing field-tested open source Au- topilot controllers and GCS softwares are readily available across the globe, it makes sense to setup a framework for Hardware in Loop simulation of cooperative missions using these community devel-oped open source hardware and software platforms as building blocks. It reduces the time and effort required for building the HILS framework for co- operative missions and makes the setup accessible to a wider community of users. Hence, the purpose of this paper is to implement a HILS framework for cooperative mission using readily available open source hardware and software platforms.

SYSTEM DESCRIPTION

As the name suggest, HILS framework involves actual hardware subsystems in the loop while creating near-realistic simulations to test the system. Cooperative missions involves multiple MAVs inter- communicating among themselves and with the GCS to accomplish a common task. The paper implements a framework for HILS simulation of upto 3 MAVs but the node based architecture of the system enables it to be extendable to perform HILS simulation of larger number of MAVs limited only by the increased communication traffic that may induce packet clashes and data loss. Each node of the HILS framework consists of a completely self contained subsystem interacting with the other nodes only through wireless modems. This enables each node to be simulated on a separate computer creating a distributed network of HILS. The different nodes of the HILS framework and there communi- cation topology for a generic cooperative mission are summed up in Fig. 2 below.

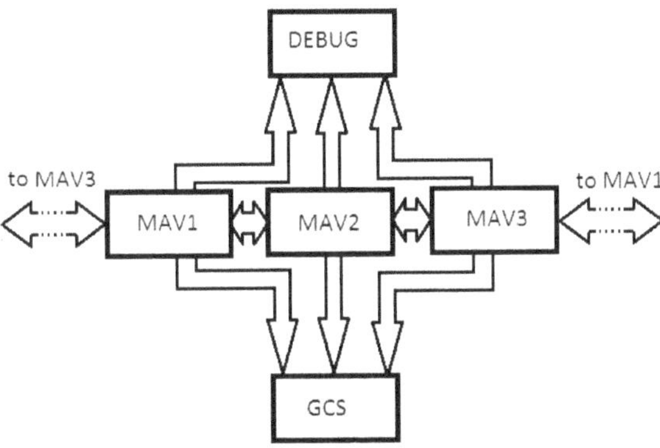

Fig. 1 Communication topology of the various nodes of HILS

A brief overview of the different hardware sub-systems and simulated components of the HILS framework are shown in the Fig. 1 below.

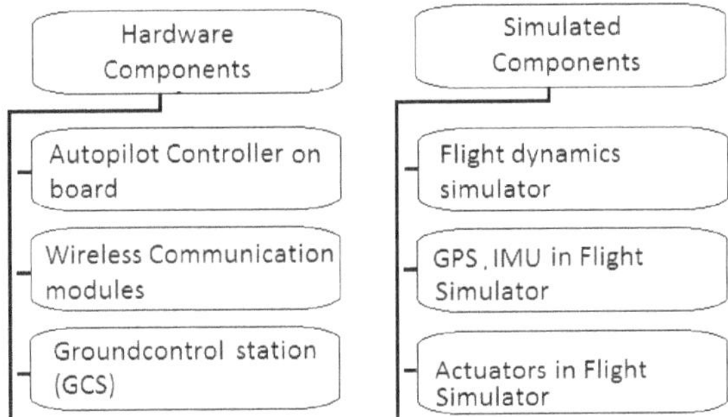

Fig. 2   Hardware and software components of the HILS framework

These various components of the HILS setup are explained in detail below.

A. Onboard Autopilot controller

The onboard autopilot controller used is Ardupi- lotMega v1.4 (APM) controller board. It is an off-the-shelf commercial autopilot board based on a 16MHz Atmega2560 processo that supports 3D waypoints and mission commands,   has   16 spare analog inputs and 40 spare digital input/outputs to add additional sensors, four dedicated serial ports for two-way telemetry (using XBee modules) and can be powered by either the RC receiver or a separate battery [1]. Full autopilot software, including IMU and groundstation/mission planning code can be downloaded online.

Interfaced with a GPS, an IMU shield and an Xbee communication module, it can be used to conduct a fully autonomous waypoint navigation mission including autonomous take-off and landing of a single MAV. The stock autopilot code support 2 way telemetry with the GCS using standard Xbee- Pro communication modules in transparent mode, effectively broadcasting the messages to any other Xbee in range. The communication protocol used is MAVlink which makes it compatible with a wide range of ground control stations like APM Mission- planner, Qgroundcontrol GCS, Happykillmore GCS etc.

The default GCS of APM i.e. APM Missionplanner provides an interface to conduct Hardware in Loop Simulations (HILS) of a single MAV running on the APM board. It supports a wide range flight simulators including Flight Gear, JSBsim, Xplane and AeroSimRC. APM Missionplanner acts as an interface between the Flight dynamics solver and the APM autopilot controller board. It uses UDP port to connect with the flight simulator and uses the onboard serial port on APM to connect with the controller. It receives the GPS information (Latitude, Longitude, Atitude, Heading etc)and IMU informa- tion(body axis accelerations and angular rates)from the flight simulator and sends them to the APM controller in the appropriate format. The controller calculates the control commands for the actua- tors(Aileron, Elevator, Rudder, throttle) based on the stabilization and navigation algorithm which are then sent to the flight simulator through UDP in the appropriate format. Fig 3. below gives an overview of this HILS setup.

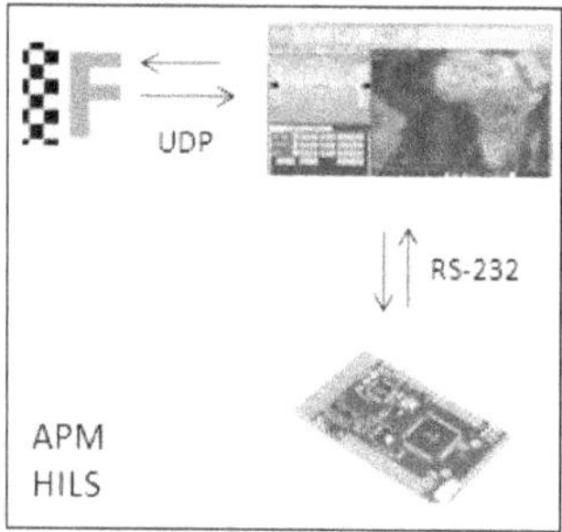

Fig. 3   ArdupilotMega HILS framework

It should be noted that this HILS setup is limited to simulations involving only a single MAV for the following reasons.

• APM controller board can process navigation commands sent only from the GCS

• APM has bidirectional communication support only with the GCS, i.e. no default support for inter-aircraft communication

• The stock GCS i.e. APM Missionplanner can monitor and control of only one MAV connected to it via MAVlink

• APM controller board and the defaut GCS Mis-sionplanner provide support for Xbee wireless telemetry only in the transparent mode i.e. broadcasting the messages

Cooperative missions involving multiple MAVs requires onboard autopilot controllers that can pro-cess navigation commands sent not only by the GCS but also other members of the cooperative system. It requires support for inter-aircraft communication. The ability to unicast messages to specific nodes rather than flooding the communication network by broadcasting information is necessary. Cooperative missions also require a GCS that can monitor mul- tiple MAVs simultaneously. Despite the above men- tioned limitations of the stock firmware, it provides a platform over which required modifications to enable cooperative mission HILS can be made.

To enable HILS of cooperative missions on APM board, a higher level of navigation logic was created above the default autopilot routine on the controller board that could process the information received from other nodes of the HILS and generate naviga- tion commands depending on the type of cooperative mission at hand for the lower level stabilization and navigation routine to execute. This is where the logic for various cooperative missions was coded. It also handled the inter-aircraft communication depending on the specific cooperative mission being simulated. Interface libraries were added to port the current communication system from broadcasting messages with Xbees in transparent mode to unicasting mes- sages using Xbees in API mode. Libraries were also added to create specific inter-aircraft communication packets and execute the communication as required by the cooperative mission logic. Finally, sticking to MAVlink potocol for aircraft-GCS communica- tion enabled a shift to Qgroundcontrol GCS as it provided support for monitoring multiple MAVs simultaneously.

Thus, the modified version of Ardupilot firmware for HILS simulation of cooperative missions could transmit and receive data packets from both GCS and other MAVs. It would then process the data obtained from the rest of the MAVs and and give appropriate navigation commands based on the co- operative mission being simulated to the autopilot routine of APM, which would then carry out the low level stabilization and navigation of the MAV and update its status to the Qgroundcontrol GCS. On the whole, an overview of the modified firmware on the APM is as shown in Fig. 4.

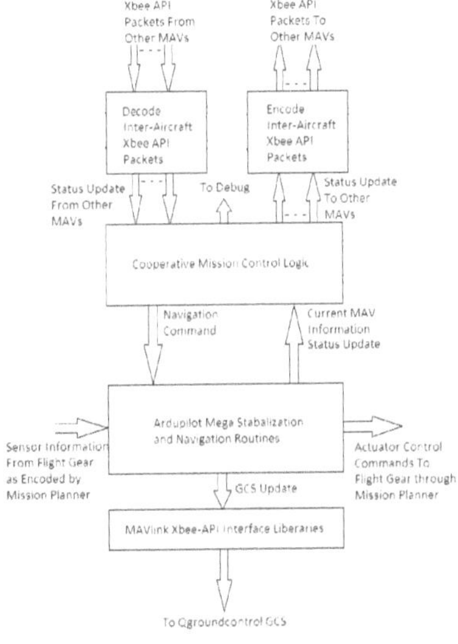

Fig. 4 An overview of the modified ArdupilotMega HILS framework

B. Communication Network

A generic cooperative mission control algorithm may require status updates from a few or all the rest of the members of the cooperative mission. Thus the ability to send and receive information among all the nodes of the simulation is paramount in enabling the implementation of any general cooperative al- gorithm. For example, communication topology for Leader-Follower mission and non linear cyclic pur- suit algorithm for 3 MAVs is as shown in Fig. 5 & Fig. 6 below.

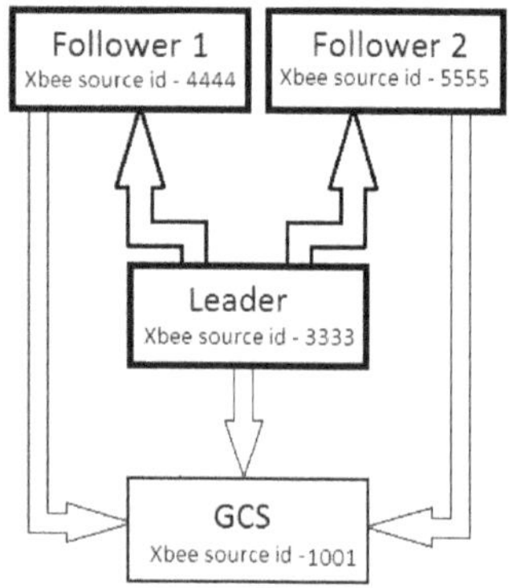

Fig. 5 Communication topology for Leader follower mission

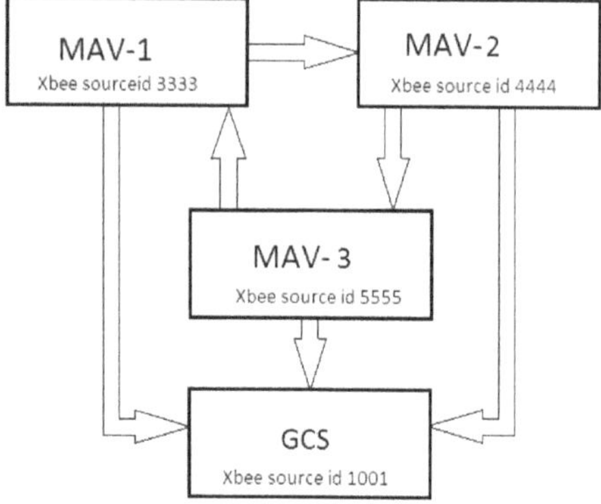

Fig. 6 Communication topology for non linear cyclic pursuit

Xbee Pro wireless modules have been used to implement the above communication strategies. Xbee modules can be programmed to either operate in transparent mode and broadcast their messages or operate in Application programming Interface mode (API) effectively unicasting their message to the source id encoded in the message packet. Transparent mode has many advantages. It does not add any overhead to the message being sent and the message need not have any specific packet structure to be delivered. The message could be received by any other Xbee module in range if it operates in the same channel. Disadvantages include that specific one-to- one messages with different receiving Xbees cannot be executed. Also, there is no acknowledgement of successful transmission of the message.

API mode has the ability to execute to one to one communication with other Xbees and receive acknowledgment of successful transmission of mes- sage packet. The message packet need to be in a particular format to be successfully transmitted and includes a checksum to verify the sanity of the received packet. This induces a small overhead of 6 bytes per packet of transmitted data. The overhead includes

- Start Delimiter

- Packet size

- Frame type

- Frame id

- Source id of destination

- [Message packet]

- Checksum

Since the ability to execute specific one-to-one communication is paramount to designing suitable communication networks for cooperative mission control, Xbees in API mode have been used on all the MAVs in HILS for inter-aircraft communication and on the GCS for MAV-GCS communication. Support for an optional debug Xbee in API mode has also been created to monitor the transfer of pack- ets in realtime on a serial communication monitors like Docklight.

C. Flight Simulator

APM Missionplanner supports HILS with various flight simulators like Flight Gear, JSBsim, Xplane and AeroSimRC. In keeping with the philosophy of designing a cooperative flight HILS based on open source hardware and software platform, Flight Gear was selected to simulated the flight dynamics. Flight Gear is a well tested open source flight simulator. It supports 3 primary flight dynamics solver JSBSim, YASim and UIUC. It works across platforms(Windows, Linus, Mac, FreeBSD, Solaris, IRIX etc), supports numerous aircraft models and has an extensive world scenery database. It supports an open aircraft modelling system that enables the users to design and modify there aircraft models. The aircraft used in this HILS setup is Rascal 110 (R/C) which is an R/C plane developed as a part of a UAV project at University of Minnesota (although support for it seems to have ended).

Fig. 7   Rascal 110 (R/C) Flightgear aircraft model [6]

D. Ground Control Station (GCS)

The Groundcontrol station used for this HILS setup is Qgroundcontrol v 0.9. It is an open source GCS software based on PIXHAWK's Groundstation and is now developed in a joint effort with community.

Fig. 8 Qgroundcontrol GCS [3]

It supports realtime monitoring and control of multiple MAVs using MAVlink 0.9 communication protocol. It supports advanced features like 2D/3D aerial maps (with Google Earth integration), realtime plotting of sensor and telemetry data, data logging, in flight manipulation of waypoints and inboard parameters, support for digital video transmission etc. Currently, Qgroundcontrol GCS is being used for one way telemetry with all three MAVs, performing data logging and realtime data monitoring operations.

On the whole, the entire 3 MAVs system for Hard- ware in Loop Simulation of cooperative mission can be looked upon as shown in Fig. 5 below. All the boxed nodes can be simulated on separate computers and all the inter-node communication is in Xbee API mode.

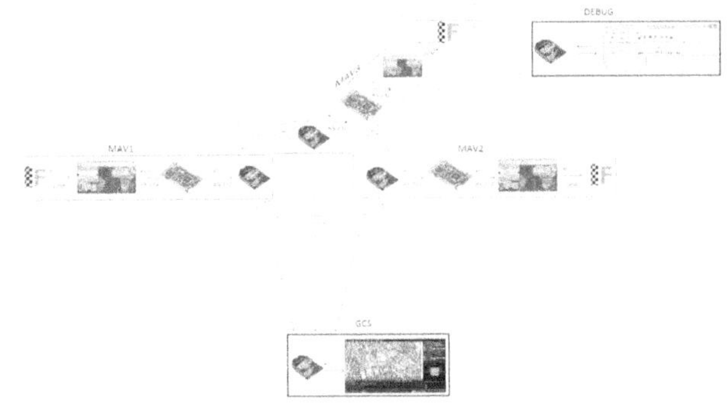

Fig. 9   Setup for cooperative mission HILS of 3 MAV system

V. RESULTS

Initially the HILS framework was verified by con- duction a HILS of autonomous way-point navigation mission using a   single MAV. The mission profile included autonomous take-off followed by a set of 6 way-points at different altitudes. It was verified that the simulated MAV closely follows the designated path connecting the way-points in a stable fashion. Fig. 10 below the path followed by the MAV as shown on the GCS.

Fig. 10 Mission 1. Autonomous way-point navigation. Note that Yellow Dots have been manually added on the MAV path(shown in blue color) to improve its visibility

The path reconstructed using the logged telemetry data as displayed on Google earth.

Fig. 11 Mission 1. Google Earth display of MAV trajec- tory(constructed from telemetry log)

After verifying that the lower level autopilot routines were functioning as desired, a cooperative flight Leader-Follower HILS mission was carried out using 2 MAVs. The Leader MAV was set to follow independent way-point navigation including autonomous take-off and landing, and send its lo- cation update to the Follower MAV at 1 Hz. The Follower MAV was set to accept the Leader's lo- cation as its target location (way-point) each time it received a fresh location update. It was observed that the Follower MAV closely followed the Leader MAV as expected. Fig. 11 below shows the screen shot of the GCS after the completion of the Leader- Follower cooperative mission.

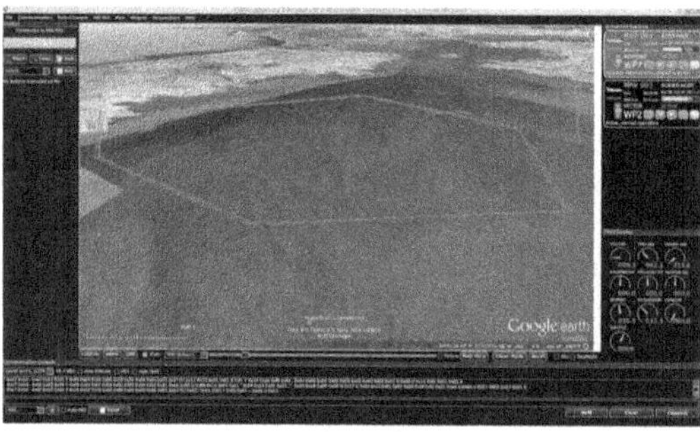

Fig. 12 Mission 2. Leader-Follower cooperative HILS

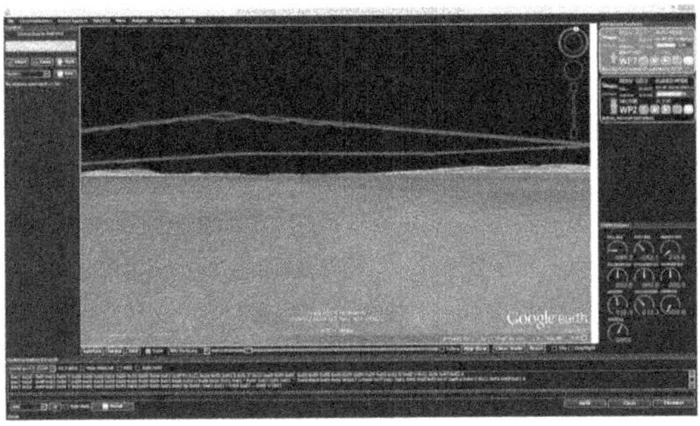

Fig. 13   Mission 2. Leader(Blue), Follower(Pink)

Fig. 14   Mission 2. Leader-Orange, Follower-Yellow

The third cooperative mission conducted involve one Leader MAV with 2 follower MAVS. The two follower MAVs were given the exact same informa- tion packets by the leader. From the screen shot of the GCS in Fig. 15 below, it can be seen that both in the firmware before the HIL simulation begun. As per the Orbit following algorithm in [11], the heading command for individual MAVs after the orbit center was set is given by π the follower MAVs followed the Leaer MAV very χd (d) = γ − − tan−1 (k(d − r))   (1) closely.

Fig. 15 Mission 3. Leader with 2 Followers. Leader-cyan, Follower 1- Blue, Follower 2-Pink
where the symbols are as shown in Fig. 17 below.

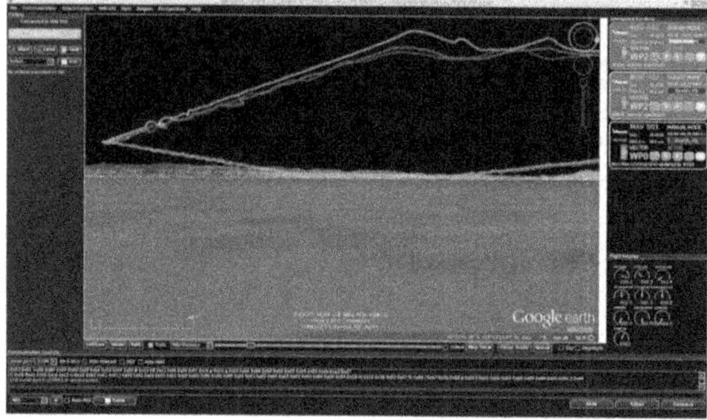

Fig. 16 Mission 3. Leader with 2 Followers. Leader-cyan, Follower 1- Blue, Follower 2-Pink

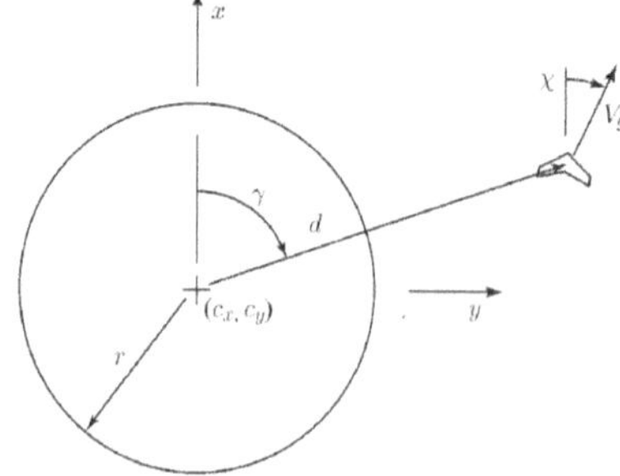

Fig. 17 Mission 4. Orbit following algorithm discussed in [?]

From the figure below, it can be observed that all the three MAVs converged into near-circular orbits around the common orbit center.

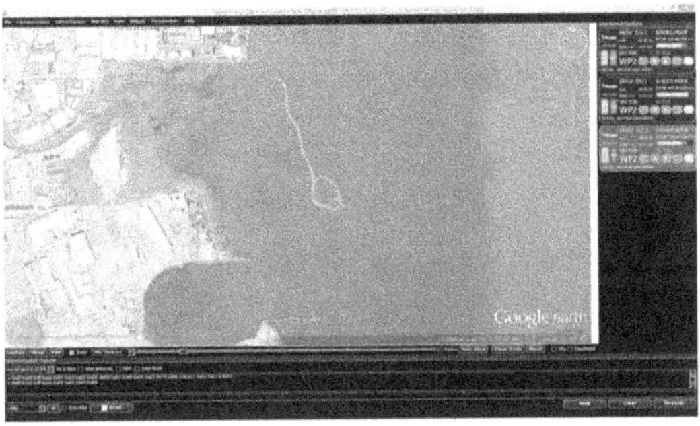

Fig. 18   Mission 4. Orbit Follower, Top view

The fourth cooperative mission conducted was Orbit follower involving 3 MAVs. The orbit center's latitude, longitude and altitude could be set in flight through the GCS on any of the 3 orbit follower MAVs. The HILS system was set such that the loca- tion of the orbit center would immediately be shared with the other two MAVs and hence all the 3 MAVs would converge into an orbit around a common center. The radius of the orbit could programmed

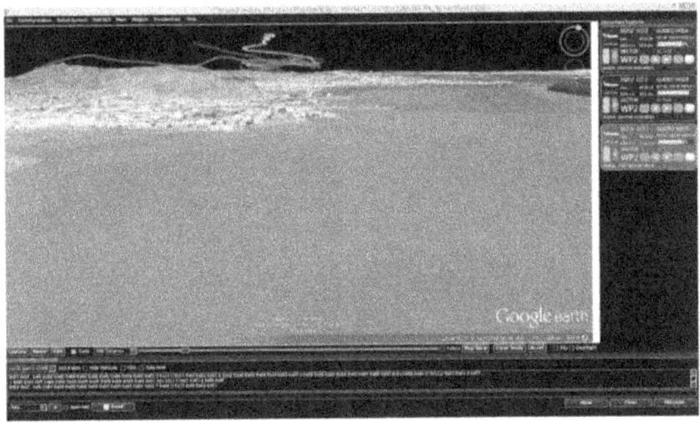

Fig. 19   Mission 4. OrbitFollower, Side view

CONCLUSION

A framework for Hardware in Loop Simulation (HILS) of various cooperative mission control algo- rithms for 3 MAVs was setup using commercially available open source Ardupilot Mega (APM) hard- ware and software platform. Open source flight sim- ulator used in the HILS was Flightgear. Qground- control GCS was used as the ground control station software to monitor and log data for all the MAVs in the simulation. The framework was verified by conducting Leader-Follower missions and Orbit fol- lower missions involving 3 MAVs.

ACKNOWLEDGEMENTS

Authors would like to thank NPMICAV for sup- port of this work.

REFERENCES

[1]   Google code wiki for ArdupilotMega, http://code.google. com/p/ardupilot-mega.

[2]   Arduino homepage, http://www.arduino.cc/.

[3]   QGroundcontrol station GCS Homepage, http://qgroundcontrol.org/.

[4]   MAVlink Developers Homepage, http://qgroundcontrol.org/mavlink/start.

[5]   Flightgear flight simulator Homepage, http://www.flightgear. org/.

[6]   Flightgear flight R/C aircraft model Rascal 110, http://wiki. flightgear.org/Rascal 110.

[7]   Shixianjun, Song Jiakun, Liu Hongxing, Hardware-in-the- loop Simulation Framework Design For a UAV Embedded Control System, Control Conference, 2006. CCC 2006. Chinese , vol., no., pp.1890-1894, 7-11 Aug. 2006.

[8]   Goktogan, A.H., Nettleton, E., Ridley, M., Sukkarieh, S., Real time Multi-UAV Simulator,Robotics and Automation, 2003. Proceedings. ICRA '03. IEEE International Confer- ence on , vol.2, no., pp. 2720- 2726 vol.2, 14-19 Sept. 2003

[9]   R. Mahajani and H. Arya MAV - hardware in loop simulation using LabVIEW, AIAA Modeling and Simulation Technolo- gies Conference, Portland, Oregon, AUGUST 2011.

[10] Dileep Krishnan, Hardware in Loop Simulator for Coopera- tive Missions, Department of Aerospace Engineering, Indian Institute of Technology, Bombay, 2012.

[11] Derek R. Nelson, D. Blake Barber, Timothy W. McLain, Se- nior Member, IEEE, and Randal W. Beard, Senior Member, IEEE Vector Field Path Following for Miniature Air Vehicles, IEEE TRANSACTIONS ON ROBOTICS, VOL. 23, NO. 3, JUNE 2007.

EFFECTS OF WIND ON COOPERATIVE MISSIONS

Aseem Vivek Borkar
Research Assistant, email: aseem.v.borkar@gmail.com

Dileep Krishnan K
MTech 2012 Batch, email: dkk.pranavan@gmail.com

Rishti Tiwari
Research Assistant, email: srishti.tiwari02@gmail.com

Hemendra Arya
Associate Professor, email: arya@aero.iitb.ac.in
Department of Aerospace Engineering, Indian Institute of Technology Bombay,
Mumbai - 400076

ABSTRACT

In order to test the cooperative control algorithms and also the on-board systems of the MAVs in realistic scenarios, a Hardware In Loop Simulator (HILS) has been presented capable of simulating cooperative missions of upto eight MAVs in real time under the influence of wind. The hardware in the loop includes the on-board computers of the MAVs simulated and servos for actuation of control surfaces. The sensor data such as the GPS, IMU and pressure sensor measurements for airspeed and altitude are all simulated in a flight simulation for each MAV running on a real time operating system called xPC Target and analog feedbacks tapped from the servos of each MAV provide inputs to the simulation. The Dryden wind model has been included in the simulation to bring in the effects of winds. Since the wind is a major factor in deciding performance of cooperative algorithms, some simulation results of multiple MAV cooperative missions with and without effects of wind have been compared and alternate trajectory tracking strategies available in literature have also been simulated under the effects of wind in real time.

KEYWORDS

Miniature Aerial Vehicles, Cooperative Missions, HILS, Autopilot, Dryden Wind Model, Cyclic Pursuit, Trajectory Tracking.

INTRODUCTION

The ability of unmanned autonomous systems to carry out dangerous and dull tasks in both military and civilian applications have made them indispens- able in our lives today. All countries strive to de- velop such platforms as they cost effective, with low risks and hence find major applications in various fields such as real time surveillance and monitoring, search and rescue,

"

etc. There is also widespread interest in making autonomous agents communicate and work as a group to accomplish these tasks effectively. The work that will be presented in this paper is the culmination of projects undertaken at the Department of Aerospace Engineering at IIT Bombay specifically addressed at confronting the challenges associated with implementing such coop- erative control theories in practice on fixed winged miniature aerial vehicles (MAVs).

In order to demonstrate cooperative control algo- rithms in flight, it is necessary to evaluate its perfor- mance in a realistic scenario, particularly under the influence of winds. Wind is a major factor mainly because small MAVs fly at velocities of around 15-20 meters/second and wind speeds can reach upto 20-60 percent of the MAVs' speed. This can result in large deviations from the desired flight trajectory, which is evident in the simulation results presented later in this paper.

In order to evaluate cooperative control strategies on the basis of their performance in wind conditions and to test the onboard systems of the MAVs such as the Onboard Computer (OBC), the communication network, etc., a HILS system has been presented in this paper for real time simulation of cooperative missions of upto eight MAVs under the influence of wind. This HILS system borrows its core features from the HILS systems presented in [1] and [2], but it has the following improvements:

- Simulation of missions involving upto eight MAVs

- Real time monitoring of flight parameters for all eight MAVS using User Datagram Protocol (UDP) communication.

- Noise filtering of servo feedbacks and analog sensor data.

- Dryden model for simulating the effects of winds on the cooperative mission.

For better understanding of the problem at hand, some literature regarding different HILS system architectures, cooperative missions and trajectory tracking strategies, the dryden wind model and the challenges faced in flight demonstration of coop- erative missions on MAVs has been reviewed and summarized in section II.

The section III explains the over all architectuere and both the hardware (OBC, communication net- work, etc.) and the software (Dryden wind model, simulated sensor data, etc.) components of the HILS system presented in this paper and also explains the information flow in the HILS system during closed loop operation.

In section IV, the simulation results of various cooperative and trajectory tracking strategies, both with and without the effects of winds, have been presented and compared.

In the final section, all the work done in this paper has been summarised and the scope for further developing the HILS system has been discussed.

LITERATURE SURVEY

HILS systems have several applications in indus- tries and academia, for simulation and testing of hardware components of different complex systems. Some literature about such systems developed in academic circles aimed at designing solutions for specific aerospace problems, particularly for Un- manned Aerial Vehicles (UAVs) and MAVs research has been summarised in this section.

The authors of [3] propose a framework design for UAV embedded systems. They have implemented a HILS system using a real time exploring and testing platform called Dspace (based on MATLAB$^®$ and Simulink. This setup has been implemented in three stages and successfully been tested by the authors. In [4] the real time simulation and testing of spacecraft on a HILS system has been discussed. The environment used includes real time control and data acquisition on Real-Time Application Interface (RTAI), and Matrix-X control system design and automation tool for real time simulation and testing on HILS.

The author of [5] has developed a HILS for MAV using RTAI for flight mechanics simulation, and Comedi drivers for configuring ADC/DAC cards in RTAI. This setup can simulate upto two MAVs on a single computer. Work done in [6] describes the development of a HILS system in LabVIEW for simulating one MAV.

In [7] the authors have presented the architecture of a TMO-HILS (time-triggered message-triggered object model) which can be used for application ranging from small rapid control prototyping sys- tems to large-scale distributed real-time simulators. Such systems can be used for validation of real- time control system like the UAV control system. The authors of [8] also propose an architecture based on TMO model as a multi-UAV simulation platform. A further improved system is presented in [9] features an improved design of the TMO HILS architecture for multiple UAVs with a new architecture which consists of network-centric distributed object models. The TMO HILS has also been successfully implemented and analysed for its real time performance.

A three degrees of freedom gimbaled platform designed to be used as a HILS system has been described in [10]. It has been used for calibration and synchronization of MAV autopilot components, and for quick and efficient verification of autopilot behaviour with different airframes under various weather conditions within a laboratory.

The authors of [11]nave proposed a fault tolerance verification on HILS based on redundancy, which in- cludes permanent faults covered by hardware repli- cation, transient faults, and fault detection which are processed by software techniques. A dual loop archi- tecture HILS for multiplication of the control system has been used for this purpose. One loop serves as active control system and the other is passive control system. Multiplication allows handling of random faults.

In [1] and [2] a HILS system based on the xPC- Target^{TM} real time operating system by MATLAB$^®$ has been presented for simulation of cooperative missions of upto three and four MAVs respectively. A TDM based communication network has also been discussed for effective cooperation among the MAVs, and some cooperative missions have been successfully simulated. The HILS system presented in the this paper is a refined extension these systems. In order to simulate multi-MAV missions on the HILS, some literature regarding cooperative control strategies and trajectory tracking schemes were also reviewed. For multi-MAV missions mainly based on LOS (line of sight) pursuit missions were chosen such as the cyclic pursuit missions and leader follower missions. The leader follower mission is very similar to the pure pursuit strategy employed in missile guidance. The non-linear cyclic pursuit (NLCP) strategy for target capture discussed in [12] has been chosen for implementation on the HILS system and has served as a basis for designing the communication network for the cooperative mission. As seen from the results obtained from actual flights in [13], winds have an adverse effect on the flight paths of MAVs. To reduce deviations from desired trajectories, implementing trajectory tracking algorithms is necessary. With this in mind some trajectory tracking schemes were also reviewed in [14], [15] and[16] and the strategy chosen for im- plementation in this paper is the vector field based strategy for tracking straight line and circular paths proposed in [16].

For incorporating the effects of wind in the HILS system, the Dryden model was chosen as it was easily available as a simulink block in MATLAB$^®$ and for including the effects of this model in the MAV's flight dynamics equations, guidelines given in [16] were used.

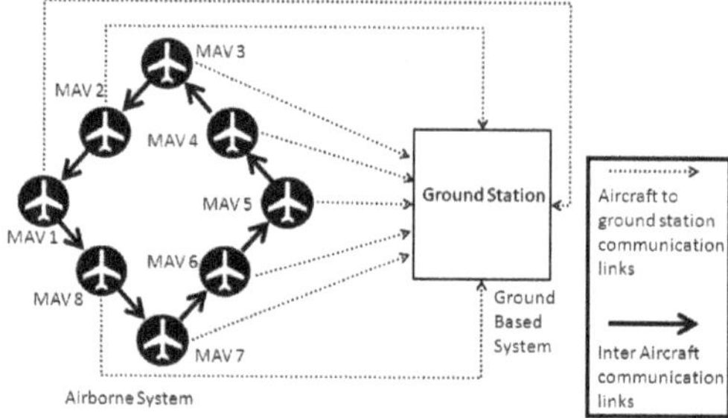

Fig. 1. Cooperative mission scenario of eight MAVs and a ground station

HARDWARE IN LOOP SIMULATOR ARCHITECTURE

One of the biggest challenges faced in implemen- tation of cooperative missions is the communication network between MAVs and the ground station.To resolve any communication issues simulation of the cooperative mission in real time with actual communication hardware in the loop is necessary. Thus the HILS system is an ideal platform for testing applicability of various time communication schemes such as Time Division Multiplexing (TDM implementation discussed in [1]) and Carrier Sense Multiple Access- Collision Avoidance (CSMA-CA implementation discussed in [2]).

Taking the non-linear cyclic pursuit (NLCP) mission for target capture discussed in [12] as a basic test case the interactions among the MAVS and the ground station have implemented as shown in figure

The HILS system can be broadly be classified into the following components, the simulation environ- ment, and the Hardware subsystems present in the simulation loop. In figure 2 the overall HILS system with its individual components is shown along with the interfacing signals between different blocks.

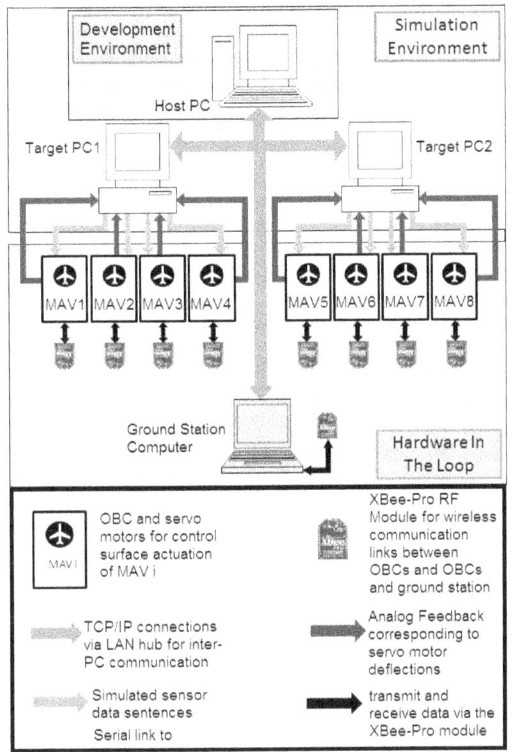

Fig. 2. Block diagram of the HILS system for for real time simulation of cooperative missions of eight MAVs.

SIMULATION ENVIRONMENT

Host PC: The Host PC is a standard desktop computer equipped with an ethernet port. The entire flight simulation program is implemented in form of two identical Simulink$^{®}$ v7.7 block diagrams, which also contain some MATLAB$^{®}$ R2011a functions for specific tasks. Each of these block diagrams simulate four MAVs each, the only difference being the initial positions of the MAVs. The host PC is used to build the block diagrams and compile them into C programs and header files using Microsoft$^{®}$ Visual C++$^{®}$ 2010. After compilation these programs are loaded on the two target computers by running a MATLAB$^{®}$ script via the ethernet connections as shown in figure 2. There is also a provision to read simulation data from the target PC via UDP packets after the simulation is completed. The flight model used for simulation has been taken from [17] in which the wind tunnel data used was obtained from National Aerospace Laboratories, Bangalore and the aerodynamic equations were derived.

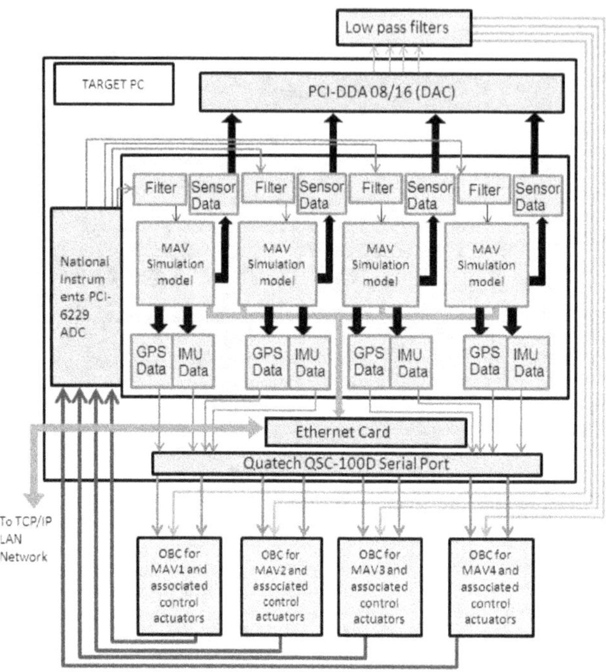

Fig. 3.   Block diagram of one of the target PCs

Target PCs: Target PCs are the computers on which the Real-Time Operating System RTOS named xPC Target® Rapid Prototyping System v5.0 is run from bootable CD-ROMs, which comes bun- dled with MATLAB® . The flight simulation for four MAVs is run on each target PC and it generates the corresponding sensor data for the OBCs in the loop in appropriate formats as shown in figure 2. The specific interface cards used to interface the information exchange between the target PCs and the OBCs are shown in figure 3.The analog feedback from the servomotors actuated by the OBCs is converted to its digital equivalent using a Analog to Digital Converter (ADC) card on each target PC. This is then given as input variables to the flight simulation of each MAV via a digital second order low pass filter (implemented in software) with 25 Hz cutoff frequency. The sensor information generated by each target PC includes the GPS and IMU data sentences (shown as GPS and IMU Data blocks in figure 3) serially conveyed via an eight port serial card on the target PC to the OBCs at correct baud- rates and after regular intervals. The pressure sensor data (sensor Data Block in figure 3)for airspeed and altitude is converted to analog voltages with proper scaling using a Digital to Analog Converter (DAC) card fitted on each target PC and this is given to the ADC inputs of the OBC via a low pass filter with a 50 Hz cut-off frequency. In the figure 3 all the green blocks are implemented as software, the blue blocks are the external hardware in loop and their interfaces, the violet blocks represent the on- board interfacing cards on the target PC. The colour conventions followed for the signals are the same as those in figure 2.

The Dryden wind model is incorporated as a simulink block as a part of the simulation and the wind velocities consisting of the steady wind and the stochastic gust component generated by the Dryden block are added to the air speed to generate appropriate ground speed for propagating position of the MAVs in the flight simulation as explained in [16].

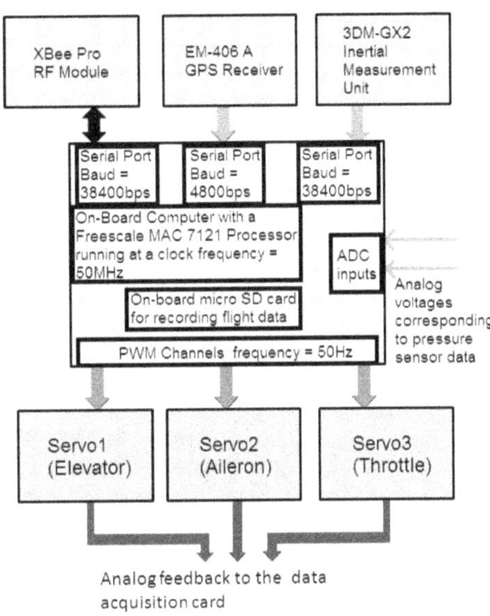

Fig. 4.   OBC block diagram with perepheral interfaces

HARDWARE IN THE LOOP

On-board Computers and its peripherals: The specifications and perepheral interfacing of the OBC used with this HILS system are as shown in the figure 4. The OBC has serial interfaces with the target computer on two of its serial ports for reading the data from GPS sentences as (sent in the NMEA format [18] by the Target PC by to mimic the EM-406A GPS receiver module)on one port, and IMU data sentences (same proprietary sentences of Microstrain® 3DM-GX2® IMU [19]) on the second port. The IMU packets, sent to the OBC every 20 milliseconds, consist of euler angles, accelerations and angular rates. The GPS packets, sent to the OBC every second, consist of current latitude and longitude, true course, GPS altitude, etc. The analog voltages from the target PC corresponding to pressure sensor data for airspeed and altitude are converted to digital values using the onboard ADC and used for processing. The onboard computer has pulse width modulated (PWM) channels by which it operates servo motors for actuation of the aircraft control surfaces. The control interrupt in the micro-controller is raised every 20 milliseconds and it keeps the states of the aircraft in check as per the mission programmed on the OBC. The OBC is also serially interfaced with the XBee-Pro® RF modules for implementation of the communication network shown in figure

5. The microcontroller used on the obc is the MAC 7121 by Freescale[TM] [20]. The flight Control System, i.e., autopilot control structures are simple PI-D control loops programmed on the OBC.

For achieving autonomous flight, the autopilots implemented on the OBC include: Altitude hold autopilot, Airspeed hold autopilot and Heading hold autopilot, the controller structures of which have been discussed in [1] and [2]. The rudder isn't used for autonomous flight and hence is maintained at 0 degrees deflection. In figure 4 the colours assigned to the signals follow the same conventions as shown in figure 2.

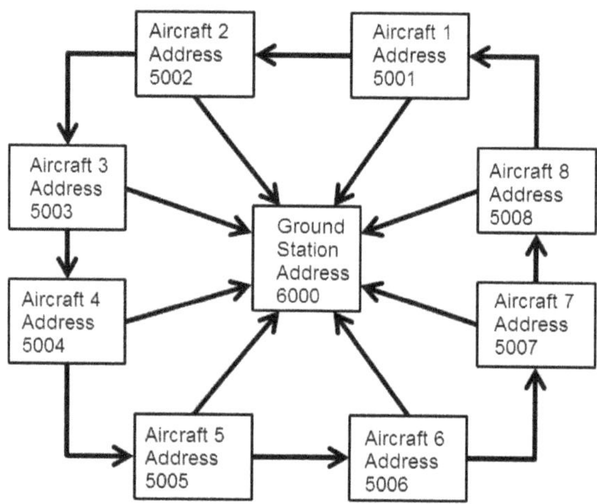

Fig. 5.   Communication Network for implementing cooperative missions of HILS

Communication Network: The network topology required for simulating the non linear cyclic pursuit mission for target capture by eight MAVs is shown in figure 5. The communication module chosen for this purpose is the XBee-Pro[®] RF module [21].

All the communication links shown in figure 5, use the API mode of packet based communication this is because since the XBee-Pro[®] module identi- fies addressed packets and verifies checksum, either the entire correct and complete packet is accepted or in the case of error or data loss, the entire packet is rejected. This avoids mixing of data from different packets at the ground station. Also in API mode, the sending XBee[®] module gets an acknowledgement from the receiving end and hence a clash or loss of data packets can be detected. It must be noted that for sending API packets the correct construction of API packet and calculation of checksum is necessary at the transmitter end.

For implementing pursuit missions of multiple MAVs, the inter MAV links shown in figure 5 com- municate packets containing current GPS latitude and longitude of the transmitting MAV. The commu- nication links from the aircraft to the ground station convey several parameters flight of the transmitting MAV for real time monitoring.

It must be noted that Xbee-Pro[®] modules supports half duplex communication only, i.e., it cannot send and receive data at the same time. Some packet collision avoidance strategy is thus necessary and some strategies such as TDM and CSMA-CA have been implemented in [1] and [2] respectively. For this HILS system the TDM scheme has been refined with a generic code which calculates the timeslots of each MAV for transmitting its data. Also all MAVs' flight times maintained by their OBCs are initialised and synced using a broadcast packet by the sent using the ground station before the mission begins. Unlike in the case of [1] and [2], currently the Futaba^{TM} Remote controllers for manually flying the MAVs in the HILS simulation have not been brought in as Hardware in-the-Loop, for the eight MAV HILS in order to reduce the communication load. This will also be done in the near future.

Ground Station: The ground station used for monitoring the MAV flight parameters during HILS simulations is a C++ application program, initially developed in [22] being run on a notebook, with Intel Core i5 processor, running at 2.4 GHz with a 4 GB RAM. The ground station application has been modified so that it is compatible with API packets containing GPS information of the MAVs, and it can plot the GPS position on receiving the packet from the MAV and verifying a correct checksum. Each MAV sends the ground station a data packet every one second and the ground station plots the current position of the MAV on the map with its corresponding coloured spot. The ground station also allows tuning gains of autopilots programmed on the OBC during the real-time HILS simulations. The ground station computer also recieves UDP data packets from the target PCs. These packets consist of the simulated flight data and states of the aircraft which can be plotted as a function of time in real time on the ground station using a simulink diagram built for this purpose.

SIMULATION RESULTS

Two types of missions have been simulated to investigate effects of wind in the HILS. The first are the pursuit based missions involving multiple MAVs, the second are trajectory tracking missions for a single MAV under the influence of winds. Missions involving multiple MAVs

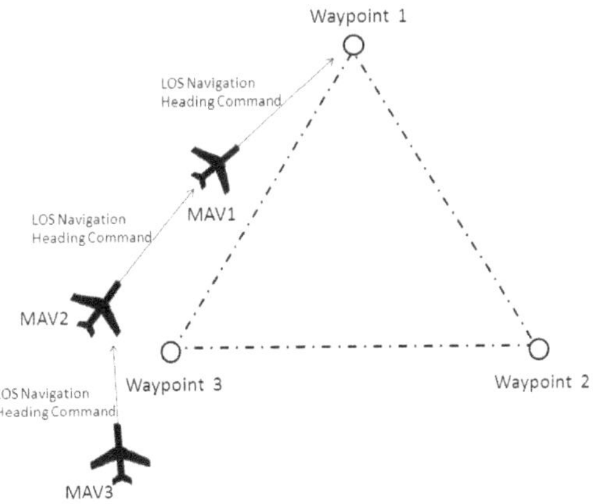

Fig. 6.   Leader follower mission for three MAVs

Leader Follower Mission: This is the first multi- MAV mission simulated on the HILS system . In a leader follower mission the leader MAV flies along a particular predefined trajectory and transmits its current GPS location to the follower MAV and the follower MAV flies along the line of sight to the leader, and hence follows the leader. In case of more than one followers each follower acts as a leader for the next follower as shown in figure 6 for three MAVs.

This mission was simulated for eight MAVs with the leader flying on a star shaped path joining five way-points by LOS navigation towards the way-points and has seven followers in successive pursuit. The results of the simulation both with and without the presence of wind are shown in figures 8 and 7 respectively. It is clear from the trajectories that in windy scenarios, there is a considerable deviation of the MAVs from the nominal path joining consecutive way-points, thought the MAVS fly to the way-points successfully.

Non-linear Cyclic Pursuit Mission for target cap- ture: This mission is a cooperative mission based

Fig. 7.   Leader following mission simulation with no winds. (Way-points are shown as yellow spots)

Fig. 8.   Leader following mission simulation with winds. (Way- points are shown as yellow spots)

on the work done in [12], where each MAV $i + 1$ communicates its current location to MAV i and the MAV i flies along a heading towards a virtual leader point, which divides the line joining its leading MAV $i + 1$ to a known target location P in a fixed ratio $\rho \in (0, 1)$ (called camouflage factor). The entire system of MAVs eventually reach a stable equilibrium where they capture the target by circling around it (as shown for three MAVs in figure 9).

This mission was simulated on the HILS system for eight MAVs both without and with the presence of winds as shown in figures 10 and 11. From figure 11 it is evident that under windy condition the MAVs fail to maintain an orbit and hence fail to capture the target. One reason for this is that the NLCP strategy assumes constant speeds of the MAV which are difficult to maintain under strong windy conditions. The winds taken for the simulation of both the multi MAV missions was a steady wind of 5 meters/second with a stochastic component, blowing at a 45° heading angle (towards the North-east).

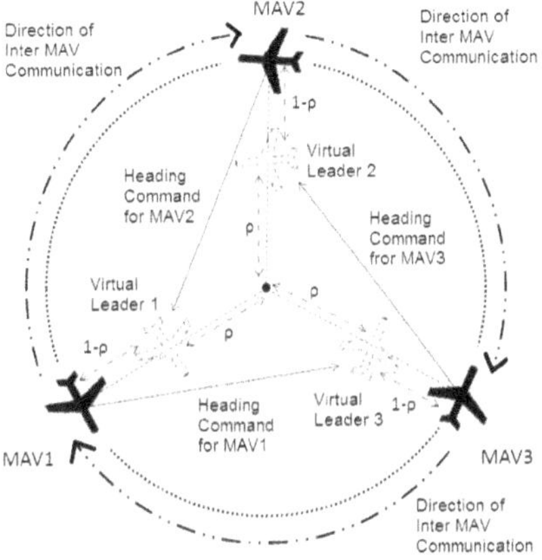

Fig. 9.   Non-linear Cyclic Pursuit mission for target capture

Fig. 10. Non-linear Cyclic Pursuit mission simulated on HILS

without winds. (The target is shown as a yellow spot)

TRAJECTORY TRACKING MISSIONS

From the above simulations it is evident that LOS navigation is not capable of preventing deviation from the desired trajectory in windy conditions. The necessity of the MAV to stick to the path planned for it is of great importance in several scenarios such as multi-MAV missions involving threat evasion, cooperative timing critical missions such as rendezvous or cooperative fly-by, etc. In this paper trajectory tracking strategies discussed in[16] for straight lines and circular trajectories has been simulated in with and without wind. Any path can be approximately implemented as a combination of straight lines and circular arcs and this can be done to meet various mission objectives such as constant path length, minimum time path, path through the way-point, etc. as explained in [14].

Fig. 11.   Non-linear Cyclic Pursuit mission simulated on HILS with winds.
(The target is shown as a yellow spot)

Straight Line Trajectory Tracking: The vector field based strategy for straight line tracking has been implemented in the HILS by issuing a heading command as shown in figute 12

The HILS simulation results of this mission with $k_{path} = 0.01$ are shown in figure 13 where the red trajectory is the nominal trajectory joining two waypoints, the green trajectory is the MAV using LOS navigation and the blue trajectory is the MAV using the vector field trajectory tracking scheme mentioned above. From figure 14 it can be seen that the lateral deviation from the nominal path is much larger in case of LOS navigation than the vector field strategy. The MAV heading in both cases is approximately towards 135° (South-East).

Circular trajectory tracking: For this purpose the vector field based orbit tracking strategy given in[14] has been chosen for implementation. In this strategy the MAV flies towards the center of the orbit from infinity and transitions to a heading perpendicular to the radius of the orbit as it reaches the perimeter of the circle as shown in figure 15 where d is the distance between the MAV and the target center and k is the transition gain.

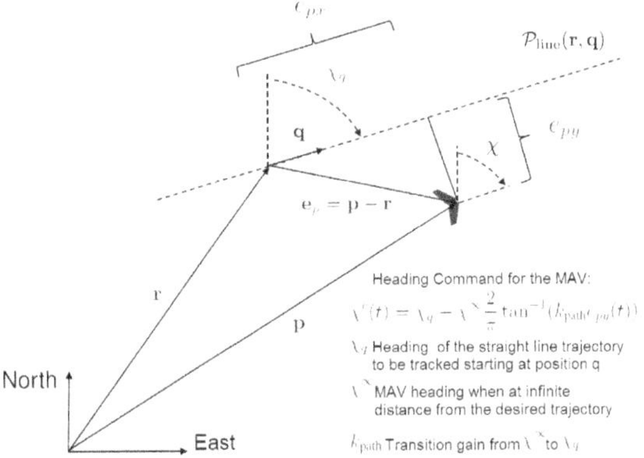

Fig. 12. Configuration of the MAV indicated by (p, χ), and the straight line path indicated by **Pline (r, q)** [16]

Fig. 13. Simulation of the vector field strategy and LOS navigation in HILS with a wind of 7.5 meters/second at a heading of 45°

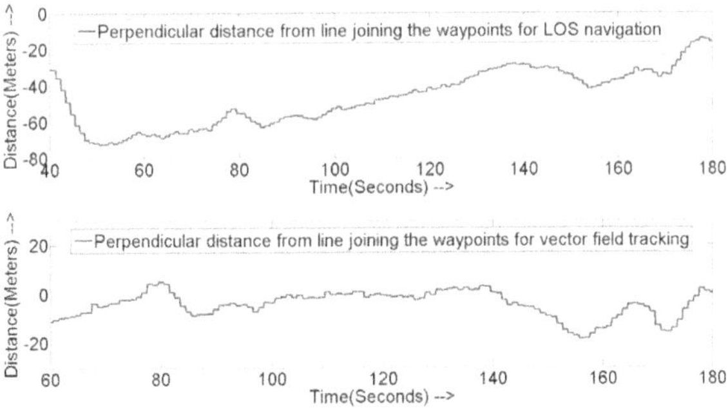

Fig. 14. Deviation form nominal trajectory for vector field strategy and LOS navigation in HILS with a wind of 7.5 meters/second at a heading of 45°

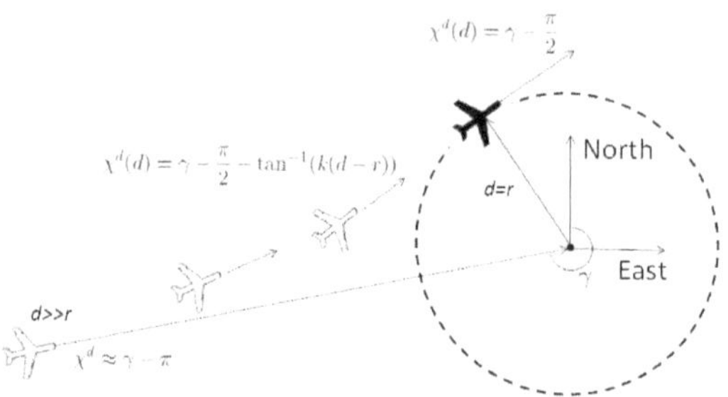

Fig. 15. Vector field based orbit following strategy for MAVs [13]

The orbit following mission was implemented on HILS for four conditions: without wind (figure 16), with a 5 meter/second wind, with a 7.5 meter/second wind (figure 17) and with a 10 meter/second wind. In all the cases k = 0.01 with different commanded radii. It was found that the orbit following strategy works better for larger radii. Also with increase in wind speed errors in radius also increase and the orbit periphery is disturbed but the MAV is still able to fly around the center or the target position as shown in figure 17. The radius of the MAV recorded onboard the OBC during the different HILS simu- lations mentioned above is shown in figure 18.

CONCLUSIONS

Thus in this paper a HILS platform capable of simulating different cooperative missions of upto eight MAVs under the effects of wind has been pre- sented. Both the leader follower and NLCP mission for target capture have been simulated on the HILS both with and without winds. Also vector field based straight line and orbit following strategies having potential applications in various cooperative mis- sions have also been implemented on HILS under windy conditions and errors in tracking trajectories recorded in simulation have also been presented.

Fig. 16. Orbit following strategy simulated on HILS for different radii with no wind. The orbit center is marked by a yellow spot

Fig. 17. Orbit following strategy simulated on HILS for different radii with no wind. The orbit center is marked by a yellow spot

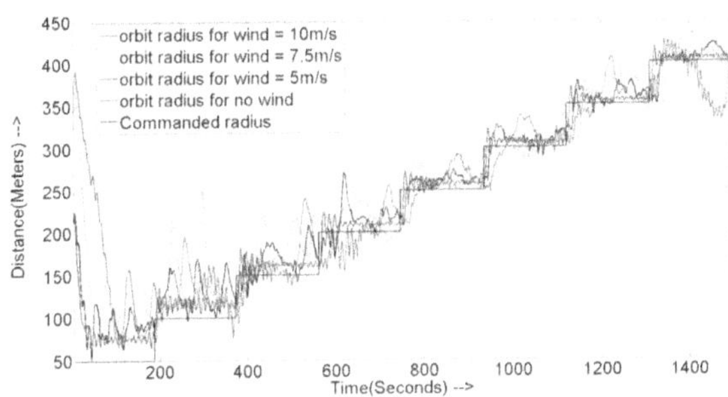

Fig. 18. Onboard recorded radii as compared to the commanded radii for different wind conditions

The Futaba^{TM} RF Remotes have not been made a part of the communication network yet, this will be done soon in the future. Also for simulating timing critical cooperative missions, (such as rendezvous) in windy conditions a ground speed control strategy is necessary. Currently an airspeed hold autopilot has been implemented on the OBC, but this will be replaced by a ground speed hold autopilot in the future.

REFERENCES

[1]   D. Krishnan, A. V. Borkar, P. Shevare, and H. Arya, "Hardware in loop simulator for cooperative missions," in Second International Conference on Advances in Control and Optimization of Dynamical Systems, IISc Bengaluru, February 2011.

[2]   D. Krishnan, A. V. Borkar, and H. Arya, "An elegant hardware in loop simulator for cooperative missions of MAVs," in AIAA Infotech@Aerospace 2012, Garden Grove, California, June 2012.

[3] Shixianjun, S. Jiakun, and L. Hongxing, "Hardware-in-the- loop simulation framework design for a UAV embedded control system," in Control Conference, 2006. CCC 2006. Chinese, pp. 1890 –1894, aug. 2006.

[4] A. Goktogan, E. Nettleton, M. Ridley, and S. Sukkarieh, "Real time multi-UAV simulator," in Robotics and Au- tomation, 2003. Proceedings. ICRA '03. IEEE International Conference on, vol. 2, pp. 2720 – 2726 vol.2, sept. 2003.

[5] T. Uppal, "Real time operating system benchmarking on mini aerial vehicle's hardwarein loop simulation," Master's thesis, Department of Aerospace Engineering, Indian Insti- tute of Technology Bombay, 2008.

[6] R. Mahajani and H. Arya, "MAV - hardware in loop simu- lation using LabVIEW," in AIAA Modeling and Simulation Technologies Conference, Portland, Oregon, aug. 2011.

[7] J. Xu, H. Park, K. Jeong, and C.-H. Chang, "TMO- HILS architecture for real-time control system," in Control, Automation and Systems, 2007. ICCAS '07. International Conference on, pp. 1855 –1861, oct. 2007.

[8] S.-H. Song, D.-H. Kim, and C.-H. Chang, "Experimental re- liability analysis of multi-UAV simulation with TMO-based distributed architecture and global time synchronization," in Object/Component/Service-Oriented Real-Time Distributed Computing Workshops (ISORCW), 2010 13th IEEE Inter- national Symposium on, pp. 211 –218, may 2010.

[9] S.-H. Song, H.-S. Park, T.-W. Kim, and C.-H. Chang, "Improved TMO HILS architecture for multi UAVs," in ICCAS-SICE, 2009, pp. 1398 –1403, aug. 2009.

[10] J. Shumaker, K. Ali, and L. Carter, "A gimbaled plat- form for MAV autopilot simulation and calibration," in Digital Avionics Systems Conference, 2008. DASC 2008. IEEE/AIAA 27th, pp. 4.C.4–1 –4.C.4–10, oct. 2008.

[11] V. Chandhrasekaran, E. Choi, and D. Min, "Multiplication of fault tolerance for unmanned aerial vehicle system using hils," in New Trends in Information and Service Science, 2009. NISS '09. International Conference on, pp. 719 –724, 30 2009-jly 2 2009.

[12] S. Daingade and A. Sinha, "Cooperative target capturing with multiple heterogeneous vehicles," in Second Interna- tional Conference on Advances in Control and Optimization of Dynamical Systems, IISc Bengaluru, February 2011.

[13] A. V. Borkar, "Cooperative control for fixed wing MAVs," Master's thesis, Department of Aerospace Engineering, Indian Institute of Technology, Bombay, July 2012.

[14] D. Nelson, D. Barber, T. McLain, and R. Beard, "Vector field path following for miniature air vehicles," Robotics, IEEE Transactions on, vol. 23, pp. 519 –529, june 2007.

[15] D. Nelson, D. Barber, T. McLain, and R. Beard, "Vector field path following for small unmanned air vehicles," in American Control Conference, 2006, p. 7, june 2006.

[16] R. Beard and T. McLain, Small Unmanned Aircraft: Theory and Practice. Princeton University Press, 2012.

[17] C. Ramprasadh and H. Arya, "Multi-stage fusion algorithm for estimation of aerodynamic angles in mini aerial vehicle," in 49th AIAA Aerospace Sciences Meeting including the New Horizons Forum and Aerospace Exposition, Orlando, Florida, jan. 2011.

[18] http://www.nmea.org.

[19] Microstrain®, Inc., Microstrain® 3DM-GX2® Data Com-munications Protocol, Jan 2010.

[20] http://www.freescale.com/webapp/sps/site/prod summary. jsp?code=MAC7121.

[21] Digi International, XBee® /XBee-PRO® RF Modules, 2009. [22] V. Prabhu, "Towards cooperative flying of miniature aerial vehicles," Master's thesis, Department of Aerospace Engi- neering, Indian Institute of Technology, Bombay, 2009.

COMPACT ACTIVE FILTER TO COMPENSATE THE HARMONICS IN AERO ELECTRIC POWER SYSTEM

Thirumoorthi P

Member, IEEE Department of Electrical and Electronics Engineering, Kumaraguru College of Technology, Coimbatore-641049, T.N., India
and

Yadaiah N

Senior Member, IEEE Department of Electrical and Electronics Engineering, JNTUH College of Engineering, Kukkatpally, Hyderabad-500085, A.P., India.

ABSTRACT

This paper proposes an effective method of compensating harmonics in aero electric power system using active filter. Aero electric power system includes number of power modulator and converter components which draw harmonic currents. The harmonic components of voltage and current lead to the overheating of electric apparatus. The harmonics create EMI. The sensitive equipments are subjected to malfunction or damage due to these harmonics. It is essential to eliminate or to compensate the harmonics to acceptable level. This proposed topology reduces the rating and size of the filter components. This proposed technique is to mitigate the distortion current by injecting equal but opposite current to shape the pulsating nature of the supply current to a sinusoidal form and in phase with the supply voltage. In the proposed work only one power switch is employed to minimize cost, reduce and switching losses. The performance of this proposed filter topology was examined with computer simulation and verified. The effectiveness of the proposed algorithm is demonstrated in results. The simulation results show that the designed filter compensate the harmonic currents produced by loads to a range of standards.

KEY WORDS: Harmonics, Active Power Filter, Pulse Width Modulation, Boost Rectifier.

INTRODUCTION

The aero electric instruments and actuators are highly sensitive and critical equipments(Fusiek,2007). The power electronic circuits and transformers draw a current which is nonlinear in nature and contains harmonics.The application of power electronic in power conversion shows drawbacks that lead to power quality problems which could relate to harmonics affecting communication interference, heating, solid-state devices malfunction, resonance and others (El-Habrouk etal,200). This problem is not solved adequately by passive filters such as LC cells due to inability to compensate random frequency variation in the source current, tuning problems etc. Active power filter is one viable solution to eliminate the harmonics, to improve the power factor, compensate the reactive current and to reduce the THD (Green and Marks, 1996)

Shunt active power filter normally operates using pulse width modulation (PWM) inverter technique to inject the required non-sinusoidal current requirements of non-linear loads but is complex with the number of switches in use (Bachry and Styczynski, 2003). Another approach is the use of series active power filter that uses basic bridge-diode circuit, boost circuit and an inductor (Salmon,1993). In this work, a new shunt active power filter is proposed with only one power switch. The proposed system introduces an active current wave-shaping technique to mitigate the distortion input current by injecting equal but opposite current to shape the pulsating nature of the input current into a sinusoidal form.

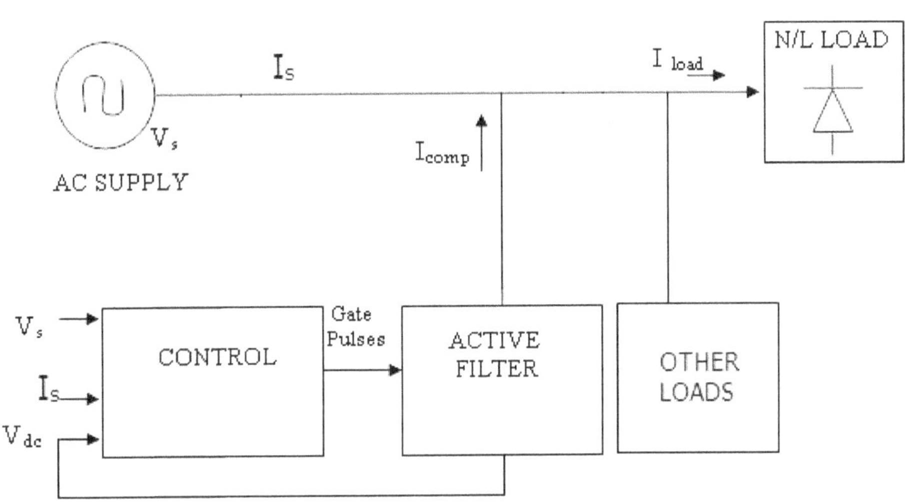

Fig. 1. Aero Electric Power System with Active Power Filter

In this paper the principle of boost technique is proposed, implemented as a new shunt active power filtering technique that could reduce input current distortions and achieve unity power factor when feeding a typical non-linear load (Athab and Shadhu Khan, 2006).

The time-domain approach is used to control the power switch of proposed active power filter during compensation process. This approach is based on the principle of holding the instantaneous current within some reasonable tolerance of a sine wave. The error is computed from the difference of instantaneous actual current signal with its reference signal, normally pure sine wave. This error is then conditioned and processed to obtain the required switching pattern known as the pulse width modulation (PWM) wave.

SHUNT ACTIVE POWER FILTER TOPOLOGIES

The active filter is connected to the supply in parallel with the load. An inverter with DC bus capacitor is used as active filter. Three main types of circuit topology are generally used for shunt active power filter (SAPF), namely; voltage source inverter (VSI), current source inverter (CSI).The VSI full and half-bridge APF circuits are as shown in Fig. 2 and 3.

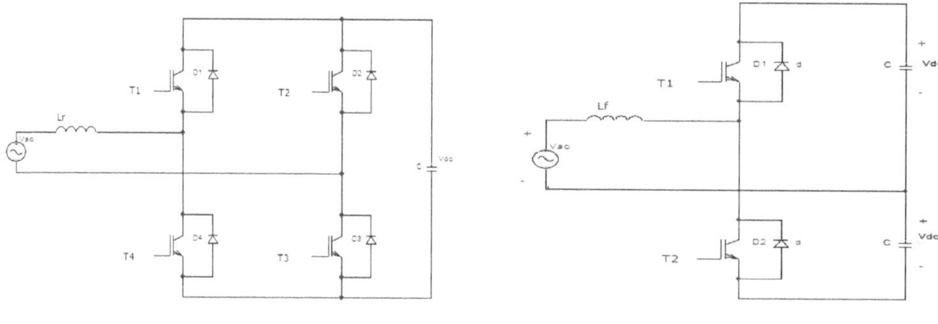

<table>
<tr><td align="center">Fig2. Full Bridge APF</td><td align="center">Fig.3. Half-bridge APF</td></tr>
</table>

BOOST RECTIFIER CIRCUIT

Typical rectifier boost converter is as shown in Figure. 4 is used extensively as a power factor correction technique and is used to increase the output voltage. The circuit uses an intermediate boost switch with inductor on the DC side. The boost topology can be operated either in the continuous or discontinuous inductor current mode and maybe switched at fixed or variable frequency. The main application of boost converter is in regulated DC power supplies and regenerative braking of DC motor.

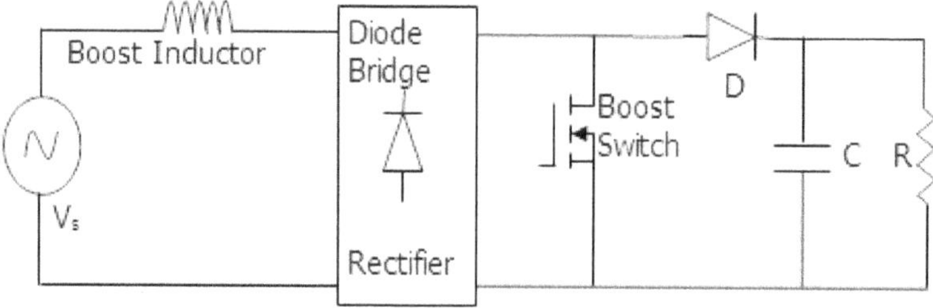

Fig.4. Rectifier boost converter circuit

PROPOSED ACTIVE POWER FILTER

The APF circuit consists of a main supply connected to a non-linear load as shown in Fig 5. RC parallel circuit connected to a single-phase rectifier circuit represents the non-linear load while the inductor represents a simple input filter. The proposed APF is placed across the main supply just before the filter. It can be clearly seen that the APF is in parallel with the non-linear load and the passive filter forming an SAPF comprising a control loop and a compensating circuit.

The compensating circuit, in turn, consists of a circuit whose components are similar to that of the non-linear load but with the addition of one active power switching device is used and RC parallel is reduced to C alone. Note that the employment of only one active power switching device enables to simplify the compensation scheme and contributes to low switch stress and losses. The IGBT is controlled by the control loop, which consists of a peak detector and a Supply Current Control Loop The peak detector is connected to a point just after the main supply terminal before the APF and the load. If the supply current is distorted, the controller will respond by providing switching signal to the IGBT that will inject the current compensation from the APF circuit to the mains to compensate the distorted supply current into a sinusoidal form. Unipolar switching is proposed due to the use of one power switch in the system.

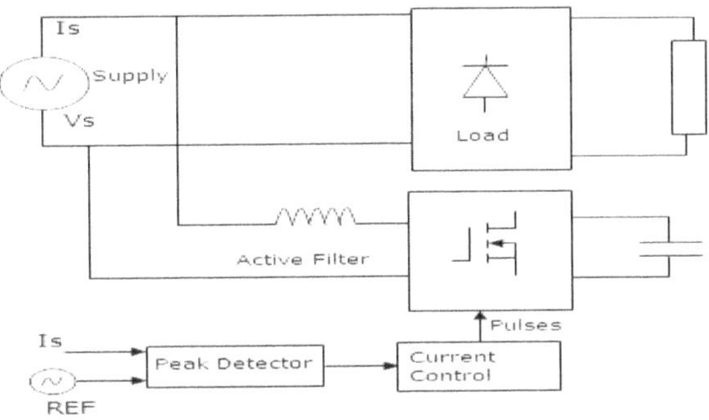

Fig.5. Proposed Active Power Filter

The proposed APF used to inject the required current into the system.

$$I_S = I_L - I_F \qquad (1)$$

When switch is turned on, diode D is reversed biased thus the output stage is isolated. The input supplies the energy to the inductor, L, causes inductor current to linearly increase. The energy stored in the inductor can be used for compensation purposes.

When the switch is turned off, there exists a change in current. Since the current in the inductor cannot change instantaneously, voltage in the inductor reverses its polarity in an attempt to maintain constant current. At this stage, the current will flow through the inductor L, diode D, and the reactive component (C) in the APF. Control is required in the design such that the inductor does not completely discharge the energy so that some residual energy remains in the inductor. When the power switch is turned on at this stage, the current ramp rides on a pedestal with a magnitude proportional to the residual energy in the core. Energy stored in the inductor is then used for charging the output capacitor and hence energy is transferred. Due to this requirement in operation, the boost voltage V_L must always be greater than the DC supply voltage V_S since the APF is intended to inject an opposite reactive current into the system.

To compensate for random variations in the waveform active pulse width modulation (APWM) technique is proposed. APWM operates by comparing the error signal that was used to determine a new magnitude with the carrier signal to produce the required PWM control. This is done by changing the modulation ratio of the PWM thus changing the width of pulse in accordance to the error detected. The higher the switching capacity of the converter circuit, the more harmonics components that could be injected thus cancelling the distortion components in the supply current. A proportional integral (PI) control algorithm is used to regulate the error.

RESULTS AND DISCUSSION

Fig.5 shows the proposed shunt active power filter using single power switch. The performance and behavior of SAPF was done by using MATLAB. The Simulink Model is as shown in Fig. 6.

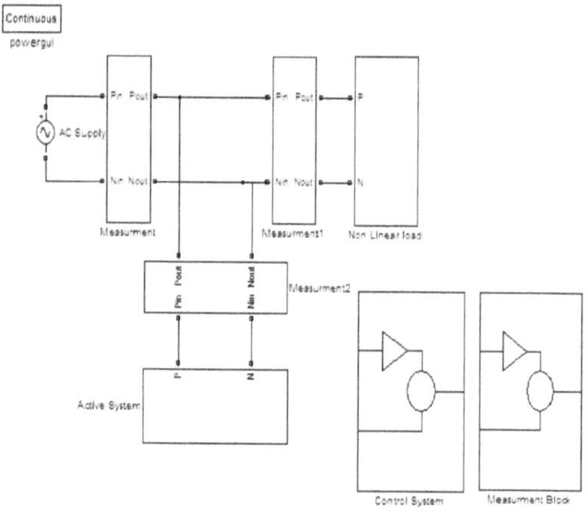

Fig .6. Simulink Model of Active Power Filter.

Results on the operation of the proposed shunt active power filter arrangement is presented.

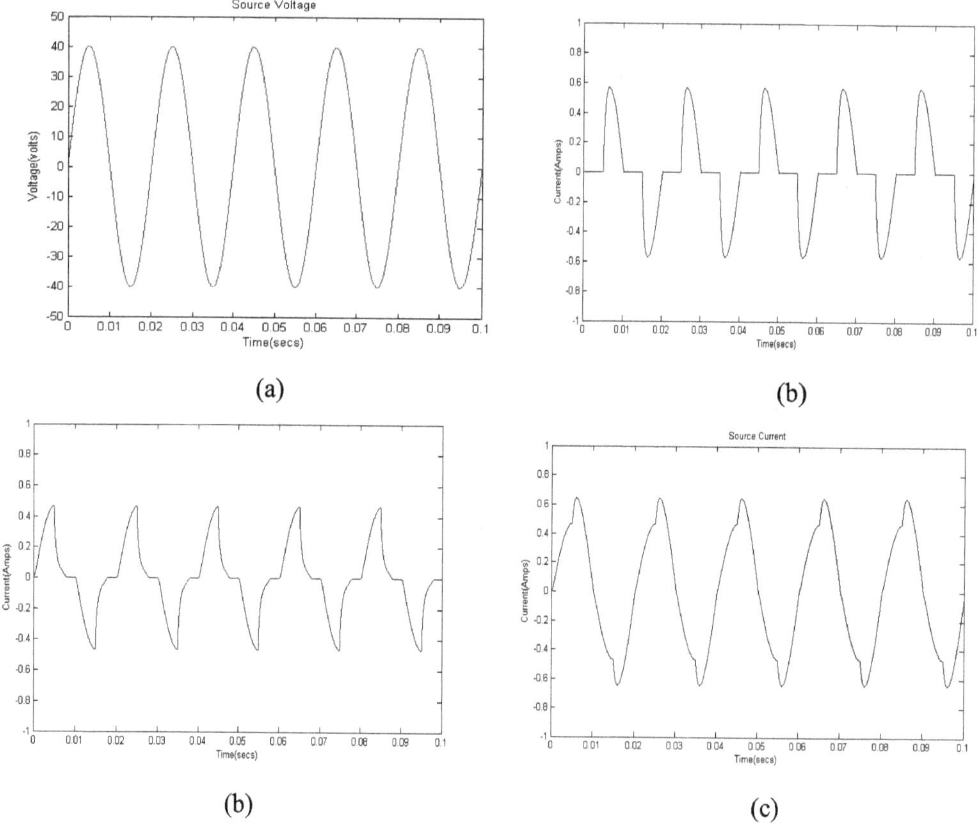

(a)

(b)

(b)

(c)

Fig.7: (a) load current (b) injected current (c) source current after compensation

From the result in Fig.7., it can be seen that the supply current was discontinuous and not in phase with the Supply voltage for uncompensated system and is measured with approximately 59.44% at a power factor of 0.88 leading due to the inductive nature of the load. During compensation, current from APF was injected into the system to mitigate the distorted current to a sine wave supply current that in phase with the supply voltage as in the result. After compensation, the supply current has a sinusoidal waveform that is in phase with the voltage supply. The THD level was reduced to 7.48% with almost unity power factor (0.99 lagging). Behaviour of system can be observed from the Fig.8 and Fig.9 and Table 1.

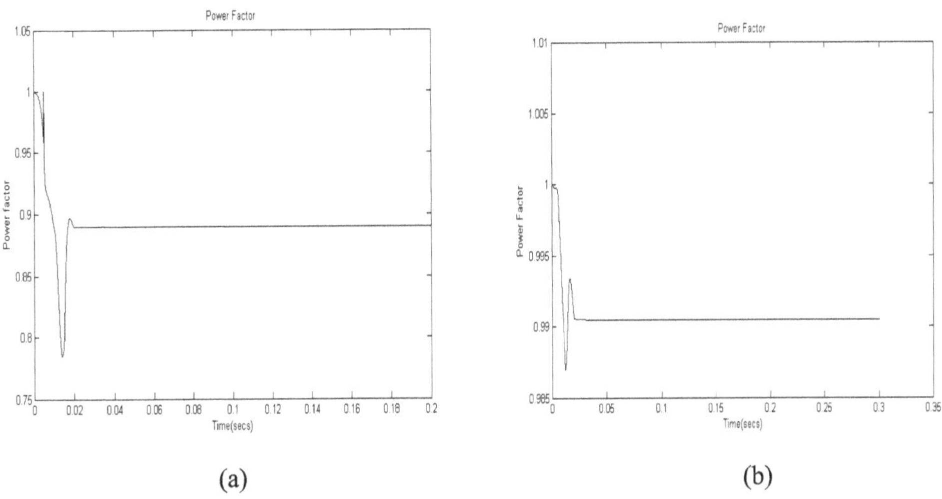

(a)   (b)

Fig.8.(a)Power factor before compensation. (b) Power factor after compensation.

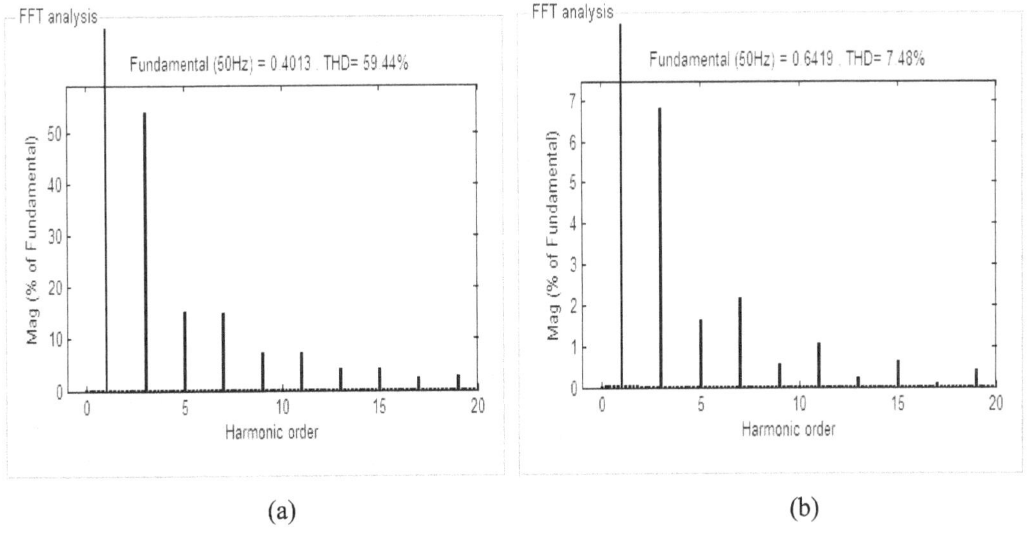

(a)   (b)

Fig.9.Total Harmonic Distortion (a) Before compensation; (b) After compensation

Table 1 Comparison of result

CASE	INPUT POWER FACTOR	TOTAL HARMONIC DISTORTION
Without active power filter	0.889	59.44%
With active power filter	0.991	7.48%

CONCLUTION

In this paper a compact active power filter for aero electric power system is presented. It uses a single-switch topology which is sufficient to reduce the input current distortions and achieve unity power factor operation in a single-phase capacitor filtered with thyristor rectifier as non-linear load. A minimal control requirement is required for achieving active current wave-shaping. This single-switch topology is used to reduce the switching stress and losses. Thus the system equipments are protected from the effects of harmonics and EMI.

REFERENCES

[1] M. El-Habrouk, M. K. Darwish, and P. Mehta, "Active power filters: A review," *Proc. Inst. Elect. Eng.—Elect. Power Appl.*, vol. 147, no. 5,pp. 403–413, Sep. 2000.

[2] T. C. Green and J. H.Marks, "Control techniques for active power filters," *Proc. Inst. Elect. Eng.—Elect. Power Appl.*, vol. 152, no. 2, pp. 369–381, Mar. 2005.

[3] J. C. Wu and H. L. Jou, "Simplified control method for single-phase active power filter," *Proc. Inst. Elect. Eng.—Elect. Power Appl.*, vol. 143, no. 3, pp. 219–224, May 1996.

[4] L. Asiminoaei, F. Blaabjerg, S. Hansen, and P. Thogersen, "Adaptive compensation of reactive power with shunt active power filters," *IEEE Trans. Ind. Appl.*, vol. 44, no. 3, pp. 867–877, May/Jun. 2008.

[5] A. Bachry; Z. A. Styczynski, "An analysis of distribution system power quality problems resulting from load unbalance and harmonics", *IEEEPES - Transmission and Distribution Conference and Exposition*, Volume 2, 7-12 Sept. 2003 pp. 763 – 766.

[6] *ww.aeroelectric.com/Reference_Docs/FAA/Part23*.

[7] Fusiek. G., "Concept Level Evaluation of the Optical Voltage and Current Sensors and an Arrayed Waveguide Grating for Aero-Electrical Systems' Applications", Instrumentation and Measurement Technology Conference Proceedings, 2007- *IEEE IMTC 2007*, 1-3 May 2007, Strathclyde Univ., Glasgow pp: 1 – 5.

[8] David Pitman, "Collaborative Micro Aerial Vehicle Exploration of Outdoor Environments" a thesis submitted to Massachusetts Institute of Technology, February 2010.

[9] John C. Salmon, "Technique for minimizing the input current distortion of current-controlled single phase boost rectifier", IEEE Transaction on Power Electronics. Vol.8.No.4. October 1993

[10] H.S. Athab and P.K. Shadhu Khan, "Single-Switch Single-Phase Boost Power Factor Correction with Harmonics Current Reduction", 1st IEEE International Power and Energy Conference, PECon '06, 28-29 Nov. 2006, pp. 447-452.

COMPARATIVE STUDY OF DIFFERENT TECHNIQUES USED FOR OPTIMIZING PARAMETERS OF FUZZY C MEANS CLUSTERING ALGORITHM

Dr.T.Satya Savithri

Professor, ECE Department, College of Engineering, JNTU Hyderabad

S.K.Chaya Devi

Professor, ECE Department, Sphoorthy Engineering College, Hyderabad.

ABSTRACT

Segmentation of medical images, particularly magnetic resonance images of brain is complex and is considered as a huge challenge in image processing, among the numerous algorithms, fuzzy c-mean algorithm is widely used in MR images segmentation .However in the cases of noisy MR images efficiency of this algorithm considerably reduces . Recently a robust segmentation technique based on an extension to the traditional fuzzy c means clustering algorithm is proposed. In this algorithm researchers have introduced two new parameters in order to improve the performance of FCM algorithm, which are calculated using neutral network in a complex and time consuming manner .These two parameters can also be calculated using Genetic algorithm, particle swarm optimization algorithm & Ant Bee Colony algorithm .with PSO&GA there is reduction in time in manipulating the two parameters ,but no change obtained in the resulted segmentation quality. Artificial Bee Colony algorithm aiming both to reduce the time and to reach a higher quality .comparison of all these algorithms presented in this paper.

INTRODUCTION

Magnetic resonance imaging is an important diagnostic imaging technique used for early detection of abnormal changes in tissue and organ[1].

Segmentation, separation of structures of interest from the back ground and from each other, is an essential analysis function for which numerous algorithms have been developed in the image processing. Fuzzy clustering can be considered the most important unsupervised learning algorithm .Accuracy of this algorithm on abnormal brains with edema,tumour etc is not efficient[11].

FCM algorithm only takes care to pixels intensity and does not consider their location (or) neighborhood properties. As a result noisy images influence effectiveness of this algorithm.MR images contain a significant amount of noise caused by operator performance, equipment and the environment ,which can lead to serious inaccuracies with segmentation. Shen[12] introduced new extension of FCM. They introduced two influential factors in segmentation ,which address issue of neighborhood attraction. First factor is the feature difference between neighboring pixels in the image and the second is the relative location of the neighboring pixels .Therefore segmentation is decided not only by the pixels intensity and considers neighboring pixels intensity and the locations.

These two parameters can be optimized using ANN ,GA,PSO and ABC algorithms. Simulation results using noisy MR images shows that the effectiveness of Ant Bee colony algorithm and robustness toward the noise.

IMPROVED FCM CLUSTERING ALGORITHM

It is assumed that number of clusters is known in advance. For MR images number of clusters is equal to four cluster, they are back ground, white matter ,graymatter, cerebrospinal fluid[24].Intensity of back ground and CSF are nearly same, therefore CSF and back ground belongs to same class and as result ,number of classes reduces to three classes. To do segmentation Fuzzy clustering use cost function and try to minimize it. The cost function is of the form

$$Jq = \sum_{i=1}^{n} \sum_{j=1}^{m} Uij^{q}\, d(xi, ej) \quad .....(1)$$

In (1) e_j represent the j-th cluster centroid ,m is number of clusters, n is the number of unknown vectors and Uij the member ship function of vector xi to the j th cluster that satisfies following conditions[20].

Uij€ [0,1],

$$\sum_{j=1}^{m} Uij = 1 \;\&\; 0 < \sum_{i=1}^{n} Uij < n \quad .....(2)$$

$d(xi, ej)$ is the dissimilarity between xi between ej .q €[1,∞],is a parameter called fuzzifier.The popular selection for $d(xi, ej)$ is

$$d(xi, ej) = (xi - ej)^{t} A (xi - ej) \quad .....(3)$$

where A is a symmetric positive definite matrix .In order to maximize equation (1) the partial derivative J with respect to Urs and ej should be performed. To do that gradient descent optimization method can be used which results

$$Uij = \frac{1}{\sum_{k=1}^{m} \left(\frac{d(xi, ej)}{d(xi, ek)}\right)^{1/(q-1)}} \quad .....(4)$$

And the equation ej is

$$ej = \frac{\sum_{i=1}^{n} Uij^{q} xi}{\sum_{i=1}^{n} Uij^{q}} \quad .....(5)$$

Note that because MR images are gray scale images ,the dimension of feature vectors is equal to one. The FCM algorithm can optimize the cost function J using equation (4)&(5) in an interactive way. To terminate the process ,an ending criterion such as

$$e(t) - e(t-1) < \varepsilon$$

Can be used . Equation(3) shows that only intensity values of pixels is important to FCM algorithm and location (or) neighboring pixels are not considered. It is caused that any changes in pixels intensity dramatically affect the clustering results. To reduce noise in MRI different dissimilarity function is used it is as follows

$$d(xi, ej) = \|(xi - ej)\|^{2} (1 - \lambda Hij) - \xi \quad .....(6)$$

Where Fij represents the distance attraction of neighboring pixels and H_{ij} is feature attraction λ and ξ have values between 0 and 1 and they can be adjusted to change the degree of neighborhood attraction.

$$H_{ij} = \frac{\sum_{k=1}^{n} u_{ik} g_{jk}}{\sum_{k=1}^{n} g_{jk}} \quad \text{......(7)}$$

$$F_{ij} = \frac{\sum_{k=1}^{n} u_{ik}^2 q_{jk}^2}{\sum_{k=1}^{n} q_{jk}^2} \quad \text{......(8)}$$

Where g_{jk} and q_{jk} are defined by:

$$g_{jk} = |x_j - x_k|, \; q_{jk} = (a_j - a_k)^2 + (b_j - b_k)^2$$

Where (a_j, b_j) denote the j-th pixel's locations. For any input image, defining an objective function and by using different optimization techniques , constant parameters λ and ξ can be computed

OPTIMIZATION APPROACH BASED ON GENETIC ALGORITHM

Domain of variables λ and ξ is {0,1]. We considered required precision is four places after the decimal points. The required bits for each variable is calculated as follows:

$$2^{m_i - 1} < 10^4 < 2^{m_i} - 1 \quad (i = 1,2)$$

Conversion from binary format to decimal is simply computed by:

$$X_j = \frac{decimal\,(binary\ string)}{2^{m_i} - 1}$$

Total bits required for simultaneous computations of λ and ξ is $m = m_1 + m_2$. If precision is set to four places after decimal point, total number of 28 bits are required ($m_1 = m_2 = 14$), the first 14 bits represents λ and next 14 bits represents ξ. We consider population size equal to 20 and then generated a random initial population cost function J_q in equation (1) is defined as fitness function and used for evaluation of initial chromosomes. In this sage some chromosomes are strong and others are weak.(some of them produce lower value for fitness function and vice versa) for selection stage a roulette wheel approach is adopted. Construction of roulette wheel is as follows [17]

PROCEDURE: ROULETTE WHEEL

Step1: Calculate the fitness value for each chromosome

Step2: Compute summations of all fitness values and calculate the total fitness.

Step3: Divide each fitness value to total fitness and get selection probability for each chromosome (noted by P_k)

Step4: Calculate cumulative probability q_k for each chromosome $q_k = \sum p_j$

In selection process, roulette wheel spin 20 times. Each time a single chromosome is selected for a new population is following Manner [18]

Procedure

selection

Step1. Generate a random number r from the rang [0, 1]

Step2. If r<=q1, then select the first chromosome,

Otherwise select the k-th chromosome such that $q_{k-1} < r < qk$.

After selection part, a new population will produce that chromosomes with higher selection probability will remain and others will clear. Now, crossover and mutation operators produce new chromosomes. Consider probability of crossover is set to 0.2 and probability of mutation is set to 0.01(it means that 4 chromosomes from total population of 20 will change using crossover operator and 6 genes from 560 (28×20) genes will converted using mutation operator).

CROSSOVER PROCEDURE CAN DESCRIBE AS FOLLOWS [17]

Procedure: cross over

Begin

$0 \rightarrow k$;

While (k<= population size) do

Random number from [0, 1] $\rightarrow$ rk

If (rk<0.2) then select k-th chromosome as one parent for

crossover.

end

$k \rightarrow k+1$;

end

end

Mutation

procedure is done as follows:

Step1.choose 6 genes from total 560 genes (we considered mutation rate is equal to 0.01).

Step2. Change current value of selected genes (if they are one, alter them to zero and vice versa)

After completion of above processes, a new population is produced and the current iteration is completed. We iterated the above procedures until a certain criterion is met. At this point, the most fitted chromosome represented the optimum values of λ and ξ.

OPTIMIZATION APPROACH BASED ON PSO

Domain of variables λ and ξ is [0,1].Total particles required for simultaneous computation of λ and ξ is 2.The first particle represents λ and next particle represents the ξ.We consider population size is equal to 20 and then generated a random initial population. Cost function Jq in equation (1)is defined as fitness function and used for evaluation of initial chromosomes. In this stage some particles are strong and some are weak. Thease particles are floated in a two dimensional space. Each particles changes its position according to own experience and its neighbours. So first we have to define a neighbourhood in corresponding population and then describe the relation between particles that fall in that neighbourhood.In this context we have many topologies such as star ,ring and wheel. Here we use the ring tropology. In ring topology each particle is related with the two neighbours and intends to move toward the best neighbour. Here neighbours can overlap, which facilitates the exchange of information between neighbourhoods and in the end convergence to a single solution. The local best algorithm is associated with this tropology[29,30].

PROCEDURE

1. Initialize a swarm of P particles in D-dimensional space, where D is the number of weights and biases.
2. Evaluate the fitness f_p of each particle p as the J.
3. .If $f_p < p_{best}$ then $p_{best} = f_p$ and $x_{pbest} = x_p$ where p_{best} is the current best fitness achieved by particle p,x_p is the current coordinates of particle p in D-dimensional weights space, and x_{pbest} is the coordinate corresponding to particle p' s best fitness so far.
4. .If $f_p < l_{best}$ then $l_{best} = p$, where l_{best} is the particle having the overall best fitness over all particles in the swarm.
5. Change the velocity V_p of each particle p

$$:\overline{V_p} = \overline{V_p}(t-1) + c_1\left(\overline{x}_{pbest} - \overline{x}_p(t)\right) + c_2\left(\overline{x}_{lbest} - \overline{x}_p(t)\right)$$

   Where c_1, c_2 are accelerate constants and rand return uniform random number between 0 and 1.
6. Fly each particle p to $x_p - V_p$.
7. Loop to step 2 until convergence.

After completion of above processes , a new population is produced and the current iteration is completed. Above procedures is iterated until a certain criterion is met. At this point ,the most fitted particle is represented the optimum values of λ and ξ.

OPTIMIZATION BASED ON ARTIFICIAL BEE COLONY ALGORITHM

Interval change of the two variables λ and ξ is between [0,1]and accuracy obtained in optimization algorithms based on the number of decimal digits of these parameters.

In each step two intervals are defined as neighbourhood for scout bees to increase search accuracy. By this way neighbourhood area for a scout bee is in far vicinity upto two decimal places and for close vicinity upto four decimal places. O/P response accuracy is controllable based on the number of decimal digits or modifying neighborhood area for scout and onlooker bees .Infact with development of neighborhood area, we attain higher accuracy and higher rate at response. First the early population of bees that consists of worker bees is constituted. The initial population of begins searching randomly the space that includes λ and ξ and the output. At the beginning, total search space has the same value for all bees and recruitment operation wont be made in this stage. After the initial search information obtained from search space considered as food source, The practice of sharing information is done in space colony. Selecting the best source of food is done regarding to cost function and the chance to select a scout bee will be regarding to as follows

$$C_i = \frac{F(\theta_i)}{\sum_{K=1}^{N} F(\theta_i)} \qquad .....(9)$$

Where Ci is the chance that bee is selected to be a scout bee and $F(\theta_i)$ is amount of nectar in θ_i food source and $\sum_{K=1}^{N} F(\theta_K)$ is amount of nectar whole around the hive. After selecting the sout bees the recruitment process to determine the neighbourhood of points λ and ξ which are selected previous step is performed .Search process to discover the best values of λ and ξ in membership function as follows.

1. Employed bees move to search randomly the space.
2. Review information obtained by the employed and onlooker bees in the colony based on cost function.
3. Selecting scout bees and going to recruitment process .
4. Search neighbours with scout bees guide.
5. Vocalize new population using selected scout bee.
6. Stop when achieving the satisfied condition otherwise go to step 2.

The search space is also repeated to obtain coefficient λ and ξ until condition of $\theta_i(t) - \theta_i(t-1) < \epsilon$ Is satisfied.

This happened in such a way that optimized values of each stagemust be located in fuzzy membership function and the algorithm is implemented and will continue until reach to minimum amount of error.

EXPERIMENTAL RESULTS

In this section performance of the GA,PSO,ABC algorithms are discussed. Fig 3 and 4 shows the images in an extracting gimproved gray matter and white matter using FCM algorithm improved with GA,PSO and ABC algorithms. ABC algorithm performs much better than other methods .Table 1 shows λ and ξ used in ABC algorithm. The number of pixels extracted using these algorithms brought in tables 2 and 3 .

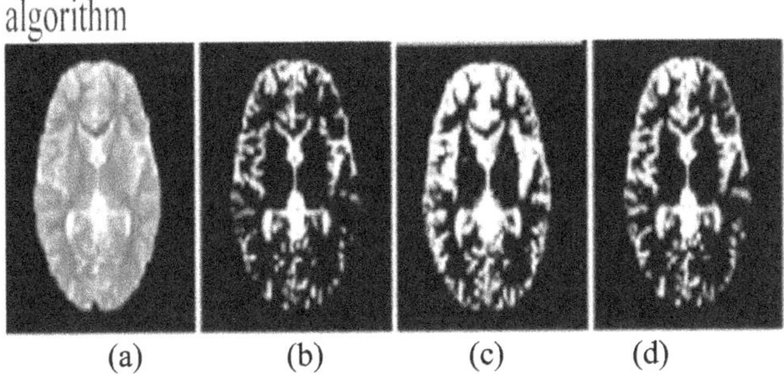

Figure1. Segmentation results on real MR image test divided into gray matter (GM) and background; (a) original image, (b) segmentation results FCM with GA algorithm, (c) segmentation results FCM with ABC algorithm

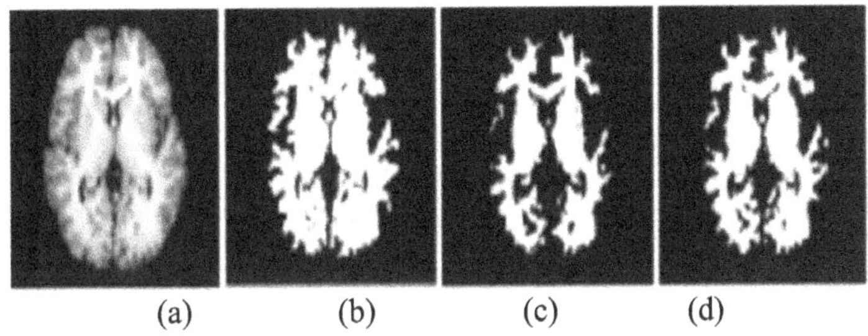

Figure2: Segmentation results on real MR image test divided into white matter (WM) and background; (a) original image, (b) segmentation results FCM with GA algorithm, (c) segmentation results FCM with PSO algorithm and (d) segmentation results FCM with ABC algorithm.

Table.1 Values of λ and ξ using ABC algorithm

	Result for gray matter (GM)	Result for white matter (WM)
λ	0.93995	0.92688
ξ	0.96884	0.94789

Table2. Number of pixels after running FCM with GA, PSO and ABC algorithm

	Figure.1	Total	Gray Matter	White Matter
FCM with GA	a	29762	15512	14250
FCM with PSO	b	29762	19048	10714
FCM with ABC	c	29762	22287	7475

Table3. Number of pixels after running FCM with GA, PSO and ABC algorithm

	Figure.1	Total	Gray Matter	White Matter
FCM with GA	a	30789	13243	17546
FCM with PSO	b	30789	19671	11118
FCM with ABC	c	30789	17224	13565

CONCLUSION

We proposed algorithms based on combining Fuzzy C mean algorithm with GA or PSO or ABC algorithms to effectively segment MR images .The comparison of ABC algorithm with PSO and GA shows that algorithm performed well interms of speed. The combination of ABC to FCM algorithm produced a method which achieves an acceptable improvement in the segmentation of images. Simulation results showed that the ABC algorithm produced optimum values of λ and ξ that are used for efficient segmentation of MR images

REFERENCES

[1] J. C. Bezdek, L.O. Hall, L. P. Clarke, "Review of MR image Segmentation techniques using pattern recognition," Med. Phys., vol. 20, No. 4, pp. 1033-1048, 1993.

[2] K. R. Castleman, Digital Image Processing. Upper Saddle River: Prentice Hall, 1996.

[3] J. K. Udupa, L. Wei, S. Samarasekera, Y. Mild, M. A. van Buchem, R. I. Grossman, "Multiple sclerosis lesion quantification using fuzzy connectedness principles," IEEE Trans. Med. Imag., vol. 16, No. 5, 1997.

[4] Y. Zhu, H. Van, "Computerized tumor boundary detection using a Hopfield neural network," IEEE Trans. Med. Imag., vol. 16, No. I, pp. 55-67, 1997.

[5] K. Lim, A. Pfefferbaum, "Segmentation of MR brain images into cerebrospinal fluid spaces, white and gray matter," J. Comput. Assist. Tomogr., vol. 13, pp. 588-593, 1989.

[6] S. R. Kannan, "Segmentation of MRI Using New Unsupervised Fuzzy C Mean Algorithm" GVIP (5), No. V2, January 2005.

[7] Zhang Y, Brady M, Smith S."Segmentation of brain MR images through a hidden Markov random field model and expectationmaximization algorithm,". IEEE Trans Medical Imaging 2001,pp. 45-57.

[8] Chen D, Li L, Yoon D, Lee J, Liang Z. "A renormalization method for inhomogeneity correction of MR images", Processing of SPIE Medical Imaging 2001, pp. 939-942.

[9] S. Theodoridis, K. Koutroumbas, Pattern Recognition, Academic Press; 2nd edition February 2003.

[10] S. Shen, W. Sandham, M. Grant and A. Ster, "MRI Fuzzy Segmentation of Brain Tissue Using Neighborhood Attraction with Neural-Network Optimization", IEEE Trans. On Information Technology is Biomedicine, vol9, No 3, September 2005.

[11]	M. C. Clark, L. O. Hall, D. B. Goldgof, "MRI Segmentation using Fuzzy Clustering Technique", IEEEEngineering in Medicine and Biology, vol. 13, no. 5, 1994.

[12]	S. Shen, W Sandham, M. Grant and A. Ster, "MRI Fuzzy Segmentation of Brain Tissue Using Neighborhood Atttraction with Neural-Network Optimization", IEEE Trans. On Information Technology Biomedicine, vol 9, No 3, September 2005.

[13]	R. C. Gonzales, R. E. Woods, Digital Image Processing Prentice Hall; 2nd edition, January 15, 2002.

[14]	D.L. Pham, "Spatial Models for Fuzzy Clustering", Computer Vision Image Understanding, Vol. 84, 2001.

[15]	R. R. Krishnapuram and J. M. Keller, "A Possibilistic Approach to Clustering" ,IEEE Trans. Fuzzy Systems, Vol I, No. 2, May 1993.

[16]	D. Goldberg, *Genetic Algorithms in Search, Optimization and Machine Learning*, Addison-Wesley,MA ,1989.

[17]	M. Mitchel, *An Introduction to Genetic Algorithms*, MIT Press, Fifth printing, 1999.

[18]	M.Gen and R. Cheng, "Genetic Algorithms and Engineer Design ", John Wiley, 1997.

[19]	A. Huang, R. Abugharbieh, R. Tam and A. Traboulsee, "Automatic MRI Brain Tissue Segmentation using a Hybrid Statistical and Geometric Model ", 3rd IEEE International Symposium on Biomedical Imaging, April 2006.

[20]	L. Lemieux, G. Hagemann, K. Krakow, and F. G. Woermann, "Fast, accurate, and reproducible automatic segmentation of the brain in T1-weighted volume MRI data," *Mag. Reson. Med.*, vol. 42, pp. 127-135, 1999.

[21]	H.Suzuki and J.Toriwaki, "Automatic Segmentation of Head MR Images by Knowledge Guided Thresholding" Computer Medical Imaging,vol 15, No 4, 1991.

[22]	K. V. Leemput, F. Maes, D. Vandermeulen, and P. Suetens, "Automated model-based tissue classification of MR images of the brain," *IEEE Trans. Med. Imag.*, vol. 18, no. 10, pp. 897-908, Oct. 1999.

[23]	R.Pohle and L.D. Toennies, "Segmentation of Medical Images using Adaptive Region Growing ", Proc, SPIE, Medical Imaging, vol. 4322, 2001.

[24]	S. Theodoridis, K. Koutroumbas, *Pattern Recognition,* Academic Press; 2nd edition February 2003.

[25]	J. C. Bezek, *Pattern Recognition with Fuzzy Object Function Algorithms* New York: Plenum, 1981

[26]	D.L Pham and L.Prince, "Adaptive Fuzzy Segmentation of Magnetic Resonance Images", IEEE trans. On Medical Imaging, vol. 18, September 1999.

[27]	B.Widrow, S. Stearns, *Adaptive Signal Processing*, Prentice-Hall, 1985.

[28]	R. C. Gonzales, R. E. Woods, *Digital Image Processing* ,Prentice Hall; 2nd edition ,January 15, 2002.

[29]	RC Eberhart, RW Dobbins and P Simpson, Computation Intelligence PC tools, Academic Press, 1996.

[30]	D Corne, M Dorigo F Filover (eds), Ne Ideas in Optimization, MeGraw Hill, 1999.

EVOLUTIONARY ALGORITHMS BASED CONTROLLER DESIGN OF LABORATORY   HELICOPTER SYSTEM

Kota Kanthi Kumar

Assist. Professor, Department of Electronics and Instrumentation Engineering
Keshav Memorial Institute of Technology, Hyderabad, India
kanthikumark@gmail.com

ABSTRACT

This correspondence presents a new approach that utilizes evolutionary computation and proportional-integral differential (PID) control to a multi-input multi output (MIMO) nonlinear system. This approach is demonstrated through a laboratory helicopter called the twin rotor MIMO system (TRMS). The goals of control are to stabilize the TRMS in significant cross-couplings, reach a desired position, and track a specified trajectory efficiently. The proposed control scheme includes four PID controllers with independent input. In order to reduce total error and control energy, all parameters of the controller are obtained by a real-value-type genetic algorithm (RGA) with a system performance index as the fitness function. The system performance index was applied to the integral of time multiplied by the square error criterion to build a suitable fitness function in the RGA. In this investigation a new method for the RGA to solve more than ten parameters in the control scheme. The initial search range of the RGA was obtained by a nonlinear control design (NCD) technique. The NCD provided a narrow initial search range for the RGA. This new method led chromosomes to converge to optimal solutions more quickly in a complicated hyper plane. Computer simulations show that the proposed control scheme conquers system nonlinearities and influence between two rotors successfully.

INDEX TERMS : Genetic algorithms, multi-input multi output (MIMO) system, proportional-integral differential (PID) control.

INTRODUCTION

In this correspondence, we investigate a control problem involving an experimental propeller setup called the twin rotor multi-input multi output system (TRMS) [1]. The TRMS is a laboratory setup designed for control experiments. In certain aspects, its behavior resembles that of a helicopter. From the control point of view, it exemplifies a high order nonlinear system with significant cross-couplings. An estimated model has been obtained by using radial basis function networks [2] and a black-box system identification technique [3]. Here, we assume that the dynamics of the TRMS are known. The control objective is to stabilize the system in a coupled condition and make the beam of the TRMS move quickly and accurately to track a trajectory or reach specified positions in 2 DOF. The control scheme is based on the PID control technique, which applies a signal to the process that is proportional to the actuating signal in addition to adding integral and derivative of the actuating signal. PID is the control algorithm most often used in industrial control [4]. It is implemented in industrial single-loop controllers, distributed control systems, and programmable logic controllers. There are two reasons why it is the most used in the industrial process. The first reason is that its simple structure and the well-known Ziegler and Nichols tuning algorithms have been developed [5], [6] and successfully used for years. The second reason is that the controlled processes in industrial plants can almost be controlled through the PID controller [7], [8]. The drawback is that the parameters of the PID controller are partially tuned by a trial-and-error process, which makes it less intelligent. Su *et al.* [9] proposed a terminal sliding mode control to the TRMS and obtained good tracking performance. But the settling time of the time response is too long. Islam *et al.* [10] designed a fuzzy logic controller for the TRMS. Trajectory tracking performance is superior to the system performance with a conventional PID or a linear quadratic controller (LQR) controller. The drawback is that the system performance depends on the number of the fuzzy membership functions. In their simulations, the tracking error of the sine wave response cannot be eliminated. In this correspondence, parameters of the proposed controller are obtained by a modified method of a real-type genetic algorithm (RGA) with a performance index as the fitness function, which can reduce control energy and total error simultaneously.

Huang and Juang [11] investigated the effect of the binary genetic algorithm (BGA) on controller tuning for improving the system performance. The gains of the PID controller are tuned by a BGA. The controller is better than what was designed by the Gauss–Seidel minimization procedure. Tsai *et al.* [12] applied the RGA on PID tuning to the TRMS control in the vertical and horizontal axes separately. In order to design a 1 DOF PID controller, the system is decoupled into two parts, that is, vertical and horizontal. They restrict the connective beam between the main rotor and the tail rotor to move only in the vertical or horizontal plane. The impact on these two rotors is ignored. The parameters of the PID controller are tuned by the RGA to reduce the total error and control energy. Fan and Juang [13] compared different kinds of fitness functions in an RGA via trial and error for tuning an optimal PID controller on the TRMS. The computer simulations show that whether the fitness function best suits the objective or not can be checked by the control response. Although these PID controllers [12], [13] can work in 2 DOF, they are still weak in tracking a desired trajectory and in reaching a specific attitude. It is difficult to stabilize the influence between two axes and a no minimum phase in a vertical plane. Liu *et al.* [14] quoted the system performance index as a part of the fitness function in an RGA. The system performance index deals with a modification of the known integral of time multiplied by square error criterion (ITSE). It is more efficient in finding the parameters of the four PID controllers. Although these controllers can reduce control force, and the total error is less than before, the trajectory tracking of a desired path in 2 DOF oscillates for several seconds. Above all, the main purpose in the control of the TRMS focuses on designing controllers to stabilize the impact between two rotors and track a desired path and specific attitude in 2 DOF efficiently. The performance of the PID control depends on the gains of the controller. Proper gains can be obtained by some optimization techniques. Among them, the genetic algorithm (GA) has recently been used very often. The GA was first proposed by Holland [15] at the University of Michigan in 1975. The searching process, based on the Darwinian "survival of the fittest" principle, allows the solution to evolve into a superior solution. It has been successfully applied to various optimization problems [16]–[19]. The GA is easily combined with intelligent control theory (mixed or hybrid) [20]–[24] and solves problems in the traditional control field. However, the disadvantage of the GA is that the searching process is time-consuming for multiple parameters with a wide search range. To overcome this problem, we propose a nonlinear control design (NCD) technique that uses a quasi-Newton method with cubic/quadratic interpolation for unconstrained conditions and a sequential quadratic programming method for constrained conditions. The NCD can provide a narrow search range to

421

the parameter searching process of the GA. In this correspondence, we investigate a cross-coupled PID controller on the TRMS through a modified method in an RGA. This approach can optimize more than ten parameters in hyper plane simultaneously and converge to a better solution. Simulations show that the performance of the proposed controller is much better than in previous works.

SYSTEM DESCRIPTION

The Twin Rotor MIMO System (TRMS) is a laboratory setup that is designed for control experiments. In certain aspects its behavior resembles that of a helicopter. As Fig.1 the TRMS mechanical unit has two rotors placed on a beam together with a counterbalance whose arm with a weight at its end is fixed to the beam at the pivot and it determines a stable equilibrium position. The TRMS consists of a beam pivoted on its base in such a way that it can rotate freely both in the horizontal and vertical planes. Either the horizontal or the vertical degree of freedom can be restricted to 1 degree of freedom using nylon screws found near pivot point. At both ends of the beam there are rotors (the main and tail rotors) driven by DC motors. This device is a multivariable, nonlinear and strongly coupled system, with degrees of freedom on the pitch and yaw angle denoted by θ_h, θ_v.

The TRMS consists of a beam pivoted on its base in such a way that it can rotate freely both in the horizontal and vertical planes. At both ends of the beam, the rotors (the main and tail rotors) are driven by dc motors. A counterbalance arm with a weight at its end is fixed to the beam at the pivot. The state of the beam is described by four process variables: horizontal and vertical angles measured by position sensors fitted at the pivot, and two corresponding angular velocities. Two additional state variables are the angular velocity of the rotors, measured by tacho generators coupled with the driving dc motors. In a normal helicopter, the aerodynamic force is controlled by changing the angle of attack.

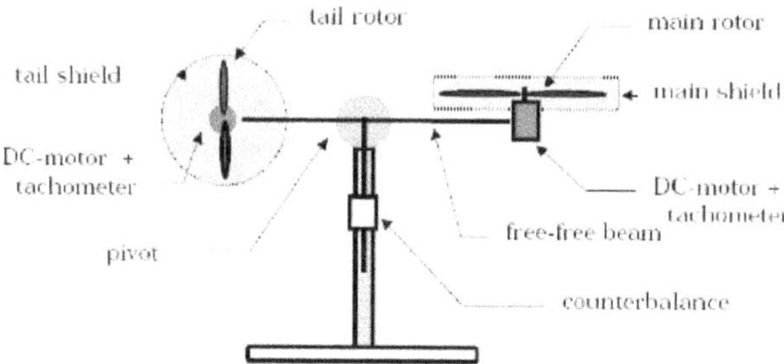

Fig.1　Aerodynamic model of the TRMS.

The laboratory setup from Fig.1 is so constructed that the angle of attack is fixed. The aerodynamic force is controlled by varying the speed of the rotors. Therefore, the control inputs are the supply voltage of the dc motors. A change in the voltage value results in a change in the rotation speed of the propeller. These further results give the change of the corresponding position of the beam.

MODELLING OF THE SYSTEM

Characteristics of the motors

It is necessary to identify the following non-linear functions:

Two non-linear inputs characteristics determining dependence of DC-motor rotational speed on voltage as follows in equation 1 respectively

$$W_m = P(U_{rr}), \quad w_t = P(U_{tt}) \tag{1}$$

Tow non-linear characteristics determining dependence of propeller thrust on DC-motor rotational speeds as follows in equation 2 respectively.

$$F_h = F_h(w_t) \tag{2}$$

The static characteristics of the propellers should be measured in the case when not delivered with the equipment documentation or if the propellers were changed by user. In this case a proper electronic balance with voltage output [0-10V] is needed. The recommended range of input force is from 0 to 2 N.

MAIN MOTOR

Having the main motor's characteristics is done by experiment as showed in Fig.2.Making the measurements is correct, first block the beam so that it can rotate around the vertical axis. Place the electronic balance under the beam in such a way that it is pulled vertically up by the propeller. To balance the beam in the horizontal position attaches the beam.

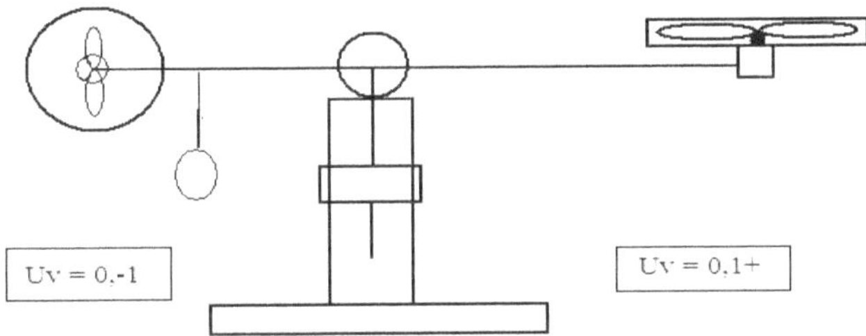

Fig.2 Main rotor Characteristics.e

Rotational speed of tail rotor is represented in below equation 3.

$$w_m(u_{rr}) = 90.99\ u_{rr}^6 + 599.73\ u_{rr}^5 - 129.26\ u_{rr}^4 - 1238.64\ u_{rr}^3$$
$$+ 63.45\ u_{rr}^2 + 1283.41\ u_{rr} \tag{3}$$

$$F_v(w_m) = -3.48.10^{-12}\ w^5{}_m + 1.09.10^{-9}\ w_m^4 + 4.123.10^{-6}\ w_m^3$$
$$- 1.632.10^{-4}\ w_m^2 + 9.544.10^{-2}\ w_m \tag{4}$$

The non linear function of aerodynamic force from main rotor as follows in Eq. (4) respectively.

TAIL ROTOR

To balance the beam in the horizontal position attaches the beam. Connect the voltage output of the electronic balance to analog to digital converter to the PLC or RT-DAC data acquisition board.

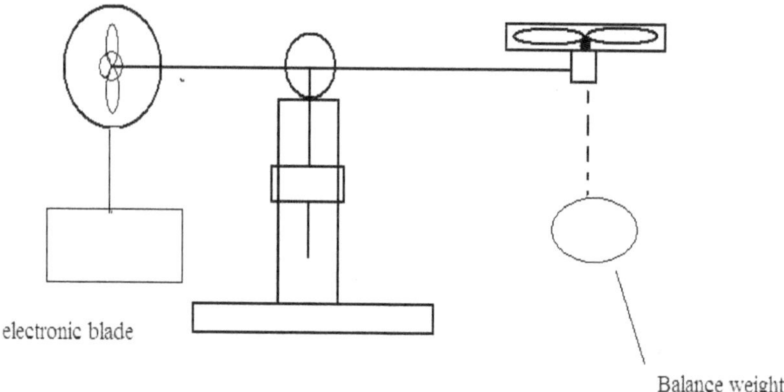

Fig.3 Tail rotor Characteristics.

The rotational speed and nonlinear functions of aerodynamic force from tail rotor is represented in above Equ. 5 and 6 the forces of the two rotors have the polynomial of the voltage. They are used in the subsystem and are the non-linear characteristics.

$$w_t(u_{tt}) = 2020\ u_{tt}^5 - 194.6\ u_{tt}^5 - 4283.1\ u_{tt}^3$$
$$+\ 262.2\ u_{tt}^2 + 3796.8\ u_{tt} \tag{5}$$

$$F_h(w_t) = -3.10^{-14}\ w_t^5 - 1.595.10^{-11}\ w_t^4$$
$$+\ 2.511.10^{-7}\ w_t^3 - 1.808.10^{-4}\ w_t^2$$
$$+\ 0.8080\ w_t \tag{6}$$

SYSTEM MODELLING

Physical model

The static characteristics of the DC motors with propellers are non-linear function which connects between a linear dynamic system and static non-linearity and the input voltage is limited to the range +/-10 volts. The linear part is in the form of first order transfer functions as follows,

$$G_h = 1/(T_h S + 1) \tag{7}$$
$$G_v = 1/(T_v S + 1) \tag{8}$$

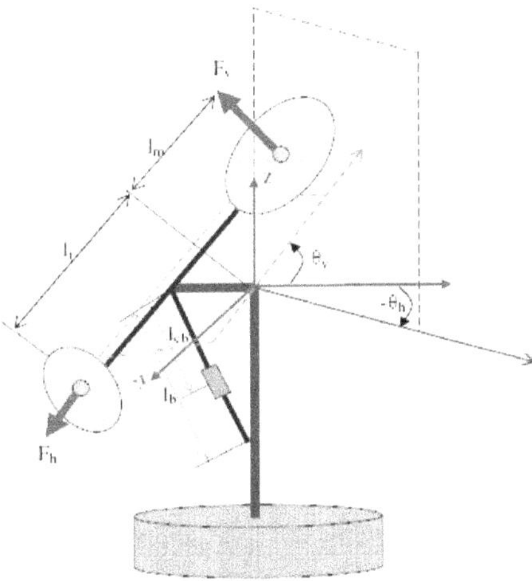

Fig. 4 The characteristic of the system.

The non-linear function u_{tt} and u_{rr} are characteristics of the DC motors. The non-linear relations between the rotor's velocity and the resulting aerodynamic force can be approximated the quadratic functions, in Eq.9and10 respectively.

$$F_h = sign(W_t)k_h.w_t^2 \qquad (9)$$

$$F_v = sign(W_m)k_v.w_m^2 \qquad (10)$$

Where k_h, k_v are positive constants.

Fig.4 represents the characteristic of the system. There are two propellers driven by DC motors at the end of There are two propellers driven by DC motors at the end of the pivot. The articulated joint allows the beam to rotate in such a way that its ends move on spherical surfaces

The physical model is developed under some simplifying assumptions: the dynamics of the propeller subsystem is first order differential equations; friction of the system is of the viscous type and propeller-air subsystem accordance with the postulates of flow theory.

MATHEMATICAL MODEL

Solving the system is considered the rotation of the beam in the vertical plane around the horizontal axis. Applying the Newton's second law of motion we obtain,

$$M_v = J_v \, d^2 \, \theta_v / dt^2 \qquad (11)$$

$$M_v = \sum_{i=1}^{4} M_{vi} , J_v = \sum_{i=1}^{8} J_{vi} \qquad (12)$$

To determine these elements on above equation, consider the figure below and these parameters are described in feedback manual [1].

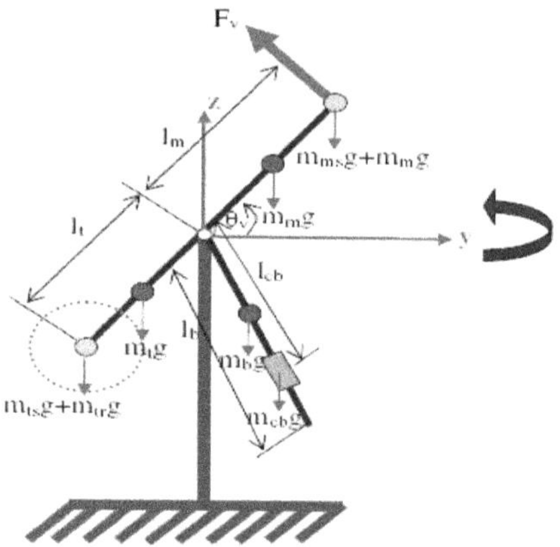

Fig.5 Side view of TRMS

From this reference manual form Feedback Company listed the parameter values. Using these values is very helpful for find the system dynamics and understands the nonlinear dynamics of the tail and main rotor. Moments of gravity forces (M_{v1}) applied to the beam and making it rotates around horizontal axis, consider the solution shown in Fig. 5 respectively.

Where, $A = (m_{t/2} + m_{tr} + m_{ts})$ (13)

$$B = (m_{b/2} l_b + m_{cb} l_{cb}) l_m$$ (14)

$$C = (m_{b/2} + m_{cb} l_{cb})$$ (15)

$$\frac{dS_v}{dt} = l_m S_f (w_m) - \Omega_v K_v$$

$$+ g((A - B) \cos \alpha_v - Sin \alpha_v)$$ (16)

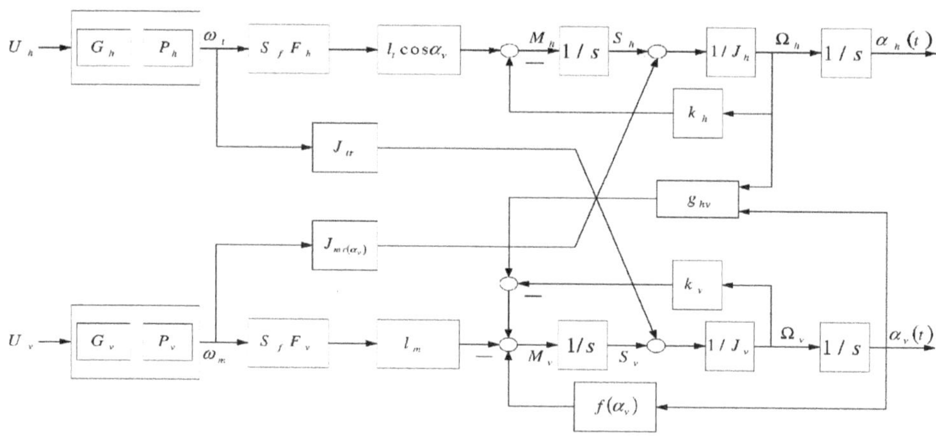

Fig.6 Blockdiagram for Twin rotor MIMO system

$$M_{v1} = \{(m_t/2 + m_{tr} + m_{ts})l_t - (m_m/2 + m_{mr} + m_{ms})l_m$$
$$-(m_m/2 + m_{mr} + m_{ms})l_m]$$
$$\cos\theta - (m_b/2l_b + m_{cb}l_{cb})\sin\theta_v\} \tag{17}$$

And the m_{mr} is the mass of the main dc motor with main rotor, m_m is the mass of the main part of the beam. m_{tr} is the mass of the tail rotor with tail rotor, m_t is the mass of the tail part of the beam. m_{cb} is the he mass of the counter weight, m_b is the mass of the counter weight beam, m_{ms} is the mass of the main shield, m_{ts} and is the mass of the tail shield. T_{mr} is the time constant of the main motor propeller system, L_b is the length of the counterweight beam, L_{cb} is the distance between the counter weight and the joint, U_{vv} is the output of the vertical dc motor G_v and g is the gravitational acceleration.

Following are the values of the model parameters:

$$l_m = 0.24\ m\ ;\quad l_b = 0.26\ m\ ;\quad l_m = 0.24\ m\ ;$$

$$l_{cb} = 0.13\ m\ ;\quad l_t = 0.25\ m\ ;$$

$$r_{ms} = 0.155m\ ;\quad r_{ts} = 0.10\ m\ ;\quad m_{tr} = 0.206\ kg\ ;$$

$$m_{mr} = 0.228\ kg\ ;\quad m_{cb} = 0.068\ kg\ ;\quad m_t = 0.0155\ kg\ ;$$

$$m_m = 0.0145\ kg\ ;\quad m_b = 0.022\ kg\ ;$$
$$m_{ts} = 0.165\ kg\ ;\quad m_{ms} = 0.225\ kg.$$

Assuming that the main rotor is an independent system.
$$d\alpha_v/dt = \Omega_v = 9.1 S_v \tag{18}$$

Similarly for tail rotor can be written as,
$$d\,S_h/dt = l_t\,S_t\,(w_t) - \Omega_h\,K_h \tag{19}$$

$$d\alpha_h/dt = \Omega_h \tag{20}$$

$$\Omega_h = 90\,S_h \tag{21}$$

Using Eqs. (18-21) simulate the system parameter values.

RGA AND SYSTEM PERFORMANCE INDEX

The GA proposed by Holland is commonly called the simple genetic algorithm (SGA). It will generate a (generally very high) number of chromosome strings at random, each representing an individual in the initial or parent population, to which the evolutionary principles of selection and mutation are applied. For the selection mechanism, the user has to provide the criterion for determining the relative fitness of every individual. This can be done by providing a fitness function that allows us to classify each individual in relation to the average fitness or error of the population and then decide which one is better. The algorithm will then favour individuals with higher fitness and selects them to be propagated into the next generation. Typically, this evaluation process consumes most of the execution time—no matter whether the fitness is determined by calculation or by experiment. The new generation is produced by applying genetic operators to the selected individuals. Some basic operators are reproduction, crossover, and mutation, which are used during the computation course. These operations and propagation are repeated until an individual matches the termination criterion. It is necessary to stress the actual solution of the problem instead of sampling solutions. However, it is important to make a fitness function more adequate by letting the population converge to a solution quickly. This correspondence uses real-value encodings of chromosomes rather than the more common binary encodings. There is no fixed rule for when a chromosome should be encoded as a bit string or as a real value. Though fixed-length, fixed-order binary encodings are the most common and best explained by the existing theoretical literature, "binary encodings are unnatural and unwieldy for many problems..., and they are prone to rather arbitrary orderings." As Mitchell [26] said, "for many applications, it is most natural to use an alphabet of many characters or real numbers to form chromosomes." The use of real-value encodings allows the development of a GA based on fitness in relation to a target chromosome. That is, instead of searching the fitness-landscape for an unknown solution, the GA is given possible solution from the beginning. The fitness of each individual is measured in comparison to the cost value of any chromosome in the context of the given problem, the goal provided at initialization. As should be apparent, this is a trivial problem in comparison to typical GA applications, where the answer is not (and sometimes cannot be) known in advance. Rather, here the answer is not of interest; what is of interest is the path that each population takes toward the known destination. The starting point is a population of chromosomes randomly generated with alleles from the specific parameters of the predefined chromosome.

PROCEDURE OF RGA

1) Population: The RGA does not work on a single individual but on a population **P** with p an individual that undergoes an evolutionary process starting with the initial population **P0**. The simplest way to create **P0** is to generate p strings or chromosomes randomly. Each gene in the chromosome represents a specific parameter of the PID controller. The population can be expressed as

$$\mathbf{P} = (\mathbf{X}1, \dots, \mathbf{X}p) \tag{22}$$

In the encoding process, we assign values for each gene within a specific range, which will be described in below section. The resolution of each gene is 0.0001 in both the encoding and decoding processes.

2) Selection and Reproduction: The selection operator S selects an individual of the given population according to its fitness value. In this correspondence, we choose the Roulette wheel selection as the selection process [25].

$$PS_i = \frac{\varsigma i}{\sum_{i=1}^{\lambda} \varsigma i} \qquad (23)$$

This method is based on the survival-of-the- fittest mechanism; the individuals with higher fitness values have higher probabilities of producing offspring. The probability is where ςi is the fitness value of each individual and λ is the population size. This operator reproduces the selected individuals to the mating pool where S' is new individuals and S is selected individuals in parent generation. R is the reproduction operator.

$$S' = R \times S \qquad (24)$$

3) Crossover: This operator exchanges the chromosome string of two selected individuals starting from a random index. Here, we use a modified crossover rule [27] for crossover process.

It is explained as follows:

$$x_{o1} = (1 - \beta) x_{pm}^{\alpha} + \beta x_{pn}^{\alpha} \qquad (25)$$

$$x_{o2} = \beta x_{pm}^{\alpha} + (1 - \beta) x_{pn}^{\alpha}$$

Where the two selected genes x_{pm}^{α} and x_{pn}^{α} are the parents

$$X_{pm} = [x_{pm}^{1} \dots \dots \dots x_{pm}^{\alpha} \dots \dots \dots x_{pm}^{u}] \text{ and}$$

$$X_{pn} = [x_{pn}^{1} \dots \dots \dots x_{pn}^{\alpha} \dots \dots \dots x_{pn}^{u}], \text{ respectively.}$$

α is the crossover position, which is an integer randomly generated within the range $[1, \mu]$, μ is the number of genes in an individual, and β is a real number randomly generated within the range $[0, 1]$.

To form the offspring chromosomes, x_{01} and x_{02} replace x_{pm}^{α} and x_{pn}^{α} respectively, then the genes on the left side of the crossover position x_{oi} are invariable, and the genes on the right side of the crossover position exchange with each other.

4) Mutation: This mutation operator creates one new offspring individual for the new population by randomly mutating a randomly chosen gene of the selected individual. For example, there are μ genes that represent d parameters in an individual $\mathbf{X} = [x_1, \dots, x_k, \dots, x_\mu]$, and the kth parameter is in the mutation position within the range $[\, UB_k \, LB_k \,]$. In (2.4), the new gene is

$$x_{\text{knew}} = LB_k + r (UB_k - LB_k)$$

Where, r *is* a real number within $[0, 1]$. The individual after mutation is

$$\mathbf{X}_{\text{new}} = [x_1, \dots, x_{\text{knew}}, \dots, x_\mu].$$

5) Fitness Function: The fitness function F is used to guide the evolution in a certain direction. In fact, the genetic algorithm is merely a method of approximating the global maximum of F in the search space of chromosome strings. The actual interpretation of this search space is packed into f and does not concern the algorithm itself.

In this correspondence, the fitness function illustrates the system performance, which will be described later. In the beginning of the searching procedure of the RGA, an initial set of solutions, called the initial population, is generated randomly. Each individual in the population is called a chromosome, which is a string composed of genes that represents encoded parameters of the solution. The chromosomes evolve through successive iterations, called generations. The first step in each generation is to decode and evaluate current chromosomes through a predefined fitness function. The offspring is a new chromosome of the next generation and is formed through the following procedures: selection, reproduction, crossover, and mutation. The chromosomes of the new generation are decoded and evaluated by the corresponding fitness function. The earlier steps are repeated for several generations, then the evolution process converges to a chromosome that has the highest fitness value, and the optimal or near-optimal solution to the problem can be obtained.

In this section, all parameters of the PID controllers will be optimized by the RGA with the system performance index. The final purpose is to build a control structure in 2 DOF. There are 12 parameters that need to be searched. In order to reach this goal, the parameters in the decoupled system are tuned first and then the results are combined for the 2 DOF systems.

PARAMETERS OPTIMIZING FOR PID CONTROLLER IN THE HORIZONTAL PLANE AND VERTICAL PLANE.

In this section, all parameters of the PID controllers will be optimized by the RGA with the system performance index. The final purpose is to build a control structure in 2 DOF. There are 12 parameters that need to be searched. In order to reach this goal, the parameters in the decoupled system are tuned first and then the results are combined for the 2 DOF system.

A. PARAMETERS OPTIMIZING FOR PID CONTROLLER IN THE HORIZONTAL PLANE AND VERTICAL PLANE.

The control system contains a PID controller and a horizontal part of the TRMS. There are ten individuals each with three genes in the initial population. The crossover rate is 0.8 and the mutation rate is 0.025. The parameters to be tuned are KP_{hh}, KI_{hh}, and KD_{hh}, within the range [0, 2], and the number of generations is 200. The computer simulation is from 0 to 50 s, with a sampling time of 0.05 s. For step response, the reference input is 1.0 rad, with the initial condition of 0.0 rad. The fitness function is

$$F = 30\,000//IPITSE. \tag{26}$$

In the procedure of the RGA, the crossover rate is 0.8 and the mutation rate is 0.025 with 500 generations. There are 12 genes in an individual generated within the range [0, 2]. Each gene represents one control parameter.

PARAMETERS OPTIMIZING FOR PID CONTROLLER IN CROSS-COUPLED CONDITION

The control system has four PID controllers in the cross-coupled condition. There are 12 parameters in the controller structure that need to be optimized in table1.

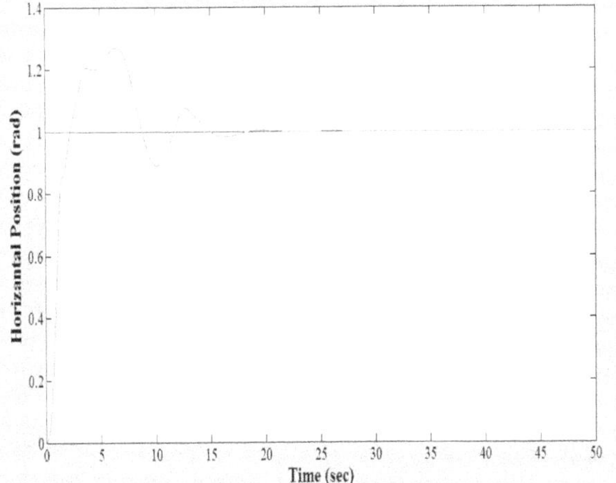

Fig.7   Horizontal position output for Cross-coupled control

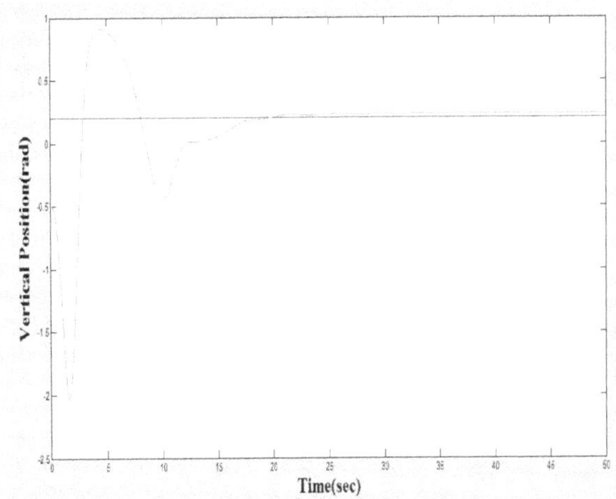

Fig.8 Vertical position output for Cross-coupled control.

In Fig.7 and Fig.8 Cross-coupled control simulation results show that the new control scheme can make the TRMS reach a specific position and track a desired path more efficiently. The total error in set point control and trajectory tracking of the sine wave and square wave are both reduced. The average computational cost of the modified RGA for cross-coupled simulation is 50 seconds over 30 runs, and each run has 500 generations. The searched solution of the RGA is related to initial search range, initial population, fitness function, and other factors.

In order to test the set point and tracking control, the computer simulation is given with step as the reference signal: Step input with 1 rad in the horizontal plane and step input with 0.2 rad in the vertical plane;In the procedure of the RGA, the crossover rate is 0.8 and the mutation rate is 0.025 with 500 generations. There are 12 genes in an individual generated within the range [0, 2]. Each gene represents one control parameter.

Table 1 Paremeters In Cross-Coupled Control

	Horizantal to Horizantal	
K_{Phh}	K_{Ihh}	K_{Dhh}
	Horizantal to Vertical	
K_{Phv}	K_{Ihv}	K_{Dhv}
	Vertical to horizontal	
K_{Pvh}	K_{Ivh}	K_{Dvh}
	Vertical to Vertical	
K_{Pvv}	K_{Ivv}	K_{Dvv}

The fitness function is shown in equation (26).The computer simulation results are shown in Fig .7&8.

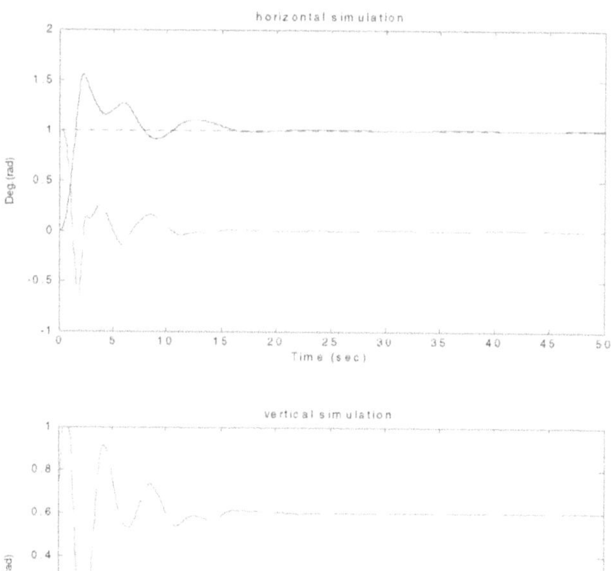

Fig.9. Step response in [12].

Table 1 shows the results of the cross-coupled control of the TRMS. The *error* of the step response in [12] is 69.09 in the horizontal plane and 34.92 in the vertical plane. The proposed controller reduces the *error* by 21% in the horizontal plane and 20% in the vertical plane and improves the output performance in 2 DOF. Fig. 9 shows the step response in [12].

Table 1   Simulation Results of Cross-coupled Control

Reference	Integral Absolute Error (IAE)	Control Energy
Step	99.9429	10.9747
Step	95.2130	23.8543
Reference	Integral Absolute Error (IAE)	Control Energy
Step	101.1744	13.9902
Step	95.4092	17.4231
Step	87.2539	26.4391
Step	99.9429	10.9747

CONCLUSION

The simulation results show that the new control scheme can make the TRMS reach a specific position and track a desired path more efficiently. The total error in set point control and trajectory tracking of the sine wave and square wave are both reduced. The average computational cost of the modified RGA for cross-coupled simulation is 50 seconds over 30 runs, and each run has 500 generations. The searched solution of the RGA is related to initial search range, initial population, fitness function, and other factors. The average computational cost of the modified RGA for cross-coupled simulation is 51 s, over 50 runs, and each run has 500 generations. The searched solution of the RGA is related to initial search range, initial population, fitness function, and other factors. In this correspondence, developed a new procedure for setting the initial search range, and propose a fitness function with a system performance index in solving large number of parameters simultaneously. The control parameters in the cross-coupled PID controller converge to satisfactory solutions quickly. Simulations show that the proposed PID control using NCD and RGA has better performance than conventional PID and GA-PID controllers

REFERENCES

[1]   *TRMS 33-220 User Manual*, 3-000M5, Feedback Company, E. Sussex,U.K. 1998.

[2]   S. M. Ahmad, M. H. Shaheed, A. J. Chipperfield, and M. O. Tokhi, "Nonlinear modelling of a TRMS using radial basis function networks," in *Proc. IEEE Nat. Aerosp. Electron. Conf.*, pp. 313– 320., Dayton, OH, 2000,

[3]   S. M. Ahmad, A. J. Chipperfield, and M. O. Tokhi, "Dynamic modelling and optimal control of a TRMS," in *Proc. IEEE Nat. Aerosp. Electron. Conf.*, pp. 391–398., Dayton, OH, 2000.

[4]   A. Odwyer, *Handbook of PI and PID Controller Tuning Rules*. London, U.K.: Imperial College Press, 2003.

[5]   B. C. Kuo, *Automatic Control Systems*, 6th ed. Englewood Cliffs, NJ: Prentice-Hall, 1995.

[6]   J. G. Ziegler and N. B. Nichols, "Optimum settings for automatic controllers," *Trans. ASME*, vol. 64, pp. 759–768, Nov. 1942.

[7]   R. A. Krohling, H. Jaschek, and J. P. Rey, "Designing PI/PID controllers for a motion control system based on genetic algorithms," in *Proc. 12th IEEE Int. Symp. Intell. Control*, Istanbul, pp. 125–130. , Turkey, 1997,

[8]   R. A. Krohling and J. P. Rey, "Design of optimal disturbance rejection PID controllers using genetic algorithms," *IEEE Trans. Evol.*, vol. 5, no. 1, pp. 78–82, Feb. 2001.

[9]   J. P. Su, C. Y. Liang, and H. M. Chen, "Robust control of a class of nonlinear systems and its application to a TRMS," in *Proc. IEEE Int. Conf. Ind. Technol.*, Bangkok, , pp. 1272–1277, Thailand, 2002.

[10] B. U. Islam, N. Ahmed, D. L. Bhatti, and S. Khan, "Controller design using fuzzy logic for a TRMS," in *Proc. IEEE Int. Multi Topic Conf.*, Islamabad, pp. 264–268., Pakistan, 2003.

[11] M. T. Huang and J. G. Juang, "Application of GA and PID control to nonlinear TRMS," in *Proc. Artif. Intell. Appl. Conf.*, Taichung, pp. 734–739., Taiwan, 2002,

[12] C. Y. Tsai, M. T. Huang, and J. G. Juang, "Application of real-type GA to TRMS position control," in *Proc. Autom. Control Conf.*, Jung-Li, pp. 1267–1273., Taiwan, 2003,

[13] J. H. Fang and J. G. Juang, "Analysis of optimal fitness function for TRMS parameter searching and its implementation on FPGA," presened at the Artif. Intell. Appl. Conf., Taipei, Paper FP4-1, Taiwan, 2004.

[14] W. K. Liu, J. H. Fan, and J.G. Juang, "Application of system-performance index based genetic algorithm to PID controller," presened at the Artif. Intell. Appl. Conf., Taipei, Paper FP4-3, Taiwan, 2004.

[15] J. H. Holland, *Adaption in Nature and Artificial System*. AnnArbor, MI: Univ. of Michigan Press, 1975.

[16] D. E. Goldberg, *Genetic Algorithms in Search, Optimization, and Machine Learning*. Reading, MA: Addison & Wesley, 1989.

[17] L. Davis, Ed., *Handbook of Genetic Algorithms*. NewYork, Boston: Van Nostrand Reinhold/Kluwer, pp. 299–305, 1991.

[18] K. F. Man, K. S. Tang, and S. Kwong, *Genetic Algorithms*. London, U.K.: Springer-Verlag, 1999.

[19] R. Subbu, K. Goebel, and D. K. Frederick, "Evolutionary design and optimization of aircraft engine controllers," *IEEE Trans. Syst., Man, Cybern. C, Appl. Rev.*, vol. 35, no. 4, pp. 554–565, Nov. 2005.

[20] G. Di Fatta, F. Hoffmann, G. Lo Re, and A. Urso, "A genetic algorithm for the design of a fuzzy controller for active queue management," *IEEE Trans. Syst.,Man, Cybern. C, Appl. Rev.*, vol. 33, no. 3, pp. 313–324, Aug. 2003.

[21] H. J. Cho, K. B. Cho, and B. I. Wang, "Automatic rule generation using genetic algorithms for fuzzy-PID hybrid control," in *Proc. IEEE Int. Symp. Intell. Control*, Dearborn, MI, pp. 271–276, 1996.

[22] B. Hu, G. K. I.Mann, and G. Raymond, "New methodology for analytical and optimal design of fuzzy PID controllers," *IEEE Trans. Fuzzy Syst.*, vol. 7, no. 5, pp. 521–539, Oct. 1999.

[23] K. Belarbi and F. Titel, "Genetic algorithm for the design of a class of fuzzy controllers: An alternative," *IEEE Trans. Fuzzy Syst.*, vol. 8, no. 3, pp. 398–405, Aug. 2000.

[24] M. Mitchell, *An Introduction to Genetic Algorithms*. Cambridge, MA: MIT Press, 1998.

[25] M. Mitchell, *An Introduction to Genetic Algorithms*. Cambridge, MA: MIT Press, 1998.

[26] A. A. Adewuya, "New methods in genetic search with real-valued chromosomes," M. S. thesis, Dept. Mech. Eng., Massachusetts Inst. Technol., Cambridge, 1996.

[27] T. Coleman, M. A. Branch, and A. Grace, *Optimization Toolbox: For UseWith MATLAB*. Natick, MA: The MathWorks, Inc., 1999.

DESIGN OF CIRCULAR PATCH ANTENNA ON METALLIC ELECTROMAGNETIC STRUCTURE FOR SURFACE WAVE SUPPRESSION

K.Praveen Kumar[1]

Ch.Raja[1]
[1]Research Scholar, Department of ECE, JNTUH, Hyderabad, India

Dr.Habibulla Khan[2]
[2]Professor, Department of ECE, K L University, Guntur, AP, India

Dr.M.Madhavi Latha[3]
[3]Professor, Department of ECE, JNTUH, Hyderabad, India
kpraveenkumar24817@yahoo.com

ABSTRACT

In this paper the performance of a micro strip circular patch antenna on different Di-electric materials is investigated on newly developed metallic electromagnetic structures characterized as (high impedance surfaces)HIS. This structure suppresses the propagation of surface waves and reflects external plane waves with zero phase reversal. Because of suppression of surface waves the power loss through Di-electric is material is minimized in this paper substrates with different Di- electric constant are considered, and the effects on resonance frequency, gain, return loss, input impedance, radiation pattern, current distribution are analyzed and presented in this paper .

KEY WORDS: MES (Metallic Electromagnetic Structure), dielectric. Surface wave suppression

INTRODUCTION

The ground plane of metallic electromagnetic surface consisting of an array of periodic planar conducting elements of hexagonal structure, characterized by high impedance surface [2, 3, 4] this structure functions like electrically thin in phase reflector. This also suppress surface wave at specific band of frequencies the currents from a nearby antenna and its images are in phase. Since surface waves are suppressed the power loss through the dielectric is minimized which in turn increases radiation efficiency. This is our key point of investigation and motive of this paper. The MES/HIS structures are configured based on texture as corrugation, mushroom type protrusions etc., In our present design we considered optically planar metallic Electromagnetic structures upon which substrate material of different Di-electric constants are positioned which is separating the circular patch antenna and his surface the dimensions of radiating element of circular patch antenna and HIS structure are maintained constant. Only the substrate materials of different Di-electric constant are changed, executed for results and the evaluation is performed using Ansoft HFSS v13 and presented in this paper.

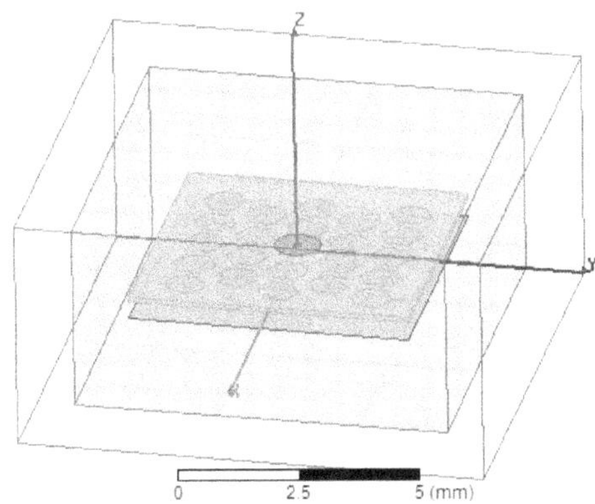

Fig (1) Circular Patch Antenna on MES

By keeping the dimensions of the radiating element of the antenna un changed we evaluated the gain and return loss for different Di-electric materials by using the Ansoft HFSS and the model of the antenna is shown in the figure (1). And the evaluated results are presented in Table (1). We observed radiation pattern, gain and return loss for materials with various relative permittivities.

Material	ϵ	Loss Tangent	Return loss (dB)	Resonant frequency(GHz)
RT Duroid 5880	2.2	0.0009	-23.1	50
Rogers Ultralam 1250	2.5	0.0015	-22.8	49
Benzocyclobuten	2.6	0	-22.9	49

Table (1) Data table of the proposed Antenna

The proposed antenna is designed with circular patch of radius 0.5mm along dimensions x-axis and y-axis. The substrate with the thickness of 0.11mm, and its dimensions along x-axis and y-axis are 6mm. feed is located at the center of it with coaxial inner radius 0.08mm, outer radius 0.12mm, and coax feed pin length is 0.079mm.

RESULTS AND DISCUSSION

The return loss curves for the proposed circular patch antenna backed high impedance surfaces for different Di-electric materials are presented in figure (2) the substrate RT-Duroid 5880(tm) is giving -23.1dB. Rogers Ultralum 1250 is giving -22.8dB. And the Benzocyclobuten is giving the return loss of -22.9dB.

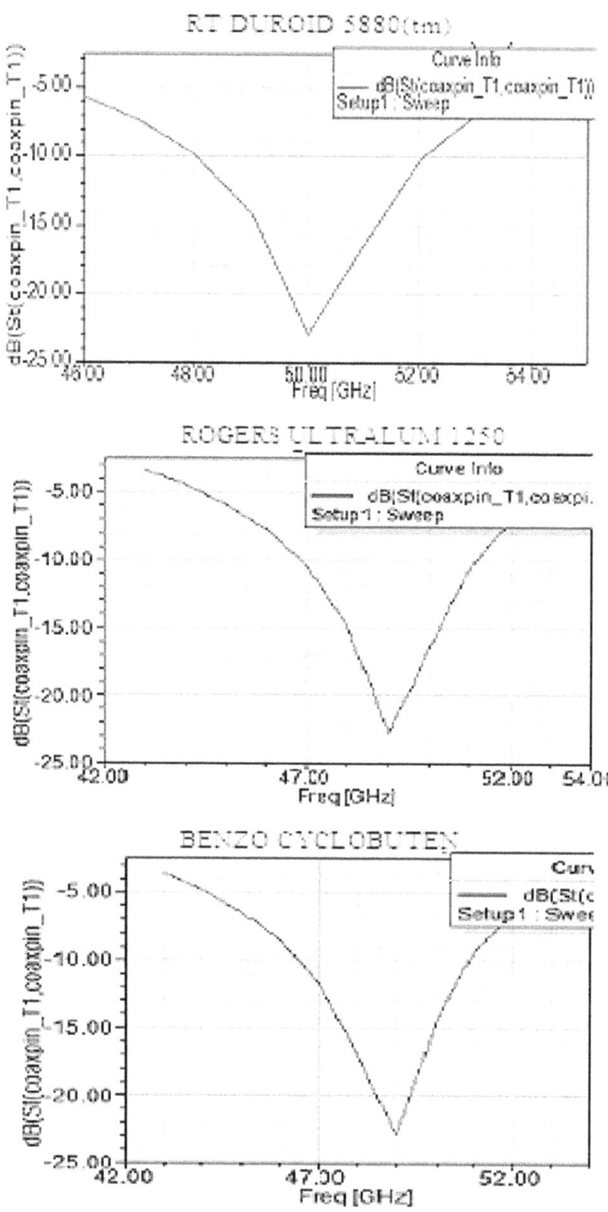

Figure (2) return loss vs frequency

The VSWR indicates the ratio of incident wave to reflected wave generally its value should lie between 1 to 2 for better results here in our present paper all the three materials obtained, the VSWR values are laying between 1 and 2 the graphical view of the results are shown in bellow figure (3)

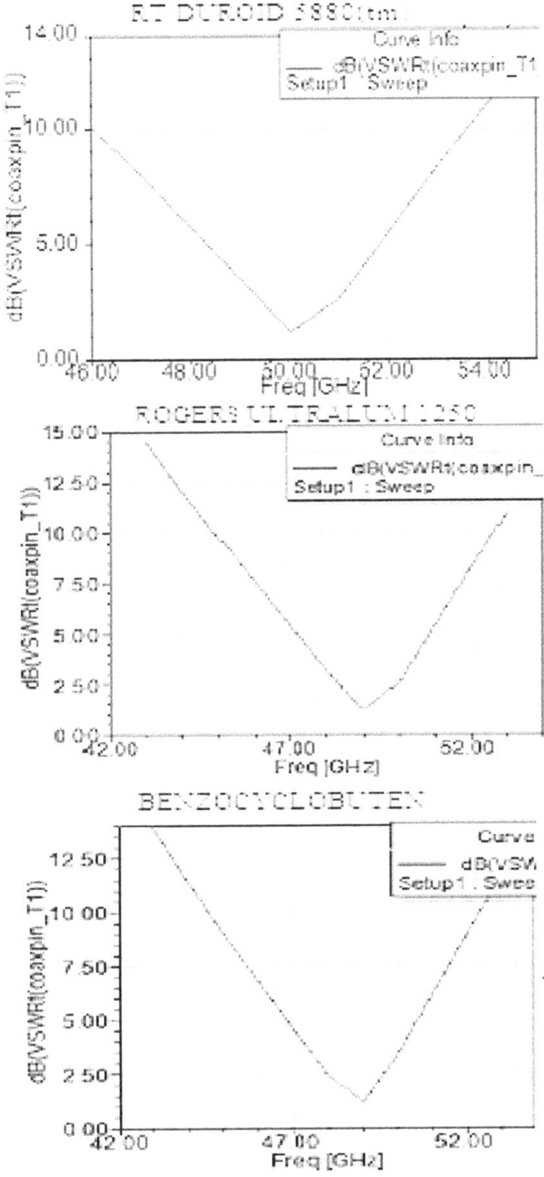

Figure (3) VSWR vs Frequency

The port is re-normalized at 55 ohm resistance for better results and maintained same for all the three materials same, the input impedance for all different Di-electric materials were presented in figure (4) as smith charts here we normalized port impedance to 55ohm for better return loss and VSWR values

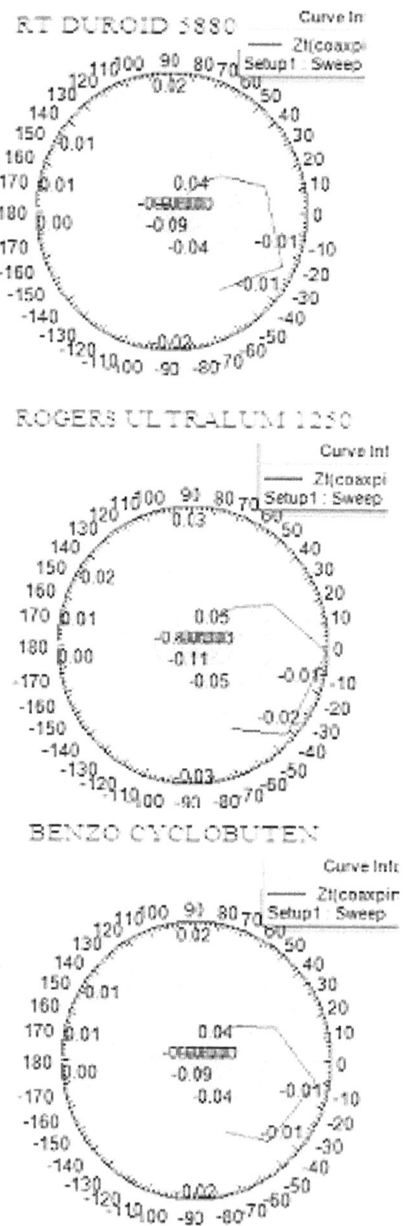

Figure (4) Impedance smith chart

The graphical representation of antenna radiation properties as a function of spherical co ordinate system is radiation pattern of the antenna in present work the radiation pattern at phi =0° and 90° is shown in the figure (5) for all the substrate materials we have taken they are almost giving same raditation on MES.

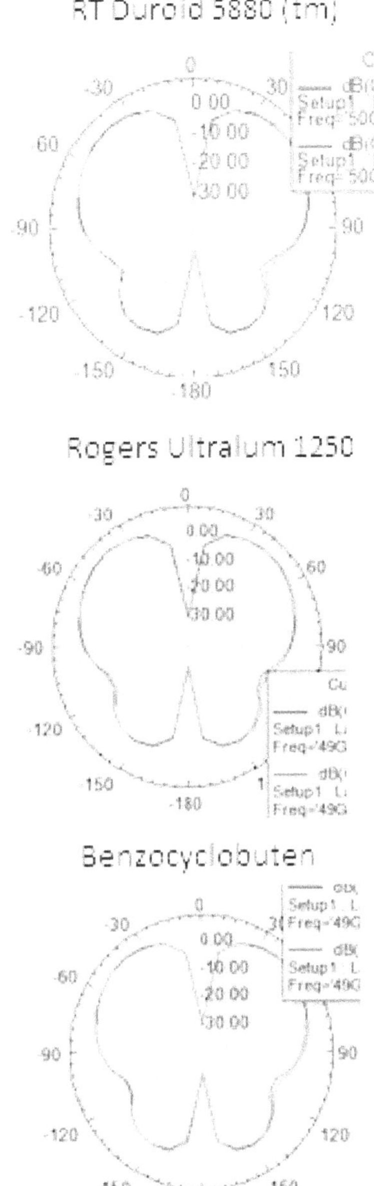

Figure (5) Radiation pattern in phi=0° and 90°

The 2D gain plots at phi=0° and 90° directions are shown in the following figure (6). The gain for Rt Duroid 5880(tm) is 3.15 dB and for Rogers Ultralum 1250 & Benzocyclobuten the gain is 4.82dB,4.91dB respectivly.

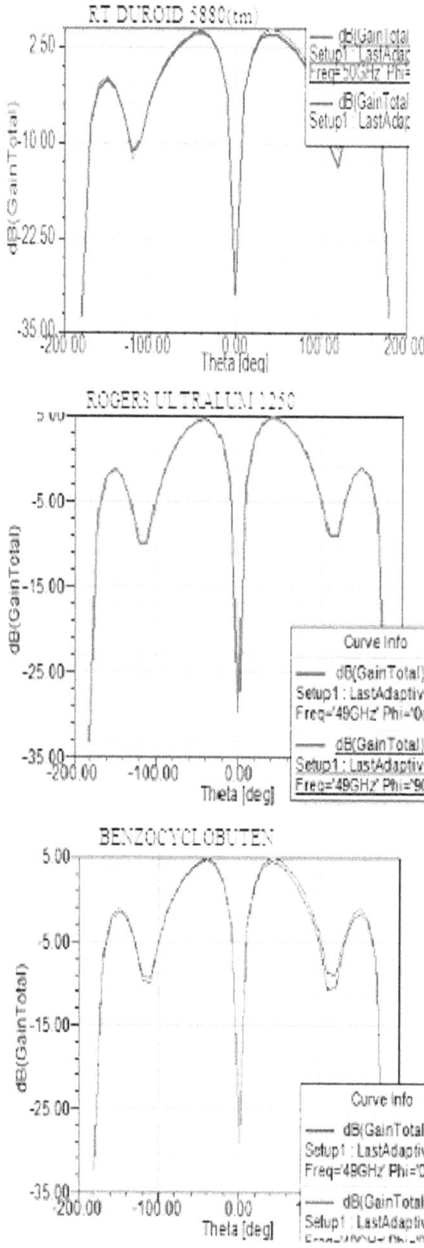

Figure (6) 2D gain in Phi=0° and 90°

The gain in three dimensional view is shown Is shown in the following figure (7) which illustrates the antenna radiation distribution among its surroundings. Its seen that the radiating element having almost the same 3D pattern for different materials we have analyzed.

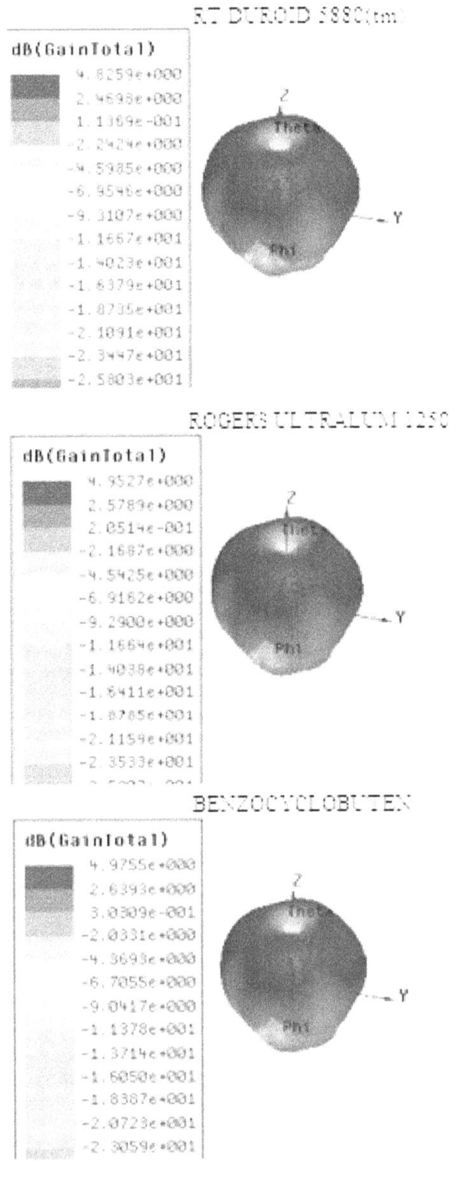

Figure (7) 3D gain

The current distribution of the designed antenna for all the substrate materials are shown in the fig (8) and we observed that it is giving some concentrated distribution of the mesh generation around the patch where the hexagonal shaped ground structure laying below and highly concentrated distribution in and around the radiating patch. This indicates that the current distribution concentration at the radiating patch for these materials based antennas are giving better results.

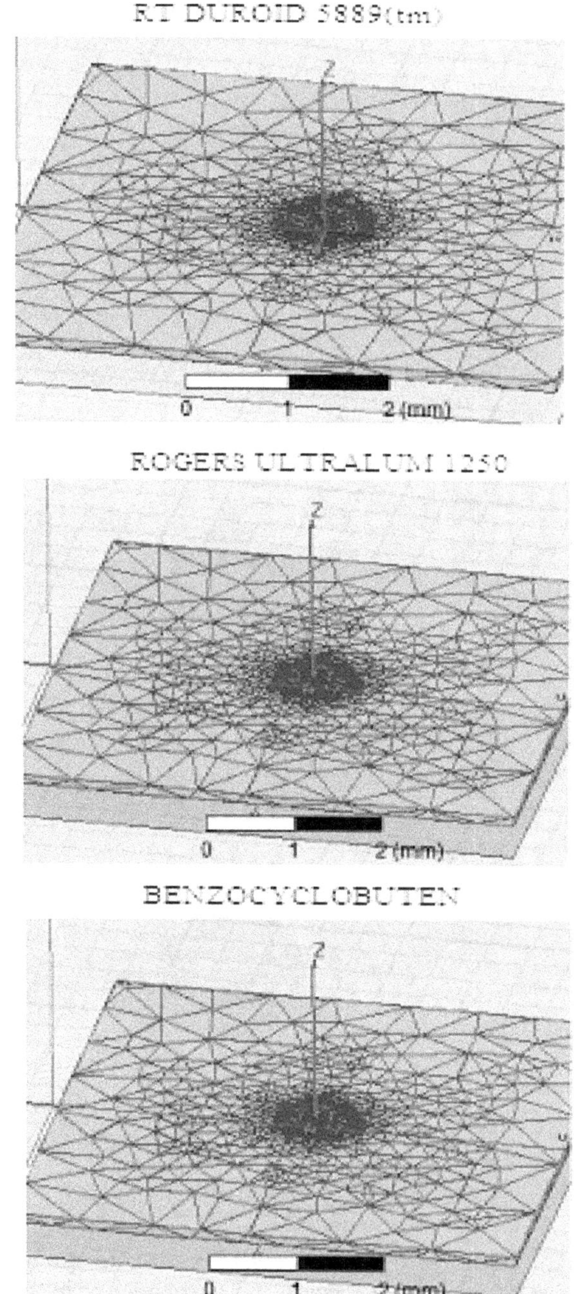

Figure (8) Current distribution of antenna

The E-field and H-field distribution at radiating patch are shown in the figure (9).

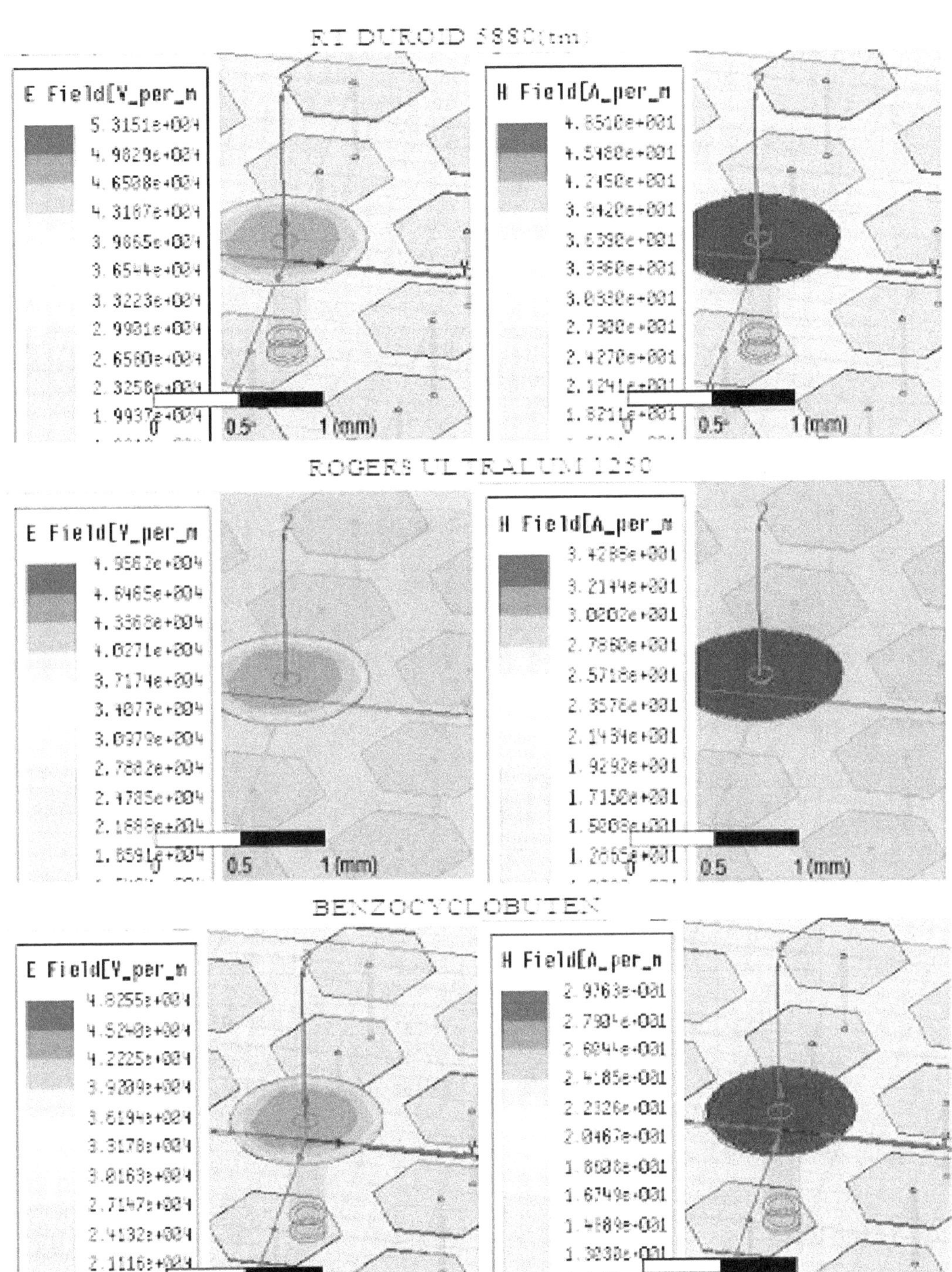
RT DUROID 5880(tm)
E Field[V_per_m
H Field[A_per_m
0.5   1 (mm)
ROGERS ULTRALUM 1250
E Field[V_per_m
H Field[A_per_m
0.5   1 (mm)
BENZOCYCLOBUTEN
E Field[V_per_m
H Field[A_per_m
0.5   1 (mm)

Figure (9) E-field and H-field of antenna

S.No	Quantity	RT Duroid 5880(tm)	Rogers Ultralam 1250(tm)	Benzocyclobuten
1	Max U	0.00495878(W/sr)	0.00506511(W/sr)	0.00455335(W/sr)
2	Peak Directivity	3.04308	3.10068	3.12418
3	Peak Gain	3.03741	3.08342	3.1522
4	Peak Realized Gain	2.08822	2.13318	1.91016
5	Radiated Power	0.0204777(W)	0.0205282(W)	0.0183153(W)
6	Accepted Power	0.020516(W)	0.0206432(W)	0.0181525(W)
7	Incident Power	0.0298414 (W)	0.0298387(W)	0.0299559(W)
8	Radiation Efficiency	0.998137	0.994433	1.00897
9	Front to Back Ratio	5.60722	4.89084	5.39212

The antenna parameters for the proposed antenna are listed in the table (2). And the antenna maximum field data is listed in table (3).

Table (2) Antenna parameters

S.No	rE Field	RT Duroid 5880(tm)	Rogers Ultralam 1250(tm)	Benzocyclobuten
		Value(v)	Value(v)	Value(v)
1	Total	1.93363	1.95425	1.8529
2	X	1.48945	1.5146	1.4232
3	Y	1.54776	1.56226	1.48438
4	Z	1.49614	1.50296	1.4225
5	Phi	0.70207	0.64221	0.659051
6	Theta	1.9335	1.95415	1.85271
7	LHCP	1.43368	1.41285	1.35237
8	RHCP	1.44755	1.43343	1.36584

Table (3) Antenna Maximum Field Data

CONCLUSIONS

In the present paper we have designed a circular patch antenna on optically planner hexagonal shaped High Impedance Electromagnetic surfaces, this structure reduce the surface current propagation via ground plane. The Di-electric constant of material is also effecting the propagation properties of antenna so in our paper I presented how the Di-electric constant of material is effecting the propagation properties like gain, directivity, return loss, efficiency was comparatively presented

ACKNOWLEDGEMENTS

The author like to express his thanks to Mr. Dr. Habibulla Khan, Prof dept. of ECE KL University for his great support during this work.

REFERENCES

[1] C. Balanis, Antenna theory, Analysis, and Design 2nd ed., John Wiley and sons, New York (1997)

[2] D. Sievenpiper, "High - Impedance EM surfaces", Ph.D. Dissertation, University of California, Los Angeles, 1999.

[3] D. Sievenpiper, E. Yablonovitch, U.S. provisional patent application, serial number 60/079953, filed on March 30,1998.

[4] B.T.P. Madhav, Prof. VGKM Pisipati, "Substrate permittivity Effects on the Performance of the Microstrip Eliptical Patch Antenna", Vol 2 No 3,2010-11.

[5] K.Praveen kumar, J. Doondi Kumar, "Microstrip GPS Patch Ceramic Antenna" *IJETAE* Vol 2, Issue 4, April 2012

PROPELLER-WING INTERACTION EFFECT ON MICRO AIR VEHICLES USING CFD FOR DESIGN ANALYSIS

Jeeva J[1], Avijit Chaterjee, H. Arya
[1] Department of Aerospace Engineering
Indian Institute of Technology Bombay
Powai, Mumbai - 400076

Abstract

The Recent interest in the development of Micro Air Vehicles (MAVs) has revealed a need for the thorough understanding of the aerodynamics of low Reynolds number flow. The present work is about an investigation into the effects of Propeller-wing interaction on Micro Air Vehicles using Computational Fluid Dynamics. While much research has been carried out on the propellers for large aircrafts, not much data exists for low Reynolds number vehicles. Air flow pass the vehicles with a propeller installed is one of the most complicated types of flow due to the interaction of the propeller flow with the wing flow. The flow around the air vehicle is affected by the strong vortices generated at the propeller blade tips. Flow interaction varies continuously with time due to the unsteady propeller motions. Therefore, it is necessary to consider the entire air vehicle, integrating its aerodynamic characteristics and its propulsion system rather than just isolated airframe or propeller.

This paper is part of a numerical study of propeller-Wing interaction effects on Micro Air Vehicles, where MAV Wing is modeled numerically and its aerodynamics characteristics are compared with the experimental results available from literature. A Zimmerman wing planform design was selected for the vehicle in the current study due to its advantage in retarding the early separation at design condition. A computational study of the MAV propeller has also been carried out by modeling the propeller NAL MAV PR 01using a Multiple Reference Frame Model, which consist propeller in rotating frame with an interface, the remaining mesh around the propeller to the stationary frame. Comparison of aerodynamics parameters of wing under propeller flow and without flow will be compared to understand the effect of interaction. The numerical simulation is done using ANSYS FLUENT.

Keywords: Micro Air Vehicles (MAVs), Propeller analysis, Zimmerman Wing analysis.

INTRODUCTION

Today, the strength of Computational Fluid Dynamics (CFD) is not in providing data, but of providing understanding so necessary for improved design. The design/optimization role of CFD has no counterpart in the wind tunnel. It has allowed us to achieve the design solutions that are otherwise unobtainable within the allowable time frame of a development program. CFD can be used where low Reynolds number flows are too difficult to investigate experimentally. CFD is also useful in extrapolating on published results when there is a gap in experimental data, or where little data is available. However, CFD typically has difficulty in predicting the location and size of the laminar separation bubble which in turn may result in poor quantitative predictions for lift, moment, and drag. Poor aerodynamic prediction may also result from the fact that low Reynolds number flows are not well understood computationally due to inherent problems in modeling thicker boundary layers where the flow may transition from laminar to turbulent.

Air flow pass the vehicles with a propeller installed is one of the most complicated types of flow due to the interaction of the propeller flow with the wing flow. The flow around the air vehicle is affected by the strong vortices generated at the propeller blade tips. Flow interaction varies continuously with time due to the unsteady propeller motions. Therefore, it is necessary to consider the entire air vehicle, integrating its aerodynamic characteristics (i.e. aerodynamic forces), and its propulsion system (such as propellers, engine inlet, etc.), rather than just isolated airframe or propeller. Micro air vehicles operate at the low Reynolds numbers, typically between 100,000 and 200,000. For such low Reynolds numbers, the flows around the airframe can involve laminar separation bubble formation, transition, and reattachments, which can all be influenced by the wake vortices, generated behind the propeller and hence influence the overall aerodynamic performance significantly. Therefore, a detailed investigation on the propeller effects is critical to understand the overall MAV performances.

This paper is part of a numerical study of propeller-Wing interaction effects on Micro Air Vehicles, where MAV Wing is modeled numerically and its aerodynamics characteristics are compared with the experimental results available from literature. A Zimmerman wing planform design was selected for the vehicle in the current study due to its advantage in retarding the early separation at design condition. A computational study of the MAV propeller has also been carried out by modeling the propeller NAL MAV PR 01using a Multiple Reference Frame Model, which consist propeller in rotating frame with an interface, the remaining mesh around the propeller to the stationary frame. Comparison of aerodynamics parameters of wing under propeller flow and without flow will be compared to understand the effect of interaction. The numerical simulation is done using ANSYS FLUENT.

MODELLING

GEOMETRIC MODEL

The Model employed by the Computational case was a Zimmerman Wing of Aspect Ratio 0.5 and 2.0 and a NAL MAV PR01 Propeller. A Zimmerman wing planform design was selected in the current study due to its advantage in retarding the early separation at design condition (Chen et al., 2010). It consists of two half-ellipses joined at the quarter-root-chord location. Fig.1 shows an illustration of how the Zimmerman planform geometries are generated. In this diagram, two half-ellipses are joined at the quarter-root-chord location. One ellipse has semi-major axis a3 and semi-minor axis a1 while the other has semi-major axis a2 and semi-minor axis a3.

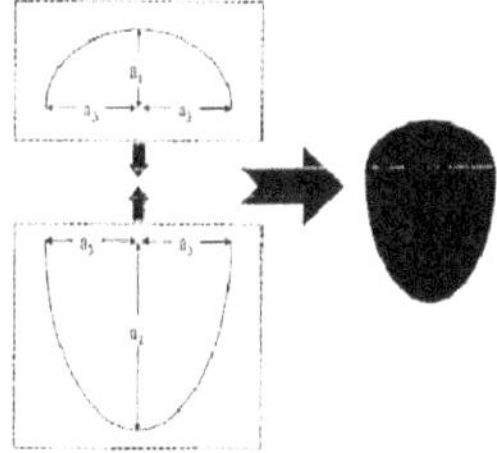

Fig.1 Formation of Zimmerman Planforms[1]

For the above planforms, two aspect ratio AR=0.5 and AR=2 having root chord of 0.2032m were selected for the numerical simulation (Mueller et al., 2001).

The models have a thickness to chord ratio of 1.96% and 5 to 1 Elliptical Leading and Trailing edge shape as shown below

Fig.2 Leading Edge Geometry: 5 to 1 Elliptical L.E[1].

As discussed above a Zimmerman planform wing is created in GAMBIT having the following dimensions given in Table.1 and Table.2

Table1 For wing having AR=0.5

Semi-major axis of Elliptical one	a3=0.039878 m
Semi-minor axis of Elliptical one	a1=0.0508 m
Semi-major axis of Elliptical two	a2=0.1524 m
Semi-minor axis of Elliptical two	a3=0.039878 m

Table 2 For wing having AR=2

Semi-major axis of Elliptical one	a3=0.159512 m
Semi-minor axis of Elliptical one	a1=0.0508 m
Semi-major axis of Elliptical two	a2=0.1524 m
Semi-minor axis of Elliptical two	a3=0.159512

The Propeller NAL MAV PR01 geometry details are given by Prathapanayaka et al (2011). The same geometry is modelled for numerical simulations using GAMBIT. The propeller was created inside two cylinders and one rectangular domain in order to create the moving reference frame model required to simulate flow around a propeller. The set-up is in a way, which the fluid interface is formed between two equally sized inner cylinders, housed inside an outside larger Rectangular domain. Using Boolean subtract operations in Gambit, the propeller geometry is subtracted from the first inner cylinder, forming the rotating zone; and the second inner cylinder (same dimensions as the first) is subtracted from the larger outer rectangular, forming the stationary zone. For a 0.1524m diameter propeller, the inner cylinders had a diameter of 0.16m and length of 0.015m, and the outer rectangular had a length of 2m, base of 1m and a height of 0.8m. This study was carried out at model scale, and the simulated working conditions were same as the ones of the experimental environment in order to compare the results between them.

MESHING AND DEFINING BOUNDARY CONDITIONS

The Unstructured mesh has been generated around the Wing and the flow domain using GAMBIT. The wing and domain has fine mesh of around 381398 nodes. A grid refinement was conducted to ensure that the results were adequately independent of the grid size. The Continuum type is specified as Fluid. Then the boundary conditions are defined as follow

Inlet face of Rectangular Domain- Velocity Inlet B.C. Outlet face of Rectangular Domain – Pressure Outlet B.C. Top, bottom and side face of Rectangular Domain – Wall B.C. Wing Body - Wall B.C

The Unstructured mesh has been generated around the propeller and the flow domain. The rotating zone has fine mesh of around 375617 nodes and the stationary zone has course mesh of 137614 nodes. A grid refinement was conducted to ensure that the results were adequately independent of the grid size. Then the boundary conditions are defined as follow

Inlet face of Rectangular Domain- Velocity Inlet B.C, Outlet face of Rectangular Domain – Pressure Outlet B.C, Top, bottom and side face of Rectangular Domain – Symmetry B.C, Two Inner Cylinder- Interface B.C and Propeller- Wall B.C

NUMERICAL SIMULATIONS

SOLVE STRATEGY

The meshed Wing design was imported into FLUENT, where the grid check has been performed. Since it is a low Reynolds number flow, the solver selected is Pressure based implicit solver with Green Gauss cell based gradient options. The flow is modeled as a viscous having laminar flow. The solution has Pressure Standard and Momentum second order upwind scheme. Energy equation is not used as flow is incompressible. The relaxation factors are all set to their default value. The reference values for calculating dimensionless parameter are defined from inlet condition. The force monitor is used to monitor the lift, drag and moment coefficient for various AOA. All simulations were converged to residual values of 10^{-3} where possible, and where the lift, drag and moment coefficient values are constant the result was assumed to be converged. This typically took between 800-1000 iterations and 2-3 hours of computation time.

Then the meshed Propeller design was imported into FLUENT, where the tetrahedral mesh was converted to polyhedral grid elements. This conversion process typically took between 1/2-1 hours, depending on the number of elements in the original tetrahedral mesh. Conversion to a polyhedral mesh has numerous advantages, such as reducing the number of elements by up to 400%, eliminating problems of excess cell skewness (which were frequently a problem with the un-polyhedral propeller mesh), and improved solution accuracy. In some cases, using a polyhedral mesh requires half the memory and 10-25% of the computing time of tetrahedral meshes to achieve the same accuracy (Peric et al., 2010). Instead of 4-6 faces, polyhedral elements have 12-14 faces, and the higher number of surrounding elements leads to better model flows in different directions through each individual element. Polyhedral cells require more computational operations per element; however, this is offset by the reduction in total number of elements. The fluid interface was designed by assigning the two inner cylinder boundary conditions to 'interface,' and then using the FLUENT interface setup tool as given by Ryan et al., 2010. The outlet was assigned as a pressure outlet boundary condition. The turbulence model used is a standard K-ε model with the enhanced wall treatment model was coupled with 2nd order momentum, turbulent kinetic energy, turbulent dissipation rate and standard pressure solution methods.

Air becomes compressible at the flow speeds of approximately 112m/s, and the maximum tip velocity of the 0.152m diameter propeller at 10,000RPM was 79.58m/s, so the energy equation was not used. Furthermore, NASA suggests that changes in density below Mach 1.0 are very small and for subsonic flow it can be ignored (NASA Glen Research Centre., 2010). The rotating fluid zone was created by assigning the cell zone condition of the inner cylinder containing the propeller to Moving Reference Frame, setting the rotational reference to the x-axis and assigning a rotation speed in rad/s. All simulations were converged to residual values of 10^{-7} where possible, and where residuals reached steady state at higher values the result was assumed to be converged. This typically took between 800-1000 iterations and 3-4 hours of computation time.

RESULTS AND DISCUSSION

This chapter details the results of the wing and propeller characteristic obtained from the numerical simulation of flow over the wing and Propeller model as discussed above. Lastly, results from the Numerical simulation are compared and analyzed to experimental data.

WING RESULTS

A selection of the results obtained from the Numerical simulation for Zimmerman Wing of AR=0.5 and AR=2 are compared with the experimental values (Mueller et al., 1998-2001). The following results correspond to numerical simulation having chord Reynolds number of 70000 and AR=0.5.

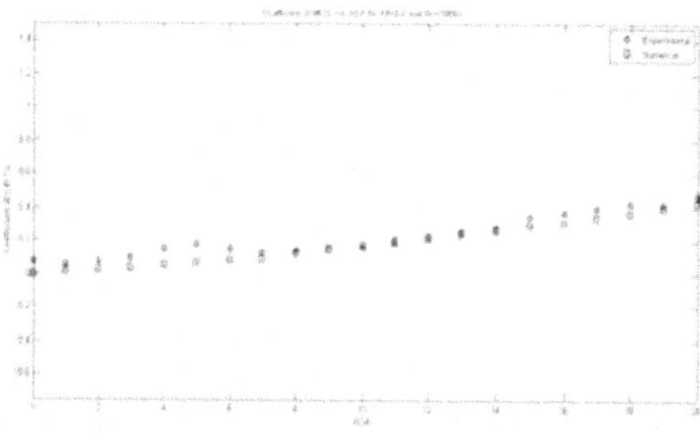

Fig.3 Coefficient of Lift (C_L) Vs AOA (α)

In the Experimental result a striking feature of the C_L Vs α curves for AR=0.5 wing, especially at Reynolds's Number of 70000 is the kink observable near α=5 deg as shown in Fig.3. The cause for the occurrence of the kink is currently being studied. But in numerical solution the C_L curve doesn't have any kink. As you can see except at the kink region the numerical solution matches closely to the experimental result. From the C_L curve we can say that the as angle of attack of increases lift coefficient increases. The stall angle for this wing is very high when compared to conventional wing which would be less than 15 deg. As angle of attack increases the laminar separation bubble increase in size and move towards the leading edge. Here for the angle of attack 20^0 the laminar separation bubble is very small hence numerically it is possible to capture the flow as of experimental.

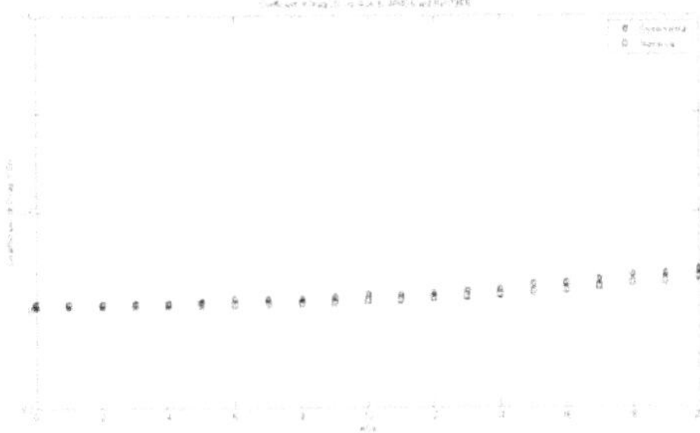

Fig.4 Coefficient of Drag (C_D) Vs AOA (α)

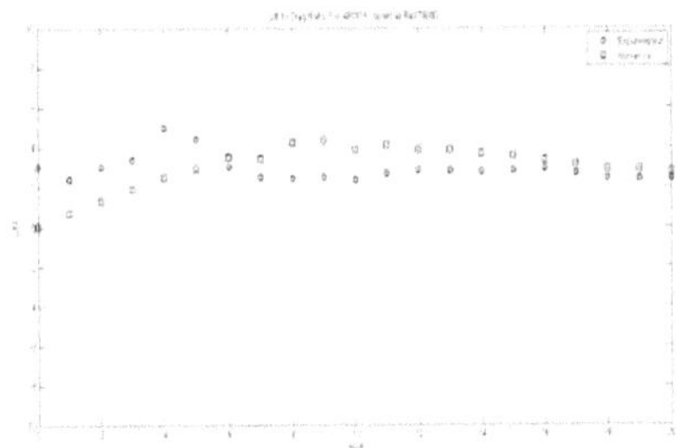

Fig.5 Lift to Drag Ratio Vs AOA (α)

The above plots give the good match of numerical and experimental results. As C_L and C_D value of numerical result is slightly lower than the experimental result, the aerodynamic efficiency L/D of numerical result is higher than the experimental result.

The following results correspond to the numerical simulation having chord Reynolds number of 70000 and AR=2. Here lift at higher angle of attack exhibit a constant lift characteristic. The lift curve is constant after 15 deg of angle of attack i.e. as angle of attack increases lift remains constant. This nature is also captured in numerical result as shown below where lift is almost constant at higher angle of attack. The Drag coefficient obtained by the numerical result is lower than the experimental result as shown in Fig.7 where the gap increases at higher angle of attack.

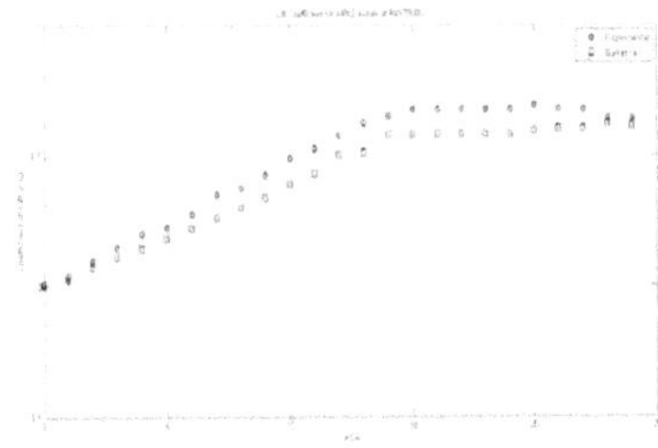

Fig.6 Coefficient of Lift (C_L) Vs AOA (α)

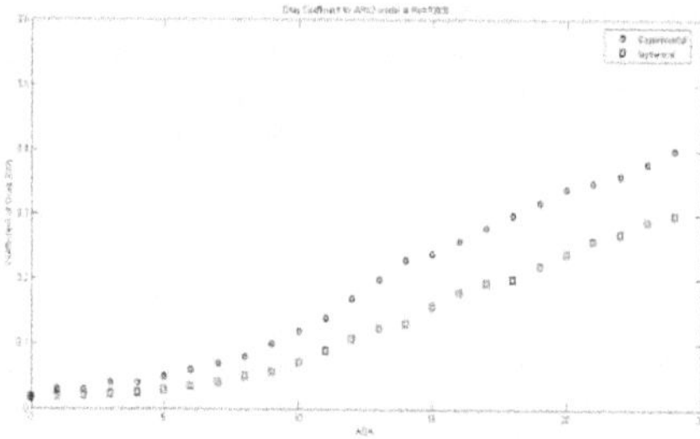

Fig.7 Coefficient of Drag (C_D) Vs AOA (α)

As drag obtained in numerical simulation is low compared to experimental, the lift drag ratio of numerical result is slightly greater than the experimental result.

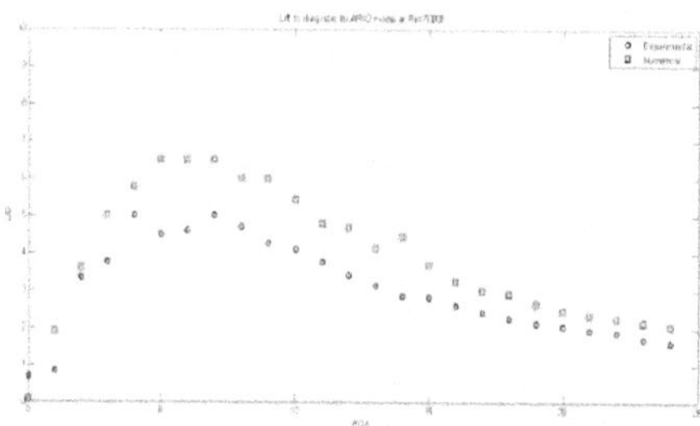

Fig.8 Lift to Drag Ratio (L/D) Vs AOA (α)

The below plot gives the co-efficient of moment for Zimmerman Wing of AR=2 and Re=140000.

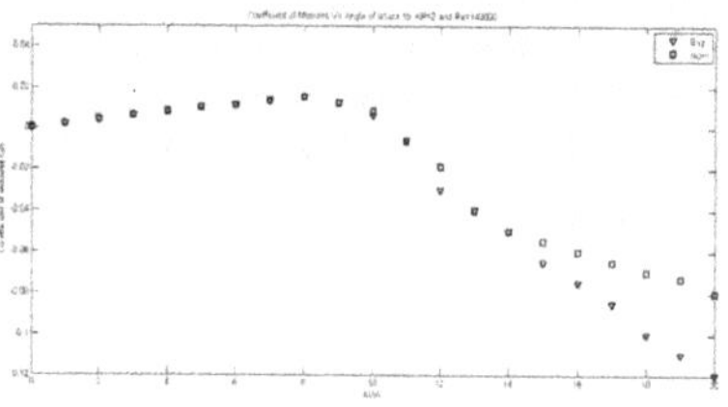

Fig.9 Coefficient of Moment (C_M) Vs AOA (α)

As the angle of attack increases, the numerical result deviating from the experimental result. The reason is, as angle of attack of increases the laminar bubble formed increases in size and move toward leading edge which has more reverse flow in domain which is not captured by the numerical solver. As bubble size less it gives better result than the larger one. It shows that numerical solution not able to capture the result at higher angle of attack which involves large separation bubble.

454

PROPELLER RESULTS

Numerical simulation were carried out for different free stream velocities ranging from 2 to 14 m/s in steps of 2 m/s for the speed of 8000 rpm in steps of 1000 rpm. The iteration is carried on by increasing the rpm in steps of 1000 till 8000 for a particular velocity. Then the coefficient of Thrust (C_T) and the Advanced Ratio (J) is calculated and plotted. Fig.10 show the comparison of coefficient of thrust v/s advance ratio obtained from CFD and Experiment for design speed of 8000rpm.

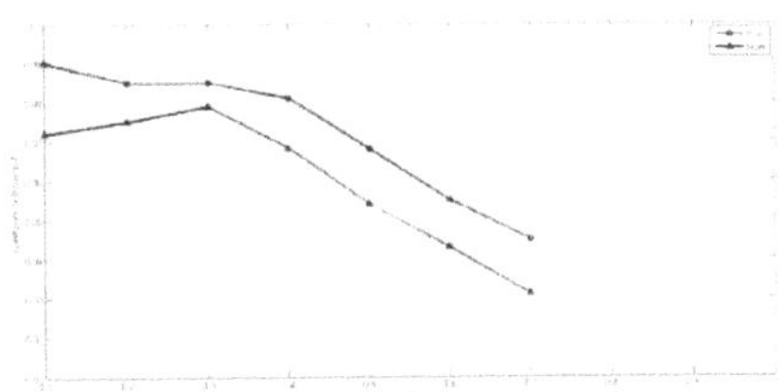

Fig.10 Comparison of coefficient of thrust

Vs ADVANCE RATIOS FOR CFD AND EXPERIMENT

The numerical result gives value less than the experimental value as shown above. However the results are close to the experimental result which follows similar trend of experimental curve.

FUTURE WORK

As it mentioned the main objective of this project is to analyze the propeller wing interaction effect on the MAVs. But so far we have discussed the flow over a Wing and the Propeller separately. In later part the aerodynamics parameters of wing under propeller flow and without flow will be compared to understand the effect of interaction.

STRATEGY TO SOLVE PROPELLER-WING INTERACTION PROBLEM

A common domain will be created for both the wing and the Propeller. The propeller is placed just ahead of the Wing. Then the control domain volume will be subtracted from the Wing and the propeller along with the cylinder, so it will be considered as a one stationary volume and one rotating volume. Then a 3D unstructured Mesh will be created over the domain. The problem will be solved using a Multiple Rotating frame (MRF) method in Fluent.

REFERENCES

1. Mueller, T.J. and Torres, G., "Aerodynamics of Low Aspect Ratio Wings at Low Reynolds Numbers with Applications to Micro Air Vehicle Design and Optimization", University of Notre Dame 1998-2001.

2. Chen, Z.J., Qin, N. and Nowakowski, A. F. "Alleviation of Flow Separation on Cambered Plate Wing Micro Air Vehicles", RAeS Aerodynamics Conference, 27-28 July 2010.

3. Chen, Z.J., "Propeller Effects on MAV Aerodynamics: A Numerical study", Department of Mechanical Engineering, University of Sheffield, UK, 2010.

4. Prathapanayaka R, Vinod Kumar N, Krishnamurthy S J,"Design, Analysis, Fabrication and Testing Of Mini Propeller for MAVs", SAROD 2011.

5. Peric, M and Ferguson, S. "The advantage of polyhedral meshes", CD-adapco [Online] 2010

6. Ryan S.A. Turner," Design and Optimization of a Propeller for a Micro Air Vehicle Using Computational Fluid Dynamics", OFFCDT, School of Engineering & Information Technology, ZEIT 4500 Aeronautical Thesis & Practical Experience 2010.

7. National Aeronautics and Space Administration, "Mach number: role in compressible flows", NASA Glenn Research Center [Online] 2008.

IMPROVED FAULT TOLERANT CONTROL OF THREE PHASE VOLTAGE SOURCE INVERTERS

R. K. Pongiannan[1], M. Sathiyanathan[2]
[1-2] Dept.of EEE, RVS College of Engineering and Technology, Coimbatore, Tamilnadu.

A. Prakash[3],
[3]R&D, LogIT (P) Ltd, Coimbatore, Tamilnadu.

M. Mubarak[4]
[4]Dept. of ECE, Shivani Engg. College, Trichy, Tamilnadu.
and

N. Yadaiah[5]
[5]Dept. of Electrical and Electronics Engg., JNTUH College of Engg., Hyderabad.
Pongiannan_rk@yahoo.co.in, sathiyangm@gmail.com, prakash@logit.co.in.

ABSTRACT

This paper proposes an improved fault tolerant control and its analysis for Voltage Source Inverters (VSIs). The line fault and line to line fault in a Three Phase system has been considered for the proposed fault tolerant control scheme. The proposed control approach minimizes the time duration between fault occurrence and its diagnosis. The detection and diagnosis of the fault is carried out in a short period which is less than 10 μs. The proposed Fault tolerant control for VSI fed induction motor drive application is modeled and simulated using MATLAB/Simulink. The effectiveness of the control approach to the fault diagnosis scheme is investigated and the simulation results are presented.

KEYWORDS- Voltage Source Inverters (VSIs), induction motor drive (IMD), Fault-tolerant VSI topology, SVPWM, Fault detection, control.

INTRODUCTION

The reliability of power electronic equipments becomes extremely important in general in industrial applications. The fault mode behavior of static converters, protection and fault tolerant control of voltage source inverter systems has been covered in a large number of papers. Most of them are focused on induction motor drive applications.

A fault-tolerant control system for a high-performance induction motor drive for an electrical vehicle (EV) is explained in Demba Diallo et al.,2004. In this a practical sensor less control scheme in Qun-Tao An et al., 2011 is developed and used within the fault-tolerant control system. Andre M. S. Mendes et al., 2007 comprises various fault modes of a voltage source PWM inverter system for induction motor drive is considered. It explains the rectifier diode short circuit, inverter transistor base driver open and inverter transistor short circuit conditions. However, it doesn't discuss about the stand by protection of the inverter topology. A fault tolerant control of induction motor drive applications using analytical redundancy, providing solutions to most frequent occurring faults is given in S. Karimi et al.,2007.

Fault detection of open-switch damage in voltage source PWM for induction motor drive systems is investigated in Karimi et al.,2007.and Mohammad-Ali Shamsi-Nejad et al.,2008. It mainly focused on detection and identification of the power switch in which the fault has occurred. In H. El Brouji el al., 2009, the utilization of a two-leg based topology when one of the inverter legs is lost explained.

Then the machine operates with only two stator windings. This paper is also proposed to modify PWM control to allow continuous free operation of the drive.

A fault tolerant shunt three phase active filter topology is explained in H. El Brouji el al., 2009, M. S. Khanniche et al., 2001 and M.Rozailan Mamat et al., 2006. It mainly details converter reconfiguration and examines a fast and reliable optimized fault diagnosis method. The design, implementation, experimental validation and performances of a fully digital fast power switch failure detection and compensation for fault-tolerant voltage source inverters (VSIs) are discussed in Karimi et al.,2007 and Alexander L. Julian et al.,2010. In this approach the time interval between the fault occurrence and its diagnosis is less than 10 ms using FPGA control.

In M. S. Khanniche et al.,2001 and M.Rozailan Mamat et al., 2006 detection and identification of transistor base drive open-circuit fault of 3-phase voltage source inverter (VSI) is investigated. Fault detection is achieved using wavelet transform. The control scheme is implemented using fuzzy logic for the reliable operation of an induction motor.

Design and testing of two-level fault-tolerant voltage source inverter (VSI) for permanent magnet drives is given in Rammohan Rao Errabelli et al.,2012. In this case of failure, a redundant leg is added that replaces the faulted leg. Faulted leg isolation and redundant leg insertion are done by using independent back-to-back-connected thyristors. In Andre M. S. Mendes et al., 2007 and 2008, fault tolerant operating strategy based on the connection of the faulty inverter leg to the dc link middle point was considered.

This present paper deals with open switch faults detection and localization in three-phase VSI fed Induction motor drives. This paper concentrates on detection and diagnosis of two leg failures at a time. The new control topology has been developed for the isolation of the faulty leg and insertion of new leg. The proposed method is simple and highly reliable.

First, an inverter based on standard three-phase power structure with the improved fault tolerant is presented in detail. Then, the Fault tolerant Controller for proposed system is clearly presented. Finally, simulation results illustrating the effectiveness of the proposed fault tolerant control topology for the induction motor drive under power switch failures.

FAULT TOLERANT VOLTAGE SOURCE INVERTER TOPOLOGY

Three phase inverter with the proposed two leg inverter system is shown in Fig.1 The fault-tolerant VSI is composed of five legs. The fourth and fifth leg formed by the switches F_1-F_2 and F_3-F_4 for the redundant legs respectively. When a fault has occurred in the semiconductor switches (S_1-S_6) or associated drivers, the fault detection scheme detects the fault occurrence and isolates the faulty leg with the help of back-to-back-connected thyristors (B_1-B_2, B_3-B_4 and B_5-B_6) are connected between output terminals of the inverter (V_a , V_b and V_c) and corresponding motor phases respectively. If the fault occurrence is an open circuit, the isolation is first implemented by removing the gate signal from the switches of the faulty leg. In the case of short circuit, the faulty leg is isolated by very fast acting fuses. The model has been implemented in MATLAB simulink environment.

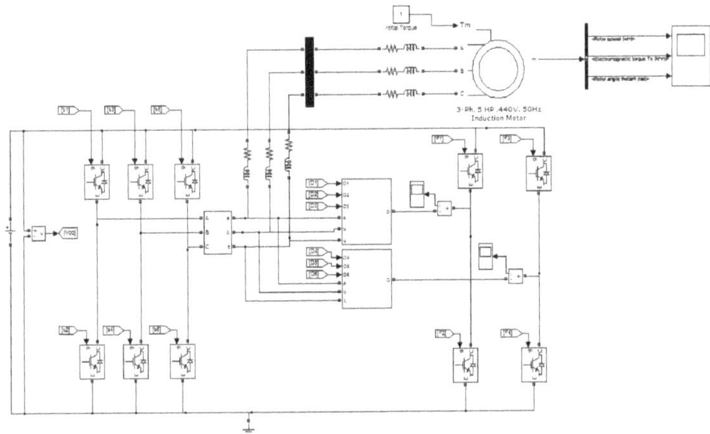

Fig.1 Fault tolerant control topology

SPACE VECTOR PULSE WIDTH MODULATOR (SVPWM)

In this section, the description of the space vector modulator is discussed in Auzani idin et al., 2009. The space vector modulator is constructed using MATLAB simulink. In The space vector modulator (SVM) dc voltage to the inverter is measured. This dc voltage value is used to compute the voltage vector applied to the motor.

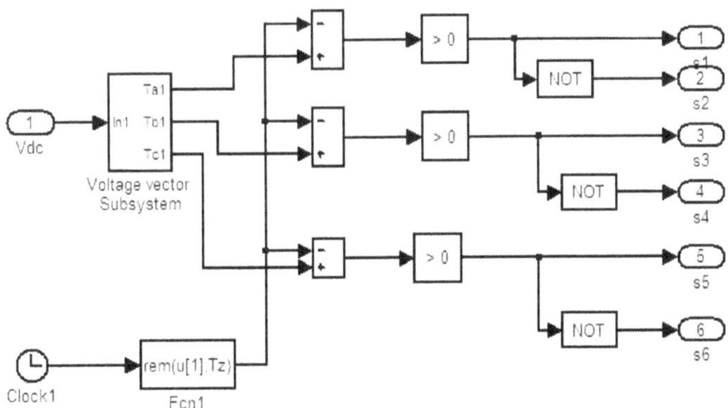

Fig.2 SVPWM controller MATLAB simulink model

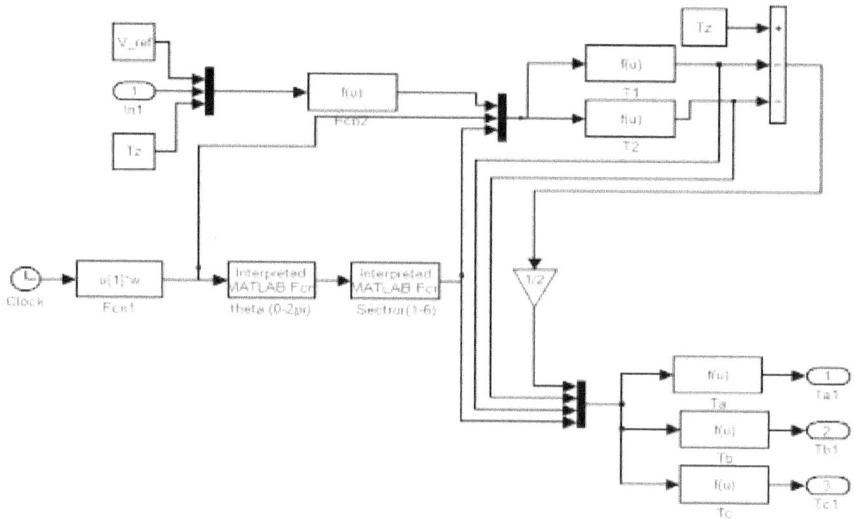

Fig.3 Voltage Vector Subsystem

SVPWM controller and voltage vector block is shown in Fig.2 and Fig.3. It will generate corresponding vectors and divides the two-axis plane into six-different sectors spaced by 60 degrees.

FAULT DETECTION SCHEME

In this paper, IGBT gate driver open-circuit fault of power converter is considered. This fault will only reduce the operating conditions of the drive without involving the short circuit protection of the system. The drive system can operate for a considerable period of time but with degraded performance and low efficiency. The injected dc offset in the machine phase currents caused by a gate drive open-circuit fault, worsen the current stress of the inverter healthy switching devices. Continuous operation in such faulty condition may lead to the terrible breakdown of the drive system. To operate the IGBT, an appropriate gate current must be applied in order to drive the power switch into the saturation mode for low on-state current. Malfunctioning of gate drive circuit can lead to the IGBT gate drive open- circuit fault.

Since the power switch S_1 has now an open-circuit fault, the phase A of the induction machine is connected to the positive dc rail through the diode D_1. The machine phase A voltage is then determined by the polarity of current and the switching pattern of the IGBT switch S_4. The phase voltage (V_a) will be clamped to the negative rail if stator current phase A, (I_a) is positive. On the other hand, the phase voltage V_a will be clamped to the negative rail when switch S_4 is switch on, and then to the positive rail when switch S_4 is off and D_4 is on, if I_a is negative.

The phase currents will be balanced and sinusoidal with a dc offset after the fault because the phase voltages (V_a, V_b and V_c) are balanced with the SVPWM signal before and after the fault explained in Auzani idin et al., 2009.

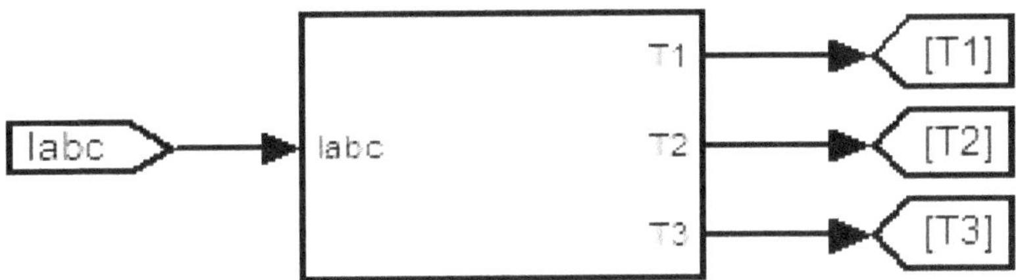

Fig.4 MATLAB Subsystem for the Fault detection controller

The proposed fault identification system has been developed in MATLAB simulink environment. The fig. and fig. shows the MATLAB simulink model of a Fault detection controller for the proposed VSI-IMD.

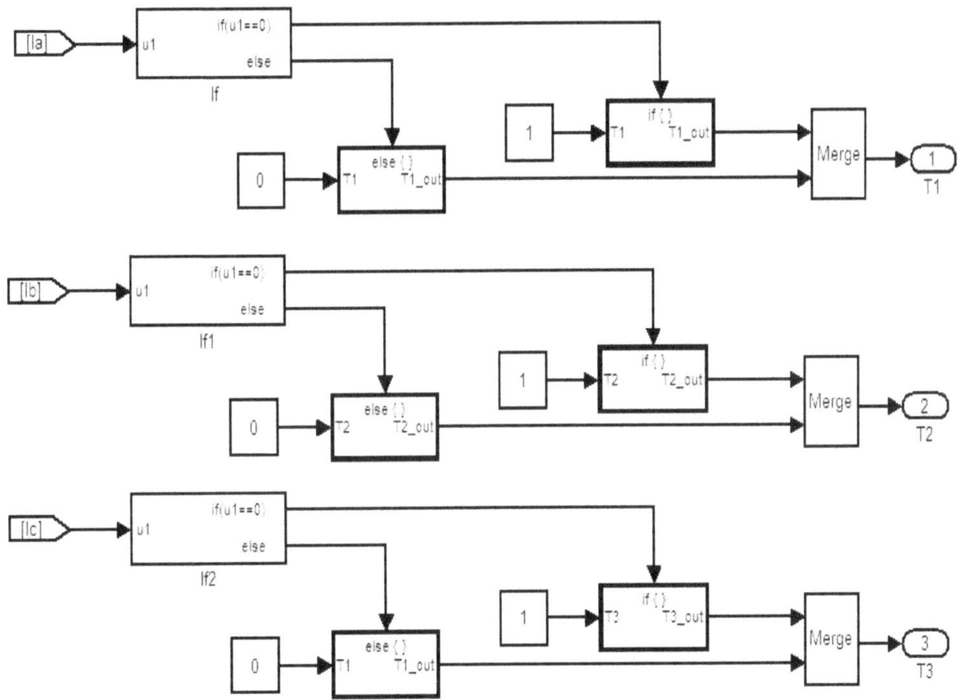

Fig.5 Fault detection controller logic

This controller is frequently monitoring the three phase input currents from the VSI-IMD system. If any one of the phase current is equal to zero, then the corresponding leg is considered as failure leg. These fault modes for three phases (A, B and C) has been considered as T_1, T_2 and T_3 respectively, which decides the corresponding control over the bypass and redundant leg switches.

PROPOSED FAULT TOLERANT CONTROL

The control concept for the proposed VSI is suitable for the effective operation of VSI under single leg or double leg failure.

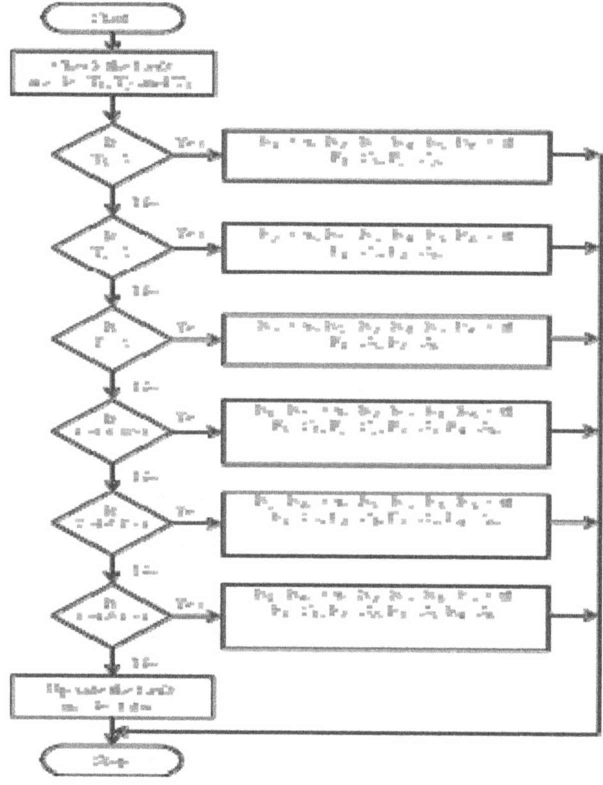

Fig.6 Flow chart for the proposed fault tolerant controlTable I. Conduction Table for Fault Tolerant VSIs

Failure Legs			Status of Bypass switches						Stand by leg switches			
A	B	C	B_1	B_2	B_3	B_4	B_5	B_6	F_1	F_2	F_3	F_4
0	0	0	Off	Off	Off	Off	Off	Off	Off	Off	Off	Off
0	1	1	On	Off	Off	Off	Off	Off	S1	S2	Off	Off
1	0	0	Off	On	Off	Off	Off	Off	S3	S4	Off	Off
1	1	0	Off	Off	On	Off	Off	Off	S5	S6	Off	Off
0	0	1	On	Off	Off	Off	On	Off	S1	S2	S3	S4
1	0	0	Off	On	Off	Off	Off	On	S3	S4	S5	S6
0	1	0	On	Off	Off	Off	Off	On	S1	S2	S5	S6

The conduction table for the proposed control concept is shown in Table I. The fault tolerant controller periodically monitors the status of an each leg and transfers the corresponding SVPWM signals to the standby leg by bypassing the legs through the back to back connected thyristors.

The flow chart for the proposed fault tolerant VSI is shown in Fig.6 It clearly shows the operating modes of a VSI-IMD under power switch failure conditions.

SIMULATION AND RESULTS

The dc voltage equal 400V, which is connected to the VSI IMD. A 5 HP, 440V, 50Hz 3-phase squirrel cage induction motor has been used for this modeling.

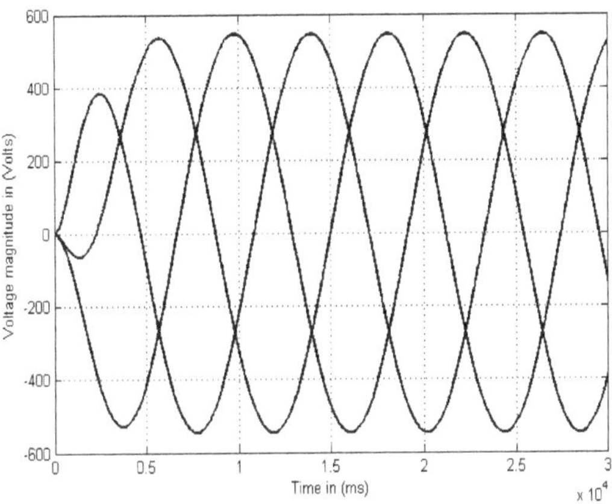

Fig.7 Output voltage of VSI-IMD under the failure of switches in leg A and B.

The fault has been identified and cleared less than 10 μs.

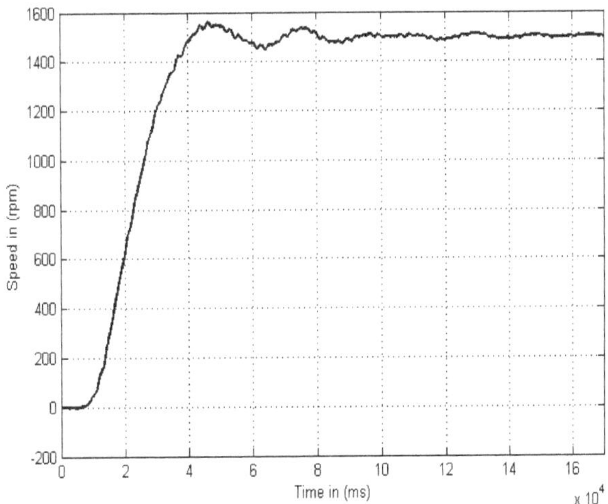

Fig.8 Speed of the Induction motor during fault

463

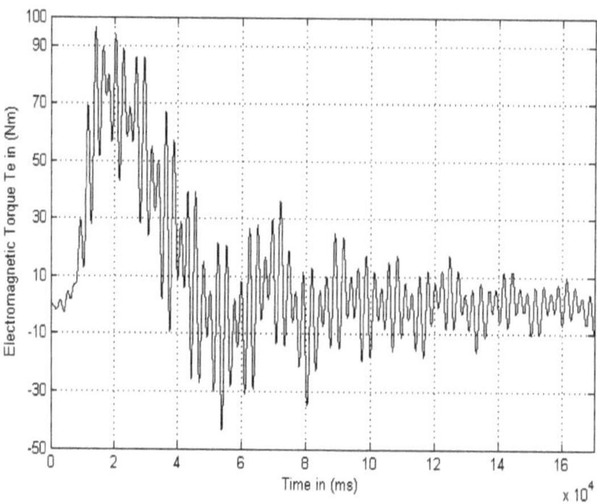

Fig.9 Electromagnetic torque (Te) under fault condition.

The performance of the Induction motor, during the fault condition is given in Fig.8 and Fig.9. The effect of Total Harmonic Distortion (THD) in the due to switching of the proposed VSI-IMD has been maintained at the standard level. The fig.10 shows the current harmonic spectrum of a VSI-IMD. The implementation of SVPWM in this fault tolerant VSI reduces the current THD into 5.41%.

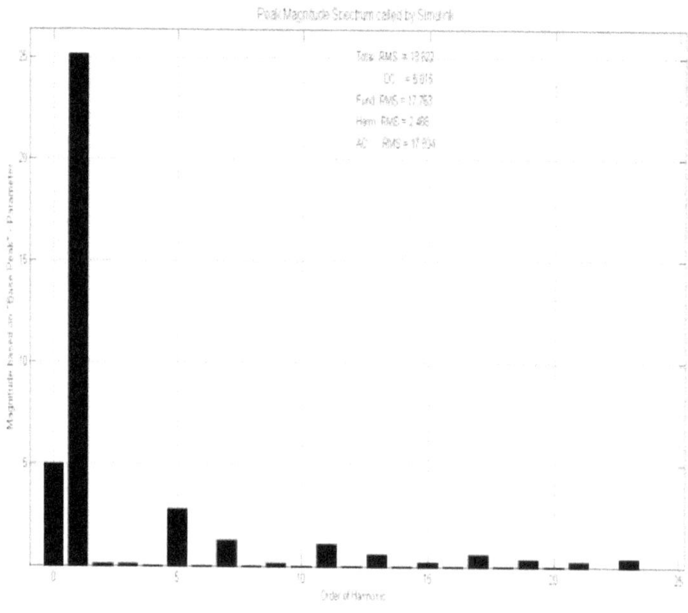

Fig.10 Spectrum of Phase current harmonics

CONCLUSION

This paper clearly explains the fault tolerant control of VSI for Induction motor drive application. The proposed system with fault tolerant control strategy has been modeled and implemented in MATLAB software. This control concept has been effectively diagnosing and controlling the VSI operation under the failure of power IGBTs in single leg or double leg. The fault has been identified and controlled less than 10 μs. The simulation results are clearly presented. It shows the effectiveness of the fault tolerant control of VSI and improves the performance of IMD under fault conditions.

FUTURE SCOPE

 The proposed fault tolerant control VSI hardware is implemented for the 100 kW grid connected Solar power plant to avoid the power oscillations during the failure of VSI power switches.

REFERENCES

[1] Demba Diallo, Mohamed El Hachemi Benbouzid and Abdessalam Makouf, 2004. "A Fault-Tolerant Control Architecture for Induction Motor Drives in Automotive Applications", IEEE Transactions on Vehicular Technology, Vol. 53, No. 6.

[2] D Kastha, B.K. Bose,1994. "Investigation of fault modes of voltage-fed inverter system for induction motor drive, IEEE Transactions on Industry Applications, 30, 4, pp. 1028–1038.

[3] H. El Brouji, P. Poure, M. C. Benhabib and S. Saadate, 2009 . "Fault tolerant shunt active power filter; topology reconfiguration using optimised fault detection algorithm", www.icrepq.com/icrepq06/444-elbrouji1.pdf.

[4] S. Karimi, P. Poure and S. Saadate, 2007. "Fast power switch failure detection for fault tolerant voltage source inverters using FPGA", IET Power Electron., Vol. 2, Iss. 4, pp. 346–354.

[5] M. S. Khanniche and M. R. Mamat-Ibrahim, 2001. "Fault Detection and Diagnosis of 3-Phase Inverter System", Rev. Energ. Ren. : Power Engineering ,pp. 69-75.

[6] M.Rozailan Mamat, M. Rizon and M.S. Khanniche, 2006. "Fault Detection of 3-Phase VSI using Wavelet-Fuzzy Algorithm", American Journal of Applied Sciences 3 (1): 1642-1648.

[7] Mohammad-Ali Shamsi-Nejad, Babak Nahid-Mobarakeh, ,Serge Pierfederici, and Farid Meibody-Tabar,2008. "Fault Tolerant and Minimum Loss Control of Double-Star Synchronous Machines Under Open Phase Conditions", IEEE Transactions on Industrial Electronics, Vol. 55, No. 5.

[8] Rammohan Rao Errabelli and Peter Mutschler, 2012. "Fault-Tolerant Voltage Source Inverter for Permanent Magnet Drives" IEEE Transactions on Power Electronics, Vol. 27, No. 2.

[9] T. Benslimane,2007. "Open Switch Faults Detection and Localization Algorithm for Three Phases Shunt Active Power Filter based on Two Level Voltage Source Inverter'', Electronics and Electrical Engineering, Signal Technology, No. 2(74.)

[10] Alexander L. Julian, Giovanna Oriti, and Stephen T. Blevins, 2010. "Operating Standby Redundant Controller to Improve Voltage-Source Inverter Reliability", IEEE Transactions on Industry Applications, Vol. 46, No. 5.

[11] Qun-Tao An, Li-Zhi Sun, , Ke Zhao, and Li Sun,2011. "Switching Function Model-Based Fast-Diagnostic Method of Open-Switch Faults in Inverters Without Sensors", IEEE Transactions On Power Electronics, Vol. 26, No. 1.

[12] Andre M. S. Mendes, Xose M. Lopez-Fernandez and A. J. Marques Cardoso,2007. "Thermal Behavior of a Three-Phase Induction Motor Fed by a Fault-Tolerant Voltage Source Inverter", IEEE Transactions on Industry Applications, Vol. 43, No. 3.

[13] Andre M. S. Mendes, Xose M. Lopez-Fernandez and A. J. Marques Cardoso,2008." Thermal Performance of a Three-Phase Induction Motor Under Fault Tolerant Operating Strategies", IEEE Transactions on Power Electronics, Vol. 23, No. 3.

[14] E.E.El-Kholy, A.El-Sabbe, A.El-Hefnawy and Hamdy M.Mharous,2004. "Three Phase Active Power Filter Based on Current Controlled Voltage Source Inverter", ACTA Electrotechnica, Vol.45, No.4.

[15] Auzani idin and Tole Sutikno,2009. "MATLAB/SIMULINK Based Analysis of Voltage Source Inverter with Space Vector Modulation", TELKOMNIKA, Vol.7,No.1. pp. 23-30.

AIRBUS 2050: VISION FOR "SMARTER SKIES"

Fakeha Azhar[1]
Dept of aero, CMR Technical Campus (Aff. to JNTUH)
Hyderabad - 500013
fakeha.aironix@gmail.com

Venkata Rama Teja S[2]
Dept of aero, CMR Technical Campus (Aff. to JNTUH)
Flat no: S-2, plot no: 24, G. V. Reddy colony,
Alwal, Secunderabad - 500010
ramteja.aironix@gmail.com

ABSTRACT

An effort was made to study the vision for the sustainable aviation in 2050 and beyond. The vision looks beyond the aircraft design to how the aircraft is operated both on the ground and in the air in order to meet the expected growth in air travel in a sustainable way.

These tough environmental targets will be met by a combination of investment in smarter aircraft design and optimizing the environment in which the aircraft operates." Our latest future by airbus smarter skies focuses on not just what we fly but how we may fly in 2050 and beyond".

The air traffic management system (ATM) and technology on board the aircraft were optimized. Assuming around 30 million flights per year, this would save around 9 million tones of excess fuel annually, which equates to over 28 million tones of avoidable CO_2 emissions and a saving of 5 million hours of excess flight time.

The smarter size concept consists of five concepts which could be implemented across all stages of an aircraft's operation to reduce waste in fuel and reduction of CO_2. These are:

A) Aircraft take-off in continuous 'eco-climb':

Aircraft launched through the assisted take-offs using renewably powered, propelled acceleration, allowing steeper climb from airports to minimum noise and reach efficient cruise attitudes quicker.

B) Aircraft in free flight and formation along 'express skyways':

Highly intelligent aircraft would be able to "self organize" and select the most efficient and environmentally friendly routes ("free flights"), making the optimum use of prevailing weather and atmospheric conditions.

C) Low noise, free-glide approaches and landings:

Aircraft allowed taking free glide approaches into airports that reduce emissions during the overall decent and reduce noise during the steeper approach as there is no need for engine thrust or air breaking.

D) Low emission ground operations:

On landing aircraft engines could be switched off sooner and runways cleared faster, ground handling emissions could be cut. Technology could optimize the aircrafts landing position with enough accuracy for an autonomous renewably powered taxiing carriage to be ready, so aircraft lands quicker, which would optimize terminal space, and remove runway and gate limitations.

E) Powering future aircraft and infrastructure:

The use of sustainable bio-fuels and other potential alternative energy sources such as electricity, hydrogen, solar etc, will be necessary to secure supply and further reduce aviation's environmental footprint in the long term. This will allow the extensive introduction of regionally sourced renewable energy close to airports, feeding both aircraft and infrastructure requirements sustainability.

INTRODUCTION

Airbus experts in aircraft materials, aerodynamics and engines came up with a concept plane design that is an "ENGINEERS DREAM". The airbus concept embodies what an aircraft looks like in the year 2050. Ultra long and slim wings, semi embedded engines, a U-shaped tail and light weight intelligent body all feature to further improve environmental performance or 'eco-efficiency'. The result is lower fuel burn, a significant cut in emissions, decreased noise pollution and greater comfort.

The Airbus Concept Cabin gives further insight into some of the innovations and technologies that will shape future passenger experiences on board. It's Concept Cabin - a whole new flying experience inspired by nature. Personalized zones replace traditional cabin classes to offer tailored levels of experience. While taking a hop between destinations, according to Airbus, passengers in 2050 could join an interactive conference; enjoy a game of virtual golf; read the kids back home a bedtime story; and recharge in a 'vitalizing seat' whilst watching the planet spread out beneath their feet.

PANAROMIC VIEW OF AIRBUS 2050

The aircraft's bionic structure mimics the efficiency of bird bone which is optimized to provide strength where needed, and allows for an intelligent cabin wall membrane which controls air temperature and can become transparent to give passengers open panoramic views. New personalized zones replace the traditional cabin classes in the Airbus Concept Cabin to offer new tailored levels of experience.

The "vitalizing zone" is all about wellbeing and relaxation allowing you to proactively recharge your batteries with vitamin and antioxidant enriched air, mood lighting, and aromatherapy and acupressure treatments whilst taking in the infinite view of the world around you. There are no limits to the kinds of social scenarios in the centre zone of the concept cabin – the "interactive zone". The virtual pop up projections in this area can transform you to whichever social scene you want to be in, from holographic gaming to virtual changing rooms for active shoppers.

The "smart tech zone" is tailored towards the more functional oriented passenger with a chameleon style offering, to meet individual needs ranging from a simple to a complete luxury service, but all allowing you to continue life as if on the ground. By offering different levels of experience within each zone, airlines would be able to achieve price differentials and give more people access to the benefits of air travel with minimal environmental impact. More than 90 percent of Airbus' annual research & development investment of over €2 billion has environmental benefits for current and future aircraft. For example, due to advances in technologies the concept cabin will be 100% recyclable. It will have self-cleaning materials made from sustainable plant fibers which reduce waste and maintenance and will harvest passenger body heat to power cabin features.

Airbus 2050 concept aircraft designed in response to user surveys about the future of air travel, the 'airbus 2050' concept craft houses replaces: first, business, and economy classes with lifestyle 'zones', all housed in a panoramic transparent body designed for both more comfortable travel and improved environmental efficiency.

In the concept cabin, hand luggage is taken at the entrance to the craft as passengers scan in via a biometric hand sensor, and is moved automatically to one's seat. The chairs themselves morph to comfortably fit passenger's bodies, and collect body heat for use in powering the cabin facilities. The airplane itself benefits from the use of lightweight 'smart' materials and composites, as well as longer and slimmer wings that improve air flow and fuel efficiency. Engines will be embedded into the rear of the craft, optimizing fuel use and reducing noise levels within.

The electrical system is self-monitoring, capable of scheduling any required maintenance. Regarding the transparent body, airbus head of engineering Charles Champion states: *'the idea is to have a technology for the fuselage that's a bit like bones of birds that allows to have large spaces that can turn transparent, in order to look outside and 'live' the panorama in which you are flying.'*

The plane's empennage (tail section) is U-shaped to reduce external noise pollution, since the more reliable engines prevent the need to have a vertical tail for directional stability. The cost of the innovative cabin designs would be defrayed by reducing maintenance costs in other areas, relying on developments in self-monitoring, 'intelligent' technology and advances in innovative materials such as dirt repellent coatings and self-healing covers.

AIRBUS CONCEPT CABIN

The Airbus Concept Cabin illustrates what the future of flight might look like from the passengers' perspective. The Concept Cabin doesn't conform to the traditional cabin classes found in today's commercial aircraft. First, Business and Economy class are replaced by zones that target more individual needs like relaxing, playing games, interacting with other passengers or holding business meetings with people on the ground. The cabin's bionic structure and responsive membrane combines panoramic views with an integrated neural network pulsing through it, which can identify and respond to the specific needs of each passenger. And the fittings and furnishings will take care of their own cleaning and repairs thanks to innovations inspired by nature, like dirt repellent coatings and self healing covers. By offering different levels of experience within each zone, airlines would be able to achieve the price differential they need to operate a successful business, give more people access to the benefits of air travel and still look after the environment. By offering different levels of experience within each zone, airlines would be able to achieve the price differential they need to operate a successful business, give more people access to the benefits of air travel and still look after the environment. Each and every passenger can now enjoy their own very unique flying experience.

VITALIZING ZONE

By the year 2050, world's population is expected to rise to 9.1 billion. But who will these people be and how will they want to fly? There will be a significant increase in female travelers as more women will travel for business than we see today. There will also be an increase in the proportion of older passengers as the number of people over 60 years old will increase by more than three times to reach nearly two billion in 2050. The vitalizing zone will both meet that demand and bring the passenger closer to nature. The bionic structure and membrane will create the perfect combination of strength, light and space, offering panoramic views of the world outside through large areas where the lining can become transparent at the wave of a hand; intelligent organically grown seats will sense your needs and adapt for the perfect fit, offering massage, drinks or vitamins as required; a gentle sea breeze or the soft aroma of a pine forest wash over you; sound showers will ease you into the perfect sleep, snug in the warm embrace of holographic shades, while the heat given out from your body is unobtrusively collected to power the cabin facilities.

INTERACTION ZONE

In the future, technology and virtual reality will be seamlessly integrated with our normal lives, services will be individually tailored and environmental needs will be addressed in everything that we do. All of which will be just as true in the air as it is on the ground. That seamless relationship between people and technology will be evident from the moment passengers' board via this zone. Touch sensitive panels will download the passenger profile and guide them through their bespoke experience.

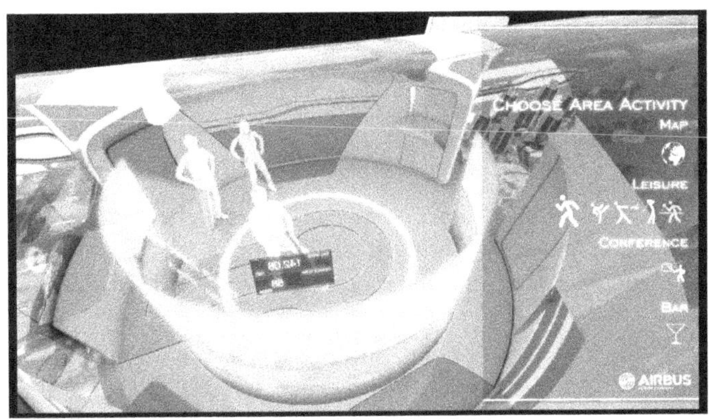

Alongside this, the crew will ensure that everything runs smoothly and provide that all important human touch. While bars on aircraft are currently only found in certain parts of a few A380s, these will be much more common in the future, providing lively areas for people wishing to interact with their fellow travelers. Intuitive technology will allow passengers to access all flight, destination and environmental information at the wave of a hand. Pop-up pods will offer more private spaces that can be used for anything from virtual business meetings or lectures to a romantic meal or reading a bedtime story to the kids back home, with global connectivity and holographic projections adapted to the needs of each user. A virtual shopping wall will project clothes directly on to passengers and the virtual gaming wall will let tennis, baseball and even golf fans get in a bit of practice, while the more adventurous will be able to try out newer options like Airbus Fusion Ball game, which lets you play catch across the skyscrapers of New York or the peaks of the Himalayas!

SMART TECH ZONE

By 2050 70% of the world's population will live in cities or towns and more of them will want and need to travel. With more people traveling and their size, shapes and needs changing, aircraft cabins need to change right alongside them. This zone will do precisely that. Morphing seats will adapt to suit passengers' budgets as well as their body, offering different levels of comfort and space for everybody; you might be traveling alone, but holographic, communication technologies and sound showers let you work with your colleagues or read a bedtime story to the kids as though they were right alongside; hand luggage is swallowed at the entrance, before reappearing beside you for easy access; all intended to provide flexible and personalized options for a productive and hassle-free flight.

FUTURE TECHNOLOGIES

While the Concept Plane shows how advanced materials can create a high performance aircraft with a more traditional look, the Concept Cabin provides a little taste of some of the alternatives.

Bionic Structures

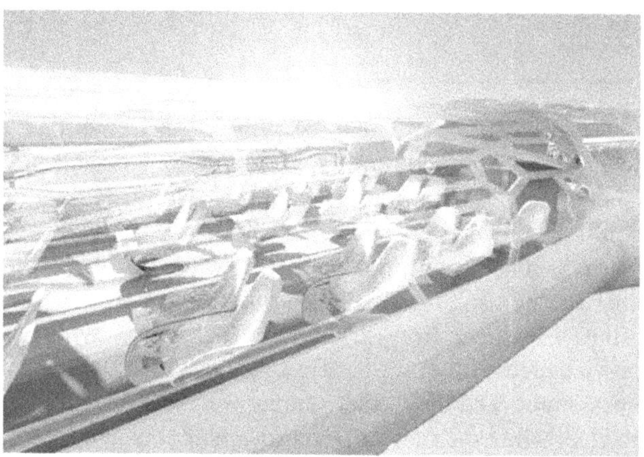

Future aircraft could be built using a bionic structure that mimics the bone structure of birds. Bone is both light and strong because its porous interior carries tension only where necessary, leaving space elsewhere. By using bionic structures, the fuselage has the strength it needs, but can also make the most of extra space where required. This not only reduces the aircraft's weight and fuel burn, but also makes it possible to add features like oversized doors for easier boarding and panoramic windows.

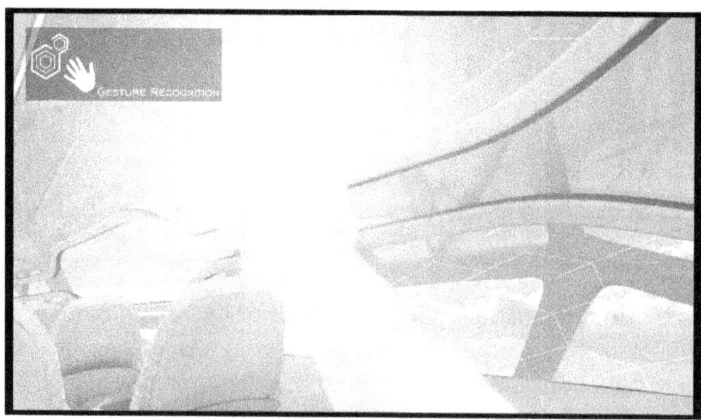

Biopolymer membrane

The cabin's bionic structure will be coated with a biopolymer membrane, which controls the amount of natural light, humidity and temperature, providing opacity or transparency on command and eliminating the need for windows. This smarter structure will make the aircraft lighter and more fuel-efficient while giving passengers 360 degree views of the skies. This will offer unparalleled, unobstructed views of the wonders of the five continents - where you will be able see the pyramids or the Eiffel Tower through the transparent walls of the aircraft.

COMPOSITE MATERIALS

Future materials may not even be the materials we see and use today. 'Composite' materials will be used - new matter made of a combination of different materials. In the future materials may not even take a solid state, but could be a composition of fluid and gas for example.

INTEGRATED NEURAL NETWORK

The cabin electrical system can be compared to the human brain, with a network of intelligence pulsating through the cabin. This network will be absorbed into the structural materials, making the hundreds of kilometers of cables and wires found in today's aircraft a thing of the past. Known as 'Smart' materials they can perform numerous functions, recognizing the passenger, so that you too are 'connected' to the plane.

MORPHING MATERIALS

Materials that change shape and return to their initial form, growing like the leaves of a plant, are a very real possibility. Morphing materials might be metals or polymers that have a 'memory'; or are covered with a 'skin' that will instigate a shape change. A memory is created using sensor and activator systems that give materials a certain level of artificial intelligence, allowing them to adapt to the passengers' needs.

Self-reliant Materials

Materials will be self-cleaning. Think of the leaves of a lotus plant, which water rolls off in beads, taking contaminants with it. Today, coatings inspired by this are used on the surfaces of cabin bathrooms. In the future they will be found on the fabric of seats and the carpets. These intelligent materials could also be self-repairing, which is already used today in surface protection. Certain paints can seal a scratch by themselves, just as the human skin does.

Ecological Materials

the future passenger cabin will be fully ecological. Fully recyclable plant fibers that can be grown to a custom shape will be sourced from responsible and sustainable practices.

Holographic Technology

Scenes showing the destination, a city skyline or a tropical forest, will be projected onto the walls. A private cabin can reflect your bedroom at home, a business conference or even a zen garden, thanks to the projection of virtual decors. Holographic technology will have advanced to such a degree that the virtual world will be indistinguishable from the real.

Energy Harvesting

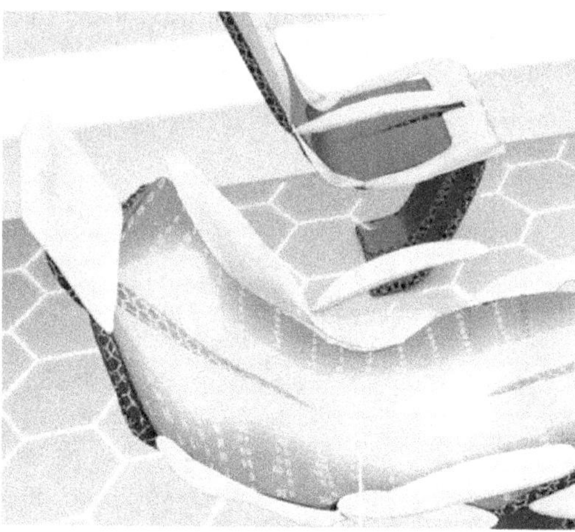

Smart energy solution such as energy harvesting will be a part of the cabin environment. The body heat you give out will be collected by your seat or pod as you relax or sleep, and combined with energy collected from other sources, like solar panels, to fuel cabin appliances.

CONFIGURED WINGSPAN

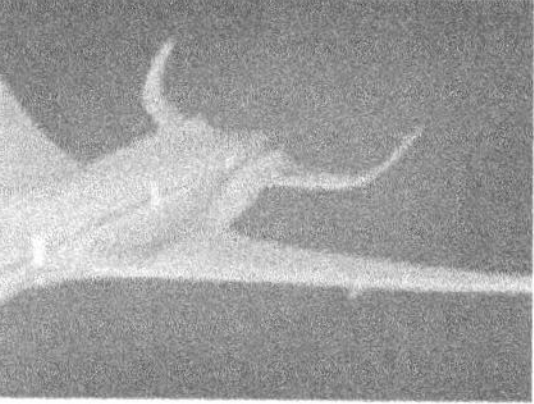

Longer and slimmer wings glide better through the skies, as the flow of air over the wing surface reduces drag and in turn, improves fuel efficiency.

EMPENNAGE

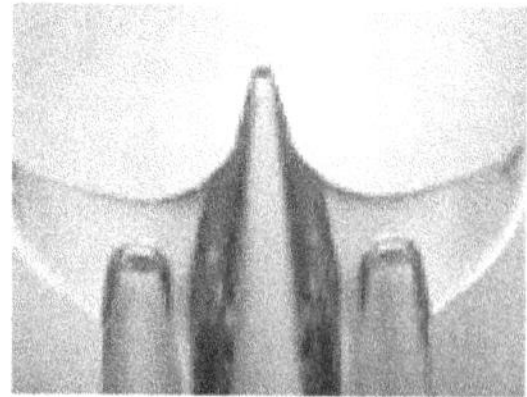

The empennage (a tail section of the aircraft) is U-shaped acting as a shield to reduce external noise pollution. The concept plane does not use the vertical tail as seen on the planes of today. Vertical tails are required when engines are installed on the wings as they provide directional stability in case of engine failure. The engines of the future will have no risk of failure, eliminating the need for a vertical plane.

DOORS

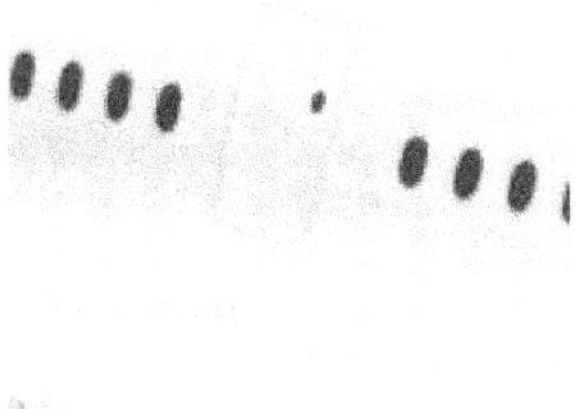

Entrance/exit doorways are double doors to allow for faster, easier boarding.

FUSELAGE

The fuselage (central body of the aircraft) is no longer a simple tube but is curved and shaped to provide more internal space for various cabin configurations, with better aerodynamics outside to improve flight. The fuselage and the entire aircraft structure is manufactured entirely from composite to take advantage of the easy to shape characteristics of the material.

MANUFACTURING METHODS

New manufacturing methods will reduce the cost of the environmental impact of building the aircraft despite the new advance materials and complex shapes.

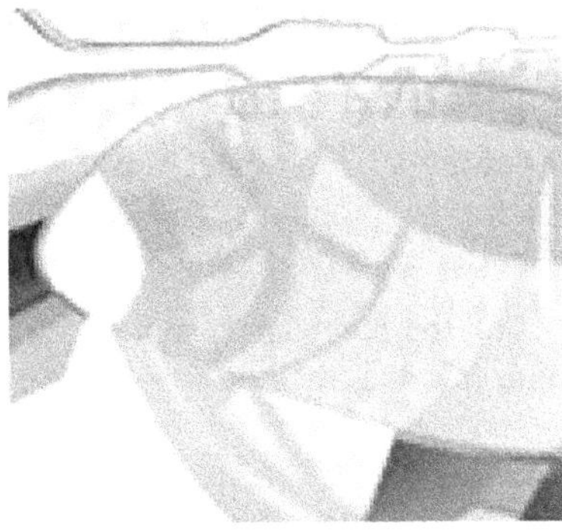

ELECTRICAL SYSTEMS

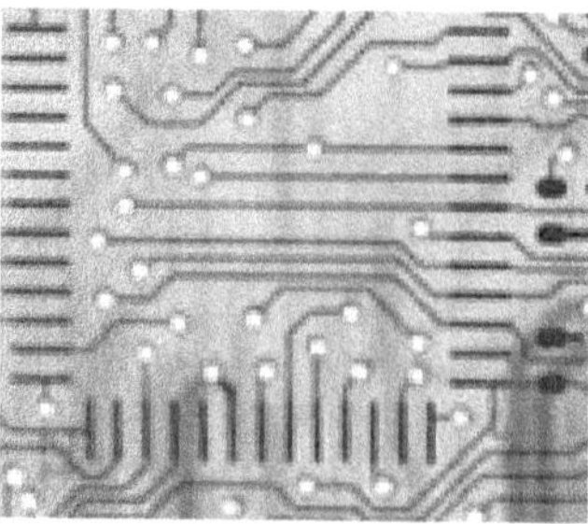

The electrical system will continuously monitor its own state of health anticipating any need for maintenance and automatically scheduling this well in advance. Electronics and other systems on board will be entirely self sufficient, requiring minimum to zero maintenance.

ENGINES

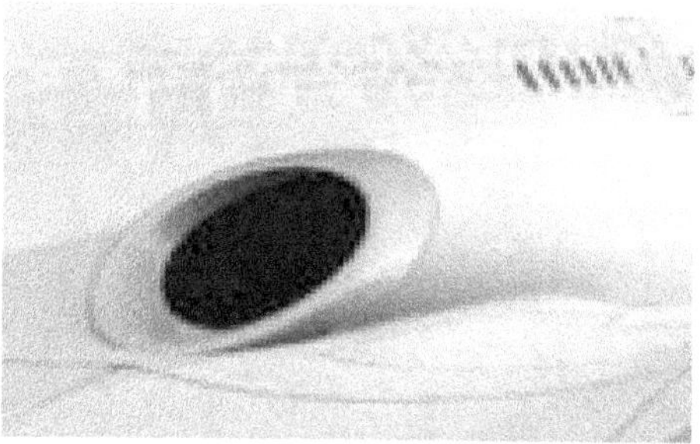

Engines will be more reliable, quieter and fuel efficient. The positioning of the engines, at the rear and semi embedded, fully optimizes the aircraft for lower fuel burn. The engine placement also boosts the cabin comfort through decreased noise levels. The engines can be incorporated into the aircraft body because technological advances will have reached such a level that superior engine reliability will diminish the need for immediate access to its components.

ECO-CLIMB

Aircraft launched through assisted takeoffs using renewably-powered, propelled acceleration will allow for steeper climb from airports to minimize noise and reach efficient cruise altitudes more quickly. As space becomes a premium and mega-cities a reality, this approach also could minimize land use, as shorter runways could be utilized.

How will it work? Aircraft could be maneuvered onto a track system and accelerated using either electro-magnetic motors built into the track or an inductive circuit within the aircraft itself. Acceptable acceleration and deceleration limits of passengers would need to be determined, but the experience would be more akin to a comfortable children's funfair ride rather than a high-octane white knuckle theme park. The ultimate, albeit it very extreme, concept is to have a system that not only launches but also captures the aircraft, removing the need for landing gear. This would require all airports to have the same system, to accommodate all routes along with alternative/diversion airports, and most likely is beyond 2050.

EXPRESS SKYWAYS: In the future, highly intelligent aircraft would be able to "self-organize" and select the most efficient and environmentally friendly routes ("free flight") - making the optimum use of prevailing weather and atmospheric conditions. High-frequency routes would also allow aircraft to benefit from -flying in formation like birds during cruise bringing efficiency improvements due to drag reduction and lower energy use.

How will it work? In a V formation of 25 birds, each can achieve a reduction of induced drag by up to 65 per cent and increase their range by 7 per cent. While efficiencies for commercial aircraft are not as great, they remain significant. Airbus is continuing to assess the feasibility for future operations, including studies in collaboration with Bristol, Cape Town and Stanford Universities. The study with Stanford was borne out of a proposal on the subject from its students that reached the final of the Airbus Fly Your Ideas challenge in 2009. It employed simulation and aerodynamics analysis to explore the optimum number of aircraft in several configurations or geometries. These included a two-aircraft formation, three-aircraft "skein" (the symmetric V-shaped formation associated with geese and ducks), an inverted-V and echelon formation. The results suggest fuel burn savings of 10-12 per cent are possible, with emissions cut by up to 25 per cent. Airbus already is looking into cooperative flight scheduling and conducting research into aircraft stability and control. In parallel, a new breed of sensors able to detect the wake of the previous aircraft and rapid state change must be developed. Avionic technologies already make this possible in principle. Lightweight remote sensing equipment such as LIDAR (Light Detection and Ranging) and Infrared cameras allow aircraft to detect the wake vortex – which is the turbulence produced by an aircraft in flight – of those ahead. For aircraft to autonomously keep station, they will need to communicate with each other. High-speed, real-time computation, communication and coordination would take inputs from all sources in the air and on the ground. High bandwidth telecommunications would cope with the increase in data being transferred around the network.

FREE GLIDE APPROACHES

Allowing aircraft to take free glide approaches into airports would lower emissions during the overall decent and reduce noise during the steeper approach as there is no need for engine thrust or air breaking. These approaches also would reduce the landing speed earlier, making shorter landing distances achievable with fewer runways needed.

A new approach to (gently) touching down: Everything is going smoothly (and quietly) as you glide back towards ground in a continuous descent. Today aircraft descend in stages and often are forced to wait in the air, circling in holding patterns to avoid congested airspace or while awaiting a landing slot. However, leveling off during descent requires an increase in thrust. That means extra fuel burn and emissions – as well as unnecessary delays for passengers. With better air traffic management, aircraft could enter a fuel-efficient descent based on when best to leave cruise level – with no risk of getting stuck in traffic. Just as when climbing, a continuous descent would see aircraft use only the minimum thrust needed. Aircraft featuring technology to optimize landing positions with pinpoint accuracy could glide smoothly into airports with their engines running in idle, for significantly reduced fuel burn, emissions and noise. Slower landing speeds would lead to shorter runways – by up to 1/3rd – a viable possibility on arrival just as with departure. And more efficient land use could see airport capacities increase or new micro-airports close to "mega-city" centers The ultimate idea, likely beyond 2050, would be to use the same renewably-powered system on landing as at takeoff, receiving aircraft and removing the need for landing gear. This would require all alternate/diversion airports to have the same system. Either way, as the aircraft touches down, kinetic energy can be captured for future use. For example, it might power on-board systems during taxiing or the ground-based propulsion system used for takeoff.

GROUND OPERATIONS

On landing, aircraft engines could be switched off sooner, runways cleared faster and ground handling emissions could be cut down.

Technology could optimize an aircraft's landing position with enough accuracy for an autonomous renewably-powered taxiing carriage to be ready, so aircraft could be transported away from runways quicker, which would optimize terminal space, and remove runway and gate limitations.

How will it work? The technologies needed to make this a reality are not as far out as might be expected. Ground vehicles that can operate without human intervention, sensing and navigating around the dynamic airport environment are more than a flight of fancy. They could use electromagnetic currents flowing through runway-installed tracks or even wireless high power, with the aircraft perhaps acting as a conduit. So too are energy storage devices capable of storing high amounts of energy received in a very short period of time. These could be discharged, either slowly or rapidly, while remaining reasonably compact and lightweight and suitable for ground use. Excess energy also could be collected during flight to power on-board systems and/or stored for use on the ground.

POWER

The use of sustainable biofuels and other potential alternative energy sources (such as electricity, hydrogen, solar and more) will be necessary to secure supply and further reduce aviation's environmental footprint in the long term. This will allow the extensive introduction of regionally-sourced renewable energy close to airports, feeding both aircraft and infrastructure requirements sustain ably.

How does it work? The potential benefits of this solution to the long-term availability and affordability of fuel means it's fast becoming a very real and viable option. 50/50 blend biofuels already are certified for commercial flights. The ultimate goal is to achieve the approval of 100 per cent blends for commercial aircraft. Airbus is acting as a catalyst for sustainable biofuels through an ambitious programmed to form regional biofuels "value chains" in every continent, using the Roundtable on Sustainable Biofuels criteria to guarantee sustainability. Five such value chains already have been established, in Australia, Brazil, Middle East, Romania and Spain. While it's not foreseeable that fuel cells would be used for commercial aircraft propulsion, they are one of the most promising "step change" technologies to power cabin operations. As hydrogen is combined with oxygen in a "cold" combustion, the only by-product is water. This could be used for the aircraft's water and waste system, saving water, weight and – in turn – fuel consumption and emissions.

CONCLUSION

Using these upcoming technologies AIRBUS intends to provide efficient and smarter air travel. These technologies may not have much efficiencies when applied to civil airline passenger planes, but the technology can be applied to concepts such smaller aircrafts where these efficiencies are significant and measurable in a huge scale.

Furthermore, as the technology tends to be enhanced from time to time, therefore it can be incorporated during the period of study of these concepts which may help in pushing the efficiencies to a higher extent.

ACKNOWLEDGEMENT

We thank our professors who enlightened us with these technologies and helped us to carry forward our thirst for knowledge by guiding us.

We wish to acknowledge AIRBUS for the availing the resources easily on the internet for everyone to know and enlisting information enough to understand the concepts. And also thanks to youtube for the free listing of videos related to different topics related to herewith of our paper which provided a great insight into the technologies being researched upon for the coming years.

REFERENCES

1. www.airbus.com
2. www.youtube.com
3. www.wikipedia.com

A NOVEL PROPULSION MECHANISM FOR MICRO AIR VEHICLES

P. Raghavendra
Assistant Professor, Sree Nidhi Institute of Science and Technology,
Hyderabad, India

ABSTRACT

Micro Air Vehicles are the application specific air vehicles which are capable of performing certain specialized tasks like accessing locations which are difficult to reach by conventional vehicles. The weight of an MAV is far less than its other counterpart, the UAV. Several measures are taken during the design phase to overcome the challenges of the low Reynolds number flight. An efficient and economical propulsion mechanism which would help the MAV to accomplish its mission is essential. Inorder to meet the unique requirements of an MAV a new kind of propulsion is proposed. The combination of a rubberband propelled motor and compressed air which would be released in the form of jets, making the MAV a thrust vector controlled vehicle. This mechanism can be applied on any fixed wing MAV with minimum of one propeller and sufficient wing span, so that the compressed air bags can be accommodated underneath. The propeller of the vehicle will be rotated using the uncoiling action of the coiled rubber band and this gives the initial thrust and lift to the vehicle. As the vehicle attains a predetermined height, the compressed air can be released so that it provides the necessary thrust in lateral plane and also the altitude is maintained. The various phases of flight are, initially the vehicle will be in the propeller mode as the propeller is made to rotate during the unwinding of the rubberband. Then the vehicle is allowed to glide till it is capable of maintaining the altitude. The compressed air is released as per the maneuvering requirements. The landing can be made by switching it to the glide mode and appropriate usage of control surfaces. The potential energy of the wound rubber band is converted to rotational kinetic energy of the propeller as the rubber band unwinds. Hence the time of propulsion depends on the quantity of potential energy in the wound rubber band.

INTRODUCTION

A novel propulsion mechanism for fixed wing micro air vehicle has been discussed in this paper. Powered flight has under gone several advancements ever since Wright brothers invented the first aircraft. The advancement lead the usage of air vehicles to a broad category of application which included air surveillance, reconnaissance, etc in military domain and disaster management, traffic management, etc in civilian domain. Widening the application base threw challenges to the aero engineers as many of these applications needed air vehicles that are as small as possible for various reasons. As a result of this the focus shifted on reducing the size of the air vehicles without compromising their mission capabilities. An air vehicle of reduced size is significantly different from its counterpart of original size and the key aspects of stability and control were totally out of bounds for small aircrafts (Scott M Ettinger, August 2001). There are several other challenges that are to be overcome by the designer of a micro air vehicle. Micro air vehicles were developed with a wing span of 15cm or less in case of fixed wing vehicles. There are also MAV's which are of

"

flapping wing type as this types of wing offers some advantages at low Reynolds number. Reynolds number, which governs the aerodynamic performance of the air vehicle is vey low for MAV's compared to their larger size counterpart, the aircraft. As the size is low, obviously the pay load capacity is also less and the inertial capability is also significantly reduced which poses a challenge to control and stabilize the aircraft. The propulsive thrust which is generally generated by an engine in a larger aircraft and electric motor in MAV is essential for forward motion of the aircraft and also contribute towards the generation of lift. In case of fixed wing MAV's the wing span, shape of the wing, including the camber in the airfoil are the governing factors for generation of lift. There are certain distinct advantages offered by fixed wing aircrafts such as adaptability to adverse weather conditions, enhanced fuel efficiency, etc (D.Blake Barber et al. 2006). As the propeller rotates the aircraft is subjected to forward motion and the body of the aircraft with wings start moving in the air. The camber in the airfoil results in a laminar flow as the air passes over the wing and generates lift. Lift is the force responsible for making the aircraft airborne and maintaining its state.

SIGNIFICANCE OF PROPULSION

An effective propulsion mechanism is necessary to propel the aircraft. The wing of a MAV is broader than the wings of the scaled down models of conventional aircrafts in order to keep the aspect ration low, to enable generation of lift and prevent stall at low air speeds while executing turns. Conventionally Lithium ion or Lithium polymer batteries (S.sankara Narayanan et.al.) are used to power the electric motor that drives the propeller. Even a slight increase in specification like increase in speed, more maneuvers etc will have a direct impact on the input power and hence more power will be drawn from the battery reducing the endurance of flight. In-order to maintain the endurance a battery of higher capacity has to be used which will increase the weight to be carried by the MAV. This paper proposes a method where in the propeller is rotated with the help of energy of a rubber band. As per the statement of law of conservation of energy, energy can neither be created nor destroyed but can be converted from one form to another. The potential energy of the wound rubber band, as it is released gets converted into the rotational kinetic energy of the propeller. There is also a historical significance to rubber band powered flight. Alphonse Penaud in 1871 used a twisted rubber band to power a flying model of a helicopter. Later a model plane Planophore was developed which was also a rubber band powered model aircraft. It is note worthy that the toy which inspired Wright brothers to build the heavier than air flying machine was also a rubber band powered toy which would reach to a particular height as the rubber unwinds and then fall down when the propeller ceases to rotate. However it was impractical to produce large thrust using this type of power source. The propulsive power required by a MAV is very low and thus electric motors are preferred to rotate the propeller. This adds up the weight on the aircraft as it has to carry a power source, the battery all through the flight. After the energy in the battery is exhausted it is just a dead load on the aircraft during the descent phase.

Rubber band powered motor – Novel approach

Rubber band based propulsion is a novel approach to propel a MAV and maintain it airborne for few minutes. The size of the propeller, propeller angle, size of the rubber band and the number of turns wound are the key factors determining the overall efficiency of this kind of a propulsion system.

Different propeller designs have different lift-drag characteristics and thus their optimal rate of advance through air also varies. So a propeller that matches the requirements of a particular aircraft has to be chosen properly. Efficient propeller delivers 80% of the motor output as thrust. The propeller should deliver its maximum efficiency when the aircraft flies at its designed glide speed. To achieve this, the motor has to deliver ideal power and torque to the propeller at that speed. This paper discusses the number and size of the rubber bands required to deliver the optimal power. It has been estimated that only 80% of the power produced by the motor can be utilized by the propeller and the rest is accounted for losses. Hence the motor should generate more power than required by the aircraft as 20% of the total power generated is wasted due to various losses. In-order to choose an optimal rubber band motor, first the power required by an electric motor used for the same purpose is calculated. Considering a medium pitch angle, the power efficiency coefficient of the propeller is 0.85 (β), the torque can be calculated by using the formula,

$$\text{Torque} = \frac{P*D}{2\pi*V*\beta}$$

P – Power required by the aircraft (Watts)

V – Natural glide speed of the aircraft (m/s)

D –Diameter of the propeller (m)

The instantaneous power and torque delivered by the motor depends on the number of winds in the rubber band and its cross-sectional area. Unlike gas or electric powered aircraft the power supplied by the motor and hence the torque of the propeller is not constant in a rubber band powered aircraft. At launch it will have more power than the one required to fly, as a result of which the aircraft climbs. The torque and output power will decrease with time. The MAV will eventually level off and cruise when the power supplied is about what is required. At the end of the flight as the available power runs out, the MAV will transition into a glide. The design of the motor should enable the propeller to operate at its peak efficiency during its cruise stage while maintaining adequate power even though considerable number of winds have been released. This improves the endurance of the aircraft. By convention a rubber band is used to 2/5 of its wind limit strength.

Several rubber band sizes and/or loops have to be tested before finalizing the rubber band configuration that provides the torque and power required by the MAV. The configuration of the rubber will affect the tightness with which the motor can be wound so that the number of wounds per flight can be decided.

Compressed air

Compressed air has been a well proven source of power but sparsely used in propulsion applications on a large scale. Compressed air, when released from an object will tend to move the object with a particular force in opposite direction. This is well supported by Newton's Third law of motion which says every action has equal and opposite reaction. Compressed air filled in small robust sacks mounted under the surface of the wings of a fixed wing MAV would serve multiple purposes. Initially it would provide the necessary inertia to the MAV to with stand the high starting torque delivered by the rubber band motor to the propeller. Secondly, it would serve as a means to control the vehicle and aid control surfaces to ease out difficult maneuvers which would otherwise make the user to depend fully on control surfaces. The air bags will release a jet of air through a proper nozzle and thus will act as booster rocket to the MAV producing additional thrust in appropriate direction.

CONCLUSION

A combination of optimally chosen rubber band motor and compressed air bags will provide adequate propulsive power to keep the MAV airborne for a long time. This reduces the dead weight on the aircraft like usage of batteries, etc and rubber band motor is a clean energy source compared to other miniature combustion engines used to power MAV. There is still a lot of scope for research on MAV propelled by rubber band powered motors. Optimization at every stage would result in a better rubber band motors and MAV's can fly with less electric power and thereby less dead load. Further development of this technology would help the military to save on their infrastructure to carry the battery and also to charge it after every flight. As this technology reaches higher level it would save enormously on cost of batteries and thus the same can be allotted for other useful payloads like onboard camera, etc.

REFERENCES

[1] Design and implementation of Autonomous vision guided MAV – Scott M Ettinger, August 2001

[2] Vision based target Geo-location using a fixed wing Miniature Air Vehicle, D.Blake Barber et.al., Journal of Intelligent Robotic Systems, 2006.

[3] Technology Driven Programme for the Development of a Fixed Wing MAV at NAL, S Sankara Narayanan,et.al.

TECHNICAL SESSION – V

SYSTEM ENGINEERING AND LEGAL ASPECTS OF MAV FLYING

ENHANCING NODE COOPERATION USING NEIGHBORHOOD WATCH IN MOBILE AD HOC NETWORKS

[1]David Solomon Raju. Y, [2]D. Vemana Chary [3] B. Ananda Krishna

[1]Assoc. Prof, Dept of ECE, Holy Mary Institute of Tech & Science, Hyderabad, AP, India.
Email: davidsolomonraju131@gmail.com
[2]Assoc. Prof, Dept of ECE, Bharath Institute of Tech & Science, Hyderabad, AP, India.
Email: vemanad@gmail.com
[3]Professor, Department of ECE, Gudlavalleru Engineering College,
Gudlavalleru, Krishna Dt., AP, India.
Email: anand_bk@rediffmail.com

ABSTRACT

A Mobile Ad hoc Network (MANET) is a self configuring network of mobile nodes and is connected by wireless links. In MANET, cooperation of nodes for transmitting packets from source to destination cannot be assumed because of the nodes selfishness and misbehavior degrades the performance of the network makes difficult in finding the routes between the nodes. Cooperation among the nodes can enhance by providing incentives and malicious nodes can eliminate by neighborhood watch are the proposed solutions. A pricing approach which is based on an auction structure is used to obtain the least cost route and after finding the required route, the misbehaving nodes should be isolated, thus making it unattractive to deny cooperation. The proposed scheme is simulated using Glomosim and the results show the improved performance.

KEYWORDS: Mobile Ad hoc networks, routing, least cost, pricing, neighborhood watch.

INTRODUCTION

A mobile ad hoc network (MANET) consists of a set of mobile hosts that carry out basic networking functions like packet forwarding, routing, and service discovery without the help of an established infrastructure. Nodes of an ad hoc network rely on one another in forwarding a packet to its destination, due to the limited range of each mobile host's wireless transmissions. Security in MANET is an essential component for basic network functions like packet forwarding and routing: network operation can be easily jeopardized if countermeasures are not embedded into basic network functions at the early stages of their design. Unlike networks using dedicated nodes to support basic functions like packet forwarding, routing, and network management, in ad hoc networks those functions are carried out by all available nodes.

In the open environments where a common authority that regulates the network does not exist, any node of an ad hoc network can endanger the reliability of basic functions like routing. The correct operation of the network requires not only the correct execution of critical network functions by each participating node but it also requires that each node performs a fair share of the functions. The later requirement seems to be a strong limitation for wireless mobile nodes where by power saving is a major concern. The threats considered in the MANET scenario are thus not limited to maliciousness and a new type of misbehavior called selfishness should also be taken into account to prevent nodes that simply do not cooperate. Also, there are situations, at which a node may not be able to cooperate though it wants to cooperate. This includes nodes running on very low battery power, nodes located at border areas with few packets to forward, nodes with a full buffer and the nodes located at a congestive point in the network. The work can be extended by including a kind of neighborhood watch, which refers that each node can be monitored and evaluated by its neighbors and can determine the condition of the node. The node cooperation can be initiated based on the condition of the node.

The rest of the paper is organized as follows. Section 2 presents the related works, Section 3 presents the problem statement and section 4 describes the proposed work with pricing method and Neighborhood watch. Section 5 gives the simulation and results and conclusion in section 6.

RELATED WORKS

There have been several works related to the pricing scheme and neighborhood watch. Michiardi and Molva, proposed a standard mechanism based on reputation to enforce cooperation among the nodes of a MANET to prevent selfish behavior [3]. Each network entity keeps track of other entities collaboration using a technique called reputation. High computation and communication overhead, as each successful request results in the adjustment of the reputation table and in propagating the success. S. Zhong, proposed Sprite, a simple, cheat-proof, credit based system for stimulating cooperation among selfish nodes in MANET [4]. It supports sender-based payment, since it avoids DoS attacks on the receiver. Buttyan and Hubaux, proposed cooperation stimulation method which is based on a virtual currency namely nuglets which are used for packet forwarding [5]. As the nuglet account balance of the destination is not considered when the packet is generated, the network can become overloaded quickly. S. Eidenbenz, introduced auction-like methods where the sender-centric Vickrey auction has been adopted to discover the most efficient routes for the cost efficient and truthful routing in MANET [6]. Z. Ji. and L.Anderegg, proposed a dynamic pricing framework to implement multistage pricing for autonomous MANETs [43][7][8]. The least cost route is selected from the available routes based on an auction. Marti et al., introduced detection-based routing protocol enhancements for wireless ad hoc networks [9]. They use a watchdog that identifies misbehaving nodes and a path rater that helps routing protocols avoid these nodes. A malicious node is possible to avoid the watchdog by dropping the packets at a lower rate than the threshold. The authors do not give the method to define an appropriate threshold.

The effectiveness of various watchdog schemes was investigated in [10]. Those schemes are suggest that watchdog schemes are indeed able to detect a number of attacks such as omissions and certain symmetric faults but expose limitations, e.g., fabrication of false route error messages. Watchdogs are not limited only to forward monitoring. In [11] watchdog was extended to an extended watchdog mechanism, it presented that implemented backward monitoring based on CTS and RTS messages at the MAC layer. By this concept of monitoring was implemented and that can be extended to neighborhood monitoring, in which groups of monitors form a neighborhood. A monitor can operate in a passive or active mode. A passive monitor just observes, but does not take action. An active monitor reacts by sending notification messages [2]. The impact of neighborhood monitoring was exploited in [12] and [9]. Authenticated neighborhoods were considered in [12]. However, the work assumes an attack-free environment during neighbor discovery, i.e., no malicious nodes exist before the completion of the neighbor discovery. Furthermore, no misrouting attacks are considered. It does not address network dynamics nor is collaboration between malicious nodes, i.e., the probability of framing due to collaborating nodes assumed small and based on statistical arguments. We diverge from this assumption and consider pathological behavior as likely. The impact of neighborhood watching was considered in [9] to analyze the impact of topology on reliability. However, this was based only on the general structure of the neighborhood, i.e., the width of adjacent neighborhoods represented as a join-graph, in order to determine the reliability and survivability of a link. In [2] k-hop monitoring is introduced for most common packet manipulations in ad hoc networks. The key issue is that no assumptions are made about the behavior of malicious nodes and possible collusions. The monitoring scheme depends on the type of faults considered, which also dictate the thresholds of monitors necessary for attack detection and survivability for topology-aware and topology-unaware implementations.

PROBLEM STATEMENT

In MANET each node acts as a router and cooperate each other for forwarding packets. In the network all nodes doesn't have same nature or working principle, for example, some nodes may be cooperative, some may be selfish and some are malicious. The malicious nodes may disturb the network performance by not forwarding control messages or data packets or modifies the original data or transmits wrong information to the neighboring nodes. The selfish nodes may stay idle without participating in the functioning of the network. Therefore the problem being addressed in this work is to find an efficient way to enhance the co-operation among the nodes and eliminate the malicious nodes. The objective of this work is to enhance node cooperation by paying incentives to eliminate selfish nodes and a keen watch on condition of all the nodes to running on very low battery power, nodes located at border areas with few packets to forward, nodes with a full buffer and the nodes located at a congestive point in the network and the malicious nodes. The figure 1 shows the block diagram of our proposed work.

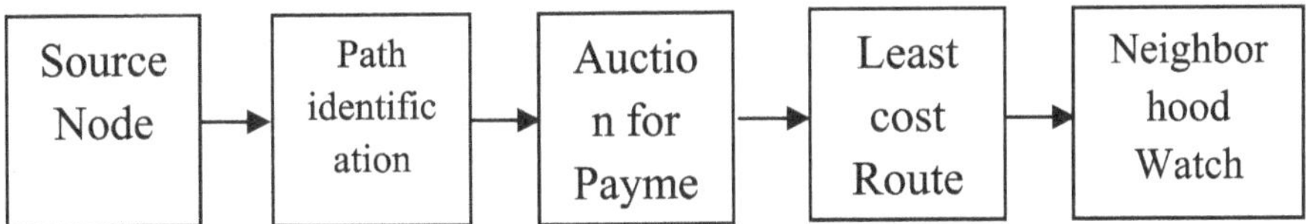

Fig.1 Block Diagram of proposed work

Selfish Nodes and Malicious Nodes

Selfishness is a passive attack in which nodes may not participate in communication due to issues like low battery, less bandwidth, energy saving, etc. *Selfish Nodes* are active in route discovery, but not in packet forwarding. They tend to drop data packets of others to save their energy so that they could transmit more of their own packets and also to reduce the latency of their packets. This type of attack comes under denial-of-service (DoS) category.

Selfish nodes, on the other hand, which cooperate during route discovery and defect during packet forwarding, need to be explored. A behavioral model that could dynamically predict the level of cooperation extended by the node towards the network functions such as routing, network monitoring and packet forwarding is therefore, crucial. Maliciousness is an active attack in which nodes may benefit, but may not transfer the data packets, may steal or modify information, manipulate the data packets. These attacks will not help the attacker saving power, but only disrupting routes [10].

PROPOSED WORK

Pricing Method

The pricing scheme is proposed in such a manner that the source, destination and the relay node gets the benefit. Packet forwarding imposes resource costs on the relay nodes. Resource costs include energy consumption, CPU usage and other factors. The forwarding cost [1] C_{rj} is a monetary resource cost, and is given by the following equation

$$C_{rj} = \mu \cdot c_{rj} \dots\dots\dots\dots\dots(4.1)$$

where c_{rj} is defined as monetary compensation for the unit resource cost. C_{rj} is the unit resource cost of forwarding one unit of packets by the node r through the path j and μ is a constant which in turn depends on the resource utilization. The pay off of a relay node corresponds to the received payment minus the incurred forwarding cost, computed as:

$$U_{rj}(p_j) = (p_j - C_{rj})Q(s) \dots\dots\dots\dots(4.2)$$

where p_j is the price requested by the relay node and $Q(s)$ is the traffic sent from source nodes. The participating nodes in the transfer of packets can be source, destination or relay node. The cost demanded by each node depends on its price function which in turn may depend on the battery level of the node, its load, or any other factor. The utility function is a function which is of importance in the pricing scheme which can be defined as

$$U_i(d) = E_i(d) - Q_i(d) \cdot r_i \dots\dots\dots\dots(4.3)$$

where $E_i(d)$ is the expected payment for the ith route, $Q_i(d)$ is the probability that the ith route will be selected for forwarding where d is vector of bidding strategies for all routes and r_i is the forwarding cost of the ith route. The auction mechanism [1] is introduced in order to find the optimum route. The type of auction is Dutch auction and is also known as descending price auction, uses a bidding process to find an optimal market price for the stock, the lowest price at which an issuing company can sell all the available shares. The multiple routes between source and destination act as bidders. The auction is performed among the multiple routes. The route with the least cost is considered as the winner of the auction. The packets are transferred through this route. There are some constraints to be considered to perform the dynamic pricing namely bandwidth B and delay T which represents the maximum number of routing stages the packets can wait. Thus we perform the proposed pricing scheme over this T period time window to obtain the least cost route.

Neighborhood watch

In this scheme, nodes in the network collectively detect and declare the misbehavior of a suspicious node. Such a declaration is then propagated throughout the network so that the misbehaving node will be cut off from the rest of the network. By neighborhood watch, each node observes its 1st hop neighbor nodes and maintains the details of the neighbor.

Discovery/Monitor:

The neighbors of the neighborhood watch can detect deviances by the next node on the source route by either listening to the transmission of the next node or by observing route protocol behavior. By keeping a copy of a packet while listening to the transmission of the next node, any content change can also be detected. In this work, we focused on the detection of observable routing and forwarding misbehavior in DSR.

Neighbor monitoring is used to collect information about the packet-forwarding behavior of the neighbors. Due to the promiscuous mode that we assume, a node is capable of overhearing the transmissions of its neighbors. With this capability, a mobile node N can maintain a *neighbor node list* (denoted by *NNLN*) which contains all of its neighbor nodes that node N learns of by overhearing. In addition, node N keeps track of two numbers, for each of its neighbor (denoted by X), as below.

- $RFN(X)$ (Request-for-Forwarding): the total number of packets that node N has transmitted to X for forwarding.

- $HFN(X)$ (Has-Forwarded): the total number of packets that have been forwarded by X and noticed by N.

The two numbers are updated by the following rules. When node N sends a packet to node X for forwarding, the counter $RFN(X)$ is increased by one. Then N listens to the wireless channel and check whether node X forwards the packet as expected. If N detects that X has forwarded the packet before a preset time-out expires, the counter $HFN(X)$ is increased by one.

Given $RFN(X)$ and $HFN(X)$, node N can create a record called *local evaluation record* (denoted by $LERN(X)$), for the neighbor node X. The record $LERN(X)$ consists of two entries, that is, $GN(X)$ and $CN(X)$, where $GN(X) = HFN(X) \ RFN(X)$ and $CN(X)$ is a metric called *confidence*, used to describe how confident node N is for its judgment on the reputation of node X. In our scheme, we set $CN(X) = RFN(X)$; that is, the more packets transmitted to X for forwarding, the better estimation about how well the neighbor X does forwarding.

Indicator

In this each node records the details of every 1st hop neighbors and distributes to its neighbors. That is each node has incoming and outgoing alert messages. Alert messages are sent by the trust manager of a node to warn others of malicious nodes. Incoming alerts originate from outside neighbors (The nodes that are available in out of range), whereas the node itself generates outgoing alerts after having experienced, observed or been reported malicious behavior.

Each route consists of its Owen black lists are maintained at each node and potentially exchanged with neighbors. The nodes can include black list in the route request to be avoided for routing, which also alerts nodes on the way. Nodes can look up senders in the black list containing the nodes with bad rating before forwarding anything for them. The problem of how to distinguish alleged from proven malicious nodes and thus how to avoid false accusations can be lessened by timeout and subsequent recovery or revocation lists of nodes that have behaved well for a specified period of time.

Punishment/Encouragement:

Once a node x classifies another node y as misbehaving, x isolates y from communications by not using y for routing and forwarding and by not allowing y to use x. This isolation has three purposes. The first is to serve as an incentive to behave well in order not to be denied service. The second purpose is to reduce the effect of misbehavior by depriving the misbehaving node of the opportunity to participate in the network. Finally, the third purpose is to obtain better service by not using misbehaving nodes on the path.

SIMULATION ENVIRONMENT AND METHODOLOGY

The routing protocols are simulated within the Glomosim library. The Glomosim library is a scalable simulation environment for wireless network systems using the parallel discrete-event simulation capability provided by PARSEC. There are N nodes, randomly deployed inside a rectangular region of 1000×1000 according to the 2 dimension uniform distribution and each node moves according to the random waypoint model. The simulation parameters are shown in Table 1.

Table 1 Simulation Parameters

Simulation Area	1000x1000m
Number of nodes	N
MAC Layer	802.11
Transport layer	UDP and TCP
Traffic	CBR
Mobility	Random
Node placement	Uniform
Routing Protocol	DSR
Simulation Time	200sec

Each source and destination pair is formed by randomly selecting two nodes in the network.

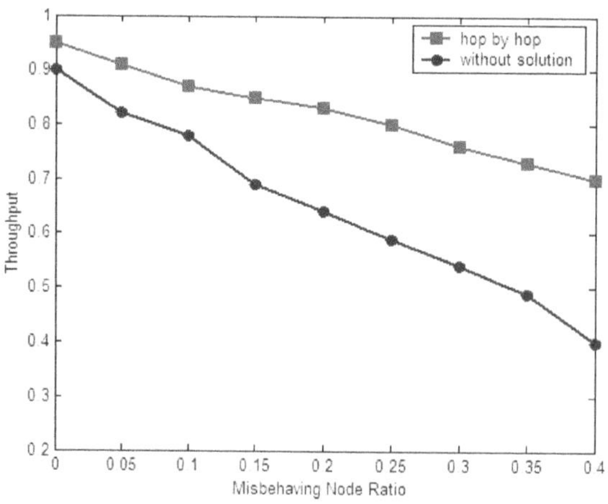

Fig. 2 Misbehaving Node Ratio Vs Throughput

From the figure 2, the first thing that we can notices is the throughput of the network degrades as the malicious nodes increase. Here we have varied malicious nodes upto 40%. With the neighborhood watch hop by hop, the number of malicious nodes can be prevented and the throughput is increased.

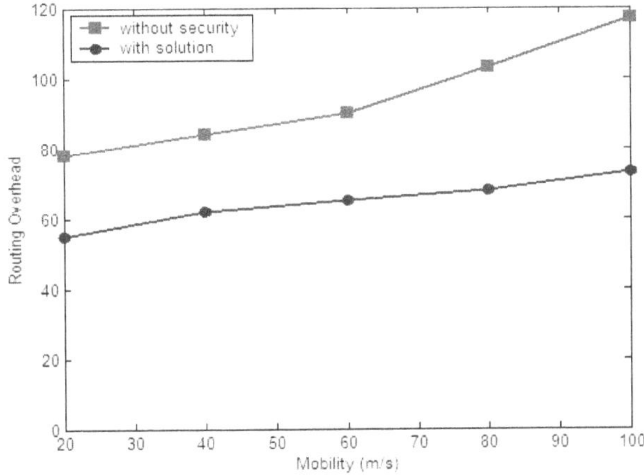

Fig. 3 Mobility Vs Routing Overhead

From figure 3, we have found that as mobility increases, the routing overhead increases due to malicious nodes and with the proposed solution, the routing overhead can be reduced.

Figure 4 shows that the number of routes available with varying cost is more with the proposed scheme than with the existing scheme.

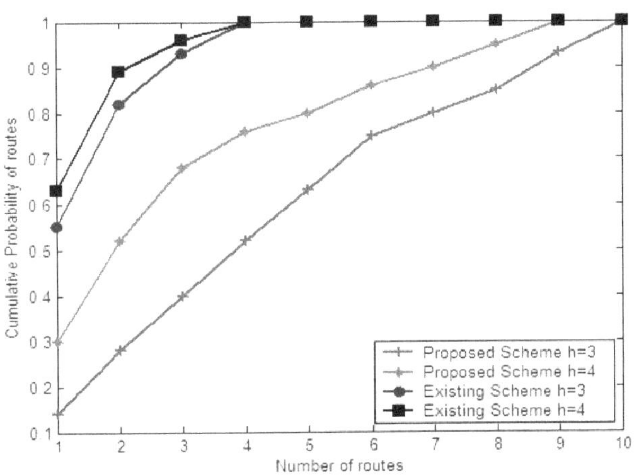

Fig. 4 Number of routes Vs Cumulative Probability of routes for
existing scheme and proposed scheme

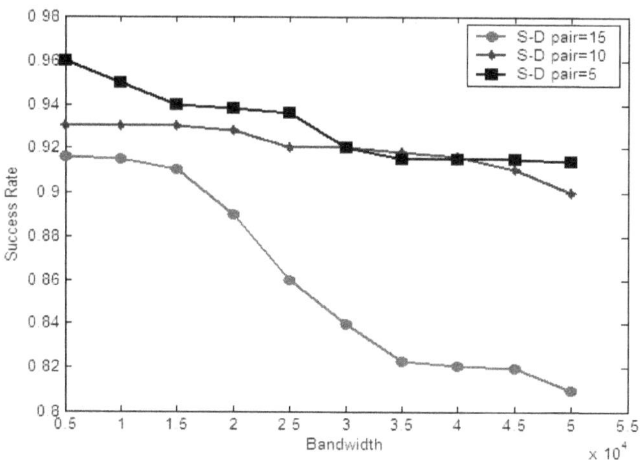

Fig. 5 Bandwidth Vs Success Rate

The success rate is defined as the ratio of the number of packets received successfully to the number of packets transmitted. The success rate is an important parameter to analyze the success of the routing scheme. Figure 5 shows the success rate versus the bandwidth. As the bandwidth increases, the numbers of packets received at the destination successfully are noted down. It is observed that as the demand on bandwidth increases the success rate decreases. The traffic rate is increased by increasing the number of source-destination pair. We have studied the simulation by varying the number of source-destination (S-D) pair as 5, 10 and 15 and it was observed that as the number of S-D pair increases the success rate decreases only by 4.2%. It was also observed that as the demand for bandwidth increases, the success rate decreases only by 3-6%. Hence it is concluded that an optimal success rate is obtained.

CONCLUSION

The proposed pricing scheme stimulates the node cooperation. The sender can transmit the packets through least cost route by spending less virtual currency. Thus the traffic control is achieved with the scheme. The network is protected by isolating misbehaving nodes using neighborhood watch. The performance of the network is increased and the work can be extended by improving Quality of Service by reducing the link failures in the network.

REFERENCES

1. Dona Mathews, "Dynamic Pricing Approach for Cooperation Stimulation and QoS in Mobile Ad hoc Networks", IEEE Computer Society, Oct 2008

2. Axel Krings, S.M.-Karrer, "Neighborhood Monitoring in Ad Hoc Networks", CSIIRW'10, April 21-23, 2010

3. P. Michiardi and R. Movla, "Simulation-based analysis of security exposures in mobile ad hoc networks," *European Wireless 2002 Conference*, Florence, Italy, Feb. 2002.

4. S. Zhong, J. Chen, and Y. R. Yang, "Sprite: A simple , cheat proof, credit based system for Mobile Adhoc Networks ", IEEE INFOCOMM, 2003.

5. L. Buttyan and J.-P. Hubaux, "Enforcing Service Availability in Mobile Ad-Hoc WANs," Proc. MobiHoc, Aug. 2000.

6. S. Eidenbenz, G.Resta, and P.Santi, "COMMIT: a sender – centric truthful and energy-efficient routing protocol for adhoc networks with selfish nodes", IEEE International Parallel and Distributed Processing Symposium, 2005.

7. Z. Ji, W. Yu, and K.J.R Liu, "Dynamic pricing Approach for self organized Mobile Ad hoc Networks", IEEE Globecom 2005.

8. Z. Ji, W. Yu, and K.J.R Liu, "An Optimal Dynamic Pricing Framework for autonomous Mobile Ad hoc Networks", IEEE INFOCOM 2006.

9. Sergio Marti, T. J. Giuli, Kevin Lai and Mary Baker, "Mitigating routing misbehavior in mobile ad hoc networks", Mobile Computing and Networking, pp. 255-265, 2000.

10. S. Buchegger, et.al., "A test-bed for misbehavior detection in mobile ad-hoc networks - how much can watchdogs really do?," Sixth IEEE Workshop on Mobile Computing Systems and Applications, WMCSA 2004, pp. 102-111, 2-3 Dec. 2004.

11. Issa Khalil, Saurabh Bagchi, Cristina N. Rotaru, Ness B. Shroff, "UNMASK: Utilizing neighbor monitoring for attack mitigation in multihop wireless sensor networks", Ad Hoc Networks, Vol. 8, pp. 148-164, 2010.

12. A. Krings, and Z. Ma, "Fault-Models in Wireless Communication: Towards Survivable Ad Hoc Networks", Military Communications Conference, MILCOM 2006, pp. 1-7, 23-25 Oct. 2006.

POSSIBLE BIOMIMICKING FOR THE DESIGN OF MICRO AERIAL VEHICLE'S CONSIDERING SPECIES SPECIFIC FEATURES OF INSECT, BIRD, AND BAT FLIGHT

N. Chari

Visiting Professor, Malla Reddy College of Engineering and Technology, Secunderabad

A. G. Sarwade

Department of Aeronautical, Malla Reddy College of Engineering and Technology, Secunderabad

C. Mohan Srikanth

Design Engineer, InfoTech Enterprises Ltd, Hyderabad,

Amit Kumar

Assistant Professor, Malla Reddy College of Engineering and Technology, Secunderabad,
InfoTech Enterprises Ltd, Hyderabad, chenji.srikanth@gmail.com

ABSTRACT

We can learn many things by observing the biological flight of insects, birds and bats since they fly with higher aerodynamic efficiency i.e., their lift to drag ratio is much higher as compared to our airplanes. They produce thrust by a complex morpho-functional flapping phenomenon. The comparison of flight features of these three types of fliers gives us some useful clues on the nature of natural wing materials having special properties (like light weight and high flexibility) which have to be considered seriously for their usage in Micro aerial vehicle's to enhance their aerodynamic efficiency. The evolutionary history of these fliers and flapping is a notable feature.

Insects are miniature fliers having low mass, resilin, and membranous chitinous wings, with high flapping frequency of 500 Hz or more. To imitate them is difficult. Birds and bat wings are modified teteapod limbs for flight. Birds have thousands of keratin feathers which cannot be replicated easily. In birds, the finger bones are reduced. However in bats, the finger bones are highly elongated (delicately) and the whole body is covered by membrane. It may be easier to imitate bat model and not its echo location. Anti stalling devices (Alula in birds) need further study. Their effortless landing and takeoff is notable for micro aerial vehicle design. Comparative moment of inertia studies give better aerodynamic information on biological wings.

These natural fliers have extraordinary sensory perception, integration, feedback mechanism and adaptive control of flight. The researchers are trying to suggest how to incorporate some of these features in micro aerial vehicle designs. In fact in the most advanced fighter aircrafts (like F – 22 Raptor of USA) some sensory features are incorporated. We can think of flexible materials for wings, like silk, thin graphite with latex support etc. The biological wings are elastic and they withstand bending and damp the vibrations.

INTRODUCTION

Biological flight of insects, birds and bats has an evolutionary history of millions of years and phylogenetically they are different in origin, structure but develop similar aerodynamic forces during flight. Biological fliers have higher aerodynamic efficiency (L/D ratio) and possess flapping flexible wings which differ in their origin and morphological structure. The biological fliers in a way can be called as MAV's keeping in mind their weight limits which are shown in Table 1.

Table1 Minimum and Maximum Weight of Biological Fliers

Parameter	Insects	Bird	Bat
Minimum Weight	$5 * 10^{-9}$	0.005	0.008
Maximum Weight	0.01	12	1.2

(All units are in SI system)

In biomimicking MAV's we have to introduce flapping flexible wings and the problems of takeoff and landing have remained unsolved. For the development of an MAV we need an interdisciplinary approach of aeronautic engineers, ornithologists and biologists. It is advisable to go for biomimicking MAV's based on insect model, bird model or a bat model. The bat model is little complex however, its elastic patagium supported by elongated delicate bones which is more ideal to copy.

The present paper aims in comparing basic aerodynamic features of an insect, bird and bat which can be the basis in detail understanding of biomimicking principles of natural fliers. All the biomimicking fliers should satisfy the conditions for hovering and successful horizontal flight.

LITERATURE REVIEW

Deakin (1970, 2010) has derived an expression for wing beat frequency by considering the dimension analysis as the basis. He considered two dimensionless ratios by using 'Bukingham Pi Theorem', which states that equations can be reduced to a simpler relationship by using dimensionless products.

Hovering is a kind of continuous power on flight where forward velocity becomes zero and the wings act as propellers of high frequency. The body is held more or less vertical and stroke plane of the wings is approximately horizontal. The wingtip traces a figure '8' and the wings move back and forth. The rate of change of momentum supports the bird weight. Small and medium size fliers such as humming birds, kingfishers practice hovering mainly for getting food. During hovering, oxygen consumption increases five times as compared to resting state. The wing stroke angle may be larger than 120^0, therefore S_d (disc area) may be taken as 360^0. The S_d concept in mass flow (1977, 2011) is replaced by S_w (small wing swept area) in Crawford theory (1971). Birds in nature hover at one spot, and the transition to horizontal flight. The wing motion in hovering is sinusoidal. There is a marked drop in power as the bird converts its position from hovering to forward flight. Big birds are not able to hover because of high energy cost and for not developing sufficient lift.

Puranik and Chari (1986) reviewed wing beat frequency in a chronological order. Pennycuick (1975) has suggested a formula for wing beat frequency of birds in steady flight. Sane (2003) has reviewed the aerodynamics of insect flight with emphasis on flapping flight. Wong (2005) has reviewed recent experimental, computational and theoretical approach to study the forces and flows around flapping wings of insects.

For the design and development of a MAV, necessary basic (Chari 2011) and derived parameters are listed in table 2.

Table 2 Species Specific Features of Insect, Bird and Bat

Parameters	Insect	Bird	Bat
Mass (M)	0.0008	0.168	0.0075
Wing Length (l)	0.0217	0.256	0.127
Wing Span (L)	0.0564	0.542	0.283
Wing Area (2A)	0.0004	0.0579	0.0114
Effective Wing Breadth, B_{eff}	0.015	0.113	0.041
Wing Loading (M/ 2A)	1.96	2.98	0.65
Wing Span Loading, (M/ L^2)	0.269	0.572	0.0936
Frequency, (v_h)	49	11	5
Induced Power (P_i)	1.19	363.6	30.65
Time (T)	0.01	0.09	0.22
Wing Hinge	Point fulcrum	Ball and Socket joint	Ball and socket joint patagium attached to body

Insect – T. Javanica (Soap nut bug), Bird – S. d, decacto (Ring dove), Bat- H. spereosis (leaf nosed bat), (All units are in SI system)

DISCUSSION AND SUGGESTIONS

The mass range as suggested for MAV's is from 0.1 to 0.2 Kg. The above mentioned fliers in Table 2 and their species specific, aerodynamic features have been very well tested in nature for millions of years under genetic control. Each flier as suggested in the Table 2 has definite environmental adaptation for flight. Our knowledge of airplane aerodynamic parameters is hardly 110 years old where we have separated lift and thrust. Therefore, scaling of biological features of insect, bird and bat flapping wing along with their body morphology might help in the design of better biomimicking MAV's. These prototypes have to be studied theoretically and experimentally.

The mass of the flier is carried on the flapping wings by developing lift and thrust at a particular frequency.

SUGGESTIONS FOR FUTURE WORK

a) For the biomimicking wing structure, pigeon (dove) or eagle (hawk) can be considered to start with. The materials should be less dense and stronger. Dragon fly or locust are the other easier insect models to mimick. Scaling is the most essential factor in biomimicking.

b) The experimental work in wind tunnels is needed to obtain the aerodynamic parameters and to validate CFD studies. Since the experimental work is tedious, CFD analysis can be carried out on much geometry after their due validation. The experimental model needs a robust flapping mechanism, for the study of unsteady aerodynamics.

c) If the flapping wing is not able to produce sufficient thrust, a stand by propeller can be thought of to generate the deficit thrust.

d) For biological fliers, the landing and takeoff are not a problem either for terrestrial, aquatic or arboreal environment. Similar one can be thought of for the terrestrial MAV's. Since present MAV's do not have adequate landing and takeoff provisions.

REFERENCES

1. Deakin, M. A. B., 1970. "The Physics and Physiology of Insect Flight", American Journal of Physics 38:1003-1005.

2. Crawford Frank S., 1972. Comments on "The Physics and Physiology of Insect Flight", American journal of physiology 39: 584.

3. Puranik, P.G., Gopal Krishna G, Adeel Ahmed and Chari N. 1977. "Wing Beat Frequency of a Flier – Mass Flow Theory", Proc. Ind. Acad. SciSoc. A., 85, 327-329

4. Puranik, P. G., and Chari N. 1986 "Bio-Aerodynamics of Fliers." National Book Enterprises, Hyderabad, A.P. India.

5. Pennycuick, C. J., 1972. "Mechanics of Flight in Avian Biology", Vol. 5 (ed. D. S. Farner and J. R. King), pp. 1-75. Newyork: Academic press.

6. Pennycuick, C. J., 1996. "Wing Beat Frequency of Birds in Steady Cruising Flight: New data and Improved Preditions", Journal of Experimental Biology, Vol. 199: 1613-1618.

7. Sane, S. P., 2003. "The Aerodynamics of Insect Flight", Journal of Experimental Biology, Vol. 209: 4191-4208.

8. Wang, Z. J., 2005. "Dissecting of Insect Flight", Annu. Rev. Fluid Mech, Vol. 37: 183-210.

9. Chari N., 2011 "Bio-Aerodynamics of Bird Flight." Vistar Graphics, Hyderabad, A.P. India, V edition.

A SOLAR POWERED HYBRID AIR VEHICLE SYSTEM FOR LONG ENDURANCE

Ganesh J[1],   **R. Krishnakumar[2],**   **Dr K. Senthil Kumar[3],**
[1]P.G Student, [2]PH.D Scholar, [3]Associate Professor
Division of Avionics, Department of Aerospace Engineering, MADRAS INSTITUTE OF
TECHNOLOGY, ANNA UNIVERSITY, Chennai-600044.

ABSTRACT

The small air vehicle usually does not have long endurance due to its size and weight restriction. With solar powered vehicle the endurance can improved much beyond the endurance of the normal battery operated air vehicle. The advantage of the tilt rotor vehicle is to take-off and land both vertically and horizontally, this phenomenon can be used in the areas of surveillance where the hover is required and also the vehicle can be landed anywhere when the supply from the solar irradiance is drought.

KEYWORDS: Aerial Surveillance, Solar power, MPPT, DC-DC converter, Hybrid aerial Vehicle

INTRODUCTION

The ability for an aircraft to fly during a much extended period of time has become a key issue and a target of research, both in the domain of civilian aviation and unmanned aerial vehicles. This latter domain takes an increasingly important place in our society, for civilian and unfortunately military applications. The required endurance is in the range of a couple of hours in the case of law enforcement, border surveillance, forest fire fighting or power line inspection. However, other applications at high altitudes, such as communication platform for mobile devices, weather research and forecast, environmental monitoring, would require remaining airborne during days, weeks or even months. For the moment, it is only possible to reach such ambitious goals using electric solar powered platforms. Photovoltaic modules may be used to collect the energy of the sun during the day, one part being used directly to power the propulsion unit and onboard instruments, the other part being stored for the night time.

In order to reach the target endurance, the design of the airplane has to be thought carefully and globally, as a system composed of many subsystems that are continuously exchanging energy.

POWER SYSTEM OF THE SOLAR UAS

The system proposed in this project uses solar power to drive the propulsion system of the UAS with which the endurance of the system can be increased upto 7 to7.5 hours, where the power system is implemented with MPPT algorithm and DC-DC converter for maximum power extraction.

Solar cell

A solar cell or photovoltaic cell is a device that converts solar energy into electricity by the photovoltaic effect. In figure 2.1, a simple silicon solar cell is represented with two doped semiconductors layers, p-type and n-type. When the sunlight strikes the solar cell surface the cell creates charge carriers as electrons and holes. The internal field produced by junction separates some of the positive charges (holes) from the negative charges (electrons). The holes are swept into the positive or player and the electrons are swept into the negative or n-layer. When a circuit is made, the free electrons have to pass through the load to recombine with the positive holes, current can be produced from the cells under illumination.

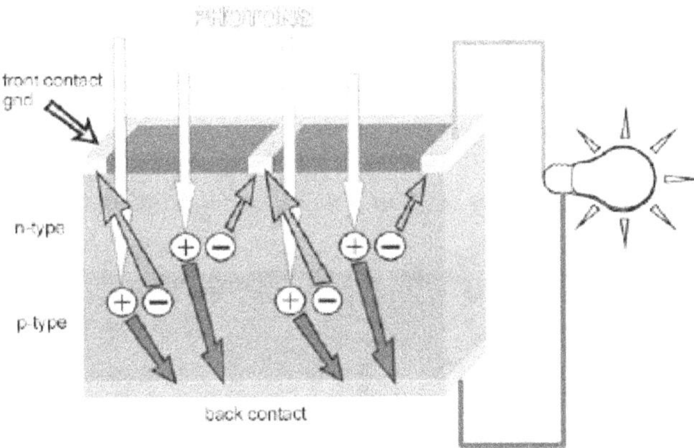

Fig 2.1 System architecture

MPPT Algorithm

A solar cell has a working point on its current to voltage curve where the power retrieved is maximal. In order to work at this point, which is continuously moving because of the constantly changing irradiance conditions, and thus get the highest amount of energy, a so called Maximum Power Point Tracker (MPPT) is required. An MPPT is basically a DC/DC converter with variable and adjustable gain between the input and the output voltage, the input being the solar panels and the output the battery. It contains electronics that monitor both the current and the voltage on each side, which allows a determination for how the gain has to be changed to ensure the best use of the solar panels. As part of the energy chain, the MPPT has to be as efficient as possible. Thus, not only the hardware design has to be optimized to minimize the losses in diodes, transistors and inductors, but also the algorithm has to be tuned to have a fast adaptation to irradiance variations and a good tracking precision. A well designed MPPT should have efficiency above 95%, but the best products reach 99 %.

DC-DC CONVERTER

DC to DC converters are used to regulate the voltage that we receive from the solar panels which increases with increase in irradiance and vice versa. They are also important in Solar UAVs', which are supplied with power from batteries primarily. Additionally, the battery voltage declines as its stored power is drained. Switched DC to DC converters offer a method to increase voltage from a partially lowered battery voltage thereby saving space instead of using multiple batteries to accomplish the same thing. Thus safe charging of battery is ensured by using such converter. Fig

2.2 shows a buck-boost configuration of DC-DC converter. Fig 2.3 shows Simulation Results of Buck –Boost converter done with MATLAB.

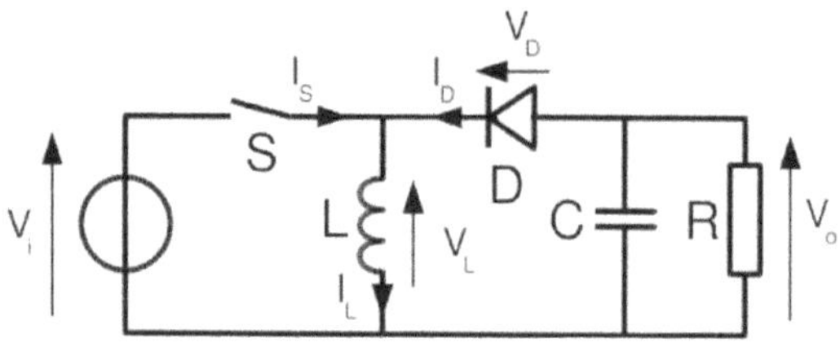

Fig 2.2 Buck-Boost Converter

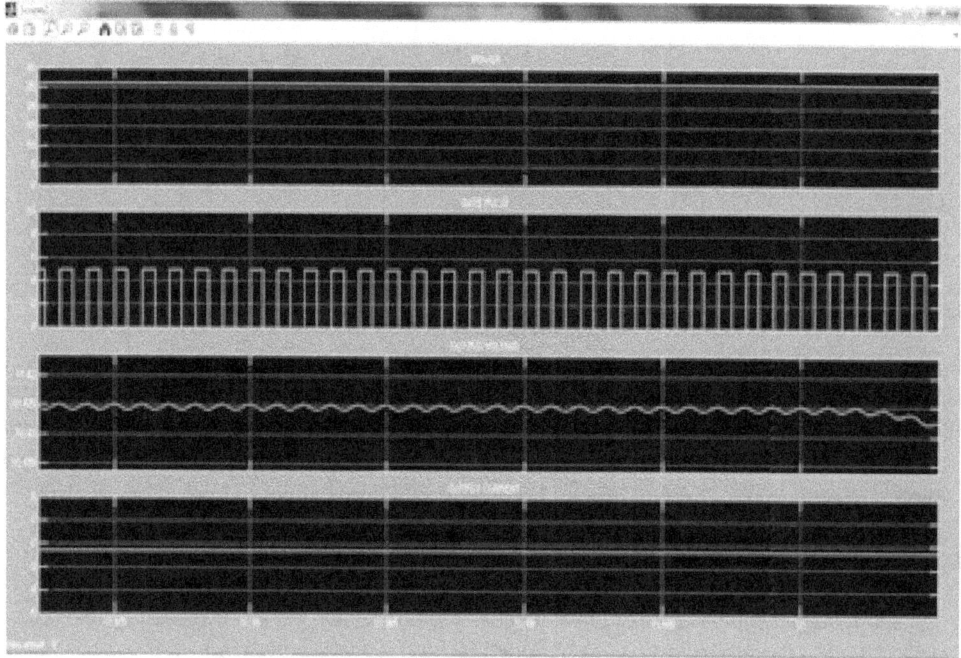

Fig 2.3 Simulation Results of Buck –Boost converter

BRUSHLESS DC MOTOR

The propulsion system of the UAV consists of brushless motor. Brushless DC motors (BLDC motors, BL motors) also known as electronically commutated motors (ECMs, EC motors) are synchronous motors which are powered by a DC electric source via an integrated inverter/switching power supply, which produces an AC electric signal to drive the motor (AC with a caveat – alternating current often implies a sinusoidal waveform; a better term would be bi-directional current with no restriction on waveform); additional sensors and electronics control the inverter output amplitude and waveform(and therefore percentage of DC bus usage/efficiency) and frequency (i.e., rotor speed).

The motor part of a brushless motor is often a permanent magnet synchronous motor, but can also be a switched reluctance motor, or induction motor.

Brushless motors may be described as stepper motors; however, the term stepper motor tends to be used for motors that are designed specifically to be operated in a mode where they are frequently stopped with the rotor in a defined angular position. This page describes more general brushless motor principles, though there is overlap.

AIRCRAFT DESIGN

A tiltrotor is an aircraft which uses a pair or more of powered rotors (sometimes called proprotors) mounted on rotating shafts or nacelles at the end of a fixed wing for lift and propulsion, and combines the vertical lift capability of a helicopter with the speed and range of a conventional fixed-wing aircraft. For vertical flight, the rotors are angled so the plane of rotation is horizontal, lifting the way a helicopter rotor does. As the aircraft gains speed, the rotors are progressively tilted forward, with the plane of rotation eventually becoming vertical. In this mode the wing provides the lift, and the rotor provides thrust as a propeller. Since the rotors can be configured to be more efficient for propulsion (e.g. with root-tip twist) and it avoids a helicopter's issues of retreating blade stall, the tiltrotor can achieve higher speeds than helicopters.

A tiltrotor aircraft differs from a tiltwing in that only the rotor pivots rather than the entire wing. This method trades off efficiency in vertical flight for efficiency in STOL/STOVL operations.

DESIGN PROBLEMS

In most tilt rotor designs, the rotors are located longitudinally near the vehicle center of gravity. This placement is typically driven by the practical requirement of mounting the motors on the front of the wings and a desire to allow the vehicle to remain in a horizontal orientation during hover. However, placement near the center of gravity results in short moment arms which limit the pitch control authority that can be achieved through rotor articulation. Typical control strategies for such vehicles utilize cyclic blade controls to augment or replace rotor articulation as the primary

Although it has been proven to be effective in several designs, this approach adds considerable mechanical complexity to the design and increases the number of moving parts and maintenance demands of the aircraft.

Fig 3.1(a) Level flight Fig 3.1(b) Take-off/Hover

Fig 3.1(a) & (b) shows the tilt rotor UAS in forward flight and vertical flight respectively.

CONCLUSION

The above UAS design increases the endurance of the mini UAV by 5 times when compared to the other normal mini UAVs. With an additional battery pack added to the system the endurance can still be extended. With this applications like aerial surveillance, etc., can be performed effectively.

REFERENCES

[1] Torabi. H.B, Sadi. M, Varjani. A.Y,"Solar Power System for experimental unmanned aerial vehicle (UAV); design and fabrication ", Power Electronics, Drive Systems and Technologies Conference (PEDSTC), 2011.

[2] Thomas L. Gibson, Nelson A. Kelly, "Solar photovoltaic charging of lithium-ion Batteries", General Motors Research and Development Center, USA, Journal of Power Sources,2010.

[3] M.H. Rashid "Power Electronics Circuits, Devices, and Applications", Third Edition, Pearson Education.Inc

[4] M. G. Villalva, J. R. Gazoli, E. Ruppert F, "Modeling And Circuit-Based Simulation Of Photovoltaic Arrays", University of Campinas - UNICAMP, Brazil, Brazilian journal of Power Electronics, 2009.

[5] Wilkins. G, Fourie. D, Meyer. J. "Critical design parameters for a low altitude long endurance solar powered UAV". IEEE Conference on AFRICON, 2009.

[6] Randall Shaffer, "Fundamentals of Power Electronics with MATLAB", Charles river media, Boston, Massachusetts, 2007.

[7] A. Noth, R. Siegwart, and W. Engel, "Design of solar powered airplanes for continuous flight", ETH Autonomous Systems Laboratory, Zurich, Tech. Rep., 2006.

[8] D. P. Hohm and M. E. Ropp, "Comparative Study of Maximum Power Point Tracking Algorithms" Prog. Photovolt: Res. Appl. 2003.

[9] Reinhardt. K.C, Lamp.T.R, Geis.J.W, Colozza.A.J, "Solar-powered unmanned aerial vehicles" Energy Conversion Engineering Conference, 1996. IECEC 96. Proceedings of the 31st Intersociety,1996.

NUMERICAL SIMULATION OF UNSTEADY FLOW OVER A FLAPPING WING

Sharanappa V. Sajjan[1] and K. Siva Kumar[2]

Computational and Theoretical Fluid Dynamics Division,
Council of Scientific and Industrial Research, National aerospace Laboratories,
Bangalore, Karnataka-560017, INDIA
Email id: svsajjan@ctfd.cmmacs.ernet.in[1], shivak@ctfd.cmmacs.ernet.in[2]

ABSTRACT

The three-dimensional unsteady viscous flow over a combined pitching and plunging (flapping) rectangular wing is simulated by using an implicit RANS solver IMPRANS. The solver used for time accurate solution is based on an implicit finite volume nodal point spatial discretisation scheme, wherein a control volume is formed by joining the centroids of the neighbouring cells around a nodal point in the computational domain. The efficiency of the solver for making time-accurate computations is enhanced by implementing an implicit dual time stepping procedure. In this approach, an equivalent pseudo steady state problem is solved at each real time step using local time stepping. The algebraic eddy viscosity model due to Baldwin and Lomax is used for turbulence closure. The computations are carried out at different reduced frequencies to study the effect on thrust generation and propulsive efficiency. The results are obtained in the form of aerodynamic coefficients, thrust coefficient and propulsion efficiency.

INTRODUCTION

Unsteady flows are common for aerospace vehicles and have strong influence over the performance of flight vehicles. These flows around an aerofoil or wing are rather complicated, because of their time-dependent nature. Significant amount of research has been carried out, both theoretically and experimentally by many scientists to understand the fluid mechanics of the flow fields around an aerofoil and wing. Also flow around oscillating aerofoils and wings have been investigated by many researchers to analyze wing flutter, aero-acoustic noise generation, and propulsion mechanism of birds, fish and insects. The detailed understanding of the flow physics behind pitching, plunging and deforming aerofoils and wings will help in the design of flapping wing Micro Air Vehicles (MAV's) and Nano Air Vehicles (NAV's).

From early times, the major idea of utilizing the thrust generated by flapping wings for the propulsion of man-made objects such as MAV's, NAV's, UAV's etc., emerged from observations of birds, insects and fish. In 1490, Leonardo da Vinci first made an attempt to explain and implement the mechanism of thrust generation by a flapping wing. At the end of the 19th century and the beginning of the 20th century, numerous attempts were undertaken to develop flight vehicles using flapping wings (Tikhomirov, 1937 and Lippisch, 1960), which were failed due to insufficient scientific research and engineering knowledge.

Flying birds or insects usually flap their wings to generate both lift and thrust. Flapping motion of birds or insects has a coupled pitching and plunging oscillation with some phase difference between them. Recent experimental and numerical studies investigated the kinematics, dynamics, flow characteristics of flapping wings and shed some light on the lift, drag and propulsive power considerations (Shyy et al. 1999 and Mueller, 2001).

The flow over two-dimensional airfoils and three-dimensional high aspect ratio wings has been computed by (Neef and Hummel, 2000) using Euler equations. The cruising flight of large birds, the reduced frequency is limited to the order of 0.1 and the pitching oscillation advances about 90° ahead of the plunging oscillation. High propulsive efficiency of the order of 90% is observed and the flow around the wing remains in the regime of attached flow.

The aerodynamics of flapping wings is unsteady and a largely unexplored area. Due to the complexity of solving Navier-Stokes equation (Hao Liu et al., 1998) for flow around flapping wings, experimental methods are widely used. Experimental investigation of fruit fly aerodynamics has been reported in Dickinson, et al. (1997).

At present, there are many scientific works are undergoing in several countries to develop small aerial vehicles such as MAV's, NAV's, UAV's, etc., which use flapping-wing systems.

The present work deals with the application of IMPRANS solver for computing unsteady viscous flow over a combined pitching and plunging rectangular wing to study the effect of reduced frequency on the thrust generation and propulsive efficiency. In brief about the IMPRANS solver is given in the next section, while the computational method is described in section III. The grid generation is explained in section IV. The wing flapping motion is given in section V. The results and discussion are presented in section VI and finally, the conclusions are summarized.

IMPRANS SOLVER

IMPRANS is an implicit RANS code developed in-house for steady / unsteady flows. It solves three-dimensional unsteady compressible viscous Reynolds-averaged Navier-Stokes equations in a moving domain. Dual time stepping approach was used with an implicit finite volume nodal point spatial discretization scheme. Inviscid flux vectors are calculated by using the flow variables at the six neighbouring points of hexahedral volume.

Turbulence closure is achieved through the algebraic eddy viscosity model of Baldwin and Lomax. For a moving body, the equations are solved in the inertial frame of reference by employing a grid which, while remaining fixed to the body, moves arbitrarily with the body. The code allows for specification of boundary conditions on the wall to account for both displacement and boundary velocity.

COMPUTATIONAL METHOD

The Reynolds-averaged Navier-Stokes equations for three-dimensional unsteady compressible flow in a moving domain in non-dimensional conservative form are given by

$$\frac{\partial \bar{U}}{\partial t}+\frac{\partial E}{\partial x}+\frac{\partial F}{\partial y}+\frac{\partial G}{\partial z}=0. \qquad \dots\dots(1)$$

Here, $\bar{U}$ is the vector of conserved variables, E, F and G flux vectors, (x, y, z) is the Cartesian coordinate system and t is the time variable.

An implicit finite volume nodal point scheme with dual time stepping approach (Dutta et al, 2003, 2005, 2007 and Sharanappa et al., 2006) is employed for solving the above governing equations (1). The dual time stepping consists of an implicit discretization in real time and the marching of solution in a pseudo time to steady state at each physical time step. Use of an implicit second order accurate backward difference formula for discretization in real time and Euler's implicit time differencing formula for pseudo time results in the following equation

$$\left[I + \frac{3\Delta t^*}{2\Delta t} I + \Delta t^* \left(\frac{\partial R}{\partial U} \right)^m \right] \Delta U^m = -\Delta t^* \left[R(U^m) + \frac{3U^m}{2\Delta t} - \frac{2\bar{U}^n}{\Delta t} + \frac{\bar{U}^{n-1}}{2\Delta t} \right] \qquad .....(2)$$

Here $U^m = U(t^*) = U(m\,\Delta t^*)$ is the solution vector at pseudo time level m, $\Delta U^m = U^{m+1} - U^m$ is the change in U^m over the time step Δt^* and Δt denotes the real or physical time step that is required to resolve the physical unsteadiness of the flow. The barred quantities denote the solution vectors at the previous real time levels n and $n - 1$ whereas R represents the spatial operators which give rise to the flux residual after a discretization in space.

This basic equation (2) of the implicit dual time stepping technique can be solved at each real time step by employing a finite volume nodal point spatial discretization. In this approach, the flow variables are associated with each mesh point (i, j, k) of the grid and the centroids of the eight neighbouring hexahedron cells surrounding the nodal point are joined to form the control volume Ω_{ijk}. To facilitate the finite volume formulation, the equations are written in integral form and the surface integrals are evaluated by summing up the contributions due to the flux terms over the six faces of the computational cell. Applying integral conservative equations to each control volume, Linearizing the changes in flux vectors using Taylor's series expansions in time, assuming locally constant transport properties, and dropping the superscript m we obtain

$$\left(I + \frac{3\Delta t^*}{2\Delta t} I \right) \Delta U_{ijk} + \frac{\Delta t^*}{\Omega_{ijk}} \sum_{m=1}^{6} \left\{ \left[\left(A - \frac{\partial E_R}{\partial x} \right) \Delta U \right]_m S_{mx} \right.$$

$$\left. + \left[\left(B - \frac{\partial F_S}{\partial y} \right) \Delta U \right]_m S_{my} + \left[\left(C - \frac{\partial G_T}{\partial z} \right) \Delta U \right]_m S_{mz} \right\}$$

$$= -\frac{\Delta t^*}{\Omega_{ijk}} \left\{ \sum_{m=1}^{6} \left[(E_I - E_V)_m S_{mx} + (F_I - F_V)_m S_{my} + (G_I - G_V)_m S_{mz} \right] \right\}$$

$$- \Delta t^* \left(\frac{3U_{ijk}}{2\Delta t} - \frac{2\bar{U}^n_{ijk}}{\Delta t} + \frac{\bar{U}^{n-1}_{ijk}}{2\Delta t} \right), \qquad .....(3)$$

Here Ω_{ijk} is the control volume surrounding the nodal point (i, j, k) of the curvilinear grid; $A = \partial E_I / \partial U$, $B = \partial F_I / \partial U$, $C = \partial G_I / \partial U$, $E_R = \partial E_{V_1} / \partial U_x$, $F_S = \partial F_{V_2} / \partial U_y$ and $G_T = \partial G_{V_3} / \partial U_z$ are the Jacobian matrices; E_I, F_I and G_I are the inviscid flux vectors and E_V, F_V and G_V are the viscous flux vectors; S_{mx}, S_{my} and S_{mz} are the x, y and z components of the surface vector corresponding to the mth surface of the control volume.

It is important to note that the terms containing inviscid flux vectors can be calculated by using the flow variables at the six neighbouring points and Taylor's series expansions can be utilised to discretize the derivatives in the viscous flux terms directly in the physical plane.

The resulting block tridiagonal system of equations is solved by using a suitable block tridiagonal solution algorithm and proper initial and boundary conditions. In order to ensure convergence and to suppress oscillations near shock waves, a blend of second and fourth order artificial dissipation terms (Pulliam, 1986) is added explicitly. Implicit second order dissipation terms are also added to improve the practical stability bound of the implicit scheme. The algebraic eddy viscosity model due to Baldwin and Lomax (1978) is used for turbulence closure. For a moving body, the equations are solved in the inertial frame of reference by employing a grid which remains fixed to the body and moves along with it. At each real time step $t + \Delta t$, starting from the solution at the previous time step t, the solution is marched in pseudo time t^* using local time stepping. Since the choice of physical time step Δt is no longer limited by stability considerations, a much larger time step, with a fixed but small number of inner iterations in pseudo time, can be used to reduce the undesirably large computational time for unsteady flow calculations. Based on this dual time stepping method an implicit Reynolds averaged Navier-Stokes solver IMPRANS has been developed at CSIR - NAL for computing a wide variety of two-dimensional and three-dimensional unsteady viscous compressible flows. This RANS solver has been extensively validated for computing unsteady flow past pitching aerofoils and wings (Sharanappa et al., 2008), plunging aerofoils and wings and combined pitching and plunging aerofoils (Siva Kumar and Sharanappa, 2010 and 2011), helicopter rotor blades (Dutta et al., 2005 and Sharanappa et al., 2006), wind turbines (Dutta et al., 2007) etc. Here, the solver has been applied for simulating three-dimensional unsteady compressible viscous flow over a combined pitching and plunging rectangular NACA 0012 wing at different reduced frequencies.

GRID GENERATION

For all present computations, a structured single block CH-type grid around a NACA 0014 rectangular wing of aspect ratio 4 was used which is shown in Fig. 2(a). The surface grid is shown in Fig. 2(b). The numbers of grid points are $247 \times 65 \times 62$ in the chord wise, normal and span-wise directions respectively. The points are clustered near the leading and trailing edges and wing tip properly to capture the flow gradients. The grid spacing normal to the wall was $2.0\times10^{-5}c$ used.

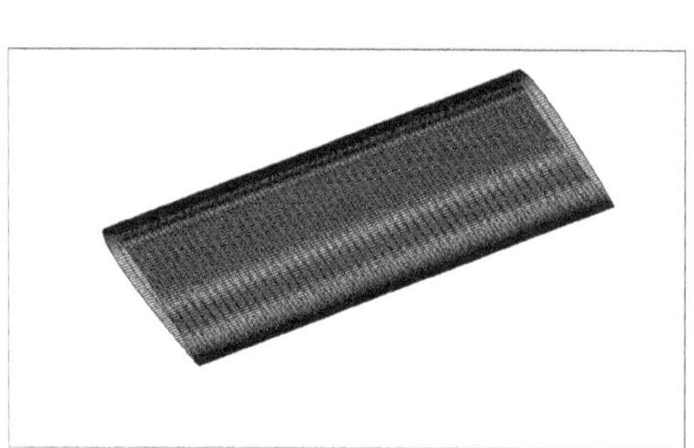

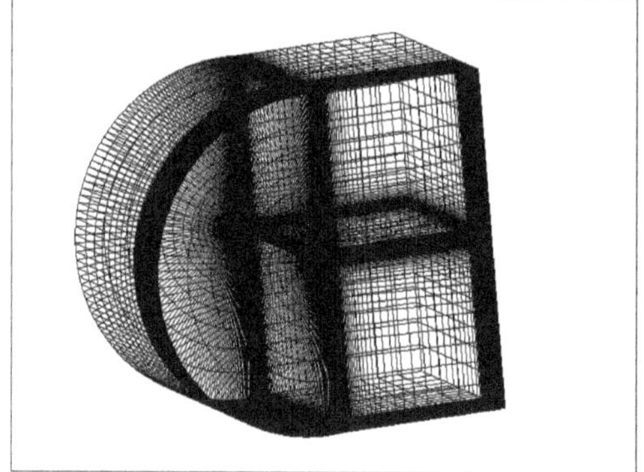

Fig. 1 (a) Surface grid (167×52) on the wing   Fig. 1 (b) C-H topology grid around the wing

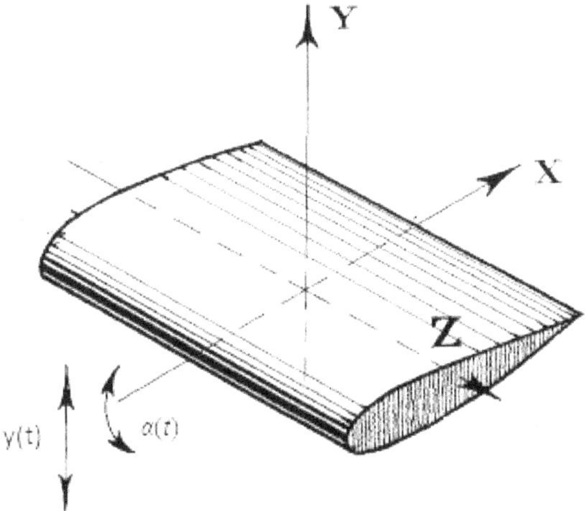

Fig. 2 Schematic diagram showing flapping motion

The sinusoidal motion of combined pitching and plunging wing is shown in Fig. 1 and is defined by the following expressions.

The motion of a pure plunge motion in normal direction is defined by

$$y\,(t^*) = h_a\,\sin\,(k\,t^*) \qquad\qquad .....(4)$$

and the plunging velocity is given by the following expression

$$\Box\,(t^*) = h_a k\,\cos\,(k\,t^*) \qquad\qquad .....(5)$$

The coupled pitching oscillation is defined as rotating about a pivot point on the aerofoil chord. The instantaneous angle measured clockwise from the mean chord is $\alpha\,(t^*)$ which is given by

$$\alpha\,(t^*) = \alpha_m + \alpha_o\,\sin\,(kt^* + \emptyset) \qquad\qquad .....(6)$$

where α_o is the amplitude of pitching oscillation, α_m is the mean angle of attack, k is the reduced frequency, h_a is non-dimensional heave amplitude y/c and $\emptyset$ is the phase angle ahead of the plunging motion.

The mean thrust coefficient and propulsion efficiency are computed using the following expressions. The mean or time-averaged thrust coefficient is defined as

$$C\Box_t = -\,C\Box_d + (C_d)_{\text{steady}} \qquad\qquad .....(7)$$

where $C\Box_d$ is the mean drag coefficient, averaged for one flapping period. $(C_d)_{\text{steady}}$ is the steady drag of the non-moving wing at its present angle of attack.

The % propulsion efficiency (η_{Prop}) can be calculated from the ratio between power output and power input. In this case, this is given by ratio of mean drag coefficient ($C\Box_d$) to mean input power coefficient. Where mean power input coefficient is calculated from the product of lift coefficient, C_l and plunging velocity, $\Box(t^*)$.

RESULTS AND DISCUSSION

The computations have been carried out for three-dimensional unsteady viscous flow over a combined pitching and plunging rectangular NACA0012 wing having aspect ratio of 4, for different reduced frequencies ($k = 0.5$ to $k = 1.0$) at Mach number of 0.3 and Reynolds number $= 1.0 \times 10^5$. For all the simulations, steady state solutions are first obtained. After steady state convergence is reached, the wing is then undergoes a prescribed sinusoidal motion, both pitching about half chord (i. e, a $= 1/2$) and plunging motion. Five consecutive cycles were computed to obtain periodic solution. The non-dimensional time step, $\Delta t = 0.005$ (approx.), the non-dimensional plunge amplitude of 1.0, the amplitude of pitching motion is 20° and the phase angle between the pitching and plunging motion 90° was used at different reduced frequencies.

The IMPRANS solver has been previously, validated for the unsteady flow over a flapping aerofoil with the experimental and numerical data available in the literature in Siva Kumar and Sharanappa (2011). The same solver has been used in the present computations.

Fig. 3 represents the instantaneous lift, thrust and drag coefficient versus instantaneous angle of attack for a combined pitching-plunging rectangular wing computed by using IMPRANS Solver at 0° mean angle of attack and at reduced frequency of 0.83. Similarly for the above coefficients are plotted versus plunging amplitude are shown in Fig. 4 respectively. The computed loops of the aerodynamic and thrust coefficients clearly demonstrate the hysteresis property existing between the up-stroke and down-stroke. The lift values are higher during down stroke than during up stroke. The thrust values are positive throughout the cycle of flapping oscillation.

The surface pressure fields on the upper and the lower surface of the wing for the same case, $k = 0.83$ are shown in Fig. 5 for 8 different instant of time while undergoing flapping motion for the final converged cycle. The corresponding vorticity plots on the upper and lower surface of the wing for the same instant are plotted in Fig. 6. From these field plots the variation of pressure and vorticity on the lower and upper surface can be observed for the complete cycle. Though there is not much change near the root of the wing as moving towards the tip, there is lot of variations in the pressure and vorticity distribution is observed.

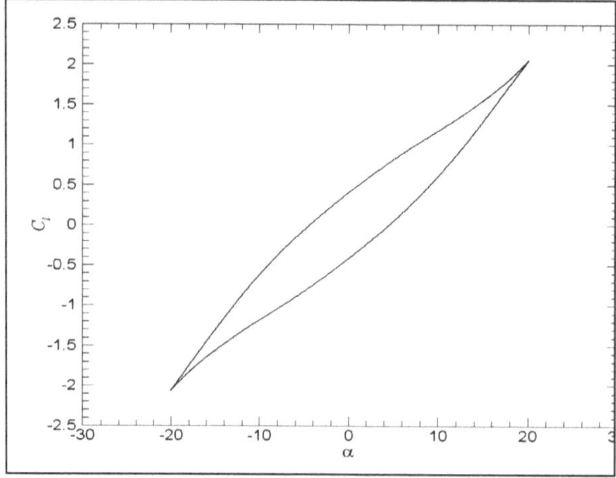

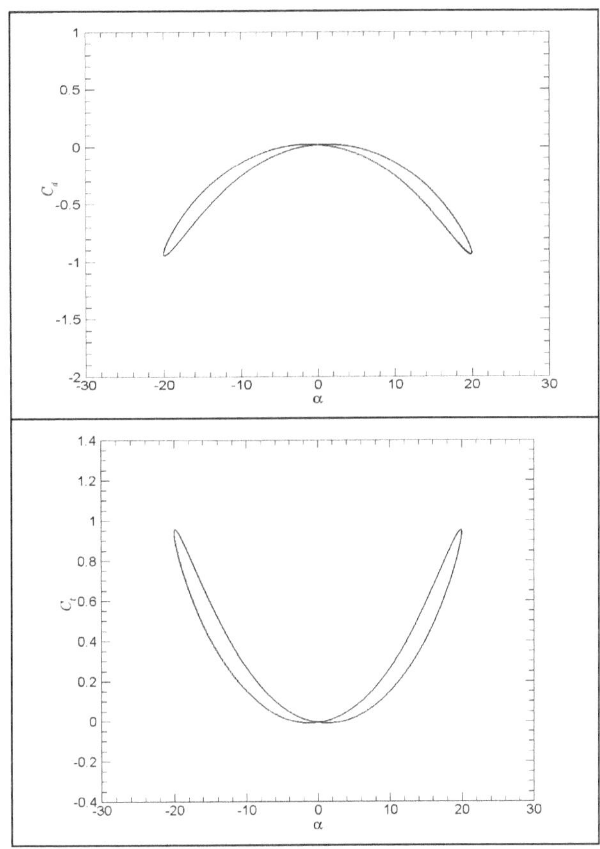

Fig. 3: The coefficient of lift, drag and thrust versus the angle of attack for the final cycle. (h_a = 1.0, α_a = 20°, a = 1/2, M = 0.3, k = 0.83, Φ = 90°)

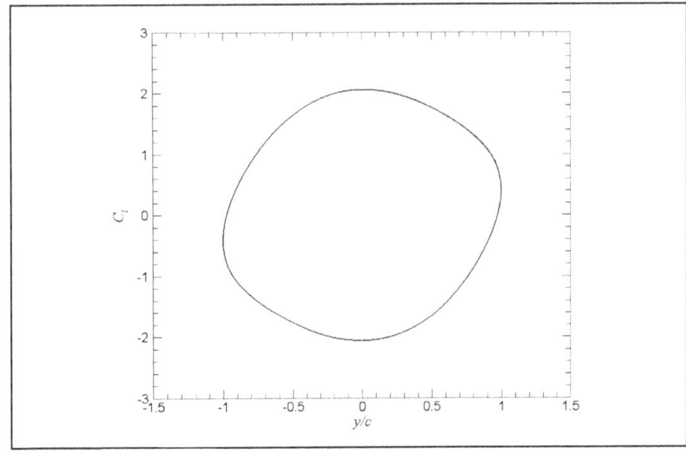

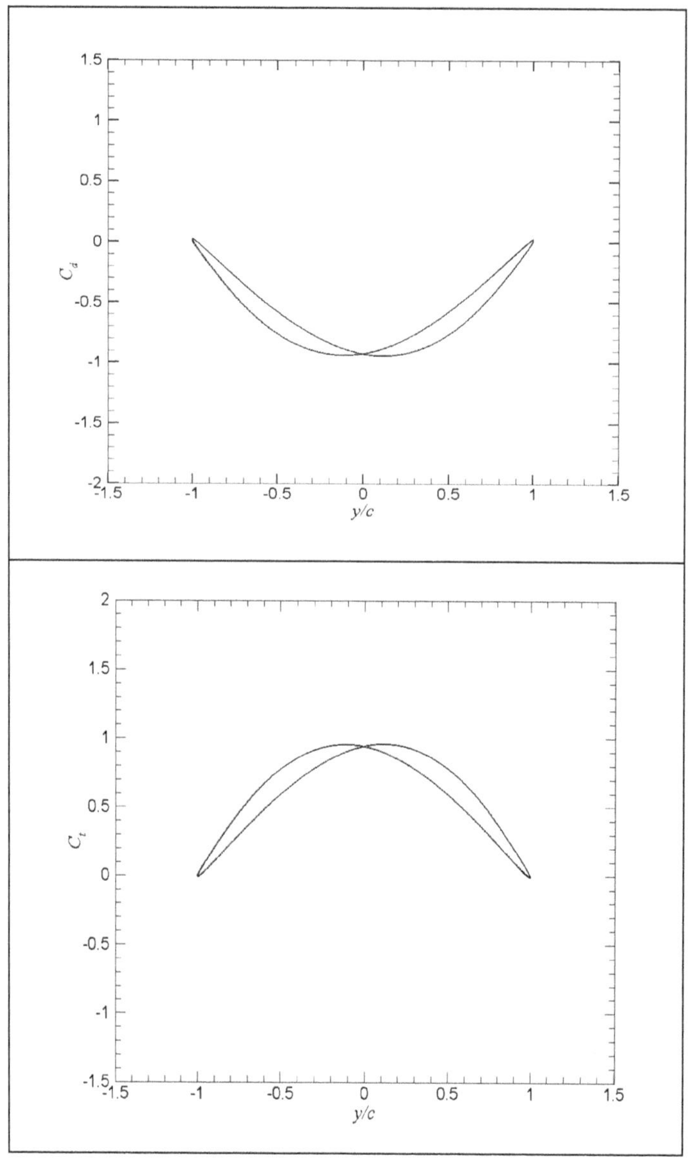

Fig. 4 The coefficient of lift, drag and thrust versus the heave distance for the final cycle.
($h_a = 1.0$, $\alpha_a = 20°$, $a = 1/2$, $M = 0.3$, $k = 0.83$, $\Phi = 90°$)

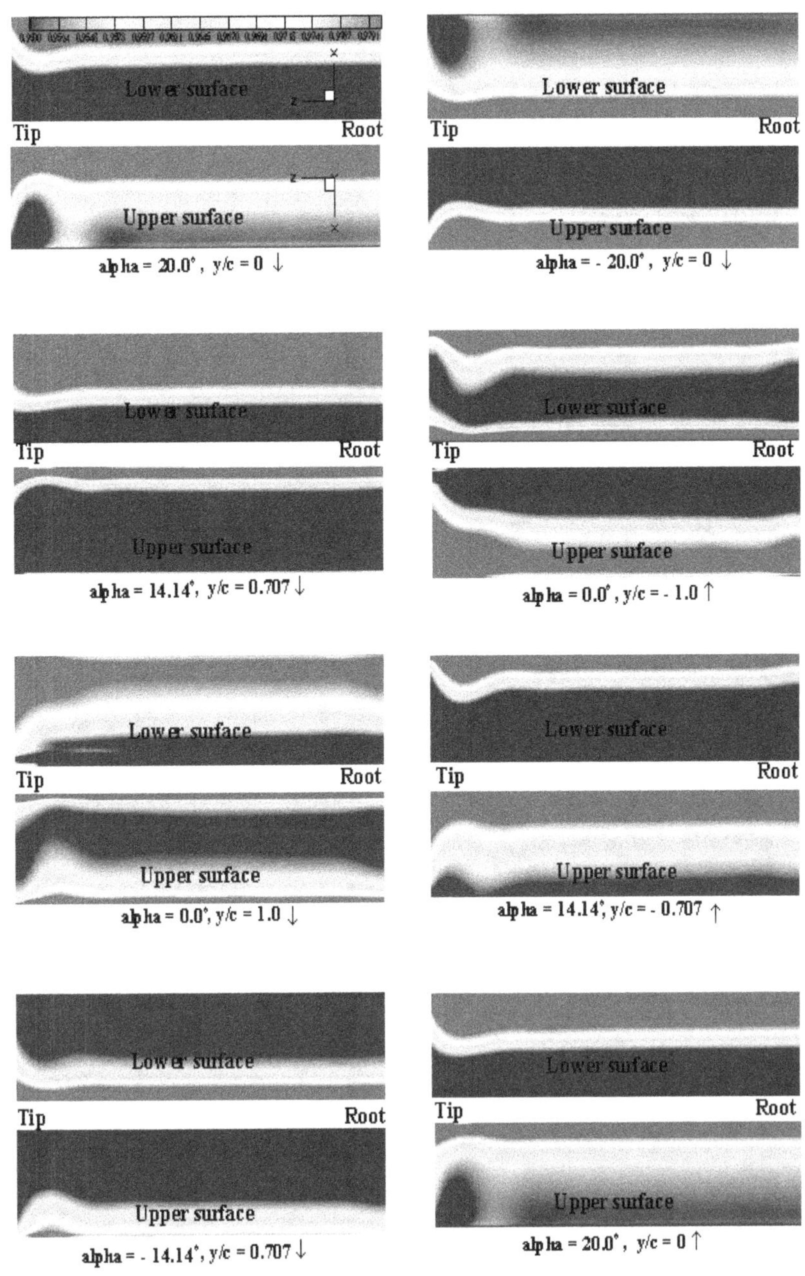

Fig. 5 Pressure fields on lower and upper surface of the wing at different instant of time for one complete flapping cycle of oscillation at $k = 0.83$

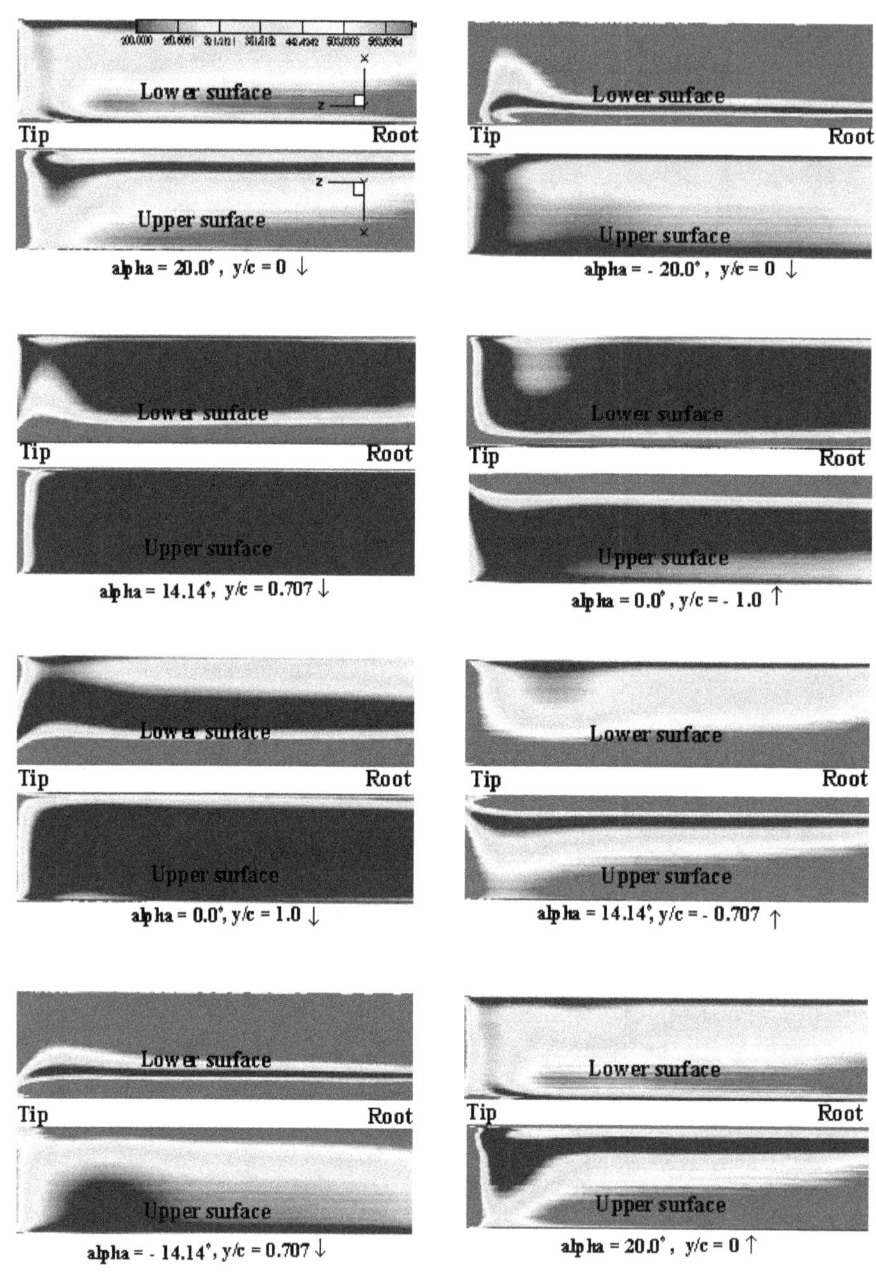

Fig. 6 Vorticity fields on lower and upper surface of the wing at different instant of time for one complete flapping cycle of oscillation at $k = 0.83$

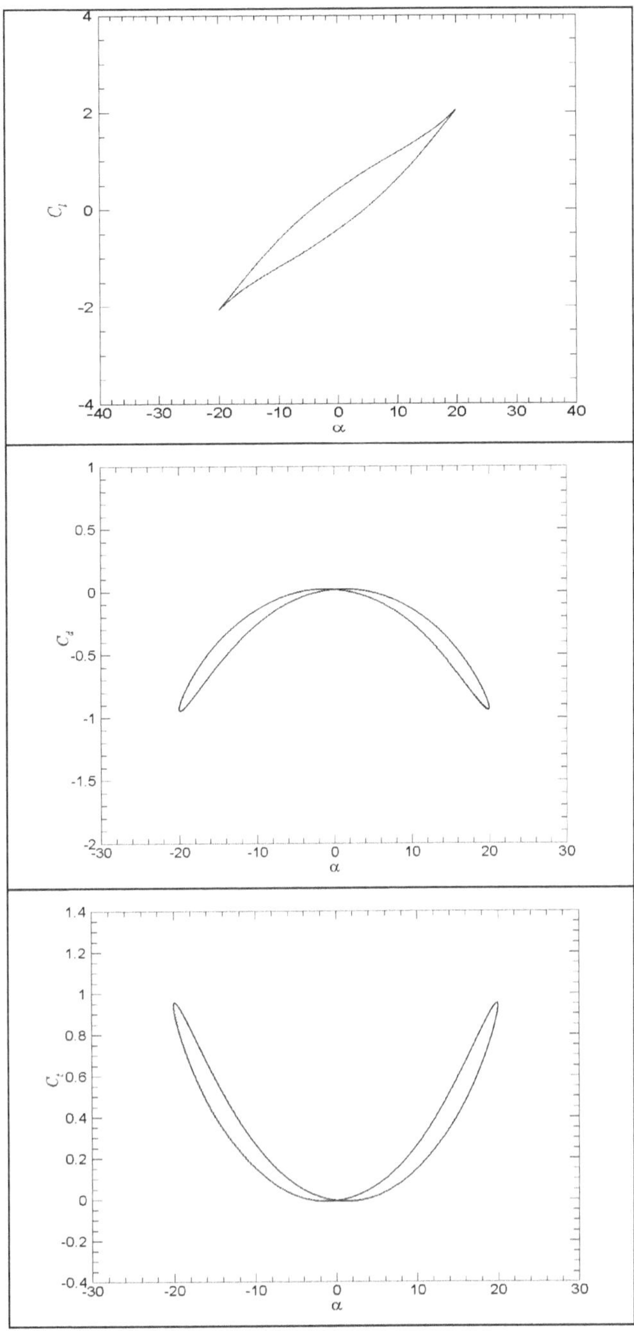

Fig. 7: The coefficient of lift, drag and thrust versus the angle of attack for   the final cycle.
($h_a = 1.0$, $\alpha_a = 20°$, $a = 1/2$, $M = 0.3$, $k = 1.0$, $\Phi = 90°$)

Similarly, the loops for lift, drag and thrust coefficients obtained at reduced frequency of $k = 1.0$ is shown in Fig. 7 for the complete converged cycle. Though the figures look similar to the one at k = 0.83 shown in Fig. 3. There is much difference in the area under the curve, which finally gives the values of thrust coefficient and the propulsive efficiency.

The mean-thrust coefficient and propulsive efficiency obtained for several cases from $k = 0.5$ to $k = 1.0$ are listed in Table 1, for Mach number = 0.5, Reynolds number = 1.0×10^5 and keeping other parameters constant (such as $h_0 = 1.0$, $\alpha_a = 20°$, $a = 1/2$, $\Phi = 90°$). From the values obtained, it can be observed that as the reduced frequency increases the thrust coefficient also increases and the propulsive efficiency decreases for range of reduced frequency values considered in the present computations. This can also be seen from the Figs. 8 and 9. In general, with the increase of reduced frequency the thrust coefficient increases but the propulsion efficiency decreases. As the flow becomes more and more unsteady with increasing reduced frequency, a large amount of vorticity is shed from the trailing edge, the thrust increases but the efficiency decreases.

Table 1 Computed Thrust and propulsion efficiency at different k values

Reduced Frequency, k	Mean-thrust coefficient, C_t	% Propulsion Efficiency, η_{Prop}
0.5	0.14076403	0.60435814
0.7	0.28639600	0.58953339
0.83	0.45177373	0.55293947
1.0	0.50836027	0.39588696

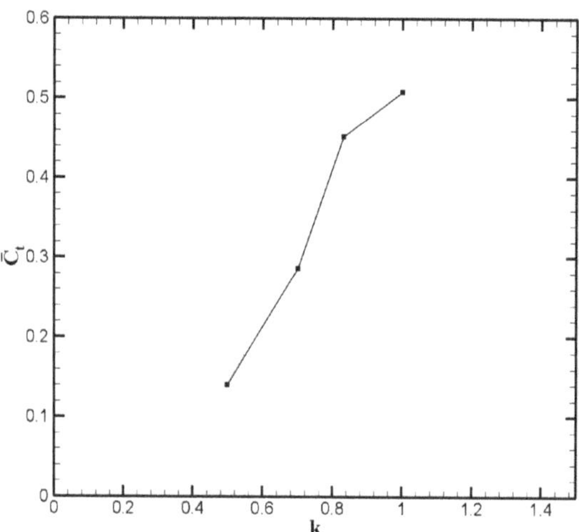

Fig. 8 Mean-thrust coefficient v/s reduced frequency for flapping wing

The simulations presented in this work may not be enough to conclude the effect of reduced frequency on the flapping wing. Further, work is being carried out at lower reduced frequencies starting with $k = 0.1$.

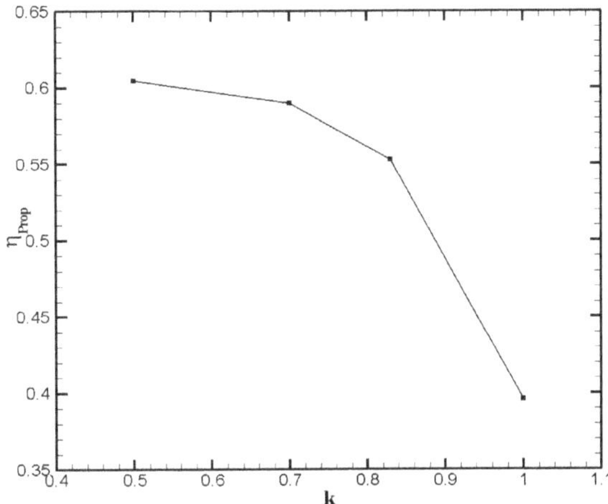

Fig. 9 Propulsion efficiency v/s reduced frequency for flapping wing.

CONCLUDING REMARKS

The three-dimensional unsteady viscous flow over a combined pitching and plunging rectangular wing has been computed using an implicit RANS solver IMPRANS. The effect of reduced frequency on the time-averaged thrust coefficient and propulsive efficiency has been studied. From the results it is observed that, as reduced frequency increases the time-averaged thrust coefficient increases and propulsion efficiency decreases for the range of values considered in the present computations. In all the cases higher thrust occurred at higher reduced frequency, while higher propulsive efficiency occurred at lower reduced frequency.

REFERENCES

[1]   Tikhomirov MK., 1937. "Flight of birds and machines with flapping wings", Moscow: ONTI.

[2]   Lippisch AM., 1960. "Man-powered flight in 1929". J R Aeronaut Soc; 64:3958.

[3]   Shyy, W., Berg, M. and Lyungvist, D., 1999. "Flapping and Flexible Wings for Biological and Micro Air Vehicles", Progress in Aerospace Sciences, Vol. 35, No. 5, pp. 455 - 505.

[4]   Mueller, T. J. (ed.) , 2001. "Fixed and Flapping Wing Aerodynamics for Micro Air Vehicles", Progress in Aeronautics and Astronautics, AIAA, Reston, VA, Vol. 195.

[5]   Hao Liu, Ellington, C. P. Kawachi Keiji, Van Den Berg and Willmott A.P, 1998. "A Computational Fluid Dynamic Study of Hawkmoth Hovering". Journal of Experimental Biology, 201, 461 - 477.

[6]   Dickinson, M.H, Lehmann, F.O., Sane, S.P, 1997. "Wing Rotation and the Aerodynamic Basis of Insect Flight".

[7]   Neef, M. F. and Hummel, D., June 2000. "Euler solutions for a finite-span apping wing", Tech. rep., Conference on fixed, flapping and rotary wing vehicles at very low Reynolds numbers, Univesity of Norte Dame, IN.

[8]   Dutta, P. K., Vimala Dutta and Sharanappa, 2003. "An Implicit RANS Solver for Unsteady Compressible Flow Computations", Proc. Seminar on State of the Art and Future Trends of CFD at NAL, NAL SP 0301, NAL, Bangalore.

[9]   Vimala Dutta, Sharanappa and P. K. Dutta, 2005. "Navier-Stokes Computations for a Helicopter Rotor Blade in Hover", Proc. Eighth Annual CFD Symposium, CFD Division of Aeronautical Society of India, Bangalore, August 11th - 13th , CP 18.

[10] Sharanappa, Vimala Dutta and Dutta, P. K., 2006. "Viscous Unsteady Flow around a Helicopter Rotor Blade in Forward Flight". Proc. Ninth Annual CFD Symposium, CFD Division of Aeronautical Society of India, Bangalore, August 11th - 12th , CP13.

[11] Dutta, P. K., Vimala Dutta and Sharanappa, 2007. "RANS Computation of Flow past Wind Turbine Blades", Proc. of 'The Seventh Asian Computational Fluid Dynamics Conference' (ACFD7), Bangalore, November 26th -30th , Paper 8.1, pp. 335 - 353.

[12] Pulliam, T. H., 1986. "Implicit Solution Methods in Computational Fluid Dynamics", App. Num. Math., (Trans. IMACS), 2, 6, pp. 441 - 474.

[13] Baldwin, B. S. and Lomax, H., 1978. "Thin Layer Approximation and Algebraic Model for Separated Turbulent Flows", AIAA Paper No. 78 - 257.

[14] Sharanappa V. Sajjan, Vimala Dutta and Dutta, P. K., 2008. "Numerical Simulation of flow over pitching bodies using an implicit Reynolds-averaged Navier-Stokes solver", Proc. of 12th Asian congress of Fluid Mechanics, Daejeon, Korea, August 18th - 21st.

[15] K. Siva Kumar and Sharanappa V. Sajjan, 10th - 14th January 2010. "Compu-tation of Unsteady Flow over a Plunging Aerofoil Using an Implicit Reynolds Averaged Navier-Stokes Solver", Proc. of 'The 8th Asian Computational Fluid Dynamics Conference (ACFD8)', Hong Kong.

[16] K. Siva Kumar and Sharanappa V. Sajjan, online available since October 2011. "Unsteady Compressible Flow over Heaving Bodies using an Implicit RANS Solver", International Journal on 'Applied Mechanics and Materials (AMM)', Vols. 110 - 116 (2012), pp 4589 - 4597.

[17] K. Siva Kumar and Sharanappa V. Sajjan, online available since October 2011. "Unsteady Flow past a Combined Pitching and Plunging Aerofoil using an Implicit RANS Solver", International Journal on 'Applied Mechanics and Materials (AMM)', Vols. 110 - 116 (2012), pp 3481 - 3488.

DESIGNING DEPENDABLE AGILE LAYERED SECURITY ARCHITECTURE SOLUTIONS – WEB SERVICES CASE STUDY

D.Vasumathi[1] D.Sravan Kumar[2] M.Upendra Kumar[3]
[1]Associate Professor CSE JNTU Hyderabad College of Engineering JNTUH A.P. India
vasukumar_devara@yahoo.co.in
[2]Principal and of Professor of CSE KITE WCPES Hyderabad A.P. India
dasojusravan@gmail.com
[3]Research Scholar CSE JNTU Hyderabad A.P. India
uppi_shravani@rediffmail.com

ABSTRACT

Our research entitled "Designing Dependable Agile Layered Security Architecture Solutions – Web Services case study", addresses the innovative idea and novel implementations of Security Engineering for Software Engineering using Agile Modeled Layered Security Architectures for Dependable Privacy requirements, with a validation of an exemplar case study of Web Services Security Architectures. Securing the Software Architecture in any application at design phase is known as Security Architectures, and we focus on authentication and authorization of the user. Now a day most of the applications are developed as a Layered Security Architecture pattern, typically we have user presentation layer, Business Logic layer and Database access layer. Now Agile Modeling is used in all applications design (but Agile Modeled architectures are given little importance) because of shortened development time, with customer collaborations with developers and importantly Test Driven development approaches. Securing Agile Modeled architectures, which being an iterative development, will provide enhanced Dependable Security Requirements in terms of Privacy of user, in its subsequent iterations. All this research paves a way for Secure Web Engineering. Finally we validate this research with an extended case study on simulated aircraft system design security.

KEYWORDS: Security Architectures, Agile Modeling, Layered Solutions design, Web Services, Web Engineering

RESEARCH METHODOLOGY ON DEGIGNING DEPENDABLE AGILE LAYERED SECURITY ARCHITECTURE SOLUTIONS – WEB SERVICES CASE STUDY

In Software Engineering terminology, a Qualitative or descriptive model will be built along with appropriate notation or tool for providing specific solution with a validation of a case study.

The Proposed Research work is for Integrating Security and Software Engineering with Design of Security Architectures. Software Engineering Problems must be treated by both theoretical and empirical methodologies. The former is characterized by abstract, inductive, mathematics-based, and formal-inference-centered studies; while the latter is characterized by concrete, deductive, data-based, and experimental-validation-centered studies. This research involves in theoretical designing of Secure UML diagrams using Agile Modeling. Also the results will be validated with experimental work on Web Services Security Architectures case study.

This research entitled "Designing Dependable Agile Layered Security Architecture Solutions" addresses the innovative idea and novel implementation of Security Engineering for *Software Engineering using Agile Modeled Layered Security Architecture for Dependable Privacy Management*, with a validation of an exemplar case study of Web Services Security Architecture. Security Architectures are those Software Architectures secured at the design phase in any application and authentication and authorization of the user have been focused. Most of the applications are developed as a Layered Security Architecture Pattern, typically have user presentation layer, Business Logic Layer and Database access layer. Model Driven Architecture based Agile modeling is used in all applications design (but Agile Modeled Architectures are given little importance) because of shortened developed time, with customer collaborations with developers and importantly with Test Driven Development approaches. Securing Model Driven Architecture based Agile Modeled architectures, which being an iterative development, will provide enhanced Dependable Security Management in terms of Privacy of user, in its successive iterations. This research paves way for Secure Web Engineering.

This research involves theoretical designing of Secure UML diagrams using Agile Modeling with experimental work on Web Services Security Architectures Case Study.

Web Service A Web Service is a method of communication between two electronic devices over a network. The World Wide Web Consortium (W3C) defines a "Web Service" as "a software system designed to support interoperable machine-to-machine interaction over a network. It has an interface described in a machine-processable format (specifically Web Services Description Language, known by the acronym WSDL). Other systems interact with the Web Service in a manner prescribed by its description using SOAP messages, typically conveyed using HTTP with an XML serialization in conjunction with other Web-related standards." The W3C also states, "We can identify two major classes of Web services, REST-compliant Web services, in which the primary purpose of the service is to manipulate XML representations of Web resources using a uniform set of "stateless" operations; and arbitrary Web services, in which the service may expose an arbitrary set of operations."

Web Services Security Development and Architecture: Theoretical and Practical issues, involves Web Services Security Engineering, *Web Services Security Architecture*, Web Services Security Standards, Web Services Security Threats and Countermeasures [5]. Web Services Security Engineering implies, Security Engineering integrated into software development which is one of the major topics developed during the last few years. Applying security engineering throughout the different steps devised by the different software development methodologies has been a major topic in both scientific and industrial literature. *Web Services Security Architecture* should define the highest level organization of the IT security infrastructure necessary to meet the security requirements specified for the systems to be built by articulating the necessary security mechanisms in such a way that reusability, manageability and (internal/external) interoperability is guaranteed. The Web Services Security Architecture, as per National Institute of Science and Technology (NIST) is a layered architecture consisting of Web Service Layer, Web Services Framework Layer and Web Server Layer [3]. The goal of the Web Services Security Architecture is to summarize out the details of message level security from the mainstream business logic [24] In the Web Services Secure application design, authentication and authorization are important research issues, pertaining to Security Architecture. Even though Web Services are existing from the year 2004 onwards, Web 2.0 had made Web as a platform, with mashup applications from the year 2009. This Web 2.0 Services Security needs to be investigated for research [11, 30] Moreover extension of these Web 2.0 Services applications in terms of Spatial Web Services Security needs to be investigated for research [15, 20]

Designing Dependable Solutions Designing Secure Solutions implies that, the task of developing Information Technology solutions that consistently and effectively apply security principles has many challenges including: the complexity of integrating the specified security functions within the several underlying component architecture found in computing systems, the difficulty in developing a comprehensive set of baseline requirements for security, and a widely accepted security design methods [14]. Dependability implies privacy management of the application [9, 27]. Securing the Software application in any application at the design phase is known as Security Architecture, with a focus on authentication and authorization. Now a days, most of the applications are developed as a Layered Security Architecture pattern, typically having layers like User presentation layer, Business Logic layer and Database access layer. Today Agile Modeling (like Test Driven Development) is used in all Web applications design (our focus on Web Services), because of shortened development time, with customers collaborations with developers (pairs) [2, 4, 6, 7]. Unfortunately Agile Modeled architecture is given less importance in literature because of quick development schedule, this research focuses on Secure Agile architecture for web services. Agile Modeling, being an iterative development approach, securing its architecture will provide Privacy information of the user, in the subsequent iterations. Our security approach is based on Model Driven Architecture (MDA) based Agile Security Modeling for Web Services. [10, 13, 25, 28, 29]

INTRODUCTION

Software Architecture: An important design artifact in any software development project, with the possible exception of very small projects, is the Software Architecture. An important part of any architecture is the set of Architectural Design Rules. Architectural Design Rules are defined as the rules, specified by the architect(s) that need to be followed in the detailed design of the system. A primary role of the architecture is to capture the architectural design decisions. An important part of these design decisions consists of architectural design rules [1].

Security: Security ensures that information is provided only to those users who are authorized to possess the information. Security generally includes the following:

Identification: This assumes that system must check whether a user really is whom he or she claims to be. There are many techniques for identification and it is also called as authentication. The most widely used is "Username/Password" approach. More sophisticated techniques based on biometrical data are like retinal fingerprint scan.

Authorization: This means that the system should provide only the information that the user is authorized for, and prevent access to any other information. Authorization usually assumes defining "user access rights", which are settings that define to which operations, data, or features of the system the user, does have access.

Encryption: This transforms information so that unauthorized users (who intentionally or accidentally come into its possession) cannot recognize it [11].

MDA: Model-Driven Development (MDD) is a modeling approach. The basic premise of Model-Driven Development is to capture all important design information in a set of formal or semiformal models, which are kept consistent automatically. To realize full benefits of MDD, formalize architecture design rules, which then allow automatic enforcement of architecture on the system model. There exist several approaches to MDD, such as OMG's (Object Management Group) MDA (Model-Driven Architecture), Domain Specific Modeling (DSM), and Software factories fro Microsoft. Model-Driven Architecture prescribes that three models or sets of models shall be developed as:

The Computationally Independent Model(s) (CIM) captures the requirements of the system.

The Platform-Independent Model(s) (PIM) captures the systems functionality without considering any particular execution platform.

The Platform-Specific Model(s) (PSM) combines the specifications in the PIM with the details that specify how the system uses a particular type of platform. The PSM is a transformation of the PIM using a mapping either on the type level or at the instance level.

MDA does not directly address architectural design or how to represent the architecture, but the architecture has to be captured in the PIM or in the mapping since the CIM captures the requirements and the PSM is generated from the PIM using the mapping [1].

Agile Methods: Over the past few years, a new family of software engineering methods has started to gain acceptance amongst the software development community. These methods, collectively called Agile Methods, conform to the Agile Manifesto, which states "We are uncovering better ways of developing software by doing it and helping others does it. Through this work we have come to value: Individuals and interactions over processes and tools working software over comprehensive documentation customer collaboration over contract negotiation responding to change over following a plan That is, while there is value in the items on the right, we value the items on the left more." The individual agile methods include Extreme Programming (XP), Scrum, Lean Software Development, Crystal Methodologies, Feature Driven Development (FDD), and Dynamic Systems Development Methodology (DSDM). While there are many differences between these methodologies, they are based on some common principles, such as short development iterations, minimal design upfront, emergent design and architecture, collective code ownership and ability for anyone to change any part of the code, direct communication and minimal or no documentation (the code is the documentation), and gradual building of test cases. Some of these practices are in direct conflict with secure SDLC processes [2].

Security Requirements: Agile information systems and software methods are characterized by nimbleness to rapid changes, multiple incremental iterations and a fast development pace. Agile development is defined as a set of principles and practices that differs as a whole from traditional planned development. The major principles for agile information systems and software methods include [9]:

Accept multiple valid approaches: A stable architecture, a tool orientation and component based development combine to enable a "fluid view" of methodology and the value of tailoring the methodology for each development project. Improvisation in development approach will help match the methodology to the constraints of the project environment.

Engage the customer: Close involvement of customers in the project enables accurate and fast requirements elicitation, and the customers again immediate satisfaction as their ideas and requirements arise in each new release.

Accommodate requirements change: Agility means that developers quickly and easily respond to the shifting requirements driven by the changing environment for which the software is intended.

Build on successful experience: The "right" people are important for project success in order to foster innovation in software development. Courage, specific knowledge, intelligence, and commitment are needed for agile development.

Develop good teamwork: The right mix of people operating with the right process framework means that the right mix of knowledge and working style will be present in the project. Agile development teams must often come together quickly and be immediately effective.

Agile practices include:

Develop in parallel: Releases may be completely developed in parallel, or staged onto the market such that design, development, and quality assurance are all taking place simultaneously, but sequentially on different releases. Coding may even begin before the requirements are declared.

Release more often: Releases are scoped to more frequently deliver small sets of new features and fixes. Constant re-prioritization of features enables responsiveness to changing requirements and enables features to easily slip from one release to the next.

Depend on tools: Heavy use of development tools and environment that speed up the design and coding process offer much of the functionality that used to be custom built. Ideally, agile developers try to avoid wasting time repetitively building features others have already developed.

Implant customers in the development environment: Fast and intimate access to customer views and opinions slashes time, and ensures the high-priority features are built first. When customers participate closely in all phases of development, cycle times shorten and teams can better chuck requirements into logical releases from customer views.

Establish a stable architecture: This anchors a rapid development process that is never quite stable, yet each release has some similarity and components reuse.

Assemble and reuse components: Never unnecessarily build software from scratch when it can be assembled from existing components. It is quicker and equally effective to acquire, integrate, and assemble components with wrappers, including business logic software, interfaces and back-end infrastructure.

Ignore maintenance: Building components for short life spans eliminates the need for documentation. Assembled software can be thrown away and reassembled with greater ease than maintaining complex and custom-build components.

Tailor the methodology daily: Operating with an overall development framework, but allowing project teams to adjust the exact approach to the daily situation, enabled teams to meet intense demands for speed by skipping unnecessary tasks or phases. Use just enough process to be effective, and no more.

Security requirements for Agile Security methods and Extant Security methods:

Requirements for security methods that are targeted to be integrated into agile software methods:

The security approach must be adaptive to agile software development methods.

They must be simple; they should not hinder to the development project.

The security approach, in order to be integrated successfully with agile development methods, should offer concrete guidance and tools at all phases of development (i.e., from requirements capture to testing).

A successful security component should be able to adapt rapidly to ever changing requirements owing to a fast-paced business environment, including support for handling several incremental iterations [10].

Key Security Elements in Agile Software Development:

The key security element stems from information security "meta-notation", or notation for notations, and database security. Apply these key security elements to a process aimed at developing secure software in an agile manner. This generic security process consists of these key security elements in different phases of software development (requirements analysis, design, implementation and testing). These steps are not necessarily sequential and in any case, every step is optional [9].

OBJECTIVES OF THE RESEARCH WORK

Software Engineering covers the definition of processes, techniques and models suitable for its environment to guarantee quality of results. An important design artifact in any software development project is the Software Architecture. Software Architecture's important part is the set of architectural design rules. A primary goal of the architecture is to capture the architecture design decisions. An important part of these design decisions consists of architectural design rules. In an MDA (Model-Driven Architecture) context, the design of the system architecture is captured in the models of the system. MDA is known to be layered approach for modeling the architectural design rules and uses design patterns to improve the quality of software system. And to include the security to the software system, security patterns are introduced that offer security at the architectural level. More over, agile software development methods are used to build secure systems. There are different methods defined in agile development as extreme programming (XP), scrum, feature driven development (FDD), test driven development (TDD), etc. Agile processing is includes the phases as agile analysis, agile design and agile testing. These phases are defined in layers of MDA to provide security at the modeling level which ensures that "security at the system architecture stage will improve the requirements for that system".

Problem Statement

This research entitled "Designing Dependable Agile Layered Security Architecture Solutions" addresses the innovative idea of Security Engineering for Software Engineering using Agile Modeled Layered Security Architectures for Dependable Privacy Requirements, with a validation of Case study of Web Services Security Architectures.

Research Questions Addressed:

1. How a failure addresses a specific security service at a specific layer impact other (interdependent) layers? Also how successful implementation of a security service had an affect on the rest of the system?

2. How can agile methods be used to generate effective security requirements? In what ways do these agile methods change the development of security requirements? How is the outcome of emergent security development different from more traditional forms?

3. web services security architectures

Web Services Security Engineering through Web Services Security Architecture

Problem Definition

This research entitled "Designing Dependable Web Services Security Architecture Solutions" addresses the innovative idea of Web Services Security Engineering using Web Services Security Architecture with a research motivation of Secure Service Oriented Analysis and Design. It deals with Web Services Security Architecture for Web Services Secure application design, for Authentication and authorization, using Model Driven Architecture (MDA) based, Agile Modeled Layered Security Architecture design, which eventually results in enhanced dependable (privacy) management. All the above findings are validated with appropriate case studies of Web 2.0 Services, its extension to Web 2.0 Mashups Spatial Web Services and various financial applications.

Research Questions addressed

1. How can Model Driven Architecture (MDA) based, Agile Modeled Layered Security Architecture design be used for Web Services Security Architecture, with a motivation of Dependable Privacy Management?

2. How can we extend the above approach for Web 2.0 Services Security Architecture?

3. How can we validate this approach for Web 2.0 Mashup Spatial Web Services Security Architecture case study?

ORGANIZATION OF THESIS

The research has been done in various stages and thesis is organized into six chapters.

In Chapter 1, introduction to Secure Software Engineering, Security Architectures Design and Development, Introduction to research title, software architecture security using Model Driven Architecture, Agile Methods, Case study Web Services Security Architectures are discussed so that the problem statement can be designed. In Chapter 1, introduction to Web Services Security Architecture Design and Development, Introduction and overview of research title, Software Architecture security using Model Driven Architecture, Agile Methods are discussed so that the problem statement can be designed.

In Chapter 2, a detailed literature survey was conducted on Secure Software Engineering, Model Driven Architecture, Agile Methodology, Security patterns for Agile Layered Security Architecture, UML 2.0, and Secure UML, Web Services Security Architecture to find out basis for the thesis. In Chapter 2, a detailed literature survey was conducted on Web Services Security

Architecture, Model Driven Architecture, Agile Methodology, Security patterns for Agile Layered Security Architecture, UML 2.0, and Secure UML to find out basis for the thesis.

In Chapter 3, we design Agile Modeled Layered Security Architectures, with a validation of case study for Web Services Security Architectures, with initial case study validations using on simple secure Web Services Design using Agile Modeled Test Driven Development. In Chapter 3. We discuss design of Model Driven Architecture (MDA) based Agile Modeled Layered Security Architecture, (for Web Services), with initial case study validations using on simple secure Web Services Design using Agile Modeled Test Driven Development. Initially we discuss about Agile Security Architecture: Software Engineering covers the definition of processes, techniques and models suitable for its environment to guarantee quality of results. An important design artifact in any software development project is the Software Architecture. Software Architecture's important part is the set of architectural design rules. A primary goal of the architecture is to capture the architecture design decisions. An important part of these design decisions consists of architectural design rules. In an MDA (Model-Driven Architecture) context, the design of the system architecture is captured in the models of the system. MDA is known to be layered approach for modeling the architectural design rules and uses design patterns to improve the quality of software system. And to include the security to the software system, security patterns are introduced that offer security at the architectural level. More over, agile software development methods are used to build secure systems. There are different methods defined in agile development as extreme programming (XP), Test Driven Development (TDD), Lean development, Scrum, Feature Driven Development (FDD) etc. Agile processing includes the phases like agile analysis, agile design and agile testing. These phases are defined in layers of MDA to provide security at the modeling level which ensures that "security at the system architecture stage will improve the privacy requirements for that system". Later on we extend this approach for Web Services, with initial case study validations using on simple secure Web Services Design using Agile Modeled Test Driven Development.

In Chapter 4, Designing Solutions using Agile Modeling for Layered Security Architectures with case study of Web 2.0 Services Security Architectures and its implementations are discussed. In Chapter 4, Designing Solutions using Agile Modeling for Web 2.0 Services Security Architecture and its implementations are discussed. Web 2.0 increases web based access to data processing particularly on the client side (AJAX Asynchronous Java Script and XML) that enables web applications which contains enriched functionality. Web 2.0 technologies have wide range of technologies and protocols which enables Web architecture to have greater access to data and functions. Traditional enterprises are skeptical in adopting Web 2.0 applications for internal and commercial use in public facing situations, with customers and partners. One of the prime concerns for this is lack of security over public networks. This chapter discusses and implements design of Web 2.0 services security architectures, for authentication over SSL/TLS.

In Chapter 5, Dependability (Privacy Requirements) for Agile Modeled Layered Security Architectures with case study of Web Services Security Architectures is discussed with its implementations of a financial application for Secure Stock Market. In Chapter 5, Dependability (Privacy Management) for Web Services Security Architecture is discussed with its implementations of a financial application for Secure Stock Market. Privacy is today an important concern for citizens, organizations and companies. We see an increasing number of organizations that collect data, very often concerning individuals, and use them for various purposes, ranging from scientific research, as in the medical data, to demographic trend analysis and marketing.

Organizations may also give access to the data they own or even release such data to third parties, the number of increased data sets that are thus available poses serious threats to the privacy of individuals and organizations. To address such concerns, several privacy techniques have been developed. Despite such a large body of work, privacy issues specific to Web Services have not been yet investigated. A very preliminary effort is represented by the identification of privacy requirements, as part of a larger set of Web Services Architecture Requirements, by a working model of World Wide Web Consortium. (W3C).

In Chapter 6, the results of the research are summarized and suggested the work for further research. In Chapter 6, a case study on Web 2.0 mashup spatial Web Services Security Architecture is carried for validating the research results. Role Based Access Control for Spatial Web Services implies that: RBAC model is a widely deployed model in commercial systems and for which a standard had been developed. The widespread deployment of location-based services and mobile applications, as well as the increased concern for the management and sharing of geographical information in strategic applications like environmental protection and homeland security has resulted in a strong demand for spatially aware access control systems. These application domains impose interesting requirements on access control systems. In particular, the permissions assigned to users depend on their position in a reference space; users often belong to well-defined categories; objects to which permissions must be granted is located in that space; and access control policies must grant permissions based on locations and user positions.   finally, the results of the research are summarized and suggested the work for further research.

Designing Dependable Web Services Security Architecture Solutions

This research on Web Services Security Architecture is done using an innovative idea and novel implementations, of design of Model Driven Architecture (MDA) based Agile Modeling, for authentication and authorization of Web Service secure application design, for privacy management. We had validated our research with implementations on Web 2.0 Services Security Design, with its extension to Web 2.0 Mashup Spatial application, and various financial applications case studies.

REFERENCES

[1] Anders Mattsson, Bjorm Lundell, Brian Lings, and Brian Fitzgerald, (2009) "Linking Model-Driven Development and Software Architecture: A Case Study", 2009, IEEE Transactions on Software Engineering, vol. 35, no. 1.

[2] "Real-time agility, the Harmony/ESW Method for Real-time and embedded Systems Development"

[3] "Design Approaches", Agile open source.

[4] Hossein keramati, Seyed-Hassan Mirian-Hosseinabadi,   "Integrating Software Development Security Activities with Agile Methodologies", 2008, IEEE.

[5] I. Lazar, B. Parv, S. Motogna, I.-G. Czibula, C.-L. Lazar, "An Agile MDA approach for Executable UML Structured Activities", Studia Univ. Bases, vol. LII, No. 2, 2007.

[6] Yann-Gael Gueheneuc, Giuliano Antoniol, "DeMIMA: A Multilayered Approach for Design Pattern Identification", 2008, IEEE Transactions on Software Engineering, vol. 34, no. 5.

[7]   Spyros T. Halkidis, Nikolaos Tsantalis, Alexander Chatzigeorgiou, George Stephanides, "Architectural Risk Analysis of Software Systems Based on Security Patterns", 2008, IEEE Transactions on dependable and secure computing, vol. 5, no. 3.

[8]   Erich Gamma, "Design Patterns".

[9]   M. Siponen, R. Baserville, T. Kuivalainen, "Extending Security in Agile Software Development Methods", pp 143-157.

[10] Johan Peeters, "Agile Security Requirements Engineering".

[11] Athula Ginge and San Murugesan, "Web Engineering: A Methodology for Developing Scalable, Maintainable Web Applications", Cutter IT Journal Vol.14, No.7 pp. 24-35, July 2001

[12] Cenzic Inc., "Web Application Security Trend Reports", 2009.

[13] David Geer, "Taking Steps to Secure Web Services", IEEE, October 2003.

[14] D.K.Smetters, R.E.Grinter, "Moving from the design of usable security technologies to the design of useful secure applications", ACM New Security paradigms workshop September 2002 pp 82 – 89

[15] Durai Pandian M et.al., "Information Security Architecture – Context aware Access control model for Educational applications", International Journal of Computer Science and Network Security, December 2006

[16] Ferda Tartanoglu et al, "Dependability in the Web Services Architecture", Architecting Dependable Systems, LNCS 2677, pp. 90 – 109, 2003

[17] Gunnar Peterson, "Security Architecture Blueprint", Arctec Group, LLC, 2007

[18] Halvard Skogsrud," Modeling Trust Negotiation for Web Services", IEEE February 2009

[19] Heiko Tillwick, Martin S Olivier, "A Layered Security Architecture: Design Issues", in Proceedings of the Fourth Annual Information Security South Africa Conference (ISSA 2004), July 2004.

[20] Jim Highsmith, Alistair Cockburn "Agile Software Development: The Business of Innovation", IEEE Computer September'2001 pp: 120:122

[21] J.J.Whitmore,"A method for designing secure solutions", IBM systems Journal, Vol 40 No 3 2001 pp. 747-768

[22] John Hunt, "Agile Software Construction", Springer Verlag publishers 2006

[23] Lorenzo D Martino, Elisa Bertino, " Security for Web Services: Standards and Research Issues", International Journal of Web Services Research, Oct-Dec 2009, pp. 48-74, Idea Group Publishing USA 2009

[24] Mark Harman, Afshin Mansouri,"Search based Software Engineering: Introduction to the special issue of the IEEE Transactions on Software Engineering", November December 2010, pp. 737 – 741

[25] Martin Naedele, "Standards for XML and Web Services Security", IEEE April 2003

[26] Massimo Barloletti, et. al." Semantics-Based Design for Secure Web Services", IEEE Transactions on Software Engineering, Vol 34, No.1, January 2008

[27] Matt Bishop, "Computer Security: Art and Science", Pearson Education, 2003

[28] NIST Draft, " Guide to Secure Web Services", September 2006

[29] Ross Anderson," Security Engineering: A guide to building Dependable Distributed Systems", Wiley publishers, 2003.

[30] Satoshi Makino, Takeshi Imamura, Yuichi Nakamura. "Implementation and Performance of WS-Security", International Journal of Web Services Research, Jan-March 2004, pp. 58-72, Idea Group Publishing USA 2004

[31] Sasikanth Avancha, "A Framework for Trustworthy Service Oriented Computing", ICISS 2008, pp. 124 – 132.

[32] Sandeep Chatterjee," Developing Enterprises Web Services an Architects Guide", Pearson, 2004

[33] Sarah Spiekermann, Lorrie Cranor,"Engineering Privacy", IEEE Transactions on Software Engineering", Vol 35 No 1 January February 2009 pp. 67 – 82

[34] Spyros T Halkidis et. al., "Architecture Risk Analysis of Software Systems based on Security Patterns", IEEE Transactions on Dependable and Secure Computing Vol 5 No. 3, July – September 2008, pp. 129 – 142

[35] Vipul Gupta, et. al., "Sizzle: A standards-based end-to-end security architecture for the embedded Internet", Elsevier, Pervasive and Mobile Computing, 2005

[36] Wei She, et. al. ,"Enhancing Security Modeling for Web Services using Delegation and Pass-on", International Journal of Web Services Research, Jan-March 2010, pp. 1-21, Idea Group Publishing USA 2010

[37] Wembo Mao, "Modern Cryptography: Theory and Practice", Pearson education, 2004

SMART PHONE CONTROLLED MINIATURE UNMANNED VEHICLES

T.Gajapriya* Dr.K.Senthilkumar*
*Division of Avionics, Department of Aerospace Engineering,
Madras Institute of Technology Campus,
Anna University. Chennai-600044. India.
Email:gajamuga1@yahoo.co.in,ksk_mit@rediff.com

ABSTRACT

The Use of Unmanned Aerial Vehicles (UAVs) which can operate autonomously in dynamic and complex operational environments is becoming increasingly more common. The Basic idea behind the project is to control Unmanned Vehicles. Unmanned Aerial and Ground vehicles are immensely useful in a variety of Surveillance & rescue applications. Vehicle's Complementary strengths provide operating teams with enhanced mission capabilities. While many of today's systems require independent control stations, necessitating arduous manual co-ordination between multiple operators, Multi-Robot Collaboration System can be used as an alternative system. Unmanned Aerial Vehicles (UAVs) can be used to cover large areas searching for targets. However, Sensors on UAVs are typically limited in their accuracy of localization of targets on the ground. To overcome this, Unmanned Ground Vehicles (UGVs) can be deployed to accurately locate ground targets. For the easy user accessibility, Smart Phone can be used to control both the vehicles UAV and UGV respectively. Unmanned Ground Vehicles have been implemented by the military to replace or assist soldiers in dangerous or unreachable environments. The Purpose of this project is to design a UGV and UAV, separately controlled by a Smart Phone. In most of the applications, it is mandatory to use a lap top or a tough book for the control of vehicles. In this design, Smart Phones are added for the manual intervention to promote multiple users.

Keywords: Unmanned Aerial Vehicle (UAV), Unmanned Ground Vehicle (UGV)

INTRODUCTION

Unmanned systems such as Unmanned Aerial Vehicles (UAV) and Unmanned Ground Vehicles (UGV) do high risk, mundane and dirty jobs in the complex battlefield environments of today and will be relied upon heavily on the battlefields of the future.

Unmanned Aerial Vehicles are predominantly deployed for military applications, but also used in a small but growing number of civil applications, such as firefighting and nonmilitary security work, such as surveillance of pipelines. UAVs are often preferred for missions that are too 'dull, dirty, or dangerous' for manned aircraft. Unmanned air vehicles are nowadays seen as an area of great importance in the aerospace industry. An essential part of these aeronautical systems is the ground control station or GCS. This is the unit on the ground that sends and receives signals from one or several airborne units. These are normally very complicated systems that require many personnel and a lot of computing power. So, as a result, Ground Control Station requires vast human intervention to navigate its flight path. To handle this, We can use Smart Phone as an alternative for Ground Control Station to have the portraits of landscape.

Unmanned Ground Vehicle is a vehicle that operates while in contact with the ground and without an onboard human presence. UGVs can be used for many applications where it may be inconvenient, dangerous, or impossible to have a human operator present. Generally, the vehicle will have a set of sensors to observe the environment, and will either autonomously make decisions about its behavior or pass the information to a human operator at a different location who will control the vehicle through tele-operation. The UGV is the land-based counterpart to Unmanned Aerial Vehicles and Remotely Operated Underwater Vehicles.

A remote-operated UGV is a vehicle that is controlled by a human operator via a communications link. All actions are determined by the operator based upon either direct visual observation or remote use of sensors such as digital video cameras. So, as Remote Control for UGV, We can use Smart Phone for easy user accessibility. Because, Smart Phones are ubiquitous, unobtrusive, intimate, sensor rich, computationally powerful, and remotely accessible.

On the other hand, Collaboration of Unmanned Aerial Vehicle, Unmanned Ground Vehicle and Smart Phone is planned, necessitating Multi-Robot Collaboration System as an alternative for arduous manual co-ordination between multiple operators, operating independent control stations.

MOTIVATION FOR PROPOSED WORK

U.S. and allied combat operations continue to highlight the value of unmanned systems in the modern combat environment. Combatant Commanders (CCDRs) and War fighters value the inherent features of unmanned systems, especially their persistence, versatility, and reduced risk to human life. The U.S. military Services are fielding these systems in rapidly increasing numbers across all domains: air, ground, and maritime. Unmanned systems provide diverse capabilities to the joint commander to conduct operations across the range of military operations: Environmental sensing and Battle space awareness; chemical, biological, radiological, and nuclear (CBRN) detection; counter-improvised explosive device (C-IED) capabilities; port security precision targeting; and precision strike. Furthermore, the capabilities provided by these unmanned systems continue to expand.

The Future UAV Mission Scenarios will be in following categories such as Stratospheric Telecommunications Airship, High-Altitude Imagery, Border Patrol, Maritime Surveillance, Environmental Sensing, Media and Traffic Reporting and Tactical Law Enforcement.

Unmanned Aerial Vehicles are an exciting field in the world of aviation, with new discoveries and proposed uses being documented daily. Over the next 16 years, UAVs will become a significant component of military, civil, and perhaps even commercial aviation. However, the very dynamic nature of the field also creates a significant amount of uncertainty. The wide range of UAV physical and performance characteristics, many of which will be very unlike any current aircraft, will place additional challenges on an air traffic management system already under great strain. However, many of the new paradigms being considered for the future NAS will likely facilitate the routine and safe entry of UAV operations into civil airspace. The information management system, through shared situational awareness, will allow all users of the NAS to know the location and intent of other aircraft (both manned and unmanned). The data provided by the system will also be a vital component to the functioning of autonomous systems embedded in UAVs and other advanced, data-dependent aircraft of the future. 4-D navigation and control will allow properly equipped UAVs to file 4-D flight plans and integrate seamlessly into the NAS. Sectorization strategies will allow controllers to segment slow or loitering UAVs and minimize their influence on surrounding aircraft. And finally, airborne separation and delegation procedures and technologies will provide a paradigm that allows UAVs to safely separate themselves from other aircraft, when appropriate.

NECESSITY OF COLLABORATIVE UNMANNED SYSTEMS

Coordination of unmanned aerial and ground vehicles (UAVs and UGVs) is immensely useful in a variety of surveillance and rescue applications, as the vehicles' complementary strengths provide operating teams with enhanced mission capabilities. Many of today's systems require independent control stations necessitating arduous manual coordination between multiple operators.

Reconnaissance using small unmanned air and ground vehicles (UAVs and UGVs) is finding widespread use in a variety of surveillance and rescue missions. Due to their small size and weight, such robotic vehicles can be deployed quickly by small teams near where they are needed, without substantial infrastructure. Their relatively low cost also permits deployment in larger numbers and riskier missions in hostile zones or challenging terrain from which retrieval is impractical. Finding a single vehicle that can fulfill all aspects of a mission, such as both searching a large area efficiently and providing high-fidelity reconnaissance data such as imagery is difficult, however. UGVs can get up close to objects or targets of interest to provide high-resolution imagery, can carry large accurate sensors, and execute long run-time missions, but they provide a narrow field of view, are relatively slow-moving, and must avoid ground obstacles or danger zones. In contrast, small UAVs are fast-moving and can cover wide areas quickly, above most obstacles, but they can provide only distant, low-resolution images of targets. Therefore, it is advantageous to use both in a system to leverage these complementary properties.

In an indoor search and rescue mission, the two main tasks are detecting the location of survivors or broken equipments and recovering them from a building full of danger. For the outdoor robots, they can use GPS to get the path information. But in a GPS-denied environment, the robots cannot get the support from satellites. To rectify this kind of situation we can opt for Air-Ground Collaborative Systems.

Unmanned aerial vehicles (UAVs) can be used to cover large areas searching for targets. However, sensors on UAVs are typically limited in their accuracy of localization of targets on the ground. On the other hand, unmanned ground vehicles (UGVs) can be deployed to accurately locate ground targets, but they have the disadvantage of not being able to move rapidly or see through such obstacles as buildings or fences. So, Co-Operative Air-Ground Surveillance will be an efficient method to overcome the above limitations.

SYSTEM ARCHITECTURE

Coordinated as a team, unmanned ground vehicles (UGVs) and unmanned aerial vehicles (UAVs) can deliver exponentially more effective operations than as separate, non-integrated systems. Critical synergies could be achieved through the combined resources from collaborative teams of semi-autonomous UAVs and UGVs. Critical improvements achievable through combined resources include information verification and acquisition through combined sensing, data transmission via combined communications and mobility, and force multiplication through combined control of vehicles.

Increased autonomy enables individual unmanned vehicles to communicate, coordinate and influence each other and eventually allows them to interact in collaborative teams of unmanned assets. However, technology in coordinated unmanned systems is still in its infancy. Prescribed sequences of maneuvers called "behaviors" have become the bread-and-butter of robotics autonomy today. Similarly, designed behaviors can be applied to implement tactical maneuvers to execute military missions. We foresee highly capable autonomous unmanned systems that execute mission tasks based on a decision architecture of tactical behavior with minimal necessary oversight.

The Minimum Capabilities of Autonomous Collaborative Unmanned Systems are as follows:
- Collaborative robotic team working together to perform a common mission.
- Demonstrate Semi-Autonomous Collaborative man / unmanned teaming.
- Demonstrate Single Operator controlling one or more UAV and Multiple Distributed UGV performing a common mission.
- Air / Ground Collaboration.
- Adaptive UGV Behaviors Enabling Minimal Operator Workload. The system should allow UGV working collaboratively to position and overcome obstacles to mission accomplishment with minimal operator intervention.
- Easily maintained and serviced. Capable of employment in all weather.

The System Architecture is depicted as follows:

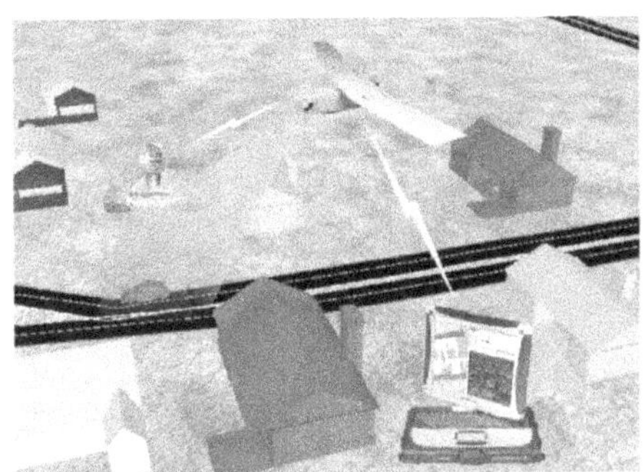

Fig 1.   System Architecture

EXIGENCY OF SMART PHONE

Since the Existing Co-Operative Air-Ground System requires Laptop or Tough book as Ground Control Station, We can propose a system which uses Smart Phone that enables GCS features. Because, Smart Phones are ubiquitous, unobtrusive, intimate, sensor rich, computationally powerful, and remotely accessible. Smart phones offer huge potential to gather precise, objective, sustained, and ecologically valid data on the real-world behaviors and experiences of millions of people where they already are, without requiring them to come into labs. Smart phones can also run controlled but perceptually and behaviorally rich surveys, tests, and experiments. Downloadable software applications for smart phones could become our central way of recruiting, obtaining consent from, observing, experimenting on, and debriefing participants—anyone, anytime, anywhere.

Each smart phone will have more computing power and memory, faster connectivity, better sensors, more input–output options, and more software apps than any current desktop computer. For many users, smart phones have already replaced a huge range of other devices: landline phones, digital cameras, photo books, video recorders, MP3 music players, radios, voice recorders, GPS navigators, handheld game consoles, watches, alarm clocks, calendars, and calculators. In psychology, smart phones could also replace a wide range of conventional research methods: paper-and pencil surveys, mail surveys, phone surveys, and, if connected to the right peripherals, many lab studies, field studies, and Internet studies.

In general, smart phones will become ever more versatile and powerful ways to run psychology studies, especially when combined with various peripherals. Their capabilities encompass and surpass most existing research methods: They can run surveys, questionnaires, field observations, and interactive experiments. Although their built-in screens and speakers are small, they can drive high-quality output to HDTVs, headphones, and speakers. Likewise, although their built-in touch screens and keyboards are small and fiddly, smart phones can receive Bluetooth input from full-sized keyboards, tablets, touch screens, game controllers, force-feedback devices, and even EEG headsets.

Smart phones uniquely combine a capacity to gather precise, objective, sustained, ecologically valid field observations of real-world behavior by very large numbers of people and a capacity to run perceptually and behaviorally rich experiments with those same people. Apart from psychology, smart phones could revolutionize empirical research in economics, political science, anthropology, sociology, geography, communication, education, medicine, management, and public policy. They also offer new possibilities for diagnosis, treatment, interventions, applications, and training, for example, in clinical, educational, health, military, organizational, and sports psychology and also in psychiatry.

Because of widespread capabilities of Smart phone, we can use the device as kind of two ways:

- Smart Phone as Remote Control.
- Smart Phone as Ground Control Station.

These two ways are applicable to both the vehicles Unmanned Ground Vehicle and Unmanned Aerial Vehicle respectively.

Current smart phones can communicate with other devices through up to seven kinds of connections: wires plugged into HDMI or Micro USB ports and radio waves at five different frequencies and ranges—near-field communication (a few centimeters' range, for example, for waving near a credit card reader), Bluetooth (around 10-meter range, for example, for connecting to laptops and heart rate monitors), Wi-Fi (around 100-meter range, for broadband access through local routers), cellular broadband (up to 50-km range, for fast Internet access through cell towers), and cellular voice (up to 50-km range, for talking and texting).

SMART PHONE AS REMOTE CONTROL

Basically, UGV will be driven using the Accelerometer Sensor (like in iphone /Android Racing Games) over Bluetooth. To integrate this sensor with the Arduino Microcontroller, an Android Application plays a vital role. With this arrangement, we can make Smart Phone as remote control to drive Unmanned Ground Vehicle. While remote-control helicopters have long been using dual, contra-rotating blades for increased stability and easier amateur piloting, the UAV goes even better and doubles that to a full four rotors. The Gyroscopes, Accelerometers and the Ultrasonic Sensors present inside a Smart phone can be used to enable it as remote control for UAVs, to make it hover and for other motions. The final key ingredient used here is Wi-Fi, which UAV use to hook up with the smart phones. Piloting the UAV requires a combination of screen taps and device movement: physically tilting the iPhone moves the quadricopter forward, back, left and right, while the right thumb controls altitude and rotation with an on-screen joystick.

SMART PHONE AS GROUND CONTROL STATION

Recent disasters have shown that wireless sensors and unmanned systems are increasingly becoming a valuable aid for first responders. Depending on the kind of incident and its extent, different assets are to be used. The more diverse these assets are, the more complex their simultaneous use and coordination. Therefore, integrated solutions are needed which comprise all necessary components such as power supply, communication infrastructure, data acquisition and processing, decision support and information dissemination. So, the Flexibility of Universal Ground Control Station can be attained by integrating Smart phone as additional mobile sensor.

In order to achieve maximum flexibility, the system is implemented open and mostly generalized so that different stationary and mobile sensors and sensor platforms can be integrated with minimal effort, establishing interoperability with existing and future assets. The system is modular and can be scaled arbitrarily or be tailored by choosing the modules suitable to the specific requirements. . The main tasks of the ground control station are to work as an ergonomic user interface and as a data integration hub between multiple sensors and a super-ordinated control center. The sensors can be stationary or mounted on moving platforms such as micro UAVs UGVs, aerostats or underwater vehicles. The system includes means to control different kinds of mobile platforms and to direct them to potentially interesting locations especially in areas with no prior sensor equipment.

As an illustration, Smart Phone as Ground Control Station can be depicted as follows:

Fig 2.   Smart Phone as Ground Control Station

Smart phones offer huge potential to gather precise, objective, sustained, and ecologically valid data on the real-world behaviors and experiences of millions of people where they already are, without requiring them to come into labs.

FIVE STEP APPROACH

There are simply five steps to make to establish Tele-operated Robot Control over Unmanned Systems with Attitude Aware Smart Phones.

1) The Major characteristics tied to the usability of an accelerometer equipped Smart phone as a controller for a Tele-operated vehicle should be studied.
2) Compare the efficiency and user feedback of the standard RC controller while viewing video feedback on the Smart phone.
3) Conduct trials using the Smart Phone as both video screen and controller with a simple push button interface to control the robot.
4) Assess efficiency and user feedback when video is viewed on the Smart Phone and the robot is controlled via accelerometer input. A dead man switch will activate accelerometers, and operators will tilt forward, backward, and side to side to control the four-wheeled vehicle. A possible variation on this could test other methods for steering control e.g., rotation vs. tilt.
5) Compare/contrast the controller described with two different "leveling" points. In one, the zero point would be a defined point in space i.e. user must hold device at a 45 degree angle to activate the dead man switch and therefore the accelerometers; alternatively, the device could self-level meaning the the zero point is wherever the user activates the dead man switch. The results of this comparison have important usability implications, especially for users that require a more heads up controller environment. It will also help inform the researchers as to how users' mental models are built and how they perceive the controller should function and react.

In tele-operated robot control, where accelerometer values are the main controller input, they should *not* be used to reorient the screen in response to device orientation. The research described below strives to identify heuristics, such as the one just defined, to inform future controller development using accelerometers in tele-operated robot control.

SMART PHONE CONTROLLED MINIATURE UGV

Unmanned ground vehicles have been implemented by the military to replace or assist soldiers in dangerous or unreachable environments. From remotely controlled bomb diffusers to completely autonomous reconnaissance vehicles, these unmanned ground vehicles have certainly played a major role in recent military history. To supplement these ground vehicles, there have also been implementations of naval and air unmanned vehicles for similar applications. These types of vehicles are becoming more popular and more complex every day.

An UGV can be controlled by a smart phone or a self-configuring ad-hoc network via a WIFI connection. This would allow ground troops to have direct control over the car from a smart phone or similar device, or would allow for a number of these UGVs to be deployed and controlled by a self-configuring network where human interaction would be minimal.

The UGV shall be able to be controlled in two different manners, either directly controlled by a human, or controlled by commands received through an ad-hoc network. The UGV will have to different operation modes under direct user control, mode number one would allow for the user to have full control over the Vehicle, much like an RC car. The second mode will allow for the user to give the UGV latitude and longitude coordinates and the UGV shall navigate by itself to the specified location. The UGV will also have a mode to receive commands over the network, this will be much like the second user mode, but the commands will be coming from a different source.

The UGV will prioritize the different commands that are issued to it, and will decide what shall be carried out. The UGV will first respond to direct control from a human operator via the smart phone platform. The UGV will consider the commands received over the network as the second priority. The UGV will also have the ability to override any of these sources if it detects a self-destructive order, such as driving off a cliff or running into a wall.

The other part is to implement a control system for the car in a smart phone. Considering that Android phones are more easily obtainable globally and their less proprietary hardware and software, we have chosen to only rely on the Android platform as a control system. Since the project requires that ground troops will be able to control this system, we need to choose a relatively small packaged device, but also allow for a large enough screen to make the system useful. Therefore we will be looking in the 4" to 4.5" screen categories of smart phones.

The control system will allow for the two different user interaction schemes outlined above. The first will allow for the user to directly control the UGV, with controls for direction, speed, and camera direction. The second mode will allow the user to direct the UGV to a location on a map. All control screens will provide sufficient location and or video feeds to allow for the user to properly and effectively control the Vehicle.

Thus, Smart phones have put video communications, computation and proprioceptive sensing (e.g. accelerometers and gyros) into the hands of hundreds of millions of consumers. These small, Micro Electro Mechanical Systems can be used in many applications, including remote control. Hundreds of millions of smart phones have been sold since their introduction, possessing impressive computational power and a suite of Micro Electro Mechanical Systems (MEMs) on board. Accelerometers, used to detect and measure motion, are one of these MEMs, and contribute to the user experience by switching orientations, self-orienting maps, and more. Developers have used the commercial availability of these sensors to tailor them for use in handheld gaming platforms, pedometers and remote control.

The Main Goals are as follows:

- Control a vehicle using a smart phone.
- Have vehicle respond to navigation commands triggered through a smart phone.
- Have vehicle avoid environment obstacles such as buildings. Maintain continuous wireless connectivity of mobile vehicle.
- Control a vehicle sensor using a smart phone.
- Have sensor respond to commands triggered through a smart phone.
- Integrate with "Re-configurable Ad-hoc Network to Track Points of Interest".

UGV – CONCEPT SKETCH

The Problem statement of the Project is that as military environments get increasingly dangerous, having a robotic scout to travel in front of the troops could be a valuable, life-saving asset. The goal for this project is to build a vehicle that can be controlled through a smart phone and can send video from a mounted camera back to the operator. This Project is also responsible for developing the smart phone application, as well as developing hardware that can be mounted on our vehicle to control it and communicate with it through the smart phone application.

Utilize a commercially available smart phone (for example a droid) to command and control a vehicle and its sensor. These capabilities are applicable to multiple scenarios including military missions, police surveillance and search and rescue activities. The prototype should be developed and demonstrated using commercially available products.

The Concept Sketch of UGV is depicted as follows:

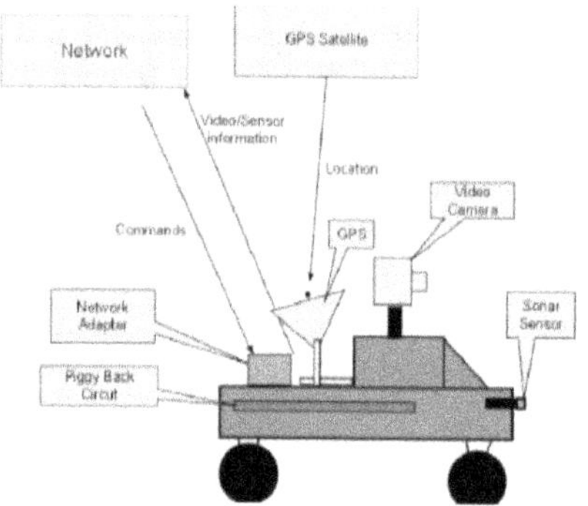

Fig.3. UGV-Concept Sketch

Navigation of the vehicle will either be manual controls or autonomous, via a GPS location. The vehicle will be equipped with WIFI so that it can receive and transmit messages. This system will be able to receive commands from both an ad-hoc network and/or personnel via smart-phone. It must maintain connection to the network and be able to reconnect if the connection is lost. The vehicle will have a camera mounted on the vehicle for gathering video and still images. The camera shall be able to pan and tilt so that it can point at an intended target. Infrared sensors should also be added to the vehicle to aid in autonomous navigation. The vehicle must be able to be deployed for at least 30 minutes of navigation, and be small enough to be easily picked up and transported.

Some Android manufactures have their devices listed as only supporting 802.11b/g, while others only list 802.11n networking chips in their models. Due to software limitations, no Android phone is natively able to support connectivity to an ad-hoc network. There are only two dual-band phones available as of this document: the Motorola Atrix and the Samsung Galaxy S2.

A development called Wi-Fi Direct is a promising contender for native ad-hoc support and is implemented into the next iteration of Android, called Ice Cream Sandwich.

CONCLUSION

UAVs and UGVs are clearly the way future when considering effectiveness on the battlefields of today and tomorrow. These Systems have proven themselves individually in limited circumstances and will be ever more powerful when combined into a family of systems.

Currently fielded UGVs and UAVs act as standalone tele operated systems with one operator per asset. Integrating UGV and UAV into single collaborative system has the potential to create dramatic improvements in situational awareness and co-ordinated control providing force multiplication in intelligence, reconnaissance and overall mission performance.

This Project is scheduled to culminate with a demonstration of the performance of the integrated team of UAVs, UGVs and Human leader in a Reconnaissance type mission. The Ground Vehicle is able to follow the path provided by the Aerial Vehicle to find the target and finish rescue mission. In the near future, the size of the vehicle team will be increased to be able to accomplish more difficult search and rescue tasks under different environmental conditions.

The collaborative framework presented provides unified control of multiple unmanned air and ground vehicles requiring minimal operator effort and maximizing situational awareness through high-level presentation of simultaneous vehicle and mission state. This is enabled by underlying algorithms implementing high-level capabilities permitting a variety of practical mission scenarios, a number of which were demonstrated in extensive field testing under varying conditions. Despite the substantial capabilities of the system, this work has barely scratched the surface of the powerful potential of air-ground collaboration and centralized multi-vehicle control. Several future directions we anticipate pursuing include cooperative localization and navigation in which vehicles observe one another to provide improved state estimates and avoid obstacles visible only from other perspectives, communication-aware vehicle motion planning to keep vehicles positioned for minimal transmission losses, more complex reasoning for pursuit and capture to derive possible escape routes to block, extending view-predictive positioning to take into account known terrain occlusions, adding additional UGV capabilities such as more general outdoor autonomous navigation, and demonstrating the system on other types of air and ground vehicles.

Future work will aim to improve performance of the capabilities enabling search and pursuit and establish the Decentralized Data Fusion for seamless collaboration to maximize state estimation. We also aim to further enhance operator situational awareness through incremental improvements to the Smart Phone to reduce cognitive load and the addition of real time geo-referenced UAV imagery to the overhead satellite map.

REFERENCES

[1]   Michael Dille, Ben Grocholsky, Stephen Thomas Nuske, Mark Moseleyy, and Sanjiv Singh, " *Air-Ground Collaborative Surveillance with Human-Portable Hardware*",2010.

[2]   Cai Luo, Andre Possani Espinosa, Alessandro De Gloria and   Roberto Sgherri, *"Air-ground multi-agent robot team coordination"* ELIOS Lab, University of Genoa, Genoa, Italy 16145-2011.

[3]   Nathan Michael,Jonathan Fink, and Vijay Kumar, , *"Controlling a Team of Ground Robots via an Aerial Robot"* University of Pennsylvania Philadelphia, Pennsylvania 19104-6228,2007.

[4]   Ben Grocholsky, James Keller, Vijay Kumar, and George Pappas , *"Cooperative Air and Ground Surveillance -A Scalable Approach to the Detection and Localization of Targets by a Network of UAVs and UGVs"* IEEE Robotics & Automation Magazine 1070-9932/06/2006 IEEE SEPTEMBER,2006.

[5]   L. Chaimowicz, A. Cowley, D. Gomez-Ibanez, B. Grocholsky, M. A. Hsieh, H. Hsu, J. F. Keller, V. Kumar, R. Swaminathan, C. J. Taylor *"Deploying Air-Ground Multi-Robot Teams in Urban Environments"* GRASP Laboratory – University of Pennsylvania Philadelphia – PA – USA,2004.

[6]   Mark B. Moseley*a, Benjamin P. Grocholskyb, Carol Cheunga, Sanjiv Singh , *"Integrated Long-range UAV/UGV Collaborative Target Tracking"* G&I Research, iRobot Corp, 8 Crosby Drive, Bedford, MA, USA 01730,   bRobotics Institute, Carnegie Mellon University, 5000 Forbes Ave., Pittsburgh, PA USA 15213-3890,2008.

[7]   R. D. Garcia, K. P. Valavanis, M. Kontitsis, *"A Multiplatform On-board Processing System for Miniature Unmanned Vehicles"* Dept. of CSE, CRASARUSF, Tampa, FL 33620, USA,2006.

[8]   William Fyfe IV Robert Johnson SAIC, *" Unmanned Tactical Air-Ground Systems Family of Unmanned Systems Experiment"* AMRDEC 4901-D Corporate Drive AMSRD-AMR-AS-CC-CO Huntsville,USA -2005.

[9]   Lee bh, *"Intelligent Web-based Remote Management for Unmanned Vehicles,"* Electrical Engineering Summer Conference 2010, 33 1, 1149-1150, 2010.

[10]   T. Kato, T. Higashi and K. Shimizu, *" Teleoperation of a robot arm system using pneumatic artificial rubber muscles: Teleoperation over the internet using UDP and a web camera"* , Proceedings of The 2010 International Conference on Broadband, Wireless Computing Communication and Applications, BWCCA 2010, 714–718, 2010.

[11]   D.I. Katzourakis, E. Velenis, D. A. Abbink, R. Happee and E. Holweg, *"Race-Car Instrumentation for Driving Behavior* Studies" , IEEE Transactions on Instrumentation and Measurement, vol. 61, pp. 462 – 474, 2012.

[12]   Huixia Xue, Lin Gao, Wenbin Li, Jun Liu, Junmei Zhang, Liu Yang, *" ARM9-based Real-time Acquisition of Orchard Soil Information and GPRS Wireless Transmission Control System"*, The 2010 International Conference on Future Computer and Communication (ICFCC 2010), The conference proceeding by IEEE, 2010, Vol(2):594-598.

[13]   W. Gasior, M. McNeely, B. Davidoff, et al *"EVDroid: Electric Vehicle Simulation and Range Finder for the Google Android Platform"*, Techniq Report, Department of Computer Science and Engineering, The University of Tennessee at Chattanooga *"Introduction to Algorithms"*, Third Edition, by Cormen, Leiserson,Rivest, and Stein 2009.

Mission to Machine: Multi-disciplinary optimization and design methodology for fixed wing MAV

Gaurav Tendolkar*, **K. Sudhakar**[†], and **Hemendra Arya**[‡]
* *Undergraduate student*, email: grvtendolkar@iitb.ac.in
† *Professor*, email: sudhakar@aero.iitb.ac.in
‡ *Associate Professor*, email: arya@aero.iitb.ac.in
Department of Aerospace Engineering, Indian Institute of Technology Bombay, Mumbai - 400076

Abstract—There is an increasing interest, in India, towards flight vehicles whose span is 30 cm or less, i.e. Mini Aerial Vehicles (MAVs). Multi-disciplinary optimization (MDO) techniques have already been used for MAVs, but there is a lack of a complete design methodology from specified mission to final MAV model. In the present study, hence, we strive to address this issue, by developing an integrated approach to demonstrate, design, and manufacture MAV that best performs a surveillance mission. To make the approach holistic, mathematical models are developed and/or adopted for various MAV subsystems, namely, video camera (payload), aerodynamics, propulsion (propeller, motor and battery), flight mechanic, structures and controls.

I. INTRODUCTION

THE development in the field of Micro air vehicles (MAVs) started in the late 90s due to scope of application and emerging technology of miniaturisation. The first feasibility study for MAVs was performed by RAND Corporation[12] followed by a more detailed study at Lincoln Laboratories[7].These studies led to further investigation in the field of MAVs. By 1997, many universities had started hosting competitions which tested performance of various MAVs in a given mission. Also, by virtue of its low cost ownership and operation many skilled hobbyists have built and successfully flown a variety of MAVs - fixed, rotary and flapping wing type. Recent advances in electronics have taken MAVs to a stage where it can perform up to 40 minutes of a surveillance mission[5]. Out of the different types of MAVs,

fixed wing MAVs have attracted much attention due its distinct advanatge of high endurance and range against other two, despite its lack of hovering capability. Benefits of formal design approaches in enhancing performance of MAV far beyond what can be achieved by off the shelf hobby items was demonstrated by [11]. This opened up opportunities for Multi-disciplinary Design optimization (MDO) techniques in the area of MAVs.

Multidisciplinary Design Optimization (MDO) can be defined as a methodology for design and analysis of complex engineering systems and subsystems which coherently exploits the synergism of mutually interacting phenomena [20]. The important distinction is that MDO is a systematic method that exploits coupling between subsystems to achieve a better overall system. MDO treats these inter disciplinary interactions as opportunities rather than nuisances or liabilities [20]. In highly coupled systems like MAVs, where everything influences everything, MDO helps in making design process objective and provides more control in design process.

In 1999 [17] first used MDO to find the best biplane configuration for an internal combustion engine MAV of maximum span of 22.8 cm (9 inches). The optimization variables included the wing geometry and relative positions of both wings and tail. The paper notably mentions that, The fuel weight and engine size are noticeably absent in the preceding design variable list. This is because, based on the available data for each engine, we know the amount of fuel an engine needs to operate at full throttle for a period of 10 min, which is thought to be sufficient to perform the specified mission.

In 2001 AeroVironment came out with Black Widow MAV which had a 15.24 cm (6 inch) span and an endurance of 30 minutes[11]. Earlier, by experimentally selecting among a range of planform shapes, they had developed a 15.24 cm (6 inch) span disk shaped MAV which gave a maximum endurance of 16 minutes without any payload. Since MAVs have to fly with payload and advanced controls, MDO was used to explore better MAV configurations. Their MDO considered variables related to propulsion system and wing geometry keeping other variables constant. The wing geometry was varied using wing tip chord keeping root chord close to 6 inch. This MDO methodology almost doubled the endurance of the MAV. Even though the MDO considered all the components of MAV, it did not consider the coupling between various components. For example, propeller wash which significantly influences lift, drag and control surfaces area depends on thrust provided by propulsion system. This important link between propulsion system, aerodynamics and control system was missing.

In 2003 [16] selected optimization variables related to the wing geometry and angle of attack only. Vortex lattice method was used to analyse wing aerodynamics, while propulsion system and payload were not considered. The objective was the lift to drag ratio subject to a constraint on static longitudinal stability. They compared optimization results using different global optimization algorithms namely GA and a non linear optimizer based on sequential quadratic programming.

In the same year MLB designed a 15.24 cm (6 inch) span MAV using MDO and followed by experimental verification[14]. The design variables included wing geometry, lift coefficient, MTOW and installed power. Optimization was done for a specific mission involving climb, cruise, loiter and descent. The propulsion system consisted of gasoline powered micro motor. Their most successful MAV named Trochoid gave an endurance of 20 minutes.

In 2008, [13] used genetic algorithm to optimize wing geometry. The only optimization variables were wing span, aspect ratio and planform type.

Four planform types were considered namely elliptical, rectangular, zimmermann and inverse Zimmermann. Fuselage drag was considered by an experimentally obtained empirical relation.

In 2009,[18] designed a tilt wing VTOL MAV with coaxial counter rotating propellers. They performed experimental analysis of aerodynamic and propulsion systems. For the first time and the coupling between propeller and planform were accounted for by momentum theory to check for the effectiveness of control surfaces in propeller wash. The report did not mention any formal optimization technique used for design process.

It is now clear that MDO can help improve performance of a system like MAV that has highly coupled subsystems. Most of the efforts in formal optimization so far have been for finding optimum wing geometry in isolation. [11] have considered propulsion system too but neglected its influence on wing aerodynamics. Recently momentum theory has been used by [18] to account for the propeller wash and its interaction with planform and control surfaces. A study that considers all subsystems of an MAV and their interactions for deciding on the globally optimum MAV for a specific mission can be of great interest. Such an MDO for MAV shall consider subsystems like, wing planform, fuselage, motor, propeller, battery, payload by using their models capturing coupled behaviour and come out with an optimum MAV in the context of its mission using a robust global optimization technique

II. Aerodynamics

Aerodynamics of different planforms and aspect ratios is characterized using Polhamus and Prandtl equations. The propeller slip stream is modelled using momentum theory and is then used to correct lift and drag.

A. *Equations of lift and drag coefficients*

The MAVs considered in this study are flying wing type. In reconnaissance flight about 95% of the time is spent in level flight. The present analysis is restricted to level flight. Thus lift equals weight and thrust equals drag of the MAV. Aerodynamics for 4 different planforms namely Zimmerman, inverse Zimmerman, elliptical and rectangular with aspect ratios ranging from 0.5 to 2.5 is available in the form of Polhamus and Prandtl equations.

$$C_l = k_p sin \propto cos^2 \propto + k_v cos \propto sin^2 \propto \quad (1)$$

$$C_d = C_{d0} + kC_l^2 \quad (2)$$

Where k_p, k_v and k are functions of planform shape and aspect ratio
$\propto$ is the angle of attack
and C_{d0} is the zero lift drag coefficient which is averaged to 0.015 [15]

B. Froude's Momentum Theory

In MAVs, the propeller dimensions are comparable to the wingspan and thus the propwash affects the aerodynamics. In present analysis, propeller only adds velocity to the flow and no other interaction effects are considered. The propeller slip stream is modelled using Froude's momentum theory and it is then used to correct lift and drag. Momentum theory gives the velocity behind a rotating propeller in terms of free stream velocity and propeller thrust as

$$\sqrt[2]{\left(V_0^2 + \frac{2T}{\rho \pi R^2} \right)} \quad (3)$$

Where V_0 is the free stream velocity
T is the thrust
ρ is the density of air
and R is the radius of propeller [18].
This corrected velocity is used to calculate lift and drag for the area of the planform which lies in the propeller's wake. Free stream velocity is used for calculations for the remaining part of the planform.

C. Equations of lift and drag

Thus, for a given forward velocity, we have lift in terms of thrust and C_l and drag in terms of C_l and thrust. The Polhamus and Prandtl equations give C_l and C_d in terms of angle of attack ($\propto$). Thus for a particular velocity, we have two equations namely lift equal to weight and thrust equal to drag and two variables, thrust and $\propto$.

$$W = \frac{1}{2}[\rho((V_0^2 + \frac{2T}{\pi \rho R^2})A^* + V_0^2(S - A^*))C_l] \quad (4)$$

$$T = \frac{[\rho((V_0^2 + \frac{2T}{\pi \rho R^2})A^* + V_0^2(S - A^*))(C_{d0} + kC_l^2)]}{2} \quad (5)$$

The equation for C_l can be replaced by equation 1

Where k_p, k_v and k are functions of planform shape and aspect ratio.
V_0 = Free stream velocity
ρ = Density of air
S = Planform area
A^* = Area of planform in the wake of the propeller
C_{d0} = Zero lift drag coefficient
C_l = Lift coefficient
T = Thrust generated by the propeller
R = Radius of the propeller
$\propto$ = Angle of attack
Solving these two nonlinear equations gives value of and thrust required to fly at that velocity for a particular planform shape and aspect ratio and at the given velocity.

III. PROPULSION

A. Propeller Model

The propeller performance curves for propellers used for 30cm MAVs are given by [19] and are generated using QPROP [9]. The thrust coefficient versus advance ratio curve is used to calculate the motor RPM since thrust and forward velocity are known.

B. Motor Model

A first order model of electric brushless motor as given in QPROP is used. The equations of motor model given in shown below are used to calculate power extracted from battery for a specific RPM [8]. This power value is used to calculate the current versus time using battery voltage.

$$Q_m(\Omega, v) = \frac{[i(\Omega, v) - i_0]}{K_Q} = \frac{\left[\frac{(v - \frac{\Omega}{K_v})}{R} - i_0\right]}{K_Q} \quad (6)$$

$$P_{shaft}(\Omega, v) = Q_m \Omega \quad (7)$$

$$\eta_m(\Omega, v) = \frac{P_{shaft}}{iv} = \left(1 - \frac{i_0}{i}\right)\frac{K_v}{K_Q}\frac{1}{1 + \frac{iRk_v}{\Omega}} \quad (8)$$

where i is the current i_0 is the zero load current
K_Q is the motor torque constant
v is input the voltage
R is the internal resistance
Q_m is the shaft torque
Ω is the rotation rate
k_v is the motor speed constant
η_m is the motor efficiency
and P_{shaft} is the shaft power

C. Battery Model

Since each phase of the mission is a constant velocity flight, the power required from the battery does not change with time. As the battery discharges, the terminal voltage of the battery drops as seen in 1. Thus the motor extracts larger current from the battery as the battery discharges. For each discharge rate, there is a particular voltage versus state of discharge (SOD) curve, which is the characteristic of the battery. Since the discharge rate increases due to drop in voltage for the same power, the voltage versus SOD curves changes. Thus to model a battery, voltage versus SOD curves for all discharge rates will be required which are not available from the manufacturer. Thus there is a need for a dynamic battery model.

1) Lithium-Ion Battery Model [10]:

1) A typical curve of battery voltage versus the depth of discharge is chosen as a reference curve near the median operating range. The equilibrium potential as a function of the state of discharge is found by excluding the internal potential losses due to ohmic-limitation, kinetic-limitation, and concentration-limitation resistances. A polynomial equation is fitted to that curve.

2) Secondly, the discharge rate (i.e., the current) for the reference curve is chosen as the reference rate. The dependence of the state of discharge on rate is then accounted for by a rate factor $\propto (i))$, which has value unity for the reference curve.

$$\propto (i) = \frac{b}{a} \qquad (9)$$

refer to figure 1

3) A linear curve for $\propto (i)$ vs discharge rate is obtained using the few points obtained from curves provided by the manufacturer as shown in figure 2

4)

$$SOD = \frac{1}{Q} \int \propto (i) i \, dt \qquad (10)$$

Instead of $\frac{1}{Q} \int i \, dt$, where Q is the total charge of the battery for the reference discharge rate.

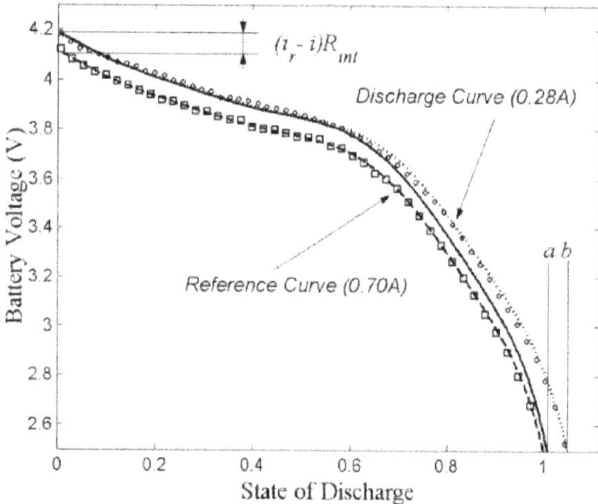

Fig. 1.   Voltage versus state of discharge curves [10]

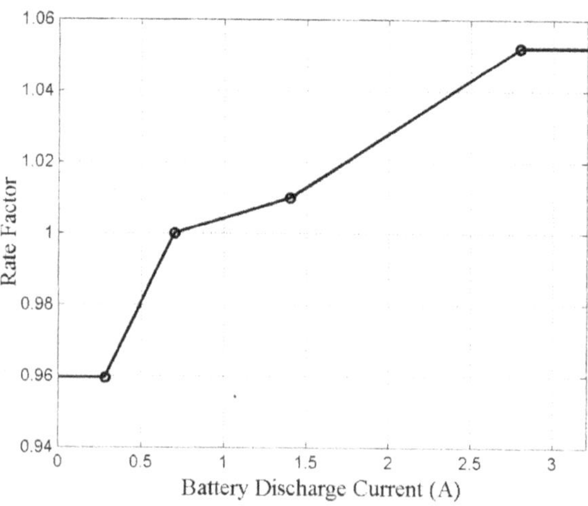

Fig. 2. $\propto$ versus discharge rate [10]

5) Gaos model also accounts for temperature effects by another factor $\beta(i)$ which has been neglected here. Capacitance is not included.

2) Voltage versus SOD Reference Curve: Li-Po batterys reference voltage versus state of discharge curve is modelled using averaged voltage versus SOD curve for low discharge rates as given in [6]. A dynamic Li-Ion battery model given in [10] is used to find the voltage versus SOD curves for different discharge rates. The endurance is calculated using total discharge time which is equal to the time taken for battery voltage to reach its cut-off voltage.

IV. MDO PROBLEM

A. Objectives to be maximized

- Endurance (E)
- Area scanned (function of V, h and E)

B. MDO variables

The variables used in MDO are listed in table I

C. Constraints

- Angle of attack not exceeding stall angle.
- Current does not exceed maximum discharge rate.

The design space could be further reduced by constraining the allowable combinations of motor and propeller. Each off the shelf motor has recommended propellers to be used with it. Using other dimension propellers might heat up the motor excessively. A literature survey of all motors (within 50 grams weight) available on HobbyKing was conducted to find out if there was any simplified relation between motor parameters and the recommended propeller dimensions. The graph shown in figure 3 shows one such relation between maximum pitch and diameter and minimum pitch and diameter of the propeller versus motor Kv. The data was collected from HobbyKing stores website where each motors specifications along with recommended propellers were given.

This data can be used to narrow the design space by adding a constraint on the motor-propeller combinations allowed.

	Variable	Nature	Range of values
1	Planform Shape	Discrete	Elliptical, Rectangular, Zimmerman, Inverse Zimmerman
2	Planform Area	Discrete	(**)
3	Aspect Ratio	Discrete	0.75 to 2
4	Propeller	Discrete	5 propellers namely 5.5x4.5, 6x3, 6x4
5	Motor	Discrete	3 motors namely Turnigy C1616, Turnigy C2020, AXI 2203-52
6	Battery	Discrete	[500mAh, 20C, 7.4V] (*)
8	Flight Velocity	Continuous	4 m/s to 30 m/s
9	Roll Rate	Continuous	5°/s (*)
11	Total Weight	Continuous	120 grams (*)
(*) Single values have been considered for the results			
(**) Calculated using span and aspect ratio			

TABLE I
MDO VARIABLES

V. MDO METHODOLOGY

Since some of the variables are discrete, a gradient based approach cannot be used to solve the optimization problem. A global optimization technique has to be used. There are many global optimization techniques that can be used to solve the

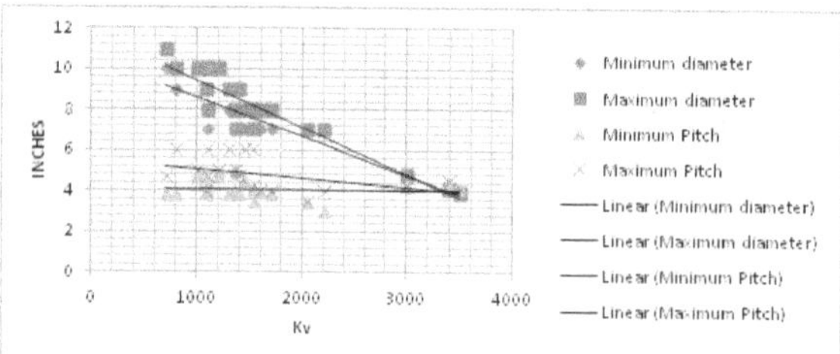

<table>
<tr><td>Planform shape</td><td>Rectangular</td></tr>
<tr><td>Aspect ratio</td><td>0.75</td></tr>
<tr><td>Motor</td><td>Turnigy C2020</td></tr>
<tr><td>Propeller</td><td>APC 5.5x4.5</td></tr>
<tr><td>Cruise velocity</td><td>7 m/s</td></tr>
<tr><td>Endurance</td><td>54 minutes</td></tr>
</table>

TABLE II
RESULTS OF MDO

Fig. 3. Recommended of the shelf propeller for of the shelf motor

Fig. 4. Endurance versus planform shape and aspect ratio

problem like genetic algorithm, simulated annealing, particle swarm, etc. For the results presented in this paper, no such global optimization technique was used. Instead the whole design space was searched systematically for the optimum configuration.

VI. MDO RESULTS

The global optimum endurance configuration obtained from the above procedure is a rectangular planform of lowest aspect ratio, which is 0.75 in this case, mounted with a 5.5x4.5 propeller and Turnigy C2020 motor. The optimum cruise velocity is 7 m/s. It should be noted that airfoil was not considered for analysis. During manufacturing, a common airfoil namely EH2070 was chosen for all configurations. The EH airfoils [3] were developed specifically for swept wing tailless aircraft. The airfoil has excellent stall characteristics and near zero pitching moments, making it ideal for this type application. The optimum configuration is sumarrized in table II. It should also be noted that weight was not considered as a variable in optimization problem. This is because the total weight of all the components is almost constant and is far greater than the weight of the planform. Thus the total weight remains almost constant irrespective of the shape and size of the planform which is why single value of total weight was considered for optimization process. The weight breakup of the MAV to its individual components is shown in table VI.

The endurance does not vary much with change in planform shape, but aspect ratio and velocity significantly influence the endurance. The graph shown in figure 4 provides an insight into the variation of endurance with respect to different variables ceteris paribus.

VII. VALIDATION OF PROPULSION MODEL

The endurance calculated by propulsion models used in this paper seem unrealistic and thus the model needs to be validated and correspondingly corrected using statistically obtained fudge factors. To obtain these fudge factors, some static thrust tests were conducted by measuring power consumption and endurance for same propeller, motor and thrust and comparing them with power consumption obtained from QPROP and endurance obtained from battery model. The actual power consumption in static conditions is compared with power consumption estimated by QPROP for static conditions in table VII.

The endurance estimated by battery model using actual power consumption values is compared with actual endurance for different values of static thrust in table VII. The endurance was for a 800mAh battery. Both tests used Turnigy C2020 motor with a 6x3 propeller.

Also, a comparison study was carried out between Turnigy C2020 motor and AXI 2203-52 motor by comparing experimental power consumption for different values of static thrusts. The results are shown in table VII

During the flight test of second model, which is discussed later in detail, it was observed that the 110g MAV flew at 60 grams thrust. We see that for static conditions, AXI 2203-52 motor consumes almost 50% less power than Turnigy C2020 motor for 60g thrust.

VIII. MANUFACTURING

The manufacturing phase began in second stage of the project. The results of MDO were used to generate digital solid model of the planform. The

Thrust (g)	Actual power consumption (Watts)	Power consumption estimated by QPROP (Watts)	Unaccounted efficiency
10	0.8	0.787	0.98
15	1.2	1.2	1
20	2	1.7	0.85
25	2.7	2.2	0.81
30	3.2	2.7	0.84
35	4.2	3.3	0.78
40	5.4	3.9	0.72
45	6.1	4.5	0.74
50	7.3	5.2	0.71
55	8.3	5.9	0.71
60	9.5	6.6	0.69
100	20	13.2	0.66

TABLE III

DERIVING FUDGE FACTOR FOR POWER CONSUMPTION

Thrust (g)	Actual endurance (secs)	Endurance estimated by battery model (sec)	Fudge factor
20	107	181	1.69
25	80	135	1.68
30	67	112	1.67
40	54	66	1.22
50	40	49	1.22
60	30	37	1.23
100	16	17	1.06

TABLE IV

DERIVING FUDGE FACTOR FOR ENDURANCE

Thrust (g)	AXI 2203-52 (Watts)	Turnigy C2020 (Watts)
10	0.3	0.8
20	1.2	2
30	2.0	3.2
40	3.0	5.4
50	4.0	7.6
60	5.1	9.5

TABLE V

DERIVING FUDGE FACTOR FOR ENDURANCE

slot for fuselage, along with motor mount coordinates, is generated based on exact dimensions of the components. The model is then fed to CNC machine to get an optimized ready to fly MAV kit for the specified mission.

A. Generating three dimensional coordinates of planform surface

Solving the MDO problem gives us the optimal planform shape and aspect ratio which corresponds to maximum endurance/range. In order for the CNC machine to manufacture this configuration, we need to provide it an STL or an IGS format file describing three dimensional geometry of the model. The whole model is divided into 3 parts namely planform, fuselage and vertical tail. This is because, as shown in Fig.20, making vertical tail and planform from a single piece would require a thick work piece and most of the material will be wasted. The fuselage shape is complicated and thus generated manually using SolidWorks.

To obtain a solid model we need to generate three dimensional coordinates of the models surface and then use the coordinates to generate the surface. The origin of the coordinate system is at the point of intersection of span line and centre chord with Y axis pointing forward and Z axis upwards.

Initially we generate the coordinates of the perimeter of top view of planform, i.e. the planform shape in X-Y plane using its parametric equation. The parametric equations of 4 planform shapes considered for this study are given below

Zimmermann,

$$x = \frac{b}{2}\cos t \tag{11}$$

$$y = \frac{b}{\pi}(2\sin t + |\sin t| - 1) \tag{12}$$

Inverse Zimmermann,

$$x = \frac{b}{2}\cos t \tag{13}$$

$$y = -\frac{b}{\pi}(2\sin 2\pi - t + |\sin 2\pi - t| - 1) \tag{14}$$

Elliptical,

$$x = \frac{b}{2}\cos t \tag{15}$$

549

$$y = \frac{2b}{\pi} \sin t \tag{16}$$

Rectangular, For $t \in [0,\pi)$

$$x = \frac{b}{2} \cos t \tag{17}$$

$$y = \frac{2b}{AR} \tag{18}$$

For $t \in [\pi,2\pi]$

$$x = \frac{b}{2} \cos t \tag{19}$$

$$y = -\frac{2b}{AR} \tag{20}$$

Once we get the perimeter of top view, we generate coordinates of surface by plotting coordinates of a specific airfoil on planes parallel to Z-Y planes. The tips of these airfoils lie on the perimeter of the top view as illustrated in 5.

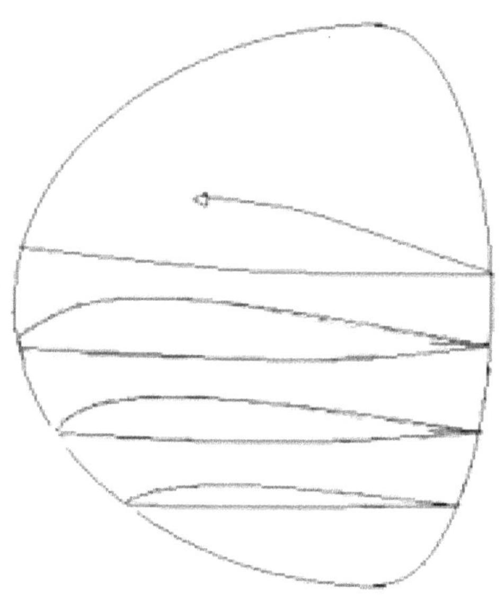

Fig. 5.   Generating coordinates of points on planform surface

To achieve that we first determine the chord length at multiple points on X axis from the top view plot. This chord length is multiplied to the y/c and z/c data of the airfoil under consideration (where z is in the thickness direction). Thus we have Y and Z coordinates for each X coordinate, but now, all the airfoils have their leading tips on the X axis. This is because the coordinate axes used by most of the airfoil data files have their origin at the leading tip of the airfoil. Thus we shift the airfoil by the positive Y coordinate value of point on the perimeter (in the top view plot) corresponding to the point on the X axis. Thus we obtain the complete set of three dimensional coordinates of points on the surface of the planform.

Since the chord at the centre of the planform is larger than at the tips, the thickness at the root is greater than thickness at the tips giving the planform a small anhedral. To make the MAV stable in roll, we may give the planform a small dihedral. Dihedral can be given by shifting the Z coordinates up by corresponding X coordinate multiplied by tan of dihedral angle.

B. Generating surface geometry

The three dimensional coordinates obtained are stored in a text file. This file cannot be used as an input to the CNC machine since the machine only accepts a .STL or .IGES file formats. The STL file format defines the surface geometry as a raw unstructured triangulated surface defined by the unit normal and vertices (ordered by the right-hand rule) of the triangles using a three-dimensional Cartesian coordinate system. To convert the text file into .STL file, we first need to generate a triangulated surface using 3 dimensional coordinates. The method adopted for this process is called Delaunay Triangulation. Using this method a matrix of faces and vertices is obtained from 3D coordinates matrix [2]. This matrix of faces and vertices is then converted to .STL file using MATLAB function STLWRITE [4]

C. Limitations on manufacturing

The optimum configuration which was a rectangular planform of aspect ratio 0.75 has dimension of 400mmx300mm when seen from top. Therefore, the size of workpiece required to make this model exceeds 400mm in one dimension. The CNC machine available at the IIT Bombay MAV lab does not accept a workpiece whose dimensions are greater than 400mm. Thus the optimum configuration could not be manufactured. Moreover, low aspect ratio models, although efficient, are laterally unstable, and since we had not arrived at the optimal control surfaces' area, we decided to manufacture and test a high aspect ratio first. After getting a qualitative idea of control, we could manufacture the low aspect ratio models.

IX. MAV COMPONENTS

Weight is a critical factor in deciding the endurance or range of MAV and thus a typical MAV uses the lightest components available in the market. The smallest and lightest off the shelf components were used for the MAV are described in this paper.

A. Planform material

The planform was made out of either blue foam or Polystyrene because of its their low density (3 gm/cm3). The disadvantage of using these materials is that the surface has a slightly rough texture even after giving a finish in CNC machine. Since the flow is mostly laminar due to low Reynolds number, skin friction drag plays a very important role and thus the surface has to be made smooth. Moreover the structure is flexible. To add stiffness and make the surface smooth, the whole planform was covered with tape or monokote. The lamination also protected the leading edge of the planform during rough landings.

B. Vertical tail

The vertical tail, if made of the same material and cut as one piece on the CNC machine, would waste a lot of material. Due to the granular structure of polystyrene, a thin vertical tail wouldnt be possible to cut on the machine. Thus it was decided to make the vertical tail and elevons out of balsa wood. Balsa wood also has a very low density (160 kg/m3). It can be cut into thin sheets which have a good strength when bent perpendicular to the direction of the threads.

C. Autopilot board

Autopilot board used was ArduPilot Mega 2.0 (APM2) [1] with uBlox GPS. The board was mainly chosen due to its light weight, small size and available documentation

D. Video camera

Since light weight video transmission system was not available, it was decided to record video digitally onboard. The system used for this is popularly called 808 car keys. It is a system which has a camera, a microprocessor which encodes and compresses the raw video, mini USB port, micro SD card slot and a single cell Li-Po battery. The whole system weighs about 8 grams and has a maximum dimension of 35mm. The system can record videos for up to 30 minutes.

E. Telemetry

Xbee module was used for telemetry of flight data and for tuning the autopilot gains. Manual control was achieved using Futabas 6 channel transmitter receiver system.

Micro servos were used to control elevon deflections. Battery used belonged to the Zippy Flightmax series. A 5V to 3.3V regulator was used to provide 3.3 volts to Xbee from 5V pin on the board. A 10 Amp Turnigy ESC was used since all motors had a maximum current of 10 Amps

F. Weight breakup

The total weight of all the components is almost constant and is far greater than the weight of the planform. Thus the total weight remains almost constant irrespective of the shape and size of the planform which is why single value of total weight was considered for optimization process. Only changing battery capacity significantly affects the weight. The weight breakup of the MAV to its individual components is shown in table VI.

X. FLIGHT TESTS AND CONTROL SURFACES

Once the planform was manufactured in CNC machine, its optimal centre of gravity position is found using glide tests. Glide tests were performed by adding weight to nose in small steps using modelling clay and observing the glide performance. Once an stable glide path is obtained, the centre of gravity position of empty planform is marked. By adjusting position of components, it is ensured that the CG position of the final model with all components is located at that mark. Results of flight tests on the model are used to decide on the planform dihedral,

Component	Weight in grams
Polystyrene/Blue foam Planform laminated with tape/monokote and balsa wood vertical tails and elevons	20 *
Motor	12 **
Propeller	1.6
Electronic speed controller	7
Camera with its own single cell battery and micro SD card	8
RC receiver	9
Xbee for telemetry	3.8
ArduPilot mega 2.0 autopilot board with GPS and IMU	20.4
Servo actuators	4
Li-Po Battery 500mAH 20C	31.5
Miscellaneous wires and normal/double sided tape	3.7
Total	120

(*) 20 grams is the target weight. The amount of tape is adjusted considering this weight constraint and stiffness and surface finish requirements

(**)Weight of all motors and propellers considered is almost constant.

TABLE VI
WEIGHT BREAKUP

control surfaces and vertical tails area to be given to all subsequent models. The first model made in CNC machine (model 1) had no dihedral. Since the airfoil coordinates are given relative to the local chord, the thickness of planform decreases as chord decreases. Thus the thickness at the tips is less than the root thickness and the shape acts as an anhedral. Moreover, since the battery (which is the heaviest component) was mounted on the top surface of planform, the CG location was slightly above the neutral point in positive Z direction in body axis frame. This made the MAV unstable in roll. These effects were only noticed after flight tests, in which the

Fig. 6.   Model 1

MAV had a high tendency to roll and fly inverted. The MAV was also slightly overpowered and used to hover at 70% of maximum thrust, although it was difficult to maintain the orientation and position in hover. Also, it was observed by the pilot that the elevons had too low surface area to control the plane because of the short span near trailing edge in Zimmerman configuration. This was the main disadvantage of using a Zimmerman planform and thus Inverse Zimmerman shape was given to the subsequent model. The plane was damaged in the numerous crashes but has been repaired to some extent. The next model (model 2) was given an 8 degree dihedral. Its elevon and vertical tail area was also increased.Figure 6. shows ready to fly model 1.

Model 2, like model 1, was meant to fly in manual mode only and therefore had only battery, receiver and ESC onboard apart from motor and propeller. Its total weight was 100 grams. For model 2, a fuselage pod was carved along with the planform in single peice. This pod housed all the electronics and thus reduced drag. It also made it possible to mount battery below the planform. The pod was designed manually in SolidWorks. Model 2 is shown in figure 7. Figure 8 shows the front view to get a better view of the 8 degree dihedral.

Model 2 flew successfully. It was stable in pitch, roll as well as yaw and flew at approximately 10 m/s ground speed on average. The ground speed

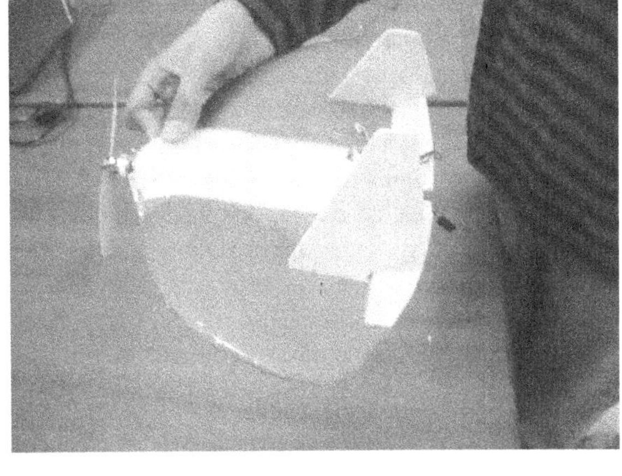

Fig. 7.   Model 2

Fig. 9.   Model 3

Fig. 8.   8 degree dihedral on model 2

increased while flying in the direction of the wind and decreased while flying against it. The control surface area was enough to control the plane. Due to fuselage, the electronic components did not block the propwash over the control surfaces giving a better control. The pilot was comfortable flying the plane. On its second flight, we conducted endurance test where we discharged the 500mAh battery from 8.4V to 6.9V. The total flight time was 11 minutes. Since the flight included many climbs, descents and turns, the endurance was reduced considerably. A pure cruise, which could be possible using autopilot aided with inertial measurement unit and GPS would increase the endurance significantly. Moreover the

test was conducted in gusty conditions and thus the pilot had to fly at a high throttle input. Model 2 was also flown manually with a 800mAh battery increasing the total weight by 10 grams. The endurance achieved in this case was 19 minutes. Both model 1 and 2 had Turnigy C2020 motor and 6x3 propeller. It was concluded that, to increase the endurance, an autopilot had to be included maintaining the weight below 120 grams. Therefore the third model was made out of polystyrene instead of blue foam, which was used to make model 1 and 2. The third model also had a longer and thicker fuselage to house the extra components. The slots were made considering the optimal CG position concluded from glide tests. The first flight of model 3 showed that the MAV was too fast and thus was difficult to control for the pilot. The model was damaged extensively during the first crash due to low stiffness of polystyrene, especially near the motor mount. A hollow fuselage further lowered the strength. Therefore the additions of carbon rods, embedded within the structure, were necessary to increase the stiffness. model 3 is shown in 9.

XI. FLIGHT TESTS WITH ONBOARD CAMERA

Model 2 was also flown with the 808 car keys camera attached to the bottom to check the quality of video. Figure 10 to figure 13 are some of the snapshots taken from the surveillance video to illustrate its quality. Figure 10 and figure 11 show snapshots taken before launching the MAV. The MAV was held in hand about a meter above the ground with camera facing downwards. Frame shown in figure 10 was taken before the motor was started while frame shown in figure 11 was taken after. The figures clearly illustrate the degradation of video quality due to vibrations caused by the propulsion system. This problem can be resolved by mounting camera on a foam mount which would absorb the vibrations.Figure 12 and figure 13 shows atheletic tracks ans basketball court of IIT Bombay respectively.

Fig. 12. Atheletics tracks at IIT Bombay

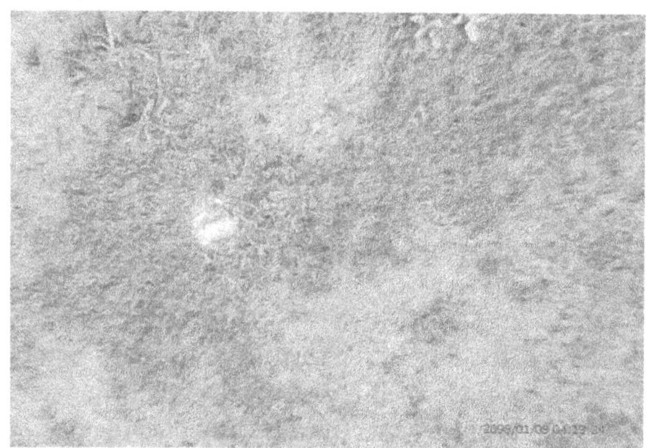

Fig. 10. Before starting the motor

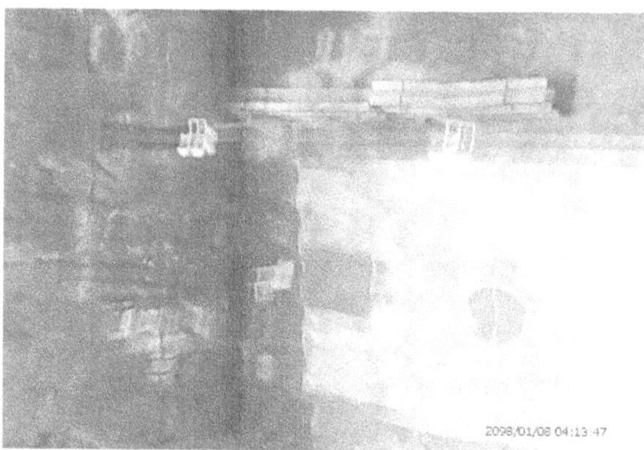

Fig. 13. Basketball court at IIT Bombay

Fig. 11. After starting the motor

developed for various subsystems. Coupling between various subsystems was accounted for. MDO was implemented to find the best MAV configuration. The models were validated and appropriate fudge factors were calculated. Some test models were manufactured and flown acheiving a maximum endurance of 19 minutes. Video quality of surveillance video from onboard camera was also analyzed. Further improvements mentioned in future work are expected to increase the endurance significantly with a better and live surveillance video quality.

XII. CONCLUSION

A detailed literature survey was carried out to understand process of design of MAVs. Models were

XIII. FUTURE WORK

Future work includes refining propulsion system models by including fudge factors derived statistically from experimental tests in static conditions. In the next model, motor will be changed to AXI

2203-52 motor from Turnigy C2020 since it consumes 40% less power as shown in VII keeping other things constant. The next model will be a rectangular planform of aspect ratio 1 with 5 degrees dihedral and 10 degree sweep. Since the aspect ratio is halved, the area will double thus giving scope for increase in weight maintaining the wing loading. The motor mount will be extended ahead of the leading edge so that there there is less flow separation at the leading edge and the whole wing experiences propwash. This feature is only limited by manufacturing capabilities of the CNC machine. A 950mAh turnigy nano-tech battery will be used which has a 18% more capacity while weighing only 10% more. A new video transmission system weighing just 3g (1g camera and 2g 200mW 5.8GHz transmitter and antenna) will be used. All these imporvements are expected to add only 30 grams more weight.

ACKNOWLEDGMENT

The authors would like to thank NPMICAV for their support. We would also like to thank members working at the MAV lab at IIT Bombay whose valuable contributions has made this project a success

REFERENCES

[1] Ardupilot mega 2. http://www.diydrones.com/profiles/blogs/apm-2-0-release.

[2] Delaunay, matlab documentation center. http://www.mathworks.in/help/matlab/ref/delaunay.html.

[3] Eh airfoils. http://www.b2streamlines.com/EH.html.

[4] Stlwrite, matlab documentation center. http://www.mathworks.in/matlabcentral/fileexchange/20922-stlwrite-write-binary-or-ascii-stl-file.

[5] Aerovironment. UAS advanced development: Hornet.

[6] Min Chen and Gabriel A. Rincon-Mora. Accurate electrical battery model capable of predicting runtime and iv performance. *IEEE Transactions on Energy Conversion*, 21(2):504 – 511, june 2006.

[7] W.R. Davis. Micro uav, presentation to 23rd. AUVSI Symposium, July 1996.

[8] Mark Drela. First-order dc electric motor model. October 2005.

[9] Mark Drela. Qprop theory document. October 2005.

[10] Lijun Gao, Shengyi Liu, and R.A. Dougal. Dynamic lithium-ion battery model for system simulation. *IEEE Transactions on Components and Packaging Technologies*, 25(3):495–505, september 2002.

[11] Joel M. Grasmeyer, Matthew T. Keennon, and Aerovironment Inc. Development of the black widow micro air vehicle. In *39th AIAA Aerospace Sciences Meeting and Exhibit*, 2001.

[12] R.O. Hundley, E.C. Gritton, United States. Advanced Research Projects Agency, National Defense Research Institute (U.S.), and Rand Corporation. *Future Technology-driven Revolutions in Military Operations: Results of a Workshop*. Rand, 1994.

[13] P. Marek. *Design, Optimization and Flight Testing of a Micro Air Vehicle*. PhD thesis, MS thesis, University of Glasgow, 2007.

[14] Stephen J. Morris. Design and flight test results for micro-sized fixed-wing and VTOL aircraft. In *In The First International Conference on Emerging Technologies for Micro Air Vehicles*, 1997.

[15] T.J. Mueller. *Introduction to the Design of Fixed-Wing Micro Air Vehicles: Including Three Case Studies*. AIAA Education Series. Eurospan Group, 2007.

[16] T.T.H. Ng and G.S.B. Leng. Application of genetic algorithms to conceptual design of a micro-air vehicle. *Engineering Applications of Artificial Intelligence*, 15(5):439 – 445, 2002.

[17] M. Rais-Rohani and G. A. Hicks. Multidisciplinary design and prototype development of a micro air vehicle. *Journal of Aircraft AIAA*, 36(1):227–234, january 1999.

[18] S. Shkarayev, University of Arizona. Dept. of Aerospace, Mechanical Engineering, and Fla.). Munitions Directorate Air Force Research Laboratory (Eglin Air Force Base. *Autonomous Micro Air Vehicles with Hovering Capabilities*. University of Arizona, 2009.

[19] Meenakshi Varyani. Modelling of propulsion unit and longitudinal dynamics of micro air vehicles. SAROD Symposium on Applied Aerodynamics and Design of Aerospace, November 2011.

[20] Thomas A. Zang and Lawrence L. Green. Multidisciplinary design optimization techniques: Implications and opportunities for fluid dynamics research. In *JAROSLAW SOBIESZCZANSKI-SOBIESKI AND RAPHAEL T. HAFTKA MULTIDISCIPLINARY AEROSPACE DESIGN OPTIMIZATION: SURVEY OF RECENT DEVELOPMENTS, 34TH AIAA AEROSPACE SCIENCES MEETING AND EXHIBIT*, pages 99–3798, 1999.

SPATIAL DATABASE INTELLIGENCE IN MAV (SDI-MAV)

Kumar Tushar Shrivastava
Software Engineer, Infotech Enterprises, Noida, U.P., India.
sritushar_GATE2k9@yahoo.co.in

Payal Sharma
Lecturer, CSIT, IST, JNTU, Hyderabad, A.P., India.
itspayalsharma@yahoo.com

J. Venkatesh
Head, CSIT, IST, JNTU, Hyderabad, A.P., India.
Csit_head@yahoo.com

Abstract

For terrestrial navigations, satellites are already playing an important role in today's traffic research and in many traffic applications such as traffic flow prediction, event detection and traffic control etc. In this paper we are concentrating on handling satellite navigations in Micro air Vehicles (MAV), using the Euclidean Distance theory to be the basis of calculation of distance covered by MAV's and point of availability of MAV through IMU GPS[2]. The above concern via spatial database with unquenchable remote sensing elicits the true potential of the location-based service in a more robust manner. The GPS market demands a breakthrough in the excellence of this technology and this attempt of combining GPS and wireless signals in a MAV for a hybrid technology is propounded as Assisted-MAV (A-MAV). This system grasps the geospatial data or location control points through A-MAV and thereby relates the attribute data residing on spatial platform/database in deriving the meaningful interpretations. Minimal fuss through navigational stuff at every unidentified place in air captivates the location of MAV in air via Euclidean Spatial Database Intelligence (ESDI). The paths follow various shortest path algorithms keeping one way paths in consideration, minimizing possibilities of mishap, availability of path at each and every step with the accurate measurements. Furthermore it also fosters the suitable conditions in case of natural disasters like floods, culverts on roads, breakdown due to bad raw material or earthquakes, bad weather conditions like fog, snow etc.

Keywords: Euclidean Spatial Database Intelligence (SDI), Air Traffic Signal Trackers and Controllers (ATSTC), Dynamic Route Guidance (DRG), Travel Time and Delay Factor (TTDF).

INTRODUCTION

Spatial database intelligence (SDI) system is management of geometry and geographic features. In this paper we are using SDI for tracking MAVs for managing the pathway without mishaps and thus getting the real time information which is required, for particular analysis and operations. The foundation of such services is clearly based upon reliable, ubiquitous and highly accurate location technology. The situation and solution has been demonstrated excellently through a model-based approach which confines all of the complimentary measures and results. The system used here for finding potential locations is A-MAV. **Assisted MAV**, generally abbreviated as **A-MAV**, is a carrier network dependent system which can, under certain conditions, improve the start-up performance of a MAV, satellite-based positioning system. It is used extensively with GPS-capability.

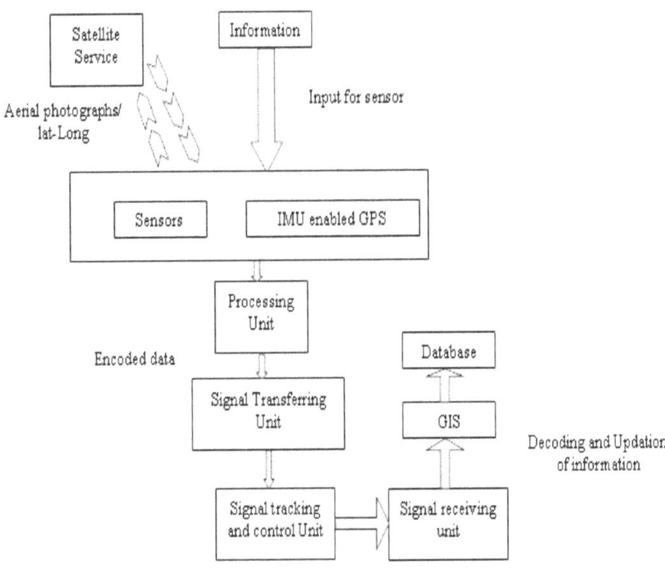

Fig. 1.1 Planning and Architecture of A-MAV

ARCHITECTURE

The architecture of SDI-MAV is shown in Fig 1.1, that includes many blocks namely, STCO (Signal tracking and control office), a high-end efficient and speedy database rich in storing spatial elements. Satellite services tracing ground control points as latitude(la) and longitude(lo), Inertial measurement unit for altitudes(al), GIS for analysis purpose, Sensors etc.

FUNCTIOALITY OF BLOCKS

Description of all the components of the architecture reprimanded through figure is given below:

Satellite Service is used for getting the information about MAV like its latitude, longitude, altitude, for sending and receiving information etc.

Information means input for sensor, used in MAV. Information is application dependent, i.e. based on the application for what purpose we are using MAV.

Sensors are equipment which can measure any kind of physical quantity and are able to convert it into the signal so that any observer may read it or any instrument can measure it. Below given diagram is showing the simple phenomena of working of a sensor.

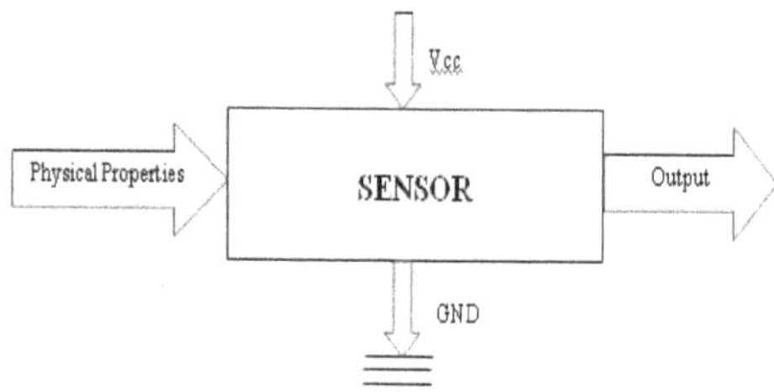

Fig 1.2 Basic working concept of sensor

IMU enabled GPS is used to make, tracking of MAV possible and moreover if GPS signals misses at home places like in tunnels, then IMU[2] enables GPS to work.

Processing Unit has always been the brain of any machine and it is not an exception here as well. It is used to process the received data, encode the data etc.

Signal Transferring Unit and **Receiving Unit** are working on the principal of mobile communication, i.e. based on GSM or CDMA, as per the importance of data and consideration of security issue.

Tracking and Control Unit is the main controller of the complete system.

GIS is used for creating, manipulating, analyzing spatial areas based on the application.

Databases are organized collection of data or in more general form it is an electronic form which responds to the query generated by user to satisfy user's requirement. Database having location related information, is known as Spatial Database.

ORDERLY APPROACH

SDI-MAV lies under the full control of Air Traffic Signal Tracker and Controller Office which gathers the requisite GPS data. Architecture is dependent on application of A-MAV. For using A-MAV, traffic and security surveillance, images sensors are required.

After selection of sensor as per the need, IMU[2]enabled GPS plays its vital role. Basic working principal of IMU[2] is to detect the rate of acceleration with the help of accelerometer. The changes in rotational attributes are measured by gyroscope. Data collected by IMU[2] is fed into processing unit, that calculates current position based on velocity and time. The disadvantage borne is, IMU[2] faces problem of accumulated error and abbe error and it increases the distance between actual location and the location where the system think it is. So, to increase the accuracy, GPS is also used with IMU[2]. As GPS is used for collecting points, processor will be programmed in a way to calculate the distance using the concept of Euclidean Distance Theory. The Euclidean distance or Euclidean metric is the "ordinary" distance between two points that one would measure with a ruler, and is given by the Pythagorean formula. By using this formula as distance, Euclidean space becomes a metric space.

The Euclidean distance between points p and q is the length of the line segment connecting them ($\overline{pq}$). The position of a point in a Euclidean n-space is a Euclidean vector. So, p and q are

Euclidean vectors, starting from the origin of the space, and their tips indicate two points. The Euclidean norm, or Euclidean length, or magnitude of a vector measures the length

$$\|p\| = \sqrt{p_1^2 + p_2^2 + \cdots + p_n^2} = \sqrt{p \cdot p}$$

Where the last equation involves the dot product.

A vector can be described as a directed line segment from the origin of the Euclidean space (vector tail), to a point in that space (vector tip). If we consider that its length is actually the distance from its tail to its tip, it becomes clear that the Euclidean norm of a vector is just a special case of Euclidean distance: the Euclidean distance between its tail and its tip. The distance between points p and q may have a direction (e.g. from p to q), so it may be represented by another vector, given by

$$q - p = (q_1 - p_1, q_2 - p_2, \cdots, q_n - p_n)$$

In a three-dimensional space (n=3), this is an arrow from p to q, which can be also regarded as the position of q relative to p. It may be also called a displacement vector if p and q represent two positions of the same point at two successive instants of time. The Euclidean distance between p and q is just the Euclidean length of this distance (or displacement) vector:

$$\|q - p\| = \sqrt{(q - p) \cdot (q - p)}. \quad (2)$$

This is equivalent to:

$$\|q - p\| = \sqrt{\|p\|^2 + \|q\|^2 - 2p \cdot q}.$$

In general, for an n-dimensional space, the distance is

$$d(p,q) = \sqrt{(p_1 - q_1)^2 + (p_2 - q_2)^2 + \dots + (p_i - q_i)^2 + \dots + (p_n - q_n)^2}.$$

So in three-dimensional Euclidean space, the distance is

$$d(p,q) = \sqrt{(p_1 - q_1)^2 + (p_2 - q_2)^2 + (p_3 - q_3)^2}.$$

Let p = p1 and q be p2 then

$$TD = d(p1\ p2) + d(p2\ p3) + \ldots\ldots$$

Total Distance(TD) =

$$\sum_{i=1}^{n} d(p_i\ p_{i-1})$$

In this proposed equational procedure, the processor will calculate the distance and for it ARM (Advanced RISC Machine) processor will be used for real time application because of its high compatibility, low energy consumption and less cost.

COMMUNICATION FLUX

For sending and receiving of data, either of the two modes (GSM/CDMA)[6] of mobile communication may be used on the basis of their requirement. The data may be transformed in encrypted format, so for more secure and faster data transfer CDMA technique may be used. The GSM as such is widely used and has nearly captured the whole market due to its freedom of using the same frequency for several calls as per requirement. So for more secured operations CDMA and for less secured and faster operations GSM can be inducted.

DATABASE INCOGNITO

Once the data will be received it will be updated in GIS as layers for visualizations and storage in spatial database for dynamic route guidance and other vast purposeful operations in future.

For spatial database, PostgreSQL bears the upper contender-ship, for its low system requirement and 36% higher speed than oracle besides being open source thus available free of cost.

DYNAMIC ROUTE GUIDANCE (DRG)

DRG is a concern that keeps on tempering the minds till the word "navigation" is associated with traffic movements on or above grounds. This not only ensures the dulcet movement of traffic but also ensures a quick and safe journey. This DRG can be planned according to the traffic and the medium of movement like earth, air, water etc. The navigations on earth are mostly required on full fledged roads, streets etc and if these navigations are planned considering congestions on roads, shortest path etc, which of course saves time, then it really outcasts the hurdles of navigations. We depend on efficient algorithms of shortest path to save time and money. These algorithms can change its shape as per their requirement in the medium. So for A-MAVs also, the DRG can play its role under its applications in various aspects like tracking the traffic on roads during its flight above the roads and the Graph Growth Algorithm can play its role in this navigation considering the height as a negligible factor. The DRG algorithms can be prepared as per its requirement in various applications.

GRAPH GROWTH ALGORITHM[7]

The Graph Growth Algorithm Implemented With Two Queues

We describe the data structures and basic procedures related to the graph growth algorithm implemented with two queues in this section. The graph growth algorithm implemented with two queues (TQQ). TQQ is an improved version of the growth graph implementation. Let us review the basic procedure in constructing a shortest path tree as shown below:

```
Procedure   ShortestPathTreeConstruction(s)
  begin
      Queue_Initialization(Q);
     for i=1 to n do
       d(i) = + infinite;
        d(s) = 0;
```

```
      while (Q != Null) do
         Queue_Removal(Q, i);
          for each successor node j of node i
do
             if d(j) > d(i) + l(i, j) then
                begin
                   d(j) = d(i) + l(i, j)
                   Queue_Insertion(Q, j)
                end
         end
```

The four basic functions involved in this procedure are:

1) Queue_Initialization(Q)   initialize queue Q;

2) Queue_Removal(Q, i)   remove node i from   queue Q;

3) Queue_Insertion(Q, j)   insert node j into   queue Q;

4) Q = Null?   Check whether queue Q is empty.

The first set of nodes is maintained by a priority queue Q. Nodes in the second set are further split into two categories: 1) the unreached nodes which have never entered Q, i.e., nodes whose distance labels are still infinite and 2) labeled nodes, i.e., the nodes that have passed through Q at least once, and the nodes whose current distance labels have already been used. A logical enhancement of the PAP algorithm is to replace the LIFO stack with a FIFO queue and construct a new data structure. This new data structure is called two-queue (Figure 3). Because both Q' and Q" are queues in the two-queue data structure, nodes can be inserted at the end of Q' and Q", and they can be removed from the head of Q' and Q".

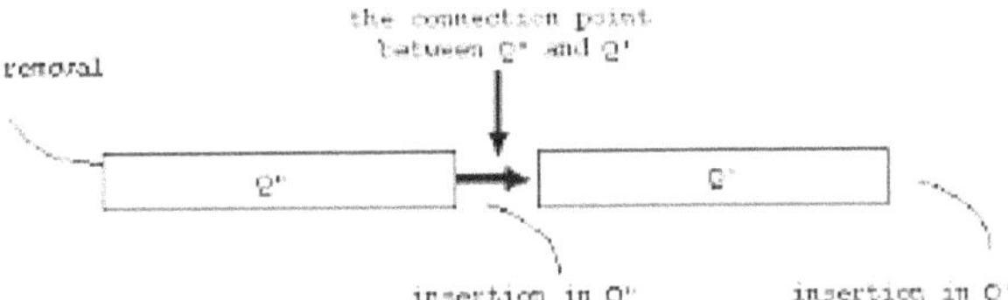

Fig. 1.3 The two-queue data structure (Q)

It follows that for any node that is not already in Q, the node is inserted at the end of Q' if it is unreached, or the node is inserted at the end of Q" if it is temporarily labeled. This leads to the following change in the Queue Insertion (Q, j) operation of the PAP implementation. Other operations remain the same.

Queue_Insertion(Q, j)   For any node j that is not already in Q, insert the node at the end of Q' if the node is unreached, i.e., if S(j) = unreached or insert the node at the end of Q" if the node is temporarily labeled.

CONCLUSION

In this paper we have concentrated on A-MAV architecture for different application. The architecture and algorithm has been discussed in detail with concept of Euclidean distance concept and with usage of GIS along with spatial Database.

REFERENCES

1) Wolf, R., Eissfeller, B., Hein, G.W., \ A Kalman Filter for the Integration of a Low Cost INS and an attitude GPS," Institute of Geodesy and Navigation, Munich, Germany.

2) Sadia Riaz, Atif Bin Asghar "INS/GPS Based State Estimation of Micro Air Vehicles Using Inertial Sensors", Innovative Systems Design and Engineering www.iiste.org ISSN 2222-1727 (Paper) ISSN 2222-2871 (Online) Vol 2, No 5, 2011.

3) Roland Brockers, Sara Susca, David Zhu, Larry Matthies "Fully Self-Contained Vision-Aided Navigation and Landing of a Micro Air Vehicle Independent from External Sensor Inputs".

4) Matthew B. Higgins, "Heighting with GPS: possibilities and limitations".

5) Ruijie He, Abraham Bachrach, Michael Achtelik, Alborz Geramifard, Daniel Gurdan, Samuel Prentice, Jan Stumpf and Nicholas Roy, "On the Design and Use of a Micro Air Vehicle to Track and Avoid Adversaries".

6) He Hai-jian, Wu Yu-hang, Chen Chun, Liu Xiang-nan , "Integration of Mobile Communication and "3S" Techniques in Electric Power Monitor System"

7) Ms. Parul Jain, Ms. Surabhi Jain, Mr. Kumar Tushar Shrivastava "Spatial Database Intelligence in Traffic Control".

FUZZY RANDOM IMPULSE NOISE REMOVAL FROM COLOR IMAGE SEQUENCES

M. Sanjeev Kumar, Sake Pothalaiah, B. Brahma Reddy

Vignana bharahti institute of technology, Hyderabad, india

ABSTRACT

Fuzzy filter plays a vital role by method of median based rank filter method known as block filtering. In this paper noise of random impulse, in which video of noise pixels are selected based on median values of noise pixels and original image pixels. This filtering frame work consists of three steps, first removal noise by step by step instead of considering whole image and retaining the details as it is. if considerable noise is already present, in that case more free neighbors to compare to available. This is based upon thresholding. Calculate for each pixel component a degree to which it is considered noise-free and a degree to which it is considered noisy. If the noisy degree is larger than the noise-free degree, the pixel component is filtered, otherwise it remains unchanged. The determination of both degrees is mainly based on temporal information (comparison to the corresponding pixel component in the previous frame). Note, however, that only innon-moving areas can large temporal differences be assigned to noise. In areas where there is motion, such differences might also be caused by that motion Only pixels that have been detect to be noisy are filtered based upon MAD(Mean absolute Difference).these simulations results are carried out using MATLAB.

INTRODUCTION

Images and videos belong to the most important information carriers in today's world (e.g., traffic observations, surveillance systems, autonomous navigation, etc.). However, the images are likely to be corrupted by noise due to bad acquisition, transmission or recording. Such degradation negatively influences the performance of many image processing techniques and a preprocessing module to filter the images is often required. Among those filters, more and more fuzzy techniques start to appear in literature Fuzzy set theory was introduced by Zadeh in 1965 [24] and is a generalization of classical set theory. A classical crisp set over a universe can be modeled by a mapping (characteristic function): an element belongs to the set or does not belong to it. Fuzzy sets are now modeled as mappings (membership functions).

There is no need to use a threshold to decide whether a difference in color component value between two pixels is large or not. Two differences that differs by   only one unit (which is not noticeable by the human eye) could then be respectively large and not large. It is better to allow a difference to be large to some intermediate degree. For a larger difference, this degree will be higher than that of a smaller difference. For an illustration of the effectiveness of fuzzy set theory in image processing, we refer to, Most filters in literature, which are developed for video, are intended for sequences corrupted by additive Gaussian noise (e.g., [3]–[7]). Only few video filters for the impulse noise case can be found, however, several impulse noise filters for still images exist. The best known among them are the median based rank-order filters .But also some fuzzy techniques can be found, Such as 2-D filters could be used to filter each of the frames of a video successively. However, temporal inconsistencies will arise due to the neglecting of the temporal correlation between successive frames.

A better alternative would be to use 3-D filtering windows, in which also pixels from neighboring frames are taken into account. The main problem in using neighboring frames is motion between them. Using pixels at corresponding spatial positions in neighboring frames for noise removal may introduce ghosting artifacts in the presence of camera and object in motion. In the method proposed in this paper, we will therefore only in non-moving areas assign a temporal impulse between two corresponding spatial positions to noise (detection phase) and for the replacement of a noisy pixel (filtering phase) motion compensation will be applied to find the most reliable pixel in the previous frame. Analogously, a distinction between filters intended for grayscale images and for color images needs to be made. Filters for grayscale images could be used for color images by applying them on each of the color bands of the image separately. In this paper, we consider the images to be modeled in the RGB color space and we thus have three color bands: red, green and blue. However, such approach will generally introduce many color artifacts, especially in textured areas, due to the neglect of the correlation between the different color bands. To incorporate this correlation, vector-based methods were introduced

To avoid blurring due to the filtering of noise-free pixels, this filtering framework has been further refined by weighted filtering techniques and switching schemes where the filter is only used for detected noisy pixels. The drawback of vector-based methods, however, is that their performance is highly reduced for higher noise levels. Consider for example a neighborhood, in which all pixels have one noisy component, and the other components are noise-free.

To preserve the details as much as possible, the noise is removed by three successive filtering steps. Only pixels that have been detected to be noisy are filtered. This filtering is done by block matching, a technique used for video compression that has already been adopted in video filters for the removal of Gaussian noise (e.g., [4]–[6]), but that has not really found its way to impulse noise filters yet. The correspondence between blocks is usually calculated by a mean absolute difference (MAD) that is heavily subject to noisy impulses. Therefore, we introduce aMAD measure that is adaptive to detected noisy pixels components. To benefit as much as possible from the spatial and temporal information available in the sequence, the search region for corresponding blocks contains pixel blocks both from the previous and current frame.

THE PROPOSED ALGORITHM

The filtering framework presented in this paper is intended for color video corrupted by random impulse noise. If we respectively denote the original (noise-free) sequence by the Io T th frame of that sequence by Io(t)and the red, green and blue component of the color $I_o(x,y,t)$of the pixel at the Xth row and Yth column in that frame by $I^R_o(x,y,t)$ $I_o^G(x,y,t)$ and $I^B_o(x,y,t)$ $I_o^G(x,y,t)$

$$I^B_o(x,y,t)(i.e: I_o(x,y,t)= I^R_o(x,y,t) \; I_o^G(x,y,t) \text{ and } I^B_o(x,y,t)$$

$$I^c_n(x,y,t)= \quad \{ \; I^c_0(x,y,t), \text{ with probability } 1\text{-}p,$$

$$I^c_n(x,y,t), \text{with probability } p,\text{-}\rightarrow \qquad\qquad .....(1)$$

The noisy sequence is determined as follows where and denotes the probability that a pixel component value is corrupted and replaced by a identically distributed independent random noise value coming from a uniform distribution on the interval of possible color component values. For the color videos used in the experiments of this paper, 8 bits are used for the storage of the color component values and we work with a uniform distribution on the interval [0, 255]. Further, the probability that a given color component value is corrupted is independent on whether the neighboring values or the values in the other color components are corrupted or not.

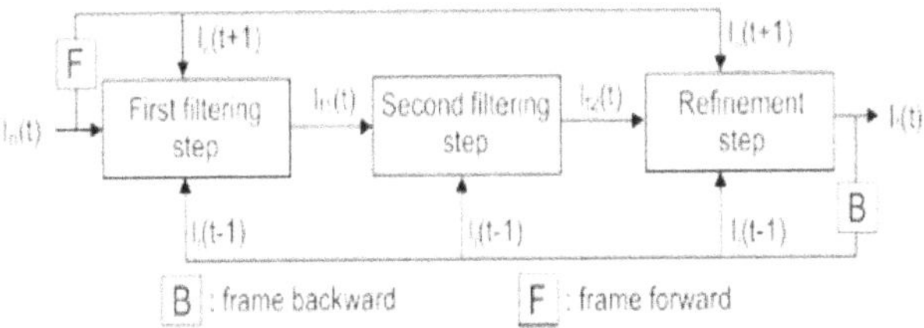

Fig.1 O overview of different steps of algorithm

The proposed filtering framework consists of three successive filtering steps as depicted in Fig. 1. By removing the noise step by step, the details can be preserved as much as possible. Indeed, if a considerable part of the noise has already been removed in a previous step, and more noise-free neighbors to compare to are available, it will be easier to distinguish noise from small details. In the first step (with output denoted by), we calculate for each pixel component a degree to which it is considered noise-free and a degree to which it is considered noisy. If the noisy degree is larger than the noise-free degree, the pixel component is filtered, otherwise it remains unchanged. The determination of both degrees is mainly based on temporal information (comparison to the corresponding pixel component in the previous frame). Note, however, that only in non-moving areas can large temporal differences be assigned to noise. In areas where there is motion, such differences might also be caused by that motion. As a consequence, and as can be seen in Fig. 2, impulses in moving areas will not always be detected in this step. They can, however, be detected in the second step (Section II-B) (output). Analogously as to the first step, again a noise-free degree and a noisy degree are calculated. However, the detection is now mainly based on color information. A pixel component can be seen as noisy if there is no similarity to its (spatio-temporal) neighbors in the given color, while there is in the other color bands. The third step (Section II-C) (output), finally, removes the remaining noise and refines the result by using as well temporal as spatial and color information. For example, homogeneous areas can be refined by removing small impulses that are relatively large in that region, but are not large enough to be detected in detailed regions and that thus have not been detected yet by the previous general detection steps. The results of the different successive filtering steps are illustrated for the 20th frame of the "Salesman" sequence in Fig. 2.

A. First Filtering Step

1) Detection: In this detection step, we calculate for each of the components of each pixel a degree to which it is considered noise-free and a degree to which it is thought to be noisy. A

Fig2:The original 20th frame of the "salesman"sequence (a),the frame rupted by 20% random impulse noise (b)(PSNR=15.05 dB) and the result after the first (c)(PSNR=23.72 dB),second (d)(PSNR=29.42dB)and refinement step (e)(PSNR=36.78dB)respectively.

Component for which the noisy degree is larger than the noise free degree, i.e., that is more likely to be noisy than noise-free, will be filtered. Other pixel components will remain unchanged. The noise-free degree and the noisy degree are determined by fuzzy rules as follows. We consider a pixel component to be noise-free if it is similar to the corresponding component of the pixel at the same spatial location in the previous or next frame and to the corresponding component of two neighboring pixels in the same frame. In the case of motion, the pixels in the previous frames can not be used to determine whether a pixel component in the current frame is noise-free. Therefore, more confirmation (more similar neighbors or also similar in the other color components) is wanted instead. For the noise-free degree of the red component (and analogously for the other components), this is achieved by the following fuzzy rule.

Fuzzy Rule1:If $((|I_n{}^R(x,y,t)- I_f{}^R(x,y,t-1)$ is NOT LARGER POSITIVE OR $((|I_n{}^R(x,y,t)- I_n{}^R(x,y,t+1)$ is NOT LARGER POSITIVE) AND there are two neighbor $(x+k,y+l,t)$ $(-2\leq k,l\leq2)$ and $(k,l)\neq(0,0)$ for which $((|I_n{}^R(x,y,t)- I_n{}^R(x+k,y+l,t)$ is NOT LARGE POSITIVE

OR (there are four neighbors $(x+k,y+l,t)$ $(-2\leq k,l\leq2)$ and $(k,l)\neq(0,0)$) for which $((|I_n{}^R(x,y,t)- I_n{}^R(x+k,y+l,t)|$ is NOT LARGE POSITIVE OR (there are two neighbors $(x+k,y+l,t)$ $(-2\leq k,l\leq2)$ and $(k,l)\neq(0,0)$ for which $((|I_n{}^R(x,y,t)- ((|I_n{}^R(x+k,y+l,t)|$ is NOT LARGE POSITIVE and

$((|I_n{}^G(x,y,t)- ((|I_n{}^G(x+k,y+l,t)|$ OR is$((|I_n{}^B(x,y,t)- ((|I_n{}^B(x+k,y+l,t)$ are NOT LARGE POSITIVE)

THEN the $I_n{}^R(x,y,t)$ red component is considered NOISEFREE. To represent the linguistic value large positive in the above rule, a fuzzy set is used, with a membership function as depicted in Fig. 3 (see Section III-A for the determination of the parameters). For the conjunctions (AND), disjunctions (OR) and negations (NOT) in fuzzy logic, triangular norms, triangular conforms and involutive negatrons [26] are used. In this paper, we will use the minimum operator, the maximum operator and the standard negatron respectively. Those operators are simple in use and yielded the best results, but the difference compared to the results for another choice of operators is neglectible. The outcome of the rule, i.e., the degree to which the red component of the pixel at position is considered noise-free, is determined as the degree to

Which the antecedent in the fuzzy rule is true:

μ^Rnoisefree(x,y,t)=max(min(max($\alpha_1(x,y,t),\alpha_2(x,y,t),M_2(x,y,t)$,max($M_4(x,y,t),M_{2b}(x,y,t)$))

where

$$\alpha_1(x,y,t)=(1-\mu_{LP}(|I_n{}^R(x,y,t)- I_f{}^R(x,y,t-1|)), \quad\quad\quad .....(2)$$

$$\alpha_2(x,y,t)=(1-\mu_{LP}(|I_n{}^R(x,y,t)- I_n{}^R(x,y,t+1|)),$$

and where $M_2(x,y,t)$ and $M_4(x,y,t)$ respectively denote the degree to which there are two (respectively four) neighbors for which the absolute difference in the red component value is not large positive, that is determined as the second (respectively fourth) largest element in the set

$$\{1-\mu_{LP}(|I_n{}^R(x,y,t)- I_f{}^R(x+k,y+l,t|)) \parallel -2\leq k,l\leq2 \text{ and } (k.l)\neq(0,0)\}$$

And $M_{2b}(x,y,t)$)denotes the degree to which there are two neighbors for which the absolute differences in the red component and one of the two color components are not large positive, determined as the second largest element in the set

$$\{\min(1-\mu_{LP}(|I_n{}^R(x,y,t)-I_n{}^R(x+k,y+l,t)|),\max(1-\mu_{LP}(|I_n{}^G(x,y,t)-I_n{}^G(x+k,y+l,t)|),$$

$$1-\mu_{LP}(|I_n{}^B(x,y,t)-I_n{}^B(x+k,y+l,t)|))) \parallel -2\leq k,l\leq2 \text{ and } (k.l)\neq(0,0)\}$$

Analogously, a degree to

Which the component of a pixel is considered noisy is calculated. In this step, we consider a pixel component to be noisy if the absolute difference in that component is large positive compared to the pixel at the same spatial location in the previous frame and if not for five of its neighbors the absolute difference in this component and one of the other two color bands is large positive compared to the pixel at the same spatial location in the previous frame (which means that the difference is not caused by motion). Further, we also want a confirmation either by the fact that in this color band, there is a direction in which the differences between the considered pixel and the two respective neighbors in this direction are both large positive or large negative and if the absolute difference between those two neighbors is not large positive (i.e., there is an impulse between two pixels that are expected to belong to the same object) or by the fact that there is no large difference between the considered pixel and the pixel at the same spatial location in the previous frame in one of the other two color bands. For the red component (and analogously the other components) this leads to the following fuzzy rule.

Fuzzy rule2:If $((|I_n{}^R(x,y,t)- I_f{}^R(x,y,t-1)$ is NOT L$((|I_n{}^R(x+k,y+l,t)$ARGER POSITIVE AND (for five neighbours $(x+k,y+l,t)$ $(-2{\leq}k,l{\leq}2)$ and $(k,l){\neq}(0,0)$) $|I_f{}^R(x+k,y+l,t-1)$is LARGE POSITIVE

$((|I_n{}^R(x,y,t)- I_n{}^R(x,y,t+1)$ AND NOT LARGER POSITIVE) AND there are two neighbor $(x+k,y+l,t)$ $(-2{\leq}k,l{\leq}2)$ and $(k,l){\neq}(0,0)$ for which $((|I_n{}^R(x,y,t)- I_n{}^R(x+k,y+l,t)$is NOT LARGE POSITIVE AND$(I_n{}^G(x+k,y+l,t)- I_f{}^G(x+k,y+l,t-1)|$ OR $|I_n{}^B(x+k,y+l,t)- I_f{}^B(x+k,y+l,t-1)|$is LARGE POSITIVE.

AND (in one of the there are four neighbors $(x+k,y+l,t)$ $(-2{\leq}k,l{\leq}2)$ and $(k,l){\neq}(0,0)$) for which $((|I_n{}^R(x,y,t)- I_n{}^R(x+k,y+l,t)|$ is NOT LARGE POSITIVE OR (there are two neighbours $(x+k,y+l,t)$ $(-2{\leq}k,l{\leq}2)$and $(k,l){\neq}(0,0)$ for which $((|I_n{}^R(x,y,t)- ((|I_n{}^R(x+k,y+l,t)|$ is NOT LARGE POSITIVE and $((|I_n{}^G(x,y,t)- ((|I_n{}^G(x+k,y+l,t)|$ OR is$((|I_n{}^B(x,y,t)- ((|I_n{}^B(x+k,y+l,t)$ are NOT LARGE POSITIVE) AND (in one of the four directions (the differences $((|I_n{}^R(x,y,t)-$ both $I_n{}^R(x+k,y+l,t)$ AND $((|I_n{}^R(x,y,t)- I_n{}^R(x-k,y-l,t)(k,l){\in}\{(-1,-1),(-1,0),(-1,1),(0,1)\})$ are both LARGE POSITIVE OR both LARGE NEGATIVE)AND the absolute difference$| I_n{}^R(x+k,y+l,t)- I_n{}^R(x-k,y-l,t)|$is NOT LARGE POSITIVE OR $(I_n{}^G(x,y,t)- ((|I_n{}^G(x,y,t-1)|$ is NOT LARGE POSITIVE OR $| ((|I_n{}^B(x,y,t)- ((|I_n{}^B(x,y,t-1)|$is NOT LARGE POSITIVE))

THEN the red component is considered noisy. Analogously to the linguistic term large positive, also large negative is represented by a fuzzy set, characterized by the membership function given in Fig. 4 (see Section III-A for the determination of the parameters). The degree to which for five neighbors the absolute differences in the red component and one of the other two components are large positive compared to the corresponding pixels in the previous frame, denoted by, is determined as the fifth largest value in the set The degree to which the absolute difference between the pixel at position and the corresponding pixel in the previous

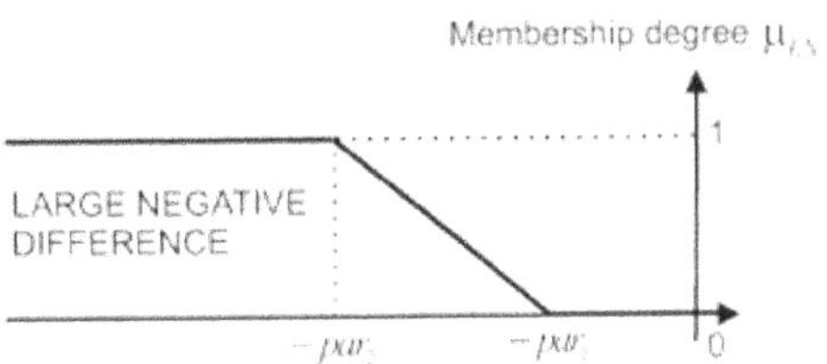

Fig. 4 The membership functions of the fuzzy set large negative.

frame is large positive and five of its neighbors do not show motion, is then given by

$$B(x,y,t)=\min\left(\mu_{LP(|I_n}{}^R(x,y,t)- I_f{}^R(x,y,t-1)|),1-t_{pos}(x,y,t)\right)$$

Further, the degree to which there is no large difference between the considered pixel and the pixel at the same spatial location in the previous frame in one of the other two color bands is given by

$$\Delta(x,y,t)=\max((1-\mu_{LP(|I_n}{}^G(x,y,t)- I_f{}^G(x,y,t-1)|),$$

$$1-\mu_{LP(|I_n}{}^B(x,y,t)- I_f{}^B(x,y,t-1)|),-\rightarrow(2)$$

Finally, the degree to which there is a direction in which the pixel at position is an impulse, denoted by, is determined as the maximum value in the set

$$\{\min\,(\max(\varepsilon^1{}_{(k,l)(x,y,t)}, \varepsilon^2{}_{(k,l)(x,y,t)}, \varepsilon^3{}_{(k,l)(x,y,t)})$$

$$\|\,(k,l)\,\varepsilon\{(-1,-1),(-1,0),(-1,1),(0,1)\}\}$$

where Combining the above, we get 2) Filtering: In this subsection, we discuss the filtering for the red color band. The filtering of the other color bands is analogous. We decide to filter all red pixel components that are considered more likely to be noisy than noise-free, i.e for which. The red components of the other pixels remain unchanged to avoid the filtering of noise-free pixels (that might have been incorrectly assigned a low noisy degree, but for which the high noise-free degree assures us that it is noise-free) and thus detail loss. On the other hand, noisy pixel components might remain unfiltered due to an uncorrected high noise-free degree, but those pixels can still be detected in the next filtering step

$$\mu^R unch(x,y,t)=\{0, \text{if } \mu R noisy(x,y,t)> \mu R noisyfree(x,y,t)$$

1,else

$$\mu^R unch(x,y,t)=(\,\mu^R unch(x,y,t),\,\mu^G unch(x,y,t),\,\mu^B unch(x,y,t)$$

is thus a vector that gives information on whether the respective color component of the color pixel should be filtered. Analogously as denotes the frame of denotes the 2-D array of vectors that gives information about the pixel components of the frame of the sequence. To exploit the spatial and temporal information in the sequence as much as possible, the filtering is performed by block matching. To do this, a noise-adaptive mean absolute difference (MAD) is used to calculate the correspondence between the color components of two blocks of image pixels (where is a general parameter that determines the block size): with and In the above equations, and are the two frames (2-D color images), to which the blocks belong, and indicate the spatial coordinates of the central pixel of the considered block and respectively stand for the vertical and horizontal coordinates of the displacement vector, i.e., the block that is considered in has as the central pixel. The binary functions and indicate whether the pixel components and are reliable and should be used (respectively) or not (respectively).

$$R_{MAD}{}^{\frac{1}{M}}(x,y,r,s,W)=\cfrac{\sum\sum \ldots}{\sum\sum \ldots}$$

Using only noise-free pixel components allows us to calculate a more reliable measure to estimate whether two blocks would correspond in the red component if they were both noise-free. If the noise adaptive MAD is assigned the value. Further, the noise adaptive MAD is not considered reliable if not for at least half of the positions in the blocks, both compared values are reliable (and) or not for half of the reliable positions the absolute difference is not large positive (i.e.,). It is also not considered reliable if both the green and blue component of the central pixels are reliable and their absolute difference is large positive (i.e., and). In these cases, the noise-adaptive changed to the value, such that the block will not be used for the filtering. For the filtering of a red component in this first step of our algorithm, we determine the displacement vectors and for the best matching block in a search region of size

$$\gamma(x,y,k,l,r,s)=\mu^R(x+k,y+l)- \mu^{\tilde{R}}(x+k+r,y+l+s)$$

$$\Phi(x,y,k,l,r,s)=|I^R(x+k,y+l)- I^R(x+k+r,y+l+s)- \rightarrow (3)$$

in respectively the previous frame and the current frame (due to large motion, sometimes no corresponding block might be found in the previous frame, but the region around the given pixel in the current frame might be similar) as follows (for the optimization of the parameters and , refer to Section III-A): The minimum value itself is denoted by . We have used the identity function Id for the binary function corresponding to the previous frame, since this frame has already been filtered and should be noise-free. The minimum value itself is denoted by .We have restricted ourselves here to pixels , for which , since only noise-free pixels should be used to replace the noisy pixel component . A pixel component , for which , is then filtered as the noise-free center of the best corresponding block in the search region, if it exists . Otherwise, a spatial filtering is performed. If the pixel component remains unchanged in this step. Summarized, the output of this first step for the red component is Given as follows. If then else if then with (in a general notation) where the spatial filtering framework is given by If which is unlikely to happen in practical situations, then the output is given by

B. Second Filtering Step

In our aim to preserve the details as much as possible, the noise is removed in successive steps. In this step, the noise is detected based on the output of the previous step . Also in this second filtering step, a degree to which a pixel component is expected to be noise-free and a degree to which a pixel component is expected to be noisy, is calculated. In the calculation of those degrees, we now take into account information from the other color bands. A color component of a pixel is considered noise-free if the difference between that pixel and the corresponding pixel in the previous frame is not large in the given component and also not large in one of the other two color components. It is also considered noise-free if there are two neighbors for which the difference in the given component and one of the other two components are not large. So, the other color bands are used here as a confirmation for the observations in the considered color band to make those more reliable. For the red component (and analogously the other color components),this gives the following fuzzy rule.

Fuzzy Rule 3:IF $|(I_f^R(x,y,t)- I_f^R(x,y,t-1)|$is NOT LARGER POSITIVE $|(I_f^G(x,y,t)- I_f^G(x,y,t-1)|$is NOT LARGER POSITIVE OR $| I_{fl}^B(x,y,t-1)- I_f^B(x,y,t-1)|$is NOT LARGER POSITIVE)

OR (for two neighbors$(x+k,y+l,t)(-1\leq k,l\leq1)$and $(k,l)\neq(0,0))| |(I_{fl}^R(x,y,t)- |(I_f^R(x+k,y+l,t)|$is NOT LARGER POSITIVE) OR $||(I_{fl}^B(x,y,t)- |(I_f^B(x+k,y+l,t)|$is NOT LARGER POSITIVE)

The degree to which the red component of the pixel at position is considered noise-free, is then

$$\mu^R_{2,noisefree}(x,y,t)=max(\zeta(x,y,t),\eta(x,y,t))$$

$$\zeta(x,y,t)=min(1-\mu_{LP(|}I_{fl}^R(x,y,t)-I_f^R(x,y,t-1|),$$

$$max(1-\mu_{LP(|}I_{fl}^G(x,y,t)-I_f^G(x,y,t-1|),$$

$$1-\mu_{LP(|}I_{fl}^B(x,y,t)-I_f^B(x,y,t-1|)),$$

given by where and is the second largest element in the set A pixel component is considered noisy if there are three neighbors that differ largely in that component, but are similar (not a large difference) in the other two components.

and $\quad\quad\quad\quad\quad\eta(x,y,t))$ is the second largest element in the set

$$\{min(1-\mu_{LP(|}I_{fl}^R(x,y,t)-I_f^R(x+k,y+l,t|),$$

$$max(1-\mu_{LP(|}I_{fl}^G(x,y,t)-I_f^G(x+k,y+l,t|),$$

$$\mu_{LP(|}I_{fl}^B(x,y,t)-I_f^B(x+k,y+l,t|))),\,\|\,-1\leq k,l\leq 1 \text{ and } (k,l)\neq(0,0)\}$$

It is also considered noisy if in the considered color band, its value is larger or smaller than the component values of all its neighbors, and this is not the case in both of the other color bands. For the red component of a pixel (and analogously for the other components), this corresponds to the following fuzzy rule.

Fuzzy Rule 4: IF (for three neighbors $(x+k,y+l,t)(-1\leq k,l\leq 1$) and $(k,l)\neq(0,0))$ $|(I_{fl}^R(x,y,t)-(I_f^R(x+k,y+l,t)|$ is LARGE POSITIVE AND $|(I_{fl}^G(x,y,t)-(I_f^G(x+k,y+l,t)|$ is NOT LARGE POSITIVE AND $|(I_{fl}^B(x,y,t)-|(I_f^B(x+k,y+l,t)|$ is NOT LARGE POSITIVE).

OR $(((\text{for all neighbors }(x+k,y+l,t)(-1\leq k,l\leq 1)$ and $(k,l)\neq(0,0))$ $|(I_{fl}^R(x,y,t)-(I_f^R(x+k,y+l,t)|$ is LARGE POSITIVE) OR (for all neighbors $(x+k,y+l,t)$ $(-1\leq k,l\leq 1)$) and $(k,l)\neq(0,0))$)

$|I_{fl}^R(x,y,t)-I_f^R(x+k,y+l,t)|$ is LARGE NEGATIVE)) AND NOT ((for all neighbors $((x+k,y+l,t)$

$(-1\leq k,l\leq 1)$ and $(k,l)\neq(0,0)$ $|I_{fl}^G(x,y,t)-|(I_f^G(x+k,y+l,t)|$ is LARGE NEGATIVE) OR (for all neighbors $((x+k,y+l,t)(-1\leq k,l\leq 1))$ and $(k,l)\neq(0,0))$ $|I_{fl}^B(x,y,t)-(I_f^B(x+k,y+l,t)|$ is LARGE POSITIVE) OR (for all neighbors $(((x+k,y+l,t)(-1\leq k,l\leq 1)$) and $(k,l)\neq(0,0))$ $|I_{fl}^B(x,y,t)-(I_f^B(x+k,y+l,t)|$ is LARGE NEGATIVE)

THEN the red component is considered NOISY. The noisy degree for the red component of the pixel at position is then calculated as follows:

$$\mu_{2,}^R noisy(x,y,t)=max(\theta(x,y,t),k(x,y,t)) \text{ where}$$

$$\{min\,(\mu_{LP(|}I_{fl}^R(x,y,t)-I_f^R(x+k,y+l,t)|),min\,(1-\mu_{LP(|}I_{fl}^G(x,y,t)-\mu_{LP(|}I_f^G(x+k,y+l,t)|),$$

$$1-\mu_{LP(|}I_{fl}^B(x,y,t)-|I_f^B(x+k,y+l,t)|)),\,(-1\leq k,l\leq 1))\text{ and }(k,l)\neq(0,0))\text{---}\rightarrow(4)$$

the third largest element in the set and where with All red (and analogously green and blue) components for which are filtered , the other red components remain unchanged : Analogously to the first step, for the filtering of the red components (and analogously the green and blue components) for which , we search for the noise-free center of the best corresponding bloc in the search region in the current and previous frame. The minimum value itself is denoted by The minimum value itself is denoted by If , the is filtered as Red pixel components that are considered noise-free remain unchanged:

C. Third Filtering Step

The result from the previous steps is further refined based on temporal, spatial and color information. Namely, the red component (and analogously the green and blue component) of a pixel is refined in the following cases:

$$\Delta(x,y,t)=(|I_{f2}^{c}(x,y,t)- I_f^{c}(x,y,t-1)|)$$
$$\sum \sum \sum_{c \in \{R,G,B\}} \Delta(x+k,y+l,t)- \sum_{c \in \{R,G,B\}} \Delta(x,y,t)/24$$

and if $\quad I_{f2}^{R}(x,y,t)- I_f^{R}(x,y,t-1)|)> par_{2}$ and

$|I_f^{R}(x,y,t-1)- I_n^{R}(x,y,t+1)|)< par_{1,\text{then the red component}} I_{f2}^{R}(x,y,t)$ is considered to be noisy

$\mu_{2,unch}^{R}(x,y,t)=0).$ The last check is to prevent noise. $\rightarrow (5)$

In non-moving areas, pixels will correspond to the pixels in the previous frame, which allows us to detect remaining isolated noisy pixels. If lies in a non-moving 3 3 neighborhood, i.e., (with) and if and , then the red component is considered to be noisy . The last check is to prevent noise propagation in the case that the pixel in the previous frame would not have been filtered correctly.

- Very small impulses might not have been detected by the algorithm. In homogeneous areas however, such impulses might be relatively large and can be detected more easily. Let and respectively denote the second largest and second smallest red component value among the eight neighbors in a 3*3 neighborhood around. If (homogeneous neighborhood) and further also or (the red component is clearly larger or smaller than the neighborhood), then the red component is considered to be noisy

- Based on color information, the red component is considered to be noisy if in a 3 3 neighborhood two neighbors can be found for which In all other cases the red component value is considered to be noise-free and should not be adapted anymore .Analogously as in the previous steps, for the filtering of the red components for which we search for the noise-free center of the best corresponding block in the search region in the current and previous frame. The minimum value itself is denoted by the minimum value it is denoted by a red component for which is filtered as Otherwise, it remains unchanged

EXPERIMENTAL RESULTS

To be able to judge the performance of the proposed method, we will use the mean absolute error (MAE),

$$MAE(I_o(t),I_f(t))$$
$$\sum_{c \in \{R,G,B\}}^{n} \sum_{x=1}^{m} \sum_{y=1}^{n} |I0c(x,y,t) - Ifc(x,y,t)|/3.n.m$$

The peak-signal-to noise ratio (PSNR)

$$MSE(I_0(t),If(t) =$$
$$\sum_{c \in \{R,G,B\}} \sum_{x=1}^{m} \sum_{y=1}^{n} (I0c(x,y,t) - Icf(x,y,t))2$$

$$/3.n.m$$

$$PSNR(I_0(t),I_f(t)=10.\log_{10} [S^{2}]/[MSE(I_0(t),I_f(t)]$$

and the normalized color difference (NCD) as objective measures of similarity and dissimilarity between a filtered frame and the original one , each containing rows and columns of pixels. The MAE is given by The lower the MAE, the more similar (less dissimilar) the images. The PSNR value is defined as where denotes the maximum possible value of a pixel component (here). The higher the PSNR value, the more similar less dissimilar) the images. Finally, the

$$\mathrm{NCD}(I_o(t),I_f(t)) = \frac{\sum_{x=1}^{M}\sum_{y=1}^{N} \| L^*_{uvw}(x,y,t) - f L^*_{uvw}(x,y,t) \|}{\sum_{x=1}^{M}\sum_{y=1}^{N} \| L^*_{uvw}(x,y,t) \|}$$

between an original and a filtered frame, is calculated as where is the Euclidean norm and and respectively denote the -transform of the original and the filtered frame. The lower the NCD value, the more similar (less dissimilar) the images. The remainder of this section is structured as follows. The parameter values for the membership functions and the window sizes are determined in Section III-A. The proposed Σfiltering framework is compared to other state-of-the-art noise reduction methods in Section III-B. Some notes on the complexity are discussed in Section III-C.

D. Parameter Selection

First the parameters and that determine the membership functions and in Figs. 3 and 4 are determined. To do this, we have fixed the window sizes and of the pixel neighborhood and the search region in the filtering as (5 5 neighborhood) and (11 11 search region) and we have let the parameters and run over a Comparison to Other State-of-the-Art Filters In this subsection, the performance of the proposed method is compared to that of the adaptive vector median filter (AVMF). the video adaptive vector directional median filter (VAVDMF) with 3-D filtering window from [30] and the 2-D fuzzy impulse noise reduction method for color images (INRC) The adaptive vector median filter orders the pixels (color vectors) in the 3-D filtering window based on increasing accumulated (Euclidean) distance to the other pixels in the window. If the Euclidean distance between the central pixel in the window and the mean of a given number of vectors that have the lowest accumulated distance, is greater than a given threshold, then the central pixel is filtered as the pixel with the lowest accumulated distance, otherwise, it remains unchanged. In the video adaptive vector directional median filter .the vectors are first ordered based on increasing angular distance. If the absolute distance between the central pixel in the window and the mean of a given number of vectors that have the lowest accumulated angular distance, is greater than a given threshold, then the central pixel is filtered as the pixel with the lowest accumulated absolute distance (magnitude), otherwise, it remains unchanged.

Matlab in combination with the mex-function and executed on an Intel® Xeon® CPU X3360 @ 2.83 GHz. Some suggestions to reduce the computation time needed by the proposed filter could be the following. First, the block matching in the filtering

CONCLUSION

In this paper, we have presented a new filtering framework for color videos corrupted with random valued impulse noise. In order to preserve the details as much as possible, the noise is removed step by step. The detection of noisy color components is based on fuzzy rules in which information from spatial and temporal neighbors as well as from the other color bands is used. Detected noisy components are filtered based on block matching where a noise adaptive mean absolute difference is used and where the search region contains pixels blocks from both the previous and current frame. The experiments showed that the proposed method outperforms other state-of-the-art methods both in terms of objective measures such as MAE, PSNR and NCD and visually.

REFERENCES

[1] R. H. Chan, C. Hu, and M. Nikolova, "An iterative procedure for removing random-valued impulse noise," IEEE Signal Process. Lett., vol. 11, no. 12, pp. 921–924, 2004. [3] S. M. M. Rahman, M. O. Ahmad, and M. N. S. Swamy, "Video denoising based on inter-frame statistical modelling of wavelet coefficients," IEEE Trans. Circuits Syst. Video Technol., vol. 17, no. 2, pp.187–198, 2007.

[2] L. Jovanov, A. Pizurica, V. Zlokolica, S. Schulte, P. Schelkens, A. Munteanu, E. E. Kerre, and W. Philips, "Combined wavelet-domain and motion-compensated video denoising based on video codec motion estimation methods," IEEE Trans. Circuits Syst. Video Technol., vol. 19, no. 3, pp. 417–421, 2009.

[3] H. B. Yin, X. Z. Fang, Z. Wei, and X. K. Yang, "An improved motion- compensated 3-D LLMMSE filter with spatio-temporal adaptive filtering support," IEEE Trans. Circuits Syst. Video Technol., vol. 17, no. 12, pp. 1714–1727, 2007.

[4] L. Guo, O. C. Au, M. Ma, and Z. Liang, "Temporal video denoising based on multihypothesis motion compensation," IEEE Trans. Circuits Syst. Video Technol, vol. 17, no. 10, pp. 1423–1429, 2007.

[5] T. Mélange, M. Nachtegael, E. E.Kerre, V. Zlokolica, S. Schulte, V. De Witte, A. Pizurica, and W. Philips, "Video denoising by fuzzy motion and detail adaptive averaging," J. Electron. Imaging, vol. 17, no. 4, pp. 043005–, 2008.

[6] R. C. Hardie and C. G. Boncelet, "LUM filters: A class of rank-order-based filters for smoothing and sharpening," IEEE Trans. Signal Process., vol. 41, pp. 1061–1076, 1993.

[7] S. Schulte, V. De Witte, M. Nachtegael, D. Van der Weken, and E. E. Kerre, "Fuzzy random impulse noise reduction method," Fuzzy Sets Syst., vol. 158, pp. 270–283, 2007.

[8] S. Schulte, M. Nachtegael, V. De Witte, D. Van der Weken, and E. E. Kerre, "A fuzzy impulse noise detection and reduction method," IEEE Trans. Image Process., vol. 15, no. 5, pp. 1153–1162, 2006.

REVIEW ON NANO COMPOSITE MATERIALS AND THEIR APPLICATION IN NANO AIR VEHICLES.

I. Rajania

Department of Physics, Joginpally M.N. Rao Women's Engineering College, Moinabad, Ranga Reddy Dist. [E-mail id: rajani.i@rediffmail.com]

V. Brahmajirao

Dept. of Nano Science and Technology, School of Biotechnology,   MGNIRSA, A unit of D. Swaminathan Research Foundation, [DSRF], Hyderabad- 500015.
[E-mail id: profvrbr@gmail.com]

C. Udaya Kiran

Department of Mechanical Engineering, JBIET, Moinabad, Ranga Reddy Dist.
[E-mail id: ukchavan@gmail.com.]

ABSTRACT

Nano composite materials possess conspicuous physical properties by virtue of their manufacturing techniques. They are very hard and preserve their structural properties, since the fillers used in their synthesis have Nano structure. Their mechanical properties suit them for extensive use in Nano Air Vehicles .This paper presents a detailed review of very recent work by established manufacturers of NAV's like Lockheed and Martin of U.S.A.

Several Current research papers & Theses about NAV's [from (1) Georgia Institute of Technology -Year 2011, (2) Massachusetts Institute of Technology –Year 2007, (3) UNIVERSITÀ DEGLI STUDI DI SALERNO-Year (2008),] are reviewed.

Reviews on: (1) Industrial applications of super hard Nano composite coatings, (2) polymer Nano fibers by electro spinning and their applications in Nano composites, (3) Nanostructured materials for advanced energy conversion and storage devices, (4) Critical issues in Nano composites research, as well as promising techniques for processing precursors for macroscopic Nano composites are discussed.

Nano composites based on Conducting Polymers and Carbon Nanotubes from Fancy Materials to Functional Applications are discussed. Magnetic and electromagnetic evaluation of the magnetic nanoparticles used in special applications of NAV's are presented. Several Industrial applications (1) of super hard Nano composite coatings, and their utilization for Novel Materials for energy applications (from 'A decade of EU-funded research') (2) of Electrical Properties of the Epoxy Nano composites are presented. Design of the SAMARAI Mono-wing Rotorcraft Air vehicle (2009), is discussed. Nano composite Coatings for Severe Applications are discussed. Information about mechanical, electrical, chemical, surface and structural properties are reviewed. Application of Functionalized Grapheme (2011) is critically analyzed.

INTRODUCTION

Nano means small (10^{-9}m) but of high potency. This emerges with large applications piercing through all the disciplines of knowledge, leading to industrial and technological growth. Precisely this implies the controlling of the morphology at Nano scale dimensions of substances or particles to produce nanostructure materials. Nanotechnology is already having its impact on products as diverse as novel foods, medical devices, chemical coatings, personal health testing kits, sensors for security systems, water purification units for manned space craft, displays for hand-held computer games, and high-resolution cinema screens. Encompassing all these overlapping fields are the tools used to measure and manipulate ultra-small structures, the Nano-scale resolution microscopes.

These advances are providing the means for nanotechnology to progress and for researchers to expand their expertise into new application fields. In current research of Nano composite material (due to some of their unique properties) which are a class of materials having promising applications., and having capabilities of replacing conventional materials, it has become critical to modify the surfaces of the Nano particles for both fundamental research and engineering applications. Nano particles exhibit novel magnetic properties; in general, the ideal response is associated with isolated particles. In practical applications, one invariably has to consider a collection or aggregate of particles forming Nano powders. Further, several applications require consolidation and sintering of Nano phase materials into solid blocks or thin films. These processing routes often lead to unavoidable formation of agglomerations and larger grains that effectively prevent the materials from attaining their full potential in terms of the desirable magnetic response associated with Nano particles, even though it is difficult to completely eliminate agglomeration in large-scale commercial synthesis of Nano materials.

A composite material typically consists of one or more fillers (fibrous or particulate) in a certain matrix. A carbon fiber composite is one in which at least one of the fillers consists of carbon fibers (short or continuous, unidirectional or multidirectional, woven or non-woven) The high performance characteristics of composites depend not only on the physical properties of the components (fiber and matrix) but also on the interfacial region (interface) that exists between these dissimilar components. Effective fillers require good bonding (chemical, mechanical, and physical) between the fibers and the matrix.

Fillers are, in general, solid substances that are embedded in polymers to reduce costs and improve performance. It can be distinguished between non-functional (extender) fillers, which are mainly used to reduce costs, and functional fillers that improve or generate new properties in the composites. Crucial parameters in determining the effect of fillers on the properties of composites are the filler geometry (size, shape, aspect ratio) and the filler matrix interactions .Generally the inclusion of fillers into polymers leads to an increase in modulus and a decrease in toughness.

Nano-particles have the ability to impart superior properties to micron-sized particles. The addition of micron-sized particles into polymer matrices leads to the enhancement of toughening properties and erosion of other mechanical properties like tensile strength, tensile modulus, and creep resistance, which can be overcome by addition of inorganic or organic Nano-sized particles. This addition of Nano-particles exhibits drastic improvement in super conductivity, magnetism, thermal stability and mechanical properties .The improvements are due to the high surface-to-volume ratio of Nano-particles and enormous interfacial adhesion between Nano-particles and pristine polymers. Magnetic polymer Nano-particles can be tailor-made depending on the final applications [15].

Polymer-coated magnetic Nano particles are of great technological interest as the coating provides a matrix for binding of the particles and also prevents grain growth and agglomeration. Coating Nano-particles with polymers affords the possibility of minimizing it to a great extent. The process of coating not only provides effective encapsulation of individual Nano-particles but also controls the growth in size thus yielding a better overall size distribution. Polymers & their composites are finding ever increasing usage for numerous industrial applications such as bearing material, rollers, seals, gears, cams, wheels, clutches and transmission belts etc. [17]. Different types of polymer show different friction and wear behavior [18-21].

Several kinds of magnetic polymer Nano-particles have been produced from natural and synthetic polymers with the intention to incorporate groups on the surface or to treat their surface to perform a certain application. The multifunctional response of these materials can have tremendous potential and lead to improved materials for applications like magneto resistive damping, mechanical and electrical devices: loudspeakers, seals, sensors, dampers, etc. All these applications depend on a core magnetic material with a modified surface resulting in additional functionality the methods of polymer modification include copolymerizing, reinforcing, and blending. Fiber reinforced polymer composites (FRPCs) are currently used for a wide variety of structural applications in many aerospace, automotive and chemical industries because of their high strength-to-weight and stiffness-to-weight ratios [22,23].Polymers and especially their blends are replacing the conventional engineering materials because of several advantages such as easy processing, improved properties, attractive appearance/color, self-lubrication and light weight, to meet the present day requirements.

NAVs are defined as small air vehicles with an operating range less than 1 km, a maximum flight altitude around 100 m, endurance less than one hour, and maximum takeoff weight (MTOW) of 25 g while MAVs are defined as 5 kg MTOW with endurance around 1 hour and an operative range around 10 km[2].

Many research institutions are actively studying and developing new air vehicles, reducing size and weight while improving performance, and adding more functionality. Examples here are Harvard Micro-robotics Laboratory in the USA [4], Department of Aeromechanics and Flying Engineering from Moscow Institute of Physics and Technology in Russia [3, 5], Aircraft Aerodynamics and Design Group at Stanford University (USA) [6, 7], the Autonomous Systems Laboratory at ETH Zurich (Switzerland) [8,9], and Department of Precision Instrument and

Mechanology at Tsinghua University in China [10]. Several companies and agencies also play an important role in the manufacturing and development of AVS. Examples here are DARPA [11] from USA, Prox Dynamics [12,13] from Norway, and Syma from USA.

NAVs reduce dimensions which are perfect for reconnaissance inside buildings, providing a very useful tactical advantage. As reported in [14], such small vehicles are currently the only way to remotely "look" inside buildings in the battlefield.

They can carry specific sensors such as gas, radiation or other sensors used to locate biological, nuclear, chemical, or other threats. They can, for instance, fly inside toxic clouds and transmit data or bring samples back to the base station, and, thus, provide vital information on the composition and extent of gaseous clouds and improve the assessment of danger.

Some of the applications described above can be extended to the civilian field. For example, the police and the fire brigade could use the capability of indoor flights for inspecting unsafe or collapsed buildings [12] in order to search for survivors or simply do a safety check of the building structure. However, even with a working prototype, there are still many technology challenges that need to be solved before NAVs are ready to take on real military missions. The key areas of ongoing research include propulsion and energy storage; aerodynamics at low Reynolds numbers; guidance, navigation, sensors, and communications systems; and advanced manufacturing techniques. Propulsion and energy storage systems for NAVs require a highly efficient power source with sufficient energy and power density to fly for sustainable time periods. Even more challenging is the requirement for dense energy storage that can be efficiently converted to thrust in order to propel the vehicle as well as power all the subsystems. Some designs employ a dual functionality power source that is also used as a structural component, thus enabling a significant reduction in size while serving to maximize flight durations [16].

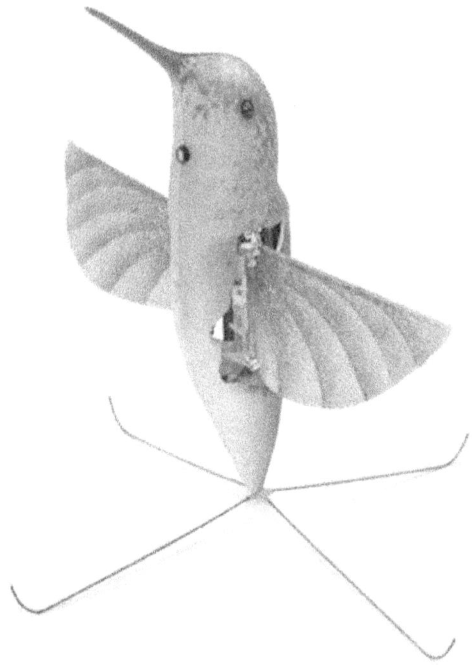

Fig:1 Nano air vehicles present engineering challenges, august 2011,Ref:[16]

There are still many technology challenges that need to be solved before NAVs are ready to take on real military missions. The key areas of ongoing research attempted to discuss include propulsion and energy storage; aerodynamics at low Reynolds numbers; guidance, navigation, sensors, and communications systems; and advanced manufacturing techniques.

PROPULSION AND ENERGY STORAGE

Propulsion and energy storage systems for NAVs require a highly efficient power source with sufficient energy and power density to fly for sustainable time periods. Even more challenging is the requirement for dense energy storage that can be efficiently converted to thrust in order to propel the vehicle as well as power all the subsystems. Some designs employ a dual functionality power source that is also used as a structural component, thus enabling a significant reduction in size while serving to maximize flight durations.

In [24],jiang.et.al developed Polypropylene/clay Nano composites used by the shell of vehicle air heater were prepared with big numerator fuse insert technics,it had high integration capability. The manufacture process of polypropylene/clay Nano composites was introduced, then we tested the capability of this material and the shell of vehicle air heater. The result indicated that polypropylene/clay Nano composites could be used by the shell of vehicle air heater. Its little filling and low density was good for the parts to reduce the weight.

Donghai Wang.et.al [25] used anionic sulfate surfactants to assist the stabilization of grapheme in aqueous solutions and facilitate the self-assembly of in situ grown Nano crystalline TiO_2, rutile and anatase, with graphene. These nanostructured TiO_2-graphene hybrid materials were used for investigation of Li-ion insertion properties. The hybrid materials showed significantly enhanced Li-ion insertion/extraction in TiO_2. The specific capacity was more than doubled at high charge rates, as compared with the pure TiO_2 phase. The improved capacity at high charge☐discharge rate may be attributed to increased electrode conductivity in the presence of a percolated graphene network embedded into the metal oxide electrodes.

AERODYNAMICS

Aerodynamic design challenges for NAVs are driven by low "Reynolds number" physics (<15,000). Natural fliers such as birds and insects operate at very low Reynolds numbers compared to a typical aircraft and have inspired research designs with flapping wings like Aerovironment's hummingbird, insects, and even maple tree seeds. The severe space limitations of the NAV make the aerodynamic challenge even greater than that overcome by the NAVs larger predecessor, the miniature air vehicle (MAV).

In Nano composites Fillers are, in general, solid substances that are embedded in polymers to reduce costs or improve performance. It can be distinguished between non-functional (extender) fillers, which are mainly used to reduce costs, and functional fillers that improve or generate new properties in the composites. Crucial parameters in determining the effect of fillers on the properties of composites are the filler geometry (size, shape, aspect ratio) and the filler matrix interactions [26].Generally the inclusion of fillers into polymers leads to an increase in modulus and a decrease in toughness.

GUIDANCE, NAVIGATION, SENSORS, AND COMMUNICATIONS

Another major technical challenge is the integration of navigation, guidance, and control sensors onto a single chip at the Nano scale to meet the restrictive size, weight, and power requirements of the NAV vehicle design. Furthermore, NAVs may be required to operate autonomously, in concert with others via a network, or travel in "swarms," making the challenge even more complex due to increased processing and sensory requirements of functions. Several kinds of magnetic polymer

Nano-particles have been produced from natural and synthetic polymers with the intention to incorporate groups on the surface or to treat their surface to perform a certain application. The multifunctional response of these materials can have tremendous potential and lead to improved materials for applications like magneto resistive damping, mechanical and electrical devices: loudspeakers, seals, sensors, dampers, etc. All these applications depend on a core magnetic material with a modified surface resulting in additional functionality [27].The methods of polymer modification include copolymerizing, reinforcing, and blending. Fiber reinforced polymer composites (FRPCs) are currently used for a wide variety of structural applications in many aerospace, automotive and chemical industries because of their high strength-to-weight and stiffness-to-weight ratios [23,28].Polymers and especially their blends are replacing the conventional engineering materials because of several advantages such as easy processing, improved properties, attractive appearance/color, self-lubrication and light weight, to meet the present day requirements.

ADVANCED MANUFACTURING

Designing and building a NAV requires revolutionary manufacturing technologies and innovative subsystem packaging and configuration layout to integrate Nano scale components into the airframe itself. This was a key component of the DARPA NAV program, which included the requirement to demonstrate a clear process to integrate other subsystems into the airframe and show the capability to manufacture the system.

CONCLUSION

The developing arena of Nano science is assuming prominent places in all the facets of life. Nano composite materials and Nano particles can create mysterious Characteristics to the specimen which have a wide variety, extensive range and imposing application when put to use as technological tools. Nano composites started its span very recently and promises extensive potential in each and every walk of life for the developing technological advancement.

REFERENCES

[1] Luca Petricca.et.al, International Journal of Aerospace Engineering Volume 2011 (2011), Article ID 214549, 17 pages doi:10.1155/2011/214549

[2] U. Yearbook, UAS: The Global Perspective, vol. 164, UAS Yearbook, 7th edition, 2009/2010.

[3] S. V. Serokhvostov, "Ways and technologies required for MAV miniaturization," in Proceedings of the European Micro Air Vehicle Conference (EMAV '08), Braunschweig, Germany, July 2008.

[4] R. J. Wood, S. Avadhanula, E. Steltz et al., "An autonomous palm-sized gliding micro air vehicle—design, fabrication, and results of a fully integrated centimeter-scale MAV," IEEE Robotics and Automation Magazine, vol. 14, no. 2, pp. 82–91, 2007. View at Publisher • View at Google Scholar

[5] S. V. Serokhvostov, "Flapping wings efficiency investigation on the basis of physical law conservation," in Proceedings of the European Micro Air Vehicle Conference (EMAV '08), Braunschweig, Germany, 2008.

[6] I. Kro, F. Prinz, and M. Shantz, "A miniature rotorcraft concept phase II final report," Tech. Rep., Stanford University, Palo Alto, Calif, USA, 2001.

[7]   I. Kroo and P. Kunz, "Development of the mesicopter a miniature autonomous rotorcraft," inProceedings of the American Helicopter Society Vertical Lift Aircraft Design Conference, American Helicopter Society, San Francisco, Calif, USA, 2000.

[8]   S. Bouabdallah, P. Murrieri, and R. Siegwart, "Towards autonomous indoor micro VTOL,"Autonomous Robots, vol. 18, no. 2, pp. 171–183, 2005. View at Publisher • View at Google Scholar• View at Scopus

[9]   S. Bouabdallah, M. Becker, and R. Siegwart, "Autonomous miniature flying robots: coming soon!," IEEE Robotics and Automation Magazine, vol. 14, no. 3, pp. 88–98, 2007. View at Publisher • View at Google Scholar • View at Scopus

[10] H. Wu, D. Sun, and Z. Zhou, "Micro air vehicle: configuration, analysis, fabrication, and test,"IEEE/ASME Transactions on Mechatronics, vol. 9, no. 1, pp. 108–117, 2004. View at Publisher •View at Google Scholar • View at Scopus

[11] T. Hylton, "Nano Air Vehicle program," http://www.darpa.mil/dso/thrusts/materials/multfunmat/nav/index.htm

[12] D. H. Paulsen, "Nano UAS- an upcoming reality," in Proceedings of the 24th International Unmanned Air Vehicles Conference, Bristol, UK, 2009.

[13] Prox Dynamics, http://www.proxdynamics.com [14] R. J. Bachmann, "Biologically inspired mechanisms facilitating multi modal locomotion for areal micro-robot," in Proceedings of the 24th International Unmanned Air Vehicles Conference, Bristol, UK, 2009.

[15] Rajani.i.et.al, International Conference on Materials Processing and Characterization March 8-10, 2012, Hyderabad Appeared in International journal of Advanced Materials Manufacturing & Characterization (2012) , [ISSN 2277-3886] , 1(1), p173-194,

[16} Nano air vehicles present engineering challenges, august 2011.

[17] PascoeM.W, (1973), Plain and Filled Plastics Materials in Bearing 6(5): 184–190.

[18] Lancaster, J. K, (1969), Abrasive Wear of Polymers, Wear, 14(4): 223–239. [19]Briscoe, B., (1981), Wear of Polymers: An Easy on Fundamental Aspects, Tribology International, 14(4): 231–243.

[20] ZumGahr K. H. (1998). Wear by Hard Particles, Tribology International, 31(10): 587–596.

[21] Liu.G, (2004), A Study on Slide Wear Mechanisms of Ultrahigh Molecular Weight Polyethylene/Polypropylene Blends, Wear, 256(11–12): 1088–1094.

[22] ASM Handbook (1992). Friction, Lubrication and Wear Technology. American Society for Metals, MetalPark, Ohio, USA.

[23] Yamaguchi.Y, (1990), Tribology of Plastic Materials.Tribology Series 16, Elsevier, New York.

[24]jiang.et.al, ChinaFAW Croup Corporation R&D Center;Chinese Academy of Science;Jilin University.

[25] Donghai Wang.et.al,ACS Nano, 2009, 3 (4), 907-914• DOI: 10.1021/nn900150y • Publication Date (Web): 26 March 2009.

[26] Hohenberger.et.al,(2001)Fillers and Reinforcements / Coupling Agents, in Plastics Additives Handbook, p. 901-943.

[27] I. Neamtu.et.al,(2005),Journal of Physics., Vol. 50, Nos. 9.–10, P. 1081.–1087.

[28] ASM Handbook (1992). Friction, Lubrication and Wear Technology. American Society for Metals, MetalPark, Ohio, USA

CLONE DETECTION USING TEXTUAL AND METRIC ANALYSIS IN JAVA

G. Anil Kumar[1], C.R.K. Reddy[2], A. Govardhan[3]

[1]MGIT, Dept. of Computer Science, Hyderabad, India
Email: anilgkumar@mgit.ac.in
[2]CBIT, Dept. of Computer Science, Hyderabad, India
Email: crkreddy@cbit.ac.in
[3]JNTUH, Dept. of Computer Science, Hyderabad, India
Email: govardhan_cse@jntuh.ac.in

ABSTRACT

Copying code fragments and then reuse by pasting with or without minor modifications or adaptations are common activities in software development This type of reuse approach of existing code is called code cloning and the pasted code fragment without is called a clone of the original One of the major shortcomings of such duplicated fragments is that if a bug is detected in a code fragment all the other fragments similar to it should be investigated to check the possible existence of the same bug in the similar fragments In this paper we compare different clone detection techniques and tools First part of this paper explains the classification of clone detection techniques and the later work done in this area and proposed method

KEYWORDS: Software clone, Clone detection, Textual Analysis, Clone Cluster, Clone Pair, Metrics computation, Abstract syntax Trees.

INTRODUCTION

Software systems provide vital support for the smooth running of an organization's business. It is the responsibility of maintainers to keep the system up-to date and functioning properly [6]. The success of free software is evident from the large and growing number of hardware devices that include free software components. Devices such as routers, televisions, set-top boxes and media players are commonly based on software such as the Linux kernel, the Samba file/print server and the BusyBox toolset [13]. Reusing code fragments by copying and pasting with or without minor changes is a common activity in software development. As a result software systems often contain fragments of code that are very similar, called code clones [8]. A code clone is a code portion in source files that is identical or similar to another. Clones are introduced because of various reasons such as reusing code by 'copy-and-paste', mental macro, or intentionally repeating a code portion for performance enhancement, etc [2]. Identifying software clones and understanding how software changes between releases are two important issues for maintainers where a text-based approach is likely to be useful. Maintenance of large software systems under pressure often leads to a phenomenon referred to as software cloning [4].

A clone is a copy of a code fragment. Usually, clones consisting out of more than 5 statements are considered interesting. Since the clone relation is symmetric we better say that the origin and the copy form a clone pair [10]. Clones are frequently introduced by code scavenging, that is, by copying existing code and modifying it. Finding clones in software systems is important in many maintenance, reengineering, and program understanding contexts [9]. Detection and removal of such clones promises decreased software maintenance costs of possibly the same magnitude [1]. One major problem in detecting a clone is that it is impossible to be absolutely certain that one section of code has been copied and pasted from another [6]. Unfortunately, a precise definition of what differentiates a clone from a non-clone is lacking. This can present problems for evaluating clone detectors [9]. A clone detector must try to find pieces of code of high similarity in a system's source text. The main problem is that it is not known beforehand which code fragments may be repeated. Thus the detector really should compare every possible fragment with every other possible fragment. Such a comparison is prohibitively expensive from a computational point of view and thus, several measures are used to reduce the domain of comparison before performing the actual comparisons [8].

A clone detection system should have ability to select clones or to report only helpful information for user to examine clones, since large number of clones is expected to be found in large software systems [2]. Although some researchers argue not to remove clones because of the associated risks, there is a consensus that clones need to be detected at least. Detection is necessary to find the place where a change must be replicated and also useful to monitor development in order to stop the increase of redundancy before it is too late [15]. An important application of clone detection is the improvement of source code quality by refactoring duplicated code fragments [7]. From the analysis of software application it appears that the inclusion of these clones results from the addition of some extra functionality which is similar but not identical to some existing logic within a system. Its seems that when presented with the challenge of adding new functionality the natural instinct of a programmer is to copy, paste and modify the existing code to meet the new requirements and thus creating a software clone [6]. Clone detection techniques attempt at finding duplicated code, which may have undergone minor changes afterward. The typical motivation for clone detection is to factor out copy-paste-adapt code, and replace it by a single procedure [5]. At the beginning of any clone detection approach, the source code is partitioned and the domain of the comparison is determined. There are three main objectives in this phase: remove uninteresting parts, determine source units and determine comparison units / granularity [8]. Code clones can be discovered manually by scavenging through the program source and identifying duplicates one by one. Depending on the size of the program, this manual process can become tedious and labor intensive. An automatic clone detection tool can be beneficial by reducing the time and effort needed to find clones [11]. A good clone detector should scale to large programs, while considering sufficient semantic-level information to detect all three types of clone. This requires that the management of necessary semantic information should be inexpensive in terms of time and memory [14].

Various approaches have been applied in practice with good results. The main technical difficulty is that duplication is often masked by slight differences: eformatting, code modifications, changed variable names and inserted or deleted lines of code all make it harder to recognize software clones [16]. Five established detection tools will be used in the evaluation process; JPlag, MOSS, Covet, CCFinder and CloneDr. JPlag and MOSS are web-based academic tools for detecting plagiarism in student's source code. CloneDr and CCFinder are stand alone tools looking at code duplication in general [6]. Problem Mining is a process change that aims at coping with the existing base of software clones in a system already in service, for which new development and maintenance is still being done [3]. The handling of duplicated code can be very problematic in many respects. An error in one component is reproduced in every copy. Since it is not documented in which places duplicates can be found, it is extremely hard to hand and remove such errors [10]. Duplicated fragments can also significantly increase the work to be done when enhancing or adapting code [12].

SHINOBI is a tool for automatic code clone detection. The main features of SHINOBI are as 1) It is highly integrated with Microsoft Visual Studio. For instance, it is implemented as an add-in of Visual Studio. A programmer can easily check and edit detected code clones. 2) SHINOBI automatically detects code clones with source code being edited. The detection process is automatic, implicit, and quick. A programmer can get a list of code clones without noticeable time penalty whenever he develops with the IDE. 3) It also highlights code clones to help recognize code clones during software maintenance tasks [18]. In the clone detection tool comparison experiment at the First International Workshop on Detection of Software Clones, clones were separated into three categories: 1) Exact copies, with no differences between them, 2) Parameterized copies, where variable and function calls can have different names and/or types have changed and 3) Modified copies, where some modification is done, such as adding or deleting lines of code [11]. Efficient token-based clone detection is based on suffix trees, originally used for efficient string search [15]. Various approaches have been applied in practice with good results. The main technical difficulty is that duplication is often masked by slight differences: eformatting, code modifications, changed variable names and inserted or deleted lines of code all make it harder to recognize software clones [16].

Hence, we propose an efficient clone detection scheme to detect all types of clones available in the source files. Here we use a hybrid technique based on textual and metric analysis to detect the duplicate codes. The rest of the paper is described as follows. A section 2 brief about the literature survey. The concept of textual and metric analysis is described in Section 3 and the proposed methodology is explained with necessary equations and diagrams in Section 4. The Results obtained in the proposed method is discussed in Section 5 and Section 6 concludes the work.

RELATED WORK

A handful of researches have been presented in the literature for the detection of Clones. Recently, utilizing artificial intelligence techniques like Abstract Syntax Trees, KClone, Substring Matching, Frequent Itemset Techniques have received a great deal of attention among researchers. A brief review of some recent researches is presented here.

Rainer Koschke *et al.* [15] proposed that, reusing software through copying and pasting was a continuous plague in software development despite the fact that it creates serious maintenance problems. Various techniques have been proposed to find duplicated redundant code (also known as software clones). This study has compared those techniques and shown that token-based clone detection based on suffix trees is extremely fast but yields clone candidates that are often no syntactic units. Current techniques based on abstract syntax trees on the other hand find syntactic clones but are considerably less efficient. It describes how they can made use of suffix trees to find clones in abstract syntax trees. That new approach was able to find syntactic clones in linear time and space. It reports the results of several large case studies in which we empirically compare the new technique to other techniques using the Bellon benchmark for clone detectors.

Stephane Ducasse *et al.* [16] proposed that, duplicated code is known to pose severe problems for software maintenance, it is difficult to identify in large systems. Many different techniques have been developed to detect software clones, some of which are very sophisticated, but are also expensive to implement and adapt. Lightweight techniques based on simple string matching are easy to implement, but how effective are they? They presented a simple string-based approach which they have successfully applied to a number of different languages such COBOL, JAVA, C++, PASCAL, PYTHON, SMALLTALK, C and PDP-11 ASSEMBLER. In each case the maximum time to adapt the approach to a new language was less than 45 minutes. In that paper, they investigate a number of simple variants of string-based clone detection that normalize differences due to common editing operations, and assess the quality of clone detection for very different case studies. Their results confirm that that inexpensive clone detection technique generally achieves high recall and acceptable precision. Overzealous normalization of the code before comparison, however, can result in an unacceptable numbers of false positives.

Chanchal K. Roy *et al.* [8] proposed that, over the last decade many techniques and tools for software clone detection have been proposed. In that paper, they provide a qualitative comparison and evaluation of the current state-of-the-art in clone detection techniques and tools, and organize the large amount of information into a coherent conceptual framework. We begin with background concepts, a generic clone detection process and an overall taxonomy of current techniques and tools. Then classify, compare and evaluate the techniques and tools in two different dimensions. First, we classify and compare approaches based on a number of facets, each of which has a set of (possibly overlapping) attributes. Second, we qualitatively evaluate the classified techniques and tools with respect to taxonomy of editing scenarios designed to model the creation of Type-1, Type-2, Type-3 and Type-4 clones. Finally, they have provided examples of how one might use the results of this study to choose the most appropriate clone detection tool or technique in the context of a particular set of goals and constraints. The primary contributions of this paper are: (1) a schema for classifying clone detection techniques and tools and a classification of current clone detectors based on this schema, and (2) a taxonomy of editing scenarios that produced different clone types and a qualitative evaluation of current clone detectors based on this taxonomy.

Yue Jia *et al.* [14] proposed that, in all applications of clone detection it is important to have precise and efficient clone identification algorithms. That work outlines a new algorithm, KClone for clone detection that incorporates a novel combination of lexical and local dependence analysis to achieve precision, while retaining speed. It also reports on the initial results of a case study using an implementation of KClone with which we have been experimenting. The results indicate the ability of KClone to find types-1,2, and 3 clones compared to token-based and PDG-based techniques, and also reports results of an initial empirical study of the performance of KClone compared to CCFinderX.

R. R. Brooks *et al.* [17] proposed that, in cloning attacks, an adversary captures a sensor node, reprograms it, makes multiple copies, and inserts these copies, into the network. Cloned nodes subvert sensor network processing from within. In a companion paper, they shown how to detect and remove clones from sensor networks using random key pre distribution security measures.

Keys that are present on the cloned nodes are detected by using authentication statistics based on key usage frequency. For consistency with existing random key pre distribution literature, and ease of explanation, the network in that paper used an Erdos-Renyi topology. In the Erdos-Renyi topology, the probability of connection between any two nodes in the network is uniform. Since the communications ranges of sensor nodes were limited, this topology is flawed. This article applies the clone detection approach from to more realistic network topologies. Grid and ad hoc topologies reflect the node connectivity patterns of networks of nodes with range limits. They provided analytical methods for choosing detection thresholds that accurately detect clones. They used simulations to verify our method. In particular they found the limitations of that approach, such as the number of nodes that can be inserted without being detected.

Shinji Kawaguchi *et al.* [18] proposed that, code clones decrease the maintainability and reliability of software programs, thus it is being regarded as one of the major factors to increase development/maintenance cost. They have introduced SHINOBI, a novel code clone detection/modification tool that was designed to aid in recognizing and highlighting code clones during software maintenance tasks. SHINOBI was implemented as an add-in of Microsoft Visual Studio that automatically reports clones of modified snippets in real time.

Nam H. Pham *et al.* [21] proposed that, Model-Driven Engineering (MDE) has become an important development framework for much large-scale software. Previous research has reported that as in traditional code-based development, cloning also occurs in MDE. However, there has been little work on clone detection in models with the limitations on detection precision and completeness. That paper presented the ModelCD, a novel clone detection tool for Matlab/Simulink models that is able to efficiently and accurately detect both exactly matched and approximate model clones. The core of ModelCD is two novel graph-based clone detection algorithms that are able to systematically and incrementally discover clones with a high degree of completeness, accuracy, and scalability. We have conducted an empirical evaluation with various experimental studies on many real-world systems to demonstrate the usefulness of our approach and to compare the performance of ModelCD with existing tools.

Kodhai. E *et al.* [19] proposed that, clone detection has considerably evolved over the last decade, leading to approaches with better results but with increasing complexity. Most of the existing approaches were limited to finding program fragments similar in their syntax or semantics, while the fraction of candidates that were actually clones and fraction of actual clones identified as candidates on the average remain similar. In that paper, a metric-based approach combined with the textual comparison of the source code for the detection of functional Clones in C source code has been proposed. Various metrics had been formulated and their values were utilized during the detection process. Compared to the other approaches, this method was considered to be the least complex and to provide a more accurate and efficient way of Clone Detection. The results obtained had been compared with the two other existing tools for the open source project Weltab.

Kodhai.E *et al.* [20] proposed that, a clone detection approach was to found out the reused fragment of code in any application to maintain .Various types of clones are being identified by clone detection techniques. Since clone detection was evolved, it provides better results and reduces the complexity. A different clone detection tool makes the detection process easier and efficiently produces the results. In many existing system, it mainly focuses on line by line detection or token based detection to find out the clone in the system. So it makes the system to take long time to process the entire system. If the fragment of code are not at exact code but the functionalities makes it similar to each other. Then existing system doesn't figure out the clone of that type of clones in it. That paper proposed the combination of textual and metric analysis of a source code for the detection of all types of clone in a given set of fragment of java source code. Various semantics had been formulated and their values were used during the detection process. Those metrics with textual analysis provides less complexity in finding the clones and gives accurate results.

Armijn Hemel *et al.* [13] proposed that, Software released in binary form frequently used third-party packages without respecting their licensing terms. For instance, many consumer devices have firmware containing the Linux kernel, without the suppliers following the requirements of the GNU General Public License. Such license violations are often accidental, e.g., when vendors receive binary code from their suppliers with no indication of its provenance. To help find such violations, they have developed the Binary Analysis Tool (BAT), a system for code clone detection in binaries. Given a binary, such as a firm ware image, it attempts to detect cloning of code from repositories of packages in source and binary form. They have evaluated and compared the effectiveness of three of BAT's clone detection techniques: scanning for string literals, detecting similarity through data compression, and detecting similarity by computing binary deltas.

TEXTUAL & METRIC ANALYSIS

In textual comparison line by line comparison is done. That is whole lines are compared to each other textually using hashing for strings. This comparison is done by string matching algorithm. The result can be plotted in a dot plot and each dot indicates a pair of cloned lines. Uninterrupted diagonals or displaced diagonals which occur in the dot plot indicate the consecutive duplicated lines.

Metric based technique gathers different metrics from a particular code fragments, such as, a function or a class, then groups these metric together into a metrics vector. After that it compares these metric vector instead of actual code directly [LPM+97, KDM+96], because this method is focused on a specific type of code fragments, it can only detect an type of high level clone, e.g. duplicated function.

Here in metric computation each code fragments are given different metric values. During comparison these metric vectors are compared instead of comparing code directly. As a hint for similar code an allowable distance can be used for these metric vectors. Text based technique is the oldest and simplest way to detect clone, which takes each line of source code as code representation. In order to increase the performance, lines are often transformed by a hash function and uninterested code, such as comments and white spaces are filtered. The result of comparison is presented in a dot plot graph, where each dot indicates a pair of cloned lines.

A clone pair can be determined as a sequence of uninterrupted diagonals line of spot. Because text based technique does not perform any syntactical or semantically analysis on source code, it's one of the fastest clone detection approaches. It can easily deal with type 1 clone, and with additional data transformation, the type 2 can also be taken care. However without information of syntactical or semantically level support, the third type of clone cannot be detected at all.

CLONE DETECTION PROCESS

A clone detector must try to find pieces of code of high similarity in a system's source text. The main problem is that it is not known beforehand which code fragments can be found multiple times. The detector thus essentially has to compare every possible fragment with every other possible fragment. Such comparison is very expensive from a computational point of view and thus, several measures are taken to reduce the domain of comparison before performing the actual comparison. Once potential cloned fragments are identified further analysis is carried out to detect actual clones. In our proposed method, a hybrid technique based on textual and metric analysis is used to detect all types of clones present in the source code

Text based clone detection technique uses the transformation such as comments removal, whitespace removal. Because text based technique does not perform any syntactical or semantically analysis on source code, it is one of the fastest clone detection approach. It can easily deal with type 1 clone, and with additional data transformation, the type 2 can also be taken care. In Metric based technique, instead of comparing the code directly, different metric of code are gathered and these metrics were compared to detect clones. Many clone detection techniques today use metrics for detecting similar codes. The proposed method is implemented as a tool in java. The system architecture of the tool is as shown in Fig. 1.

Our proposed approach use metric based and text based technique to detect clones and divided into two stages. In the first stage metric based technique is used for the selection on potential clone. Potential clones are selected on the basis of metric match and after this potential clones are further processed with text based technique. The potential clones are compared line by line to determine whether two potential clones really are clones of each other. The tool developed initially parses through the given input source code and identifies the various methods present.

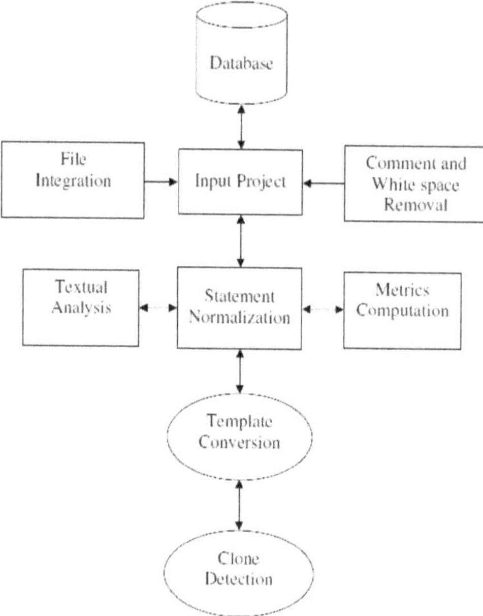

Fig. 1 Clone Detection Architecture

Clone detection process has been divided into number of phases. As shown in the fig.1 the phases include input and pre-processing, template conversion, metrics computation and finally detecting the clone types. The pairs that show similar in textual comparison are listed as the clones. The detection tool thus developed does not employ any external parsers. It requires only less overhead compared to other methods [22].

Preprocessing and input Selection

All the source code uninteresting to the comparison phase is filtered out in this phase. This phase also includes file integration, source code standardization and the normalization. File integration involves the grouping of all the files of the same project into a single large file for external parsing. This phase includes file integration, source code standardization and the normalization. File integration involves the concatenation of all the files of the same project into a single large file for external parsing. Here it includes the removal of whitespaces, comments and pre-processor statements. After removing the uninteresting code, the remaining source code is partitioned into a set of disjoint fragments called source units. These units are the largest source fragments that may be concerned in direct clone relations with each other. Source units can be at any level of granularity, for example, files, classes, functions/methods, begin-end blocks, statements, or sequences of source lines.

Source units may require to be further partitioned into smaller units depending on the comparison technique used by the tool. For example, source units may be subdivided into lines or even tokens for comparison. Comparison units can also be derived from the syntactic structure of the source unit. For example, an if-statement can be further partitioned into conditional expression, then and else blocks. The order of comparison units within their corresponding source unit may or may not be important, depending on the comparison technique. Source units may themselves be used as comparison units. For example, in a metrics based tool, metrics values can be computed from source units of any granularity and therefore, subdivision of source units is not required in such approaches. The source code is re-structured to a standard format to establish the similarity between the cloned fragments.

These steps are very similar to normalization procedures and produces gain in the recall. Almost all approaches disregard whitespace, although line-based approaches retain line breaks. Some metrics-based approaches however use formatting and layout as part of their comparison. Most approaches remove and ignore comments in the actual comparison. Most approaches apply identifier normalization before comparison in order to identify parametric Type-2 clones. In general, all identifiers in the source code are replaced by the same single identifier in such normalizations.

Template Conversion

Template conversion is the process of transformation of the input source code into a pre-defined set of statements or conversion into a standard intermediary form. For example, renaming of data types, variables, function names etc as shown in fig. 2. This type of format used in textual analysis is called 'template'. The textual comparison of the selected candidates while detecting the type-2 cloned methods where as per the definition, function identifiers, variable names, types etc., are edited during the cloning process and mere textual comparison would not suffice. Once the template conversion is over, the source file and the template file is stored in the database for applying metrics. This transformation can vary from very simple e.g., just removing the white space and comments to very complex e.g., generating PDG representation and/or extensive source code transformations. Metrics-based methods usually compute an attribute vector for each comparison unit from such intermediate representations.

SOURCE CODE	*TEMPLATE*
int templconv(ptra, buff1,leng, buff2)	*DAT FUN_NAME(S,S,S,S)*
char buff1[];	*DAT S;*
int leng;	*DAT S;*
int ptra;	*DAT S;*
char buff2[];	*DAT S;*
{	*{*
int i;	*DAT S;*
int j;	*DAT S;*
While(i<=leng)	*LOOP*
{	*{*
If(buff1[ptra+j]!=buff2[ptrb+j])	*IF*
return TRUE;	*RETURN;*
};	*};*
i++;	*ASSIGNMENT STATEMENT;*
j++;	*ASSIGNMENT STATEMENT;*
tembuf[ptra]='\0';	*ASSIGNMENT FROM FUNCTION CALL*
return TRUE;	*RETURN;*
}	*}*

Fig. 2 Example for template conversion

Metric Computation

A set of 12 existing method level metrics are used for the detection of type-1, type-2, type-3 and type-4 clone methods. They are as follows:

1) No. of effective lines of code in each method
    - Get the number of lines of code
    - Subtract white space lines
    - Subtract comment lines
    - Subtract the lines that contains only block constructs (for example in C# begin block construct is the character '{' while end block construct is the character '}'.

2) No. of arguments passed to the method
    - Calling the function involves specifying the function name, followed by the function call operator and any data values the function expects to receive. These values are the arguments for the parameters defined for the function, and the process just described is called passing arguments to the function.

3) No. of function calls in each method
    - A function call is an expression containing a simple type name and a parenthesized argument list. The argument list can contain any number of expressions separated by commas. It can also be empty.

4) No. of local variables declared in each method
- A variable declared as local is one that is visible only within the block of code in which it appears. It has local scope. In a function, a local variable has meaning only within that function block.

5) No. of conditional statements in each method
- In computer science, conditional statements, conditional expressions and conditional constructs are features of a programming language which perform different computations or actions depending on whether a programmer-specified Boolean condition evaluates to true or false.

6) No. of looping statements in each method
- A looping statement is one in which you want to execute a statement (or many) as many number of times you want. It is useful when you want to check some constraints with a specific value.

7) No. of return statements in each method
- A return statement ends the processing of the current function and returns control to the caller of the function. A value-returning function should include a return statement, containing an expression.

8) No. of function calling in each method
- Once a function has been declared and defined, it can be called from anywhere within the program: from within the main function, from another function, and even from itself. Calling the function involves specifying the function name, followed by the function call operator and any data values the function expects to receive.

9) No. of inheritance in each method
- Inheritance is a way to compartmentalize and reuse code by creating collections of attributes and behaviors called objects that can be based on previously created objects.

10) No. of virtual functions in each method
- A virtual function or virtual method is a function or method whose behavior can be overridden within an inheriting class by a function with the same signature.

11) No. of overloading constructor in each method
- Overload constructor is multiple constructors which differ in number and/or types of parameters.

12) No. of overriding functions in each method
- Function over loading means two functions will have same name but they differ in the number or type of arguments.

For each of the methods identified the metrics are computed and the corresponding values are stored in a database Table I shows the metric values for the code fragment in fig 2. After computing the metric values, the method pairs with equal or similar set of values are identified by comparing the records in the database. The short-listed set of candidates is then textually compared to be confirmed as clone pairs.

Table 1 Metric values for the example in fig. 2

Sl. No.	Metrics	Value
1.	No. of lines of code	18
2.	No. of arguments passed	4
3.	No. of local variables declared	6
4.	No. of function calls	1
5.	No. of conditional statements	1
6.	No. of looping statements	1
7.	No. of return statements	2

Finding Clone Types and Clone Pairs

By taking up a line by line comparison of the standardized and normalized source code for type-1 clone method the identification of the potential clone pairs is done. That is identical code fragments are selected except for variations in whitespace, layout and comments. For type-2 clone comparison of templates are done. Here syntactically identical fragments except for variations in identifiers, literals, types, whitespace, layout and comments are taken. In the fragments there is some modifications except there is some similarities means it must be declared as type-3 by matching template with the exact code. Copied fragments with further modifications such as changed, added or removed statements, in addition to variations in identifiers, literals, types, whitespace, layout and comments can be said as type-3 clones.

It's declared as type-4 clone when the fragments are completely different but produce similar output. If the functionalities of the two code fragments are identical or similar and referred as Type IV clones. That is when two or more code fragments that perform the same computation but are implemented by different syntactic variants are said to be type-4 clone. The identified cloned methods are then clustered separately for each type and the clusters are uniquely numbered. Clustering gives a clear image of how the methods were cloned and helps to provide an easier review process.

RESULTS AND DISCUSSION

The proposed software clone detection system has been implemented in the working platform of JAVA (version JDK 1.6). Here we use the source code with more than 500 LOC. The main aim of the proposed method is to identify all the four clone types in the source code. This can be achieved by the combining both textual analysis and metrics. The results obtained from the proposed method described as follows.

(i) *Initial Process* represents the initial screen obtained in the clone detection process which shows various fields and buttons.

(ii) *Loading the database* represents set of source programs to be loaded in to the system from a memory device.

(iii) *Selecting the source code* represents the code fragment that has to be compared with and obtain the results

(iv) *Textual analysis,* For detecting the clones in the input files, initially, the textual analysis is performed in the preprocessed codes. The textual analysis finds 2 types of clones such as type I and type II. It is presented in Fig. 3.

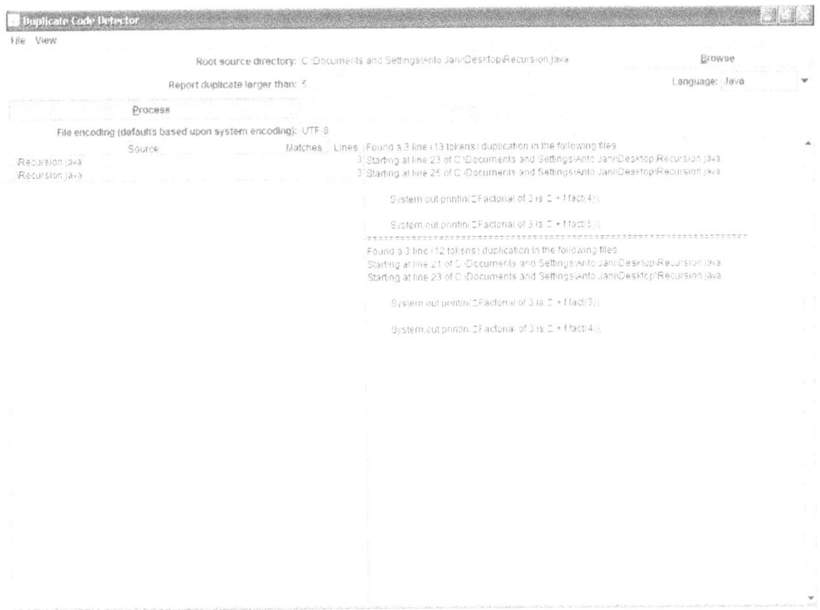

Fig. 3 Textual Analysis

(v)   *Metrics computations* are performed to detect the remaining clones in the source files. The metric analysis finds the remaining clones which are shown in Fig.4.

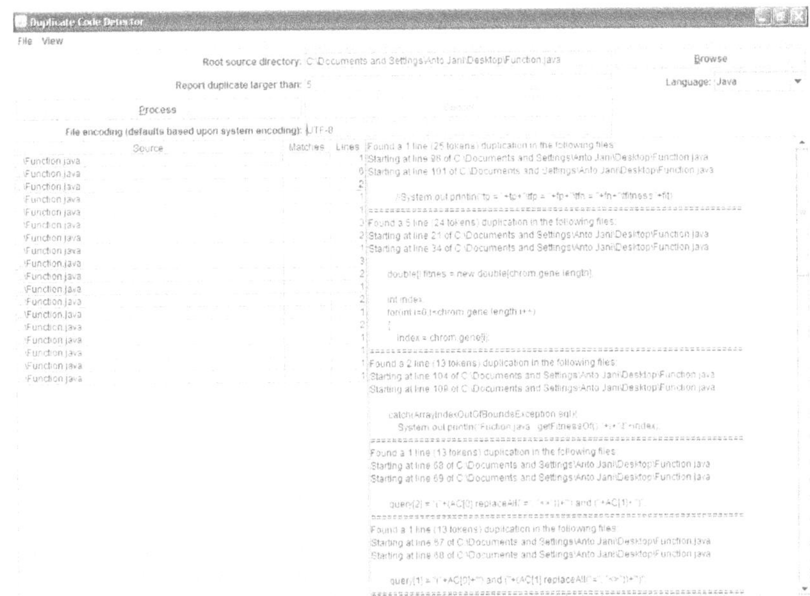

Fig. 4 Metrics computation

(vi)   *Clone Detection process* represents the clones available in the source files are detected in the efficient manner and the final output is presented in Fig. 5.

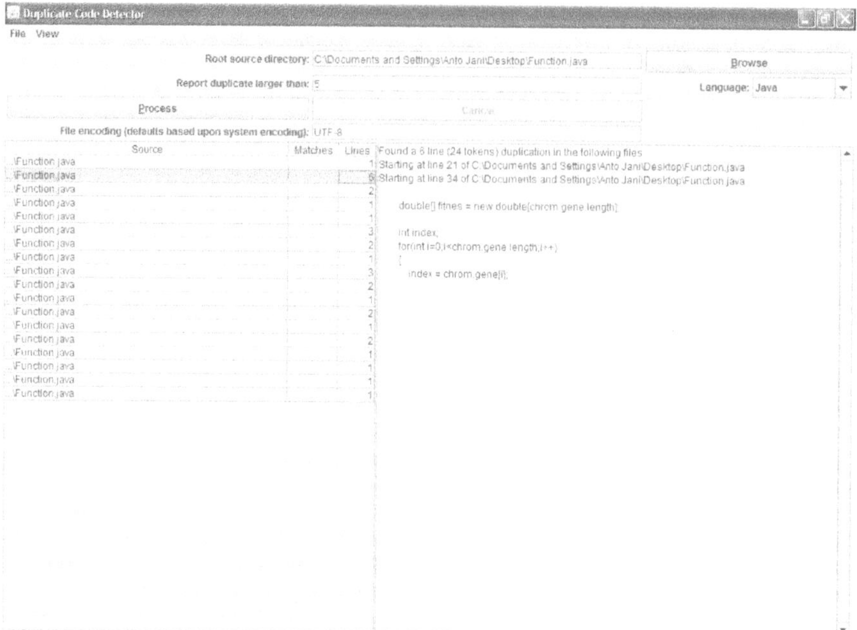

Fig. 5 Clone Detection Process

CONCLUSION

The paper has proposed a light-weight technique to detect functional clones with the computation of metrics combined with simple textual analysis technique. With the usage of metrics the existing exponential rate of comparison overhead. Since the string matching/textual comparison is performed over the shot listed candidates, a higher amount of recall could be obtained. The Proposed work is divided into two stages selection of potential clones and comparing of potential clones. The proposed technique detects exact clones on the basis of metric match and then by text match. Potential clones are compared line-by-line to determine whether two potential clones really are clones of each other. The early experiments prove that this method can do at least as well as the existing systems in finding and classifying the function clones in Java. The Precision and Recall plot describes the efficiency of the proposed work.

REFERENCES

1. Ira D. Baxter, Andrew Yahin, Leonardo Moura, Marcelo Sant'Anna and Lorraine Bier, "Clone Detection Using Abstract Syntax Trees," In Proc. of the International Conference on Software Maintenance, Bethesda, MD, pp. 368 - 377, Nov 1998.

2. Toshihiro Kamiya, Shinji Kusumoto and Katsuro Inoue, "CCFinder: A Multi-Linguistic Token-based Code Clone Detection System for Large Scale Source Code," IEEE Transactions on Software Engineering, Vol. 28, No. 7, pg. Software Engineering, Jul 2002.

3. Bruno Laguë, Daniel Proulx, Ettore M. Merlo, Jean Mayrand and John Hudepohl, "Assessing the Benefits of Incorporating Function Clone Detection in a Development Process," In Proc of the 1997 International Conference on Software Maintenance (ICSM '97), Washington, DC, 1997.

4. J Howard Johnson, "Substring Matching for Clone Detection and Change Tracking,," In Proc of the International Conference on Software Maintenance (ICSM), Victoria, British Columbia, pp. 120–126, Sep 1994.

5. Magiel Bruntink, Arie van Deursen, Remco van Engelen and Tom Tourwe, "An Evaluation of Clone Detection Techniques for Identifying Cross-Cutting Concerns," In Proc. of the 20th IEEE International Conference on Software Maintenance, Washington, DC, 2004..

6. Elizabeth Burd and John Bailey, "Evaluating Clone Detection Tools for Use during Preventative," In Proc. of the Second IEEE International Workshop on Source Code Analysis and Manipulation (SCAM'02), Montreal, Canada, Oct 2002.

7. Magiel Bruntink, Arie van Deursen, Remco van Engelen, and Tom Tourwe, "On the Use of Clone Detection for Identifying Crosscutting Concern Code," IEEE Transactions on Software Engineering, Vol. 31, No. 10, pp. 804 - 818, Oct 2005.

8. Chanchal K. Roy, James R. Cordy and Rainer Koschke, "Comparison and Evaluation of Code Clone Detection Techniques and Tools: A Qualitative Approach," Science of Computer Programming, Vol. 74, No. 7, Feb 2009.

9. Andrew Walenstein, Nitin Jyoti, Junwei Li, Yun Yang, and Arun Lakhotia, "Problems Creating Task-relevant Clone Detection Reference Data," In Proc. of the 10th IEEE Working Conference on Reverse Engineering, Victoria, Canada, Nov 2003.

10. Vera Wahler, Dietmar Seipel, Jurgen Wolff V. Gudenberg, and Gregor Fischer, "Clone Detection in Source Code by Frequent Itemset Techniques," In Proc. of the Fourth IEEE International Workshop on Source Code Analysis and Manipulation, Chicago, IL, pp. 128 - 135, Sep 2004.

11. Robert Tairas and Jeff Gray, "Phoenix-Based Clone Detection Using Suffix Trees," In Proc. of the 44th annual southeast regional conference, New York, NY, 2006.

12. Chanchal K. Roy and James R. Cordy, "Scenario-Based Comparison of Clone Detection Techniques," In Proc. of the 16th IEEE International Conference on Program Comprehension, Washington, DC, 2008.

13. Armijn Hemel, Karl Trygve Kalleberg, Rob Vermaas, and Eelco Dolstrac, "Finding Software License Violations Through Binary Code Clone Detection," In Proc. of the 8th working conference on Mining software repositories, New York, NY, May 2011.

14. Yue Jia, David Binkley, Mark Harman, Jens Krinke and Makoto Matsushita, "KClone: A Proposed Approach to Fast Precise Code Clone Detection," In Proc. of the Third International Workshop on Detection of Software Clones (IWSC 2009), pp. 12-16, 2009.

15. Rainer Koschke, Raimar Falke and Pierre Frenzel, "Clone Detection Using Abstract Syntax Suffix Trees," In Proc. of the 13th Working Conference on Reverse Engineering, Benevento, pp. 253 - 262, Oct 2006.

16. Stephane Ducasse, Oscar Nierstrasz and Matthias Rieger, "Research On the effectiveness of clone detection by string matching," Journal of Software Maintenance and Evolution: Research and Practice, Vol. 18, No. 1, pp. 37-58, 2006.

17. R. R. Brooks, P. Y. Govindaraju, M. Pirretti, N. Vijaykrishnan and M. Kandemir, "Clone Detection in Sensor Networks with Ad Hoc and Grid Topologies," International Journal of Distributed Sensor Networks, Vol. 5, pp. 209–223, 2009.

18. Shinji Kawaguchi, Takanobu Yamashinay, Hidetake Uwanoz, Kyhohei Fushida, Yasutaka Kamei, Masataka Nagura and Hajimu Iida, "SHINOBI: A Tool for Automatic Code Clone Detection in the IDE," In Proc. 16th Working Conference on Reverse Engineering, pp. 313 - 314, Oct 2009.

19. Kodhai. E, Kanmani. S, Kamatchi. A, Radhika. R and Vijaya Saranya. B, "Detection of Type-1 and Type-2 Code Clones Using Textual Analysis and Metrics," In Proc. of the 2010 International Conference on Recent Trends in Information, Telecommunication and Computing, Washington, DC, pp. 241-243, 2010.

20. Kodhai.E, Perumal.A, and Kanmani.S, "Clone Detection using Textual and Metric Analysis to figure out all Types of Clones," In Proc. of the International Joint Journal Conference on Engineering and Technology (IJJCET 2010), pp. 99 - 103, 2010.

21. Nam H. Pham, Hoan Anh Nguyen, Tung Thanh Nguyen, Jafar M. Al-Kofahi and Tien N. Nguyen, "Complete and Accurate Clone Detection in Graph-based Models," In Proc. of the 31st International Conference on Software Engineering, Washington, DC, 2009.

22. G. Anil Kumar, C.R.K. Reddy, A. Govardhan, "An efficient method-level code clone detection scheme through textual analysis using metrics," Proc. of the International Journal of Computer Engineering & Technology (IJCET 2012), Volume 3, Issue 1, pp. 273 - 288, January- June 2012.

The Development of Unmanned Aerial Vehicles and Emerging Legal Issues

Prof. (Dr.) V. Balakista Reddy
Professor of International Law and Head, Center of Air and Space Law (CASL)
NALSAR University of Law, Hyderabad

Law follows technology or technological developments necessitate the legal regulations. It is true with the aviation technology. In a span of over a century, aviation technology has expanded its leaps and bounds. Since then, aviation technology has grown at an exponential pace, with the first propeller-driven planes giving way to jets, which have now been surpassed by planes that can fly into space.The ongoing development and use of unmanned aerial vehicles (UAVs) illustrates well the observation that "law lags science; it does not lead it." UAVs serve myriad military missions; have also commercial and civilian applications.Military uses include national defense, disaster response, intelligence gathering, surveillance, reconnaissance etc,.UAVs have proven their worth on the battlefields of Iraq, Afghanistan, and Lebanon. UAVs offer a relatively low-cost, low-risk alternative to manned aircraft in the military setting.

In the coming years, law enforcement agencies will seek to use UAVs to police borders, control crowds, track criminals, detect illegal narcotics activities, and spot crime. Civilian applications include traffic surveillance, weather monitoring, communications relay, border management, maritime patrol, crime prevention, forest fire monitoring, drug interdiction, soil studies, seed fields, and dust crops. In spite of the myriad uses of UAVs, the existing legal regime is in adequate to and this paperwill discusses the legal hurdles inoperating UAVs in the national airspace systems. Due to the lack of regulations, there are many questions pending, pertaining to complex questions such as their airworthiness certification, personnel licensing, classification, registration, collision avoidance, traffic management, communication protocols, security issues, environmental protection, legal responsibility, insurance etc. The absence of a distinctive body of rules and regulations integrating UAV flight into the national airspace, coupled with an existing time-consuming certification process for UAV is an impediment and constraint for present and future UAV development will be also discussed in this paper.

This paper reviews laws and policies applicable to UAV operations inthe national airspace and calls for definite legal parameters for thisemerging sector of aviation. Part II provides a brief background of the wayin which the law has historically dealt with air and land rights relative tonew and unprecedented developments in aviation. Part III explores theoperative regulatory regime that exists today, evaluates its fitness in theUAV context, and introduces the development of UAV-related laws inforeign jurisdictions. Ultimately, there is nothing necessarily objectionableabout having the law follow technology in terms of UAV development.That is, while policymakers should aggressively encourage UAV productionand use in the civil, commercial, and military realms, they shouldapproach the laws regulating UAV operations conservatively, integratingdifferent UAV assets into the national airspace in-step with improvementsin UAV technological reliability. At the same time, given the actual proliferation of UAVs in the commercial and military markets, the time forlawmakers to more directly address UAV integration into the NAS.